2011-2012 ANNUAL REVIEW OF CHINESE ARCHITECTURAL DESIGN WORKS

中国建筑设计作品年鉴【下册】

《中国建筑设计作品年鉴》编委会　北京主语空间文化发展有限公司　编

江苏人民出版社

图书在版编目（C I P）数据

2011～2012中国建筑设计作品年鉴 / 《中国建筑设计作品年鉴》编委会，北京主语空间文化发展有限公司编. -- 南京 : 江苏人民出版社，2012.11
ISBN 978-7-214-08814-7

Ⅰ. ①2… Ⅱ. ①中… ②北… Ⅲ. ①建筑设计－作品集－中国－2011～2012 Ⅳ. ①TU206

中国版本图书馆CIP数据核字(2012)第238109号

2011–2012中国建筑设计作品年鉴（上、下）　《中国建筑设计作品年鉴》编委会 北京主语空间文化发展有限公司 编

责任编辑：刘　焱
特约编辑：卫　星
责任监印：彭李君
装帧设计：张　萌
排版设计：郝泽星　彭　丹　贾　妍　郑忠月
出版发行：凤凰出版传媒股份有限公司
江苏人民出版社
天津凤凰空间文化传媒有限公司
销售电话：022-87893668
网　　址：http://www.ifengspace.cn
经　　销：全国新华书店
印　　刷：利丰雅高印刷（深圳）有限公司
开　　本：1020mm×1420mm　1/16
印　　张：63.5（上：31.5，下：32）
字　　数：1000千字
版　　次：2012年11月第1版
印　　次：2012年11月第1次印刷
书　　号：ISBN 978-7-214-08814-7
定　　价：980.00元（上、下册）
(USD 178.00)
（本书若有印装质量问题，请向销售部调换）

主　　管：中华人民共和国住房和城乡建设部
主　　办：中国建筑文化中心
编　　辑：《中国建筑设计作品年鉴》编委会
北京主语空间文化发展有限公司
主　　编：陈建为
执行主编：肖　峰
编辑部主任：肖　措
编　　辑：代　君　郭　晨　金　峰　冷娜英　李　悦　林　朋　刘笑艳　欧　阳　齐　阳　王　伟　邢丽丽　张　澜　张　辉　朱英杰

编辑部地址：北京市海淀区三里河路13号
中国建筑文化中心712室
邮　　编：100037
电　　话：+86-10-88151966
传　　真：+86-10-88151958
监督电话：13910120811
网　　址：www.archrd.com
邮　　箱：zgjsmail@126.com

特邀编委
Contributing
Editor

宗白华先生言：一切艺术综合于建筑，而礼乐诗歌舞剧之表演，亦与建筑背景协调成为一片美的生活，所以每一文化的强盛时代莫不有伟大的建筑计划以容纳和表现这一丰富之生命。先生此言发表半个多世纪以前，今日我们更应致力于这一和谐的美的实现早日来临。

吴良镛

2011—2012 中国建筑设计作品年鉴

谨以此年鉴献给为中国建筑设计行业发展而辛勤付出的设计师们！

《2011—2012中国建筑设计作品年鉴》（以下简称《年鉴》）收录了国内外180余家设计机构的近1600件优秀建筑设计作品，这些不同类型、不同设计理念的作品基本反映了当前建筑设计行业多元化的发展状况和实践成果，从中可以直观地领略到中国建筑设计行业融汇古今、贯通中西的设计思想。

《年鉴》对于各地建设主管部门和从事城市规划、建筑设计、景观设计、工程建设、房地产开发人员以及相关科研教育机构掌握建筑设计行业发展趋向、了解新的设计理念，会有很好的参考、借鉴作用。

《年鉴》在编辑过程中，得到了各地建设主管部门和众多设计院所、设计师的大力支持和帮助，在此，谨向为《年鉴》提供帮助和支持的单位和个人表示衷心的感谢！

《年鉴》在国内公开发行，并委托中国图书进出口（集团）总公司在国外和国内的港、澳、台地区发行。在《年鉴》的编辑过程中，难免有疏漏或错误之处，敬请读者指正，并提出宝贵意见，以便我们不断提高《年鉴》的编辑水平，满足读者的需求。

《中国建筑设计作品年鉴》编辑部

2012年9月

目录 Contents

建筑设计单位
Architectural Design Units

上 册

下 册

设计作品分类 (按项目类型排序)
Contributing Editor

办公建筑

城市规划设计、景观规划设计

工业建筑

交通建筑

教育建筑

居住建筑

旅游、休闲建筑

商业建筑

体育建筑

文化建筑

医疗建筑

园林、公园、生态园

展览 展示

综合建筑

其他

主要建筑设计作品

MAIN Architectural Design Works

立地分析\项目定位
规划设计\建筑设计
园林环艺\装饰装修
智能顾问\电力服务
照明设计\工程监理

嘉博筑业集团始创于1993年，是一家为房地产投资商和开发商提供以设计为核心的集全程专业技术服务（立地分析、项目定位、规划设计、建筑设计、园林环艺、装饰装修、智能顾问、电力服务、照明设计、工程监理）于一体的房地产建设开发全程设计集成服务集团。

发展历程

1993年，香港嘉图设计工程有限公司在福州投资创办了“嘉博（福建）联合设计有限公司”。经过近20年的发展，嘉博（福建）联合设计有限公司凭借骄人业绩，跻身中国民用建筑设计50强企业，并经由专业服务延伸发展构建了房地产建设开发全程设计集成服务集团企业——“嘉博筑业”。

为实现“为投资商创造项目投资价值最大化”的服务宗旨，并成为中国大陆最专业的房地产建设开发全程设计集成服务运营商，嘉博筑业以“嘉博（福建）联合设计有限公司”为专业核心基础，先后创办了装饰装修设计、工程建设监理、智能化顾问、电力工程顾问、景观设计、照明设计、项目定位及市场研究等专业机构，形成了可为房地产投资商、开发商提供自土地评估至建成回报全过程“一条龙”的设计集成服务，以专业设计水准确保项目品质，以集成服务提升项目价值。

经营团队

嘉博筑业集团拥有一批专业水平高、经验丰富、协调能力强并极具敬业精神的项目设计及开发管理人员、建筑工程和专项设计技术专家，在与国内外知名投资及开发企业进行开发合作以及进行项目统筹运营管理服务与专业资源整合服务方面具有丰富的经验与能力。集团现有各类专业开发管理、技术人员500多人（其中具有本科以上学历的人员占总人数的90%以上），并已在香港、上海、福州、厦门四地设立了相应的管理及业务执行机构。

案例与业绩

嘉博筑业集团自始创以来已完成开发类管理专项咨询业务及专业技术服务项目超过1000个，总建筑面积超过8000万m^2，与万科、华润、世纪金源、中新、城开、融侨、建发、泰禾、永辉、福晟、阳光城、冠亚、中庚、群升、融晟、金辉、中茵、汇诚、永荣、轩辉、永嘉、东榕等诸多全国性、区域性主流房地产开发商建立了长期的开发管理及设计技术服务合作伙伴关系，市场已遍及福建全省各地市及上海、浙江、江西、山东、河北、河南、江苏、安徽、湖北、湖南、广东、广西、陕西、云南等十多个省市自治区。

核心能力

以设计为核心，以集成为特色，嘉博筑业集团业务已辐射全国各地。开发及设计技术咨询的项目种类涉及商业、住宅、酒店、工业地产、城市综合体、写字楼等多种物业。

福州：

福建省福州市鼓楼区东大路36号花开富贵A座29楼
电话：+86-591-87430666
传真：+86-591-87430680

厦门：

福建省厦门市思明区槟榔西里148号天湖大厦B座14楼
电话：+86-592-5391010
传真：+86-592-5391020

上海：
上海市卢湾区复兴中路1号申能国际大厦2006-2008A室
电话：+86-21-63919361
传真：+86-21-63919152

立地分析\项目定位
规划设计\建筑设计
园林环艺\装饰装修
智能顾问\电力服务
照明设计\工程监理

In 1993, Good Broad Industry Group was founded as a real estate construction and development full-process integrated design service group to provide real estate investors and developers with design-based full-process professional technical services, including site analysis, project orientation, planning design, architectural design, gardening & environmental arts, decoration, intelligent consulting, electrical services, lighting design and project supervision.

Development Journey

In 1993, Grotto Fine Art Ltd. (Hong Kong) founded Good Broad Architects Association Co., Ltd. in Fujian province. After a near-20-year development, Good Broad Architects Association Co., Ltd. with proud performance is listed top 50 Chinese civil building design companies. And with its extended professional services,it develops to a real estate construction and development full-process integrated design service group, namely Good Broad Industry Group.

In order to fulfiu our service tenet of "maximizing investment value for project investors", and become the most professional real estate construction and development full-process integrated design services operator in Mainland China, Good Broad Industry Group on the basis of Good Broad Architects Association Co., Ltd. has founded professional institutes in decorative design, project supervision, intelligent consulting, electrical engineering consulting, landscaping design, lighting design, project orientation, market research,etc. to form "one-stop" integrated design services from land evaluation to project completion for real estate investors and developers, with professional design level to ensure project quality, and with integrated services to increase project value.

Operating Team

Good Broad Industry Group has a group of top professional, experienced, coordinative, and committed project design & development managers, engineering and project design technicians, with rich experiences and capabilities of cooperative developments, project operation & management services and professional resource integrated services with domestic and foreign famous investment and development enterprises. At present, it has more than 500 development managers and technicians, over 90% of whom hold a bachelor's degree or above, and it sets up operation executive offices in Hong Kong, Shanghai, Fuzhou and Xiamen with management & .

Cases and Achievements

Good Broad Industry Group since its inception has completed more than 1,000 development & management consulting services and professional & technical service projects, with total Building area exceeding 80,000,000 m². It has established long-term development management and design technical service partnerships with Vanke, CRC, Century Golden Resources Group, Neo-China Land Group, UDCN Corporation, Rongqiao, C&D, Thaihot, Yonghui, FullSun Group, Yango, Guanya Group, Zhonggeng Group, ChinSun Group, Rongsheng, KamFei Group, Join•In Group, Huicheng, EverSun, Xuanhui, Yongjia, Dongrong among many other national or regional mainstream real estate developers, and it has market shares in all cities of Fujian province, Shanghai, Zhejiang province, Jiangxi province, Shandong province, Hebei province, He'nan province, Jiangsu province Yas well as in Anhui province, Hubei province, Hunan province, Guangdong province, Guangxi province, Shanxi province and Yunan province and other provinces autonomous regions, and municipalities.

Core Competencies

Based on design, and featured in integration, Good Broad Industry Group operates has extended its business across the country. Its development & design technical consulting projects cover many kinds of properties, including commercial building, residential building, hotel, industrial property, urban complex, and office building.

Fuzhou Base:

29th Floor, Tower A, Huakaifugui Building, No.36 Dongda Road, Gulou District, Fuzhou City, Fujian Province
Tel: +86-591-87430666
Fax: +86-591-87430680

Xiamen Base:

14th Floor, Tower B,Tianhu Plaza, No.148 Binlang Xili, Siming District, Xiamen City, Fujian Province
Tel: +86-592-5391010
Fax: +86-592-5391020

Shanghai Base:

Room 2006-2008A, Shenneng Internationan Plaza, No.1 Fuxing Zhonglu, Luwan District, Shanghai
Tel: +86-21-63919361
Fax: +86-21-63919152

福州大学城永嘉天地
Yongjia World, Fuzhou University Town

项目地点：福建 福州
用地面积：73 000 m²
建筑面积：180 000 m²

Location: Fuzhou, Fujian
Site Area: 73,000 m²
Building Area: 180,000 m²

永嘉天地位于福州智慧引擎——福州大学城核心。该项目以城市生活广场理念打造，规划有商业与住宅两大组团。商业部分拥有购物中心、独栋商业别墅、SOHO办公楼、酒店、地上地下复合步行街，塑造立体式的消费空间，能够全面满足消费者在永嘉天地内餐饮、购物、休闲、娱乐与商住的全面需求。住宅组团以独立小区形式嵌入商业组团中，共规划7栋ART–DECO风格精工美宅，真正做到“入则宁静，出则繁华”，其中一栋住宅楼全部为4.75 m挑高复式结构，创造截然不同的居住体验。该项目除住宅部分销售外，主力商业全部由开发商持有，并由永辉超市、台湾太平洋百货等一线主力商家定制开发，保证了空间布局的科学性。同时，由专业的商管公司及物管公司对项目进行全程服务，保证了项目的可持续发展与商业繁荣。

Yongjia World is located in the core of Fuzhou University Town, the intelligent engine of Fuzhou city, which is the project is built in the idea of urban life plaza, with both commercial and residential clusters. The commercial cluster includes shopping center, independent commercial villas, SOHO office building, hotel, aboveground/underground composite pedestrian street, and 3D consumption spaces, which can full satisfy consumers' overall catering, shopping, leisure, entertainment and commercial residential requirements within Yongjia World. The residential cluster is incorporated into the commercial cluster by independent community, with 7 ADECO-style exquisite residencial building, tranquil inside and prosperous outside, including a building in 4.75m hign lofty composite structures, providing a distinguishing living experience. Except for residential buildings for sales, main commercial buildings are all held by the developer, and custom developed by Yonghui Superstores, Pacific Department Store and other tier-1 merchants, ensuring the scientific layout of spaces. Meanwhile, the project is maintained in full process by a professional commercial management company and a professional property management company, ensuring sustainable development and business prosperity.

漳浦永嘉天地

Yongjia World, Zhangpu

项目地点：福建 漳州
用地面积：20 000 m²
建筑面积：90 000 m²

Location: Zhangzhou, Fujian
Site Area: 20,000 m^2
Building Area: 90,000 m^2

漳浦永嘉天地位于漳州市漳浦县中心。作为漳浦的标志性建筑和新商业中心，漳浦永嘉天地不仅满足漳浦县居民吃、穿、玩、乐、购、品、赏、趣等日常需求，也是年产值达千亿元的古雷港经济开发区目前最重要的商业服务配套设施。项目以科学、紧凑的商业布局囊括百姓超市、美食天地、时尚服饰、精品百货、休闲地带、商务办公、商务酒店等全方位商业配置，是漳浦首座全功能、一站式城市商业综合体。项目以围合式街区模式设计，两栋百米双子星商务大楼采用ART-DECO风格，联排商业别墅注入闽南"飞檐"的建筑元素，将现代与经典完美结合，在具有浓厚文脉特征的县级城市成功打造"上海新天地"式的浪漫风情。

The project is located in the center of Zhangpu county town of Zhangzhou City. As the landmark building and new business center of Zhangpu county, it not only satisfies local residents' daily demands for food, fashion, entertainment, recreation, shopping, etc. but also serves as the most important commercial service support for Gulei Habor Economic Development Zone, with annual output value exceeding RMB100 billion. It has a scientific and compact commercial layout, including People's Supermarket, Catering World, Fashion, Exquisite Department Store, Leisure Zone, Office, Hotel and other commercial supplies, serving as the first multi-functional one-stop urban commercial complex in Zhangpu county. The project is designed in a mode of enclosing blocks, with the two 100 m high Gemini Office Buildings in ADECO style, townhouses incorporating South Fujian architectural element of "overhanging eaves", which combines modern and classical beauties perfectly, and successfully creates "Shanghai Xintiandi" style romantics in this county with immense local characters.

贵安新天地（贵祥苑、贵荣苑、贵邸苑、贵阁苑）
Gui'an Xintiandi(Luck Garden, Glory Garden, Mansion Garden, Loft Garden)

项目地点：福建 福州　　Location: Fuzhou, Fujian
用地面积：348 900 m^2　　Site Area: 348,900 m^2
建筑面积：945 000 m^2　　Building Area: 945,000 m^2

项目位于福州市北郊的贵安温泉旅游度假区内，原生态景观资源丰富，山水环境优美，是世纪金源集团集20年之大成打造的“海西首席旅游商住综合体”，包含超五星级温泉商务酒店、五星级城堡酒店、五大旅游娱乐公园、学校、海峡传统文化街、百姓文化长廊、勉斋书院、商住基地、低密度住宅、花园洋房、高层公寓等多种业态，总用地面积达467 hm^2，气势磅礴非凡。其高层公寓中的贵祥苑、贵荣苑、贵邸苑、贵阁苑包含了87幢18～32层融汇了新古典建筑风格和半坡面经典语汇的高层住宅、35幢1～3层沿街商业及2幢3层幼儿园。

This project is located within Gui'an Hot Spring Resort in the north suburb of Fuzhou City. It enjoys rich ecologic resources and beautiful landscapes, as the "leading tourist commercial residential complex west of Taiwan Strait" built by Century Golden Resources Group with 20 years' mastership, including over-5-star business hotel with hot spring, 5-star castle hotel, 5 big tourist resort parks, school, Taiwan Strait cultural street, civic cultural corridor, Study College, commercial residential base, low-density residential buildings, garden houses, hi-rise apartments and other industrial structures, covering a total land area of 4,670,000 m^2, with uncommon grandeur. Its hi-rise apartments includes 87 high-rise 18 to 32-floor residential buildings integrating neoclassical style and semi-slope typical language, 35 roadside 1 to 3-floor commercial buildings, and two 3-floor kindergartens.

凯旋国际
Hopsca

项目地点：福建 莆田　　Location: Putian, Fujian
用地面积：48 000 m^2　　Site Area: 48,000 m^2
建筑面积：213 000 m^2　　Building Area: 213,000 m^2

项目由8幢33层融汇了ART–DECO建筑风格的单元式住宅和1幢2层商业建筑组成，以围合式大景观社区为规划手法，在有限的空间中做大、做足景观面的同时，通过各建筑错位布置、底层架空等设计，努力使景观相互延伸，致力于营造一种精神家园的归属感，将社区的住户景观受众面做到极致，使其成为当地注重社区整体居住品质及文化氛围的精品商住小区。

This project includes eight apartment buildings of 33 floors in Art-Deco style and one commercial building of 2 floors, enclosing a large landscaping community. This planning manner maximizes landscaping exposure in limited space, while extending the landscapes by dislocation of buildings and overhead ground floor and other design means, to create a spiritual homeliness, so that the residents will receive the most landscapes, and the community will become an exquisite commercial/residential community featuring both entire habitat quality and cultural atmosphere.

万科 · 大樟溪岸
Vanke · Dazhangxi River Bank

项目地点：福建 福州
用地面积：392 000 m²
建筑面积：803 300 m²

Location: Fuzhou, Fujian
Site Area: 392,000 m²
Building Area: 803,300 m²

项目地处永泰温泉生态区，山水资源丰富，素有福州市"后花园"之称，是福建省的重点旅游基地。项目依托大樟溪岸的良好生态和优美的自然环境资源，规划包括：独栋别墅、爬山叠拼低密度"花"型合院多层住宅、高层酒店式公寓、精品温泉酒店、水岸林荫院落商业街区等丰富的建筑产品，以"最大化地把自然还原给空间"为理念，将该项目打造成福州最具有度假、养生特色的温泉社区。

The project is located in Yongtai Hot Spring Eco-Zone, rich in landscape resources, well-known as "backcourt" of Fuzhou City, serving as an important tourist base of Fujian Province. Based on the good ecology of Dazhangxi River Bank and beautiful nature, the planning covers: villas, overlap low density "flower" pattern multi-floor residential buildings, high-rise apartment hotel, exquisite hot spring hotel, waterside tree-lined courtyard commercial block and other building types, to "maximize the restoration of nature into building spaces", and build this project to be the most typical hot spring community featuring resort and lifestyle of Fuzhou City.

莱芜世纪城
Laiwu Century Town

项目地点：山东 莱芜
用地面积：387 000 m²
建筑面积：1 200 000 m²

Location: Laiwu, Shandong
Site Area: 387,000 m²
Building Area: 1,200,000 m²

项目紧扣城市副中心及区域中心的主题，充分利用其周边配套完善、环境优美的地段优势，引入独特的公园居家理念，将规划建设的高层住宅、高档别墅、五星级酒店、商业步行街、商务公寓、教育设施、休闲娱乐场所及业主专属会所等建筑，通过一条景观中轴线、五大分区组团、数条中心放射状路网的规划设计，共同构筑出空间、色彩、造型、肌理、天际线全面丰富的具有ART–DECO风格的大型社区，成为莱芜市首个城市综合体。

The project is keynoted by city sub-center and regional center, to make full use of peripheral supports and beautiful environment, introduce unique park housing idea, and construct high-rise residential buildings, high grade villas, 5-star hotels, commercial pedestrian street, commercial apartment, educative facilities, recreational venues, residents exclusive club and other buildings into a large community in Art-Deco style rich in space, color, shape, texture and skyline. Through a landscaping central axis, 5 large zoning groups, and several central radiating roadwork. This community will become the first urban complex in Laiwu City.

百联华府
Bailian House

项目地点：山东 临沂
用地面积：70 700 m²
建筑面积：240 000 m²

Location: Linyi, Shandong
Site Area: 70,700 m²
Building Area: 240,000 m²

项目依托周边珍贵的环境资源，规划设置了具有经典ART–DECO风格的10幢24～32层高层住宅以及19幢以英伦风情为基调的低层联排住宅，外围利用商业街与点式高层把城市的喧嚣隔离在外，社区配套双会所、幼儿园、架空泛会所、奢华双大堂装修入户、英式皇家园林等，通过红墙、绿树、钟塔、连廊等立面元素结合建筑形体上的退台、坡屋、老虎窗等营造出原汁原味的英伦生活方式，使之成为未来临沂地区的高端人居象征。

Based on peripheral valuable environmental resources, this project plans ten high-rise residential buildings of 24-32 floors in classic Art-Deco style, and 19 townhouses in English style. The peripheral commercial street and point-style high-rise buildings insulate the project from city noises. The community is equipped with clubs, kindergarten, aerial multi—functional club, luxury dual-lobby renovated to house, English-style royal garden etc. Red wall, green trees, bell tower, corridor and other façade elements, together with backset terrace, sloping roof, dormer windows and other structures, create an original English lifestyle, enabling the community to become future symbol of hi-end habitat in Linyi City .

金域华府
Jinyu House

项目地点：山东 昌邑
用地面积：145 600 m²
建筑面积：381 000 m²

Location: Changyi, Shandong
Site Area: 145,600 m²
Building Area: 381,000 m²

项目利用地块的地理位置优势，规划了景观及商业主轴，并以此为核心，院落式地布局了6幢6层、1幢10层、2幢11层、2幢15层、3幢16层、3幢17层、12幢18层住宅及1幢25层公寓、13幢1～2层沿街商业和1幢3层16班幼儿园，各幢融汇了西班牙和新古典风格建筑语汇的单体通过具有尊贵品味和成熟气质的四坡屋面形成高低错落有致的形体美感表现，并通过比例、细节和材料的有机、巧妙运用，充分表现了项目的尊贵与典雅。

The project plans landscape and commercial main axis following the geographic advantages. Based on this, the project includes 6 residential buildings of 6 floors, 1 residential building of 10 floors, 2 residential buildings of 11 floors, 2 residential buildings of 15 floors, 3 residential buildings of 16 floors, 3 residential buildings of 17 floors, 12 residential buildings of 18 floors, and 1 apartment building of 25 floors, 13 roadside commercial buildings of 1-2 floors, as well as 1 kindergarten of 3 floors with 16 classrooms. All the buildings integrate Spanish and neo-Classical architectural language, and the noble and mature hip roofs form a beautiful figure in order. The organic and wise use of scales, details and materials fully represent the nobility and elegance of the project in the direction of urban development.

万业·锦江城
Wanye · Jinjiang Town

项目地点：福建 长乐
用地面积：80 000 m²
建筑面积：281 000 m²

Location: Changle, Fujian
Site Area: 80,000 m²
Building Area: 281,000 m²

项目由15幢17～30层高档精品住宅、局部一层地下车库及1万多 m²商业建筑所组成。西侧为长乐市气象观测站，在气象限高的不利条件下，通过科学、合理的规划和布局，不仅将"大景观、大视野、大中心"的设计理念做到极致，而且在户型设计上充分尊重当地人居生活习惯，采用了从115～180 m²以及200 m²以上的复式住宅等各类户型的平面设计，极大地满足了市场需求。在立面造型上，通过新古典及ART－DECO等建筑语汇与细节的巧妙运用，不仅彰显了项目的尊贵与典雅，扣合了案名的大气，更使其成为长乐市具有地标性意义的新楼盘。

The project includes fifteen high-end exquisite residential buildings of 17-30 floors, with underground 1-floor garage in some part and more than 10,000 m² commercial structures. The west lies Changle Meteorological Station. In unfavorable conditions of meteorological height limitation, with scientific and reasonable planning and layout, the project not only designs "big landscape, big vision, and big center" perfectly, but also respects local habitual lifestyle in unit design, adopting composite residential units with floor area 115-180 m² or above 200 m² among other unit designs which have greatly satisfied market demands. The façade is shaped by skillful use of neo-classical, Art-Deco and other architectural languages and details, not only highlighting the project elegance and nobility, in line with the greatness of project name, but also making it a landmark new building in Changle City.

凯龙城
Keylong Town

项目地点：云南 龙陵
用地面积：386 000 m²
建筑面积：730 000 m²

Location: Longling, Yunnan
Site Area: 386,000 m²
Building Area: 730,000 m²

项目在设计规划上引进沿海一线城市先进的开发理念、成熟的社区模式，产品囊括多层、板式小高层、高层电梯房、商业步行街、3500 m²高标准幼儿园、3600 m²商业综合楼（星级休闲会所）、6500 m²产权式酒店公寓（产权式酒店别墅）以及龙陵县唯一的总长900 m的商业步行街，同时还配建龙陵县唯一的4万 m²的五星级大酒店，其现代都市化的建筑风格、完善的周边配套与交通设施、精致典雅的城市花园，将式就龙陵县首席、大型融酒店餐饮、商务接待、娱乐休闲、购物消费、居住于一体的复合型社区。

The project introduces the leading development ideas and mature community models from tier-1 costal cities, whose products include multi-floor, middle-rise, high-rise, commercial pedestrian street, 3500 m² high-standard kindergarten, 3600 m² commercial comprehensive building (star-level leisure club), 6500 m² apartment hotel (villa hotel) and 900 m long commercial pedestrian street, the only one in Longling County, also equipped with 5-star grand hotel, also the only one in Longling County. Its modern metropolitan architectural style, complete peripheral supplies and traffic facilities, exquisite and elegant city garden, will together create a leading composite community, integrating hotel, catering, business reception, entertainment & leisure, shopping, residence, etc.

该项目是泰禾集团进军高端商业地产的开篇巨著，由底商、商务办公及集中商业三大功能块组成，包括4幢22层商务办公、1幢17层商务办公、1幢7层集中式商业及带地下一层局部集中式商业，拥有近4000 m²的市民休闲广场，配置机动车停车位数量达1300多个。建成之后，该项目将成为一个集购物、休闲、娱乐、餐饮、商务办公为一体的福州城北唯一的大型商业综合体，犹如一颗璀璨的钻石镶嵌在福州北大门，成为福州城市新的名片及都会时尚新地标。

This project,as the first masterpiece of Thaihot Group to enter hi-end commercial real estate, includes three major functions, namely, four commercial office buildings of 22 floors, one commercial office building of 17 floors, one collective commercial building of 7 floors with underground floor collective commercial structures in some part. There is also a 4,000 m^2 civic recreational plaza, equipped with more than 1,300 parking places. Upon completion, the project will be the only large commercial complex in Northern part of Fuzhou City, which integrates shopping, leisure, entertainment, catering and commercial office functions, resembling a bright diamond embedded in the north gate of Fuzhou City, serving as a new name card of Fuzhou City and a new landmark of metropolitan fashion.

五四北泰禾城市广场
The Thaihot City Plaza, Wusibei District

项目地点：福建 福州
用地面积：48 000 m^2
建筑面积：300 000 m^2

Location: Fuzhou, Fujian
Site Area: 48,000 m^2
Building Area: 300,000 m^2

永荣城市广场
Eversun City Plaza

项目地点：福建 长乐
用地面积：27 000 m^2
建筑面积：73 000 m^2

Location: Changle, Fujian
Site Area: 27,000 m^2
Building Area: 73,000 m^2

该项目由1幢单层面积8000 m²的四层大体量商业综合体和一条“[”形的两层商铺所组成，两者用天桥相连接，形成商业人流的互动，满铺一层地下车库 。在“[”形的二层商铺中部顶上设置了三层的公寓式商务办公楼。项目建成后，将引进大型百货、国际连锁超市、梦幻影院、精品零售、高档餐饮、儿童欢乐天地、休闲文化娱乐等业态，成为长乐市首占新区乃至整个长乐市的首席国际购物天堂，带给长乐人民前所未有的娱乐、休闲、购物新体验。

This project includes one large commercial complex of 4 floors with each floor area of 8,000 m^2 and one "[" shaped 2-floor stores connected by a flyover, to form commercial pedestrian interaction. In the middle of the "[" shape stores, on the second floor sets 3-floor apartment commercial office building. Upon completion, the project will introduce large department store, international chain supermarket, dream theatre, boutique retail, hi-end catering, children entertainment, leisure and cultural recreation etc. This project will become the primary international shopping heaven in Shouzhan New Zone of Changle City or even the whole city, bringing Changle citizens unprecedented new experience of entertainment, leisure and shopping esperience.

唐乾明月景观设计
Moon Beside the Lake Landscape Design

项目地点：福建 福州
用地面积：80 000 m^2
建筑面积：80 000 m^2

Location: Fuzhou, Fujian
Site Area: 80,000 m^2
Building Area: 80,000 m^2

项目位于福州市"后花园"的大樟溪旅游圈核心，周边旅游景点星罗棋布，地块原生态植被保留良好，是福州市首个旅游地产，也是首个直接定位为第二居所的别墅项目。根据其得天独厚的自然生态环境以及低密度、低容积率、新中式建筑风格、以联排别墅为主的建筑规划和50%绿化率的要求，项目以水系为景观主轴，因地制宜，结合道路、建筑、院落等空间的景观要求，不仅做出了户户亲水、移步换景的效果，而且营造出了恬淡幽静、绿意流动的诗意环境，确保了该项目成为福州市低密度高尚住宅的典范。

The project is located in the core Dazhangxi River, the backdoor of Fuzhou City, dotted with tourist destinations. The project maintains the ecological vegetations on the plot, and is the first tourism-themed real estate, also the first villa project positioned directly to be secondary habitat of Fuzhou City. According to its unrivalled natural ecology and low-density, low-plot ratio, neo-Chinese style, townhouse-keynoting architectural planning, and 50% greening ratio requirements, the project follows the water system as its landscaping main axis, together with roads, buildings, courtyards and other space expressions, not only to create the effects of water approaching every unit and different sight at every step, but also to create a poetic experience of tranquility and greenness, which ensures the project to become the model of low density noble residential community in Fuzhou City.

福州博林环境艺术设计顾问有限公司

福州博林环境艺术设计顾问有限公司系嘉博筑业集团旗下专业从事环境景观规划设计、景观工程施工、监理及咨询的专业性设计咨询公司。自2004年成立以来，公司在追求创新、时尚、永恒、人性化设计理念的同时，积极探寻每一个作品的时代亮点，力求完美地体现作品艺术性、实用性与经济性的和谐统一。在概念设计、扩初设计、施工图设计、招投标及施工督造等各个环节，凭借自身及嘉博筑业集团丰厚的技术资源，在景观规划、种植设计及景观小品、景观结构构造、景观照明及水景工程等专业领域为客户提供更专业、更协调、更精细的服务，从而最大限度地提升客户的项目投资价值。

地址：福建省福州市鼓楼区东大路36号花开富贵A幢20层
电话：+86-591-87430671
传真：+86-591-87430669

福州火车南站广场景观照明设计
Landscaping and Lighting Design of Fuzhou Railway South Station Square

项目地点：福建 福州
用地面积：470 000 m²

Location: Fuzhou, Fujian
Site Area: 470,000 m²

福州火车南站是一座现代化的铁路客运特等站，全国十大区域性客运交通枢纽之一，汇集高速铁路、城市轨道交通、地铁、城市公交、公路客运、大型停车场及商业购物中心于一体。总体布局为“五区”，分别为站房区、广场区、市政配套区、商业开发区及自然山体景观区。

Fuzhou Railway South Station is a modern railway special station, one of the top 10 regional passenger terminals in China. It integrates railway, urban traffic, subway, shuttle bus, long-distance bus, large parking lot and shopping center. The general layout comprises of “five zones”, namely station zone, square zone, municipal facilities zone, business development zone and natural mountain landscaping zone.

福州博维斯照明设计有限公司

福州博维斯照明设计有限公司是嘉博筑业集团旗下专业从事照明设计、照明技术咨询服务与研究的服务机构。公司引入境外先进的照明设计理念，以全新的视野、专业的服务精神，为客户提供众多照明解决方案，包括城市照明规划、各类室内空间照明设计、建筑外观照明设计、各类景观照明设计，并为客户提供各类照明顾问、咨询与培训以及企业标准的制定，力求用最少的光，设计出最好的照明效果。另外从概念规划、方案设计、后期施工及营运管理等方面为客户提供更合理的照明解决方案，为客户的投资价值最大化及品牌提升起到积极的推动作用。

地址：福建省福州市工业路523号福州大学东门行政办公楼附属楼二层
电话：+86-591-87430670（总机）
传真：+86-591-87430672

福州嘉景装饰工程有限公司

福州嘉景装饰工程有限公司系嘉博筑业集团旗下从事建筑室内外装饰装修工程设计及施工的专业公司，具有建筑装饰设计乙级及施工二级资质，现有一批经验丰富、高素质的装饰设计师、专业工程师、施工管理人员及各工种的技术工人，形成各专业配套齐全、组织机构完善的设计及施工实体，已成功地组织了近百个大中型室内外装饰工程的设计及施工，特别在房地产项目营销中心装修、星级酒店装修、商业用房装修、样板房及精装修等业务上积累了丰富的业绩和成功经验。

地址：福建省福州市鼓楼区东大路36号花开富贵A座23层
电话：+86-591-87851116
传真：+86-591-87859335

融侨水乡酒店装饰设计
Decorative Design of Rongqiao Riverview Hotel

项目地点：福建 福州 Location: Fuzhou, Fujian
建筑面积：20 000 m^2 Building Area: 20,000 m^2

该项目是福州唯一一家具有闽江江景的五星级酒店，拥有189间各式豪华客房及套房、可同时容纳数百人用膳及会议需求的680 m^2无柱式国际宴会厅以及一流的休闲健身设施，包括温热按摩浴池、理疗室、桑拿蒸汽浴室、室内恒温游泳池、室外网球场等。其24h纯天然温泉、法国Groupe GM系列高端品牌豪华洗浴用品、电动冲洗坐便器、电动静音罗马窗帘、柔软舒适的雪仑尔面料床上用品、全套欧式风格家居及装饰，尽显尊荣华贵。

This project is the only 5-star hotel with Min River view. It has 189 luxury guestrooms and suites, and a columnless international banquet hall which can accommodate hundreds of guests for banquet or conference, as well as first-class recreational and fitness facilities, Such as warm/hot massage bath, therapy room, sauna steam bathroom, indoor heated swimming pool, outdoor tennis court and other facilities. Its 24-hour natural hot spring, Groupe GM series hi-end luxury bath supplies, electric flush toilet, electric mute Roman curtains, soft and comfortable SHLR fabric bedding, full set of European-style decoration, show outstanding nobility and luxury.

扫描查看更多信息

福州市建筑设计院
Fuzhou Architectural Design Institute

福州市建筑设计院成立于1956年3月，是全国第一批勘察设计甲级设计院，具有国家甲级建筑行业建筑工程设计、工程勘察专业类岩土工程、工程咨询、工程监理、工程项目管理和房建一类施工图审查等多种资质。设计院技术力量雄厚，是福建省勘察设计骨干单位。

设计院现有员工230多名，各类专业技术人员近200人，其中教授级高工、高级职称者约占40%，并有一位工程勘察设计大师；拥有注册执业资格的各类专业技术人员近100人，其中一级注册建筑师和一级注册结构工程师42人，注册设备师、岩土师、监理师、建造师、造价师等50多人。

设计院秉持“优质、诚信、创新、高效”的企业精神，坚持“品质成就理想，服务提升价值”的质量方针，锐意进取、开拓创新，勘察、设计、监理、图审项目遍及八闽大地，服务范围辐射相关省份。先后有100多项勘察设计成果荣获全国优秀工程勘察设计行业奖和福建省优秀勘察设计奖或科技进步奖，连续多年获得“重合同、守信用”企业、行业文明单位等荣誉称号.设计院勇于担当社会责任，在社会取得较好口碑，在业界享有较高声誉。

地址：福建省福州市鼓楼区津门路32号
电话：+86-591-87554471/87556803
传真：+86-591-87616491
邮箱：fadi@fzjzy.com
网址：www.fzjzy.com

Add: No.32, Jinmen Road, Gulou District, Fuzhou City, Fujian Province
Tel: +86-591-87554471/87556803
Fax: +86-591-87616491
E-mail: fadi@fzjzy.com
Web: www.fzjzy.com

金顶国际城市综合体
Jinding International Urban Complex

项目地点：福建 霞浦
建筑面积：324 280 m²

Location: Xiapu, Fujian
Building Area: 324,280 m²

金顶国际城市综合体工程设计的项目总建筑面积约324 280 m²，包括高级酒店、商业写字楼及高档住宅。设计方案立足项目整体，结合城市设计理念，力图将项目打造成一座集“绿色生态、节能环保、特色水景、国际品质、主题商街、示范新城、和谐生活”为一体的宜居住区。

海峡出版物博览交易中心
Strait Publications Expo Trading Center

项目地点：福建 福州
用地面积：1.5 hm²
建筑面积：72 000 m²

Location: Fuzhou, Fujian
Site Area: 1.5 ha
Building Area: 72,000 m²

项目构思以“图书”这一传统的出版物为原型，试图运用形体的错动以及不同材料透光性能的排列组合，使建筑让人产生类似“条形码”式的感受，探索信息时代文化建筑的设计语言，实现新华发行集团企业文化和建筑设计理念的对应。

南平武夷名仕园
Wuyi Nobleman Garden, Nanping

项目地点：福建 南平
用地面积：130 274 m²
建筑面积：450 000 m²

Location: Nanping, Fujian
Site Area: 130,274 m²
Building Area: 450,000 m²

项目集五星级酒店、办公、公寓、商业及住宅为一体，通过综合体的功能组织实现项目整体价值提升。用地临江同时又有80 m的场地高差，设计充分利用高差形成层次丰富、逐级打开的山水步行主轴，将高差、挡墙、室外竖向观景电梯以及天桥等山地建筑要素有效组织起来，形成富有特色、现代化的城市住区。

家天下A二期、B一期地块三盛中央公园
Home World Plot A Phase II, Plot B Phase I Sansheng Central Park

项目地点：福建 福州
建筑面积：140 000 m²

Location: Fuzhou, Fujian
Building Area: 140,000 m²

建筑布置与用地条件相协调，采用南低北高的错落体量布置，在导入阳光通风的同时，形成柔和、渐变、开阔的居住空间，与周边建筑形态融合。中央景观区与主入口连接，成为视觉的焦点，给居者良好的空间感受。建筑体量及造型风格协调统一，建筑材料选用恰当，营造了居住环境的尊贵感。

厦门合道工程设计集团有限公司

Xiamen Hordor Engineering Design Group

厦门合道工程设计集团由有着40多年历史的原厦门市建筑设计院改制发展而来。1960年厦门市建筑设计院成立，2003年经改制成立厦门市建筑设计院有限公司。2007年8月厦门市建筑设计院有限公司及下属十余家参控股公司，本着“道同而融合，循道而成业”的事业愿景，正式成立“厦门合道工程设计集团”（下简称“合道设计集团”）。

合道设计集团具有建设部颁发的建筑工程设计甲级资质，国家发改委颁发的工程咨询甲级资质以及城市规划、市政工程（景观）、建筑智能化系统工程设计资质等。

合道设计集团拥有规范的现代企业制度、科学有效的管理机制和灵活的市场应变能力，集团立足设计主业，专注打造设计品牌，业务涉及建筑规划、建筑设计、景观设计、施工图审查、智能化设计、工程咨询及经济咨询、房地产开发与策划等领域，客户遍及北京、上海、深圳、福州、武汉、大连、青岛，南昌、昆明、成都、济南等地，迄今已与美国、加拿大、意大利、日本、英国、德国、西班牙、澳大利亚、新加坡、中国香港、中国台湾等国家和地区的设计机构建立了广泛的合作关系。此外，集团始终致力于人才成长平台的研究和打造，培育了一支功底扎实、思想敏锐、敬业守纪的专业技术队伍，其中教授级高工、高级工程师、工程师人数占公司人员总数的44%以上。至今集团已有百余个工程项目获得部省级优秀设计奖项，赢得中国建筑设计行业分会华东联席会“2001–2005年度改革与发展最具创意奖”，并荣获中国勘察设计协会“2006年优秀勘察设计企业”、福建省勘察设计协会“2006年度优秀勘察设计单位”、“2007年中国十大民营建筑设计企业”、“2010年中国最具影响力品牌设计机构”、“2011年中国十大民营建筑设计企业”等荣誉称号。

汗水与荣誉，变革与坚守，50多年风雨历程，合道工程设计集团历久弥新。今天，他们一如既往地关注着您的生活，构筑着城市的蓝图，他们正以专业的精神打造百年合道品牌。

地址：厦门市湖滨南路107号
电话：+86-592-2298061
传真：+86-592-2298062
邮箱：rencai@hordor.com
网址：www.hordor.com

Add: No.107 Lakeside South Road, Xiamen City
Tel: +86-592-2298061
Fax: +86-592-2298062
E-mail: rencai@hordor.com
Web: www.hordor.com

厦门高崎国际机场T4航站楼

Terminal 4 of Gaoqi International Airport, Xiamen City

项目地点：福建 厦门
用地面积：34 343.7 m²
建筑面积：74 092.6 m²
建筑密度：62%

Location: Xiamen, Fujian
Site Area: 34,343.7 m²
Building Area: 74,092.6 m²
Building Density: 62%

厦门高崎国际机场新建T4航站楼位于机场东北端，航站楼用地北侧为现有跑滑系统及与新建T4航站楼同期建设的机坪，南侧为太古七期、八期规划用地，西侧为机场规划建设的东扩二段机坪，东侧为集美大桥下穿隧道。

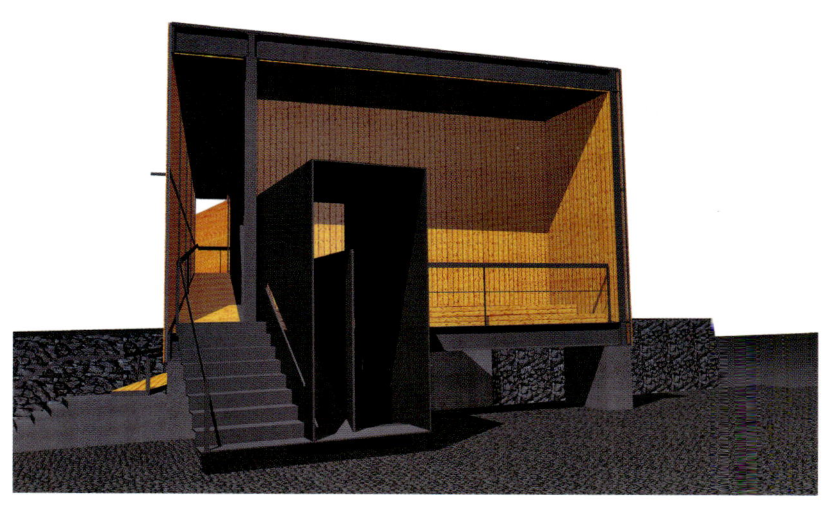

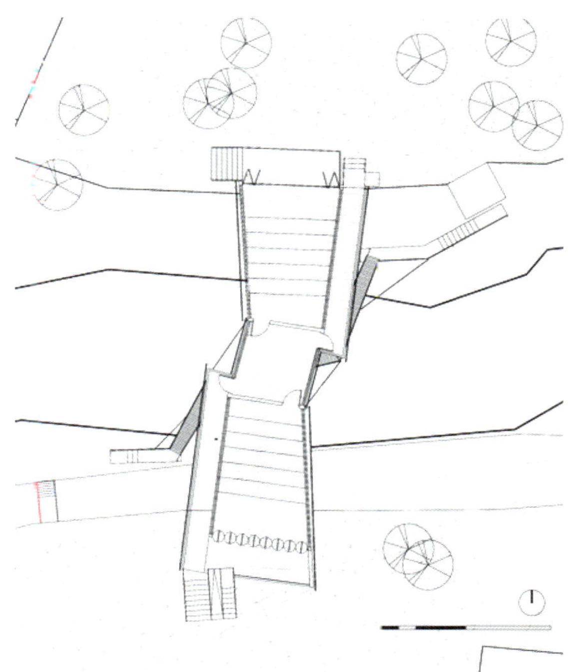

崎岭乡下石村桥上书屋

Xiashi Village Bookstore on Bridge Qiling Town

项目地点：福建 漳州	Location: Zhangzhou, Fujian
用地面积：15 507 m²	Site Area: 15,507 m²
建筑面积：2407 m²	Building Area: 2,407 m²
容 积 率：1.33	Plot Ratio: 1.33
建筑密度：62%	Building Density: 62%

项目位于福建省漳州市平和县崎岭乡下石村，在两座保留相对完好的土楼中间。功能非常简单，两个阶梯教室，1个小图书馆，1个小便利店。建筑师将3个功能块全部安置在桥上，教室单侧的走廊通向中间的图书馆，两个教室在两端，分别朝向两个土楼。书屋不仅解决河两岸的交通联系，成为两个村子共同的公共空间和交流场所，还是孩子们用以学习的教室和玩乐的场所。整个建筑采取钢桁架结构，桁架内部的整体空间作为主要的教室用途，两个教室中间设置公共空间。教室外侧设置走道，在教室和表皮之间增加一道观景通廊。下部用钢丝悬吊过河的公共桥梁，桥梁为“Z”字折线形，建筑师用朴素的语言保持着一贯的简洁和优雅。

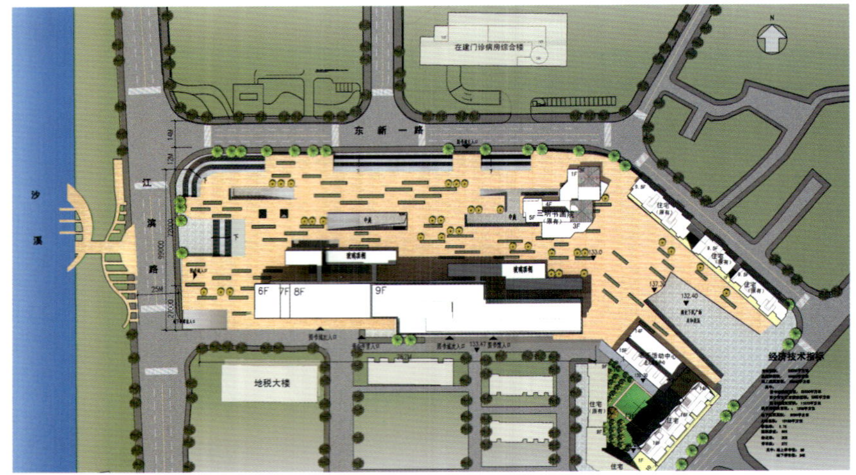

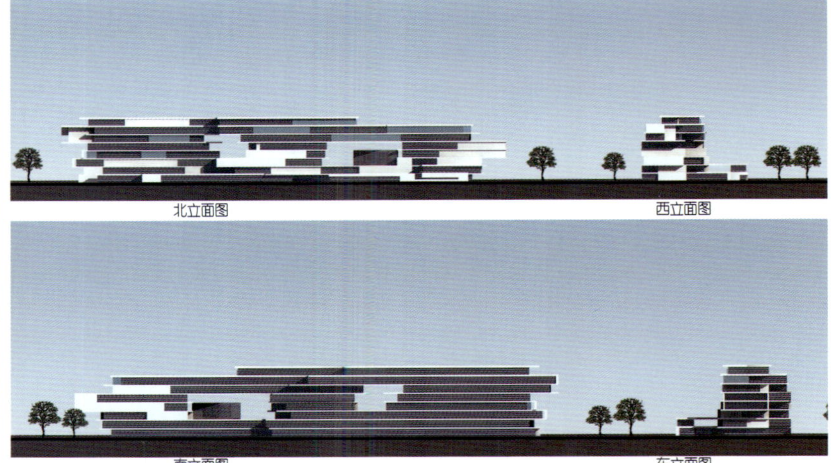

三明市图书馆、图书城、青少年宫改造项目

Renovation of Library, Book Town and Youth Palace, Sanming

项目地点：福建 三明
用地面积：23 000 m²
建筑面积：49 260 m²

Location: Sanming, Fujian
Site Area: 23,000 m²
Building Area: 49,260 m²

项目主要由三个部分组成：市图书馆新馆、闽中客家文化出版交流中心、青少年宫。图书馆的主要结构包括阅览大厅、采编室、办公、会议室、报告厅、书库；闽中客家文化出版交流中心包括闽台图书发行中心、客家文化出版展示中心、图书交流会展中心、学术报告厅、图书流转配送中心；青少年宫包括展厅、教室、办公区、青少年阅览室、动漫知识厅、会议室、活动中心、多功能厅、报告厅等。

平面功能的设置遵循以下原则：各功能分区均分别设置出入口，避免不同人流的交叉干扰。同时在设置功能时注意功能的竖向分区，人流多的设置在较低层，而人流少的设置在较高层。

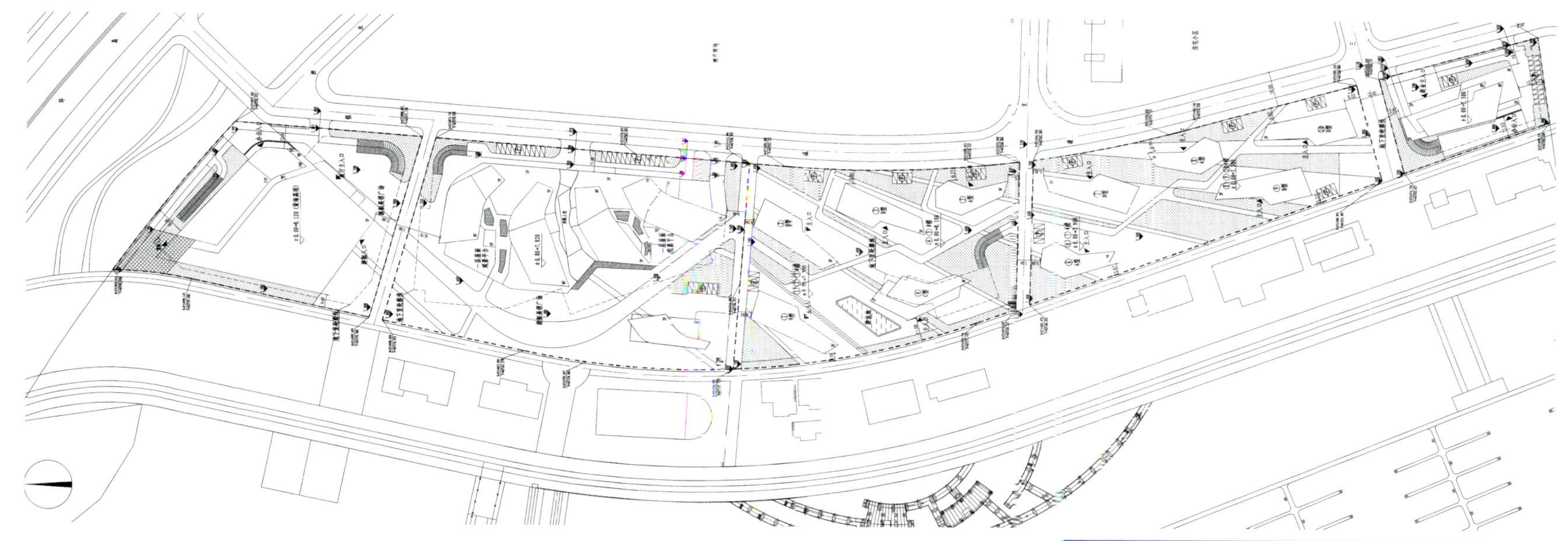

厦门五缘湾游艇帆船国际展销中心

China International Boat Show Center, Wuyuan Bay, Xiamen

项目地点：福建 厦门　Location: Xiamen, Fujian
用地面积：55 937.066 m²　Site Area: 55,937.066 m²
建筑面积：83 696 m²　Building Area: 83,696 m²

该项目位于厦门五缘湾游艇帆船港腹地，包含主题展览馆区、配件展示区（含保税仓库）、游艇品牌展示区、游艇商务办公区4个功能区。

项目的设计构思结合了游艇、帆船元素，充分发挥海景优美的地理优势，创意新颖、造型独特。为了营造海的场所感，规划以水纹理为主概念，主轴线以流动的曲线为线索，以沿环岛路开口道路的延长线，串联5个地块，融合放射性的有机线性序列，创造流动性的空间；立面概念源于风吹水动的涟漪的痕迹，以增强区域场所感。因而，从建筑的各角度观望，尤其是从五缘湾大桥远观，弧形的建筑以扇形舒展有张力的群体布局，高低错落有序，从北往南的富有张力的动感天际线，形成良好的建筑形象，成为五缘湾片区标志性的建筑群。

1

2

3

4
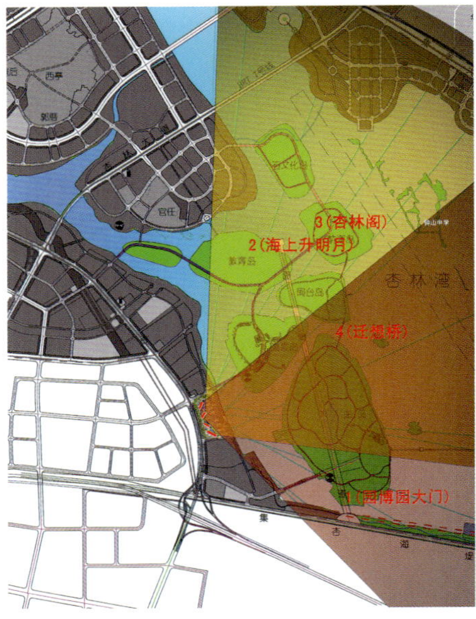

景观视线区域分析

总平面图

园博新天地
Garden New Field

项目地点：福建 厦门
用地面积：42 231 m²
建筑面积：183 340 m²

Location: Xiamen, Fujian
Site Area: 42,231 m²
Building Area: 183,340 m²

园博新天地位于集美杏林湾片区，项目力图构筑绿色生态，国际高端物业型豪宅社区。建筑风格是现代滨海建筑风格，个性鲜明，标志感强，通过现代感十足的立面风格打造杏林湾片区的标志性建筑。

该设计把握空间态势，大胆引入水系，形成“两轴”、“一心”、“四区”脉络清晰的规划组织结构。此外景观方面，最大化利用杏林湾片区及园博园景观，提升和放大整个区域的价值。注重中央水系和各组团中心的景观联系，观赏性景观和绿化有机结合，形成连续景观空间序列。车行系统独立，绕场地建筑外围连接南北两车行出入口。临水建筑、各栋别墅都有游艇码头，由社区可直通杏林湾及广大的海域。沙滩，健康步道贯穿社区南北，创造人性化的纯步行景观系统。

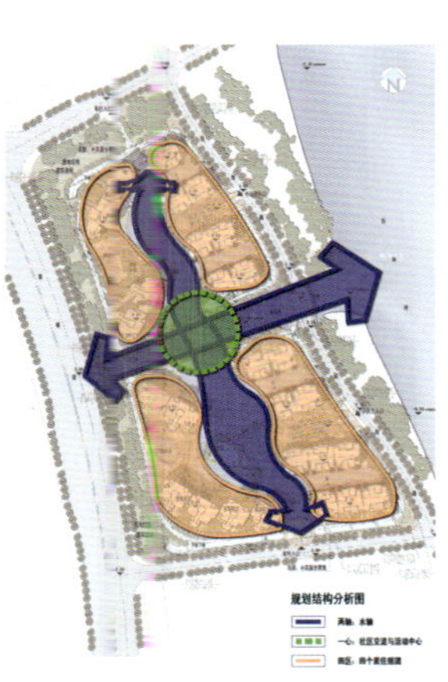

扫描查看更多信息

广州市纬纶建筑设计有限公司
Win-land Architecture Design Co.,Ltd.

········设 计 无 限 可 能········

广州市纬纶建筑设计有限公司是一个具备雄厚实力的设计团队，旗下的智海建筑设计公司及艺筑建筑设计公司分别拥有建筑行业甲级及乙级资质。纬纶专注于房地产开发建筑设计，在全国完成多项工程，对房地产项目总体规划、市场状况、小区空间环境及住宅户型有着专业研究，向客户提供商业、零售中心、公建、酒店、住宅等多个领域的建筑设计服务。
纬纶以设计“无限可能”为核心理念，以精益求精的态度完成每一项设计作品。创立14年来，累计为客户完成建筑过2000万m²，为60万人提供了优质舒适的居住环境，以精品的设计和优质的服务在客户群中树立了极佳形象。现已拥有合景泰富、万科集团、新鸿基地产、美林基业、天伦控股、苏宁集团、佛奥集团、保利集团、佛山能兴地产、兴发铝业等知名集团为长期客户。项目辐射到山东省及佛山、茂名、惠州、湛江、梧州、贵港、南宁、南昌、武汉、南京、郑州、海口等地，其中合景泰富·科汇金谷大型项目、佛奥·星光广场商住项目、佛山宾馆、万科·天景花园、美林湖畔别墅、美林海岸住宅小区、天伦林和村城中村改造、天伦·郑州龙湖小区、苏宁·天润城等重点项目更在业界获得一致好评。“正·大·光·明”是纬纶作品的最大特色，大气的规划、户型，宽敞的空间，最优的自然采光、明快开敞的建筑感受，令地产商及住户的多层次需求得到充分满足，销售业绩不断创出新高。
在建筑业风起云涌的今天，纬纶坚持“无限可能”的设计理念，独特的建筑设计特色，以专业的触觉及市场敏感度，致力于打造出符合客户和市场需求的设计精品。

Win land is a resourceful design team, with subsidiaries Guangzhou Zhihai Architectural Design Co., Ltd. and Guangzhou Yizhu Architectural Design Co., Ltd. licensed for class-A and class-B building industry qualifications. Win-land focuses on real estate development and architectural design, has finished many projects across China. It has done with precise researches on real estate project general planning, market conditions, community space environment and residential unit, and offers customers with architectural design services in commercial complex, retailing center, public buildings, hotels, residence, planning among many other fields.
Win-land follows the core philosophy of “design of limitless possibilities” to finish every design work with elaborate attitude. During the 14 years since its establishment, it has completed the construction of floor area totaling 20,000,000 m², providing high quality and comfortable residential conditions for 600,000 persons, and its exquisite design and quality services have won good reputation among customers. Now it has many famous long-term group customers, including KWG Property, Vanke Group, SHK Properties, Mayland, Tianlun Holding, Suning Group, Foao Group, Poly Group, Foshan Nenking Group, Xingfa Aluminum Holdings. Its projects radiate to Shandong Province, Foshan, Maoming, Huizhou, Zhanjiang, Wuzhou, Guigang, Nanning, Nanchang, Wuhan, Nanjing, Zhengzhou, Haikou and other places. Of its projects, KWG International Creative Valley project, Foao Star Plaza commercial & residential project, Foshan Hotel, Vanke Skyview Garden, Mayland Lakeshore Villas, Mayland Coastal Viuas Residential Community, Tianlun Linhe Village Urbanization Project, Tianlun Zhengzhou Longhu Community, Suning Spain Town and other key projects have won good reputation in the industry. “Right, Large, Bright, Shining” are the biggest features of Win-land products. Its magnificent planning and unit pattern, broad space, optimal natural lighting and bright feelings, greatly satisfy real estate developers and users' multi-level requirements, creating new records of sales performance.
In the surging architectural industry, Win-land follows the design philosophy of “limitless possibilities”, with unique architectural design features, professional sense and market sensitivity, committed to creating exquisite designs meeting the requirements of customers and the market.

地址：广东省广州市海珠区琶洲大道8号701-708室
邮编：510335
电话：+86-20-89890986
传真：+86-20-37621177
邮箱：winland@win-land.com
网址：www.win-land.com

Add: Room 701, 8 Pazhou Avenue, Haizhu District, Guangzhou City, Guangdong Province
P.C.: 510335
Tel: +86-20-89890986
Fax: +86-20-37621177
E-mail: winland@win-land.com
Web: www.win-land.com

扬翔办公楼

小龙·盛世华府

天津星光天地
Star World, Tianjin

项目位置：天津
用地面积：17 515 m²
建筑面积：173 676 m²

Location: Tianjin
Site Area: 17,515 m²
Building Area: 173,676 m²

佛奥地产携手天津金盛置业发展有限公司在天津市中心地铁上盖物业合作经营大型商业地产项目。项目将打造成一座规模宏大、功能齐全的现代型综合购物中心，集购物、游览、美食、娱乐、休闲、商务、广告、信息、展览、康体等多功能于一体，在当地开创一种全新的消费概念，带动周边地带的繁荣，使之成为天津现代商业的标志，成为市民和旅游宾客购物观光休闲的首选场所。

天津星光天地项目原名金盛广场，由A、B两座高层公寓和6层的商业裙楼组成，原商场平面是以百货业态为基础的设计，中庭及商业动线布局不能满足新时代商场的发展需求，需要进行改造，具体改造的内容有：

1. 3个中庭重新布局设计，使商业动线流畅、连续并富有生气；
2. 调整原有电动扶梯位置，新增电动扶梯及跨层扶梯，引导人流到达各层商铺；
3. 西北角增加一个商场出入口，并加强出入口的空间、造型的设计，吸引人流；
4. 重新设计外立面，加强广告、橱窗展示空间的设计，展示现代高档新形象；
5. 将所有结构、梁柱重新改建并满足消防规划要求。

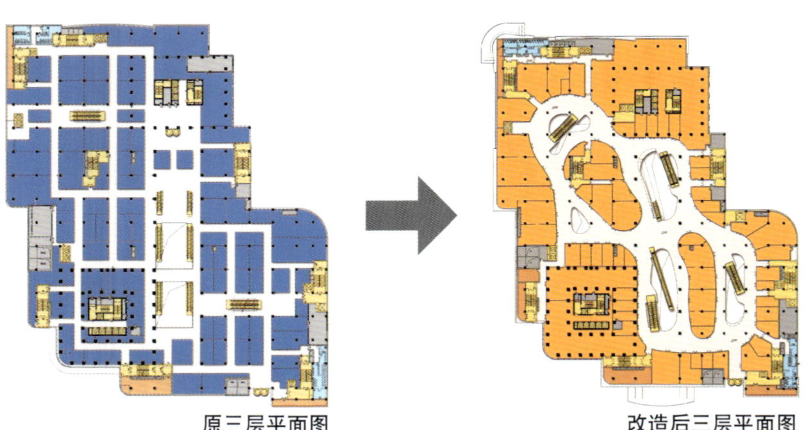

原三层平面图　　改造后三层平面图

改造前

改建后效果图

广州沙步社区城中村改造项目

佛山奥园第五期住宅小区

南京苏宁·天润城

扫描查看更多信息

广州珠江外资建筑设计院有限公司
Guangzhou Pearl River Foreign Investment Architectural Design Institute

广州珠江外资建筑设计院是华南地区著名的国有综合性建筑工程甲级设计院。
设计院和位于珠江之滨的白天鹅宾馆一起诞生，伴随着祖国改革开放的大潮崛起，记载了南粤大地建设发展的辉煌业绩。
建院以来，设计院坚持技术进步、科学管理、繁荣创作，高质量完成了广州及全国各地面积达上千万平方米的工程设计项目，项目涵盖公共建筑、居住建筑、景观园林和室内装饰等领域。作品创新意识浓郁，设计风格独特，既有岭南文化的韵味，又具有现代建筑的风采。众多项目荣获国家、部、省、市级优秀设计及科技进步奖，赢得了客户及社会各界的广泛赞誉。

地址：广州市环市东路360号珠江大厦东16楼
邮编：510060
电话：+86-20-83843732
传真：+86-20-83846074
邮箱：mail@pearl-river.com/zjtb16@126.com
网址：www.pearl-river.com

Add: 16th Floor East Tower, Pearl River Building, 360 East Huanshi Road, Guangzhou
P.C.: 510060
Tel: +86-20-83843732
Fax: +86-20-83846074
E-mail: mail@pearl-river.com/zjtb16@126.com
Web: www.pearl-river.com

广州亚运媒体村项目
Asian Games Media Village, Guangzhou

项目地点：广东 广州
建筑面积：422 332 m^2

Location: Guangzhou, Guangdong
Building Area: 422,332 m^2

总体构思——内纳外延，动静皆宜；文脉传承，生态簇团；包容开放，内敛高效。
设计在研究岭南水乡聚落特色基础上，追寻“依山而建、活水穿村”的布局特点。布局上，住宅整体上采用外围行列式布局，内部点式或拼式结合，让更多的住宅享受区内安静、优美的环境。呈开放姿态的中央景观与组团景观互相渗透，形成居住区内建筑与自然的流动空间。
设计将地块划分为3个生态组团，在住区组团内利用花带、草坡等景观元素模拟水乡河网的肌理，将便捷的人行交通系统、曲线状的休憩小径叠加在原生态组团中。同时，利用地块现有的景观优势并加以整合，造景与借景设计手法并用，公共景观、组团景观和庭院景观层次分明、此起彼伏，形成一个建筑外密内疏、高低错落、山环水绕、绿树成荫、空间流动的新式岭南水乡社区及亚运媒体居住区。

General conception – inclusive yet extensive, dynamic yet static, cultural, ecologic, open and efficient.
Based on southernmost Chinese village features, the project seeks for a layout of “following the hilly landform with a river penetrates the village”. The layout of a residential buildings are linear outside, and point or mix type inside, to make more residential buildings enjoy a quiet and beautiful environment. The open central landscape and cluster landscape penetrate each other, forming a flexible space between residential community and the nature.
The design divides the plot into three ecologic clusters. The residential building cluster uses flower strip, grass slope, and other landscaping elements to mimic the texture of a waterside village's river network, and add convenient pedestrian traffic system and curved leisure path to the primitive ecologic cluster. It integrates the existing landscapes, manages to design or borrow scenery, arranges public landscape, cluster landscape and courtyard landscape in distinct layers, up and down, so as to form a new southernmost Chinese waterside village and Asian Games media residential community with buildings intense outside and loose inside, of different heights in order, with embracing mountains and circling waters, with green trees and flexible spaces.

广州辛亥革命纪念馆

The 1911 Revolution Memorial Museum, Guangzhou

项目地点：广东 广州	Location: Guangzhou, Guangdong
建筑面积：18 652.1 m²	Building Area: 18,652.1 m²

“石”——艰难险阻。
建筑造型如一方石块，平静地置于历史的长河（水）之中。石块已被凿开，隐喻辛亥时期英烈们开天劈地之勇气和艰苦卓绝的历程。
“足迹”——寻迹之旅
广场设计线条流畅，富于指向性的地景灯带和纪念馆中庭内墙上的条形灯带将人流自然引入建筑主题，步入百年辛亥的寻迹之旅。
“路”——共和之路。
建筑入口是一座跨越水面，插入石穴的“路”——寓意“共和之路”。孙中山先生领导的辛亥革命英烈在走向共和之路上前仆后继，百折不挠。
9.6 m 宽的道路从缺口穿入，笔直、向上。道路上随机陈列着辛亥英烈铸铜雕像，似在沿路而行。作为参观流线的主轴，当人们穿梭于英烈雕塑丛中，仿佛时空又回到了辛亥革命那激情燃烧的岁月。沿路穿越昏暗的石穴，尽端豁然开朗，象征着辛亥革命推翻了清朝帝制，走向共和，为中国资产阶级民主革命迎来了胜利的曙光。中山先生背影雕像矗立在路的尽头，又是路的最前方——面朝蓝天，心系远方，似在思索着中国的未来。

“Rock” – Hardships
The architectural shape is like a rock, lying in the long river of history. The rock is rived, implying the pioneering courage and painstaking journey of The 1911 Revolution.
“Footprint” –Journey of Findings
The square is designed in smooth lines, where the indicative landscaping light poles and the light trip on the atrium interior wall introduce the pedestrian flow into the building theme, and go to the journey of findings of The 1911 Revolution.
“Path” – Path to Republic
The entrance is a path spanning water surface and inserting into a stone cave, meaning the “path to republic” that Dr. Sun Yat-sen was leading with a great number of followers unremitting and battling.
The path is 9.6 m wide, penetrating from the opening straightly to the upward. On the path display a lot of bronze busts of martyrs in that revolution, and they seems walking along the path. It serves as the main axis of sightseeing flow, so the pedestrians can experience the revolutionary age in that space. Passing through the dark cave, the pedestrians finally come out and embrace brightness, symbolizing the fact that the revolution overthrew the Manchurian imperial monarchy, and transformed China into a republican democracy, bringing sunlight of victory to Chinese bourgeois democratic revolution. The back sculpture of Dr. Sun Yat-sen was at the end of path, also the foremost front of the path – facing the sky, and looking in the far, he is pondering on the future of China.

柳州工业博物馆
Liuzhou Industrial Museum

项目地点：广西 柳州
建筑面积：62 307 m²

Location: Liuzhou, Guangxi
Building Area: 62,307 m²

设计理念："追忆激情燃烧的火红年代"
红色是柳州工业历史发展中的一个印迹深刻的颜色，也象征着柳州人激情澎湃、敢为人先的工业精神。设计以红色作为主线，串联起一个个熟悉的场景，唤起柳州人民对那段火红的年代的回忆。

Design philosophy: "the memory of passionate red age"
Red is a footprint color in the industrial development history of Liuzhou City. It symbolizes the energies of Liuzhou people, and their pioneering industrial spirits. The design is clued by the red color, to line up a string of familiar scenes, inspiring citizens to call back that red passionate age.

武汉万达中央文化区秀场
与六星酒店
Wanda Central Culture
District & 6-star Hotel,
Wuhan

广州东风中路 S8 项目
S8 Project, Dongfeng Middle Road, Guangzhou

广州市气象监测预警中心

Guzngzhou Metrological Surveillance & Pre-alarming Center

项目地点：广东 广州
建筑面积：9597.7 m²

Location: Guangzhou, Guangdong
Building Area: 9,597.7 m²

总体构思：和谐地景——嵌入自然环境中的地景式建筑形象。
规划设计上设计师将建筑作为大地景观的元素，结合场地的现状条件，利用开挖地下室的土方，采取削高填低的平衡土方的方法，通过地景式的建筑形象，创造性地还原场地的丘陵自然地貌。建筑通过不同标高的屋顶花园与自然坡地相连，建筑内部的步道延伸进自然环境中，自然环境通过条状的绿化屋面和坡道将自然景观延伸进建筑里面，最终达到建筑与环境的交融。建筑犹如从山坡上生长出来，造就出从基地环境自然生长而出的地景式建筑，借此营造节能、生态、环保的低碳型建筑。

General conception: harmonious landscape – a landscaping architectural image integrated into natural environment.
In the planning and design, architecture is regarded as an element of landscape. In accordanct with the condition, we use the earthworks escavating the basement, and creatively reproduce the hilly topography by lowering and filling to balance earthworks and the layout of landscaping architectural images. The architecture connects with natural slope through rooftop gardens in different elevations. The walkway inside the building extends into natural environment, while natural landscape extends into the architecture by strip greening roof and ramp, to finally integrate the architecture and the nature. The architecture is growing from the slope, to create a landscaping architecture growing from the base environment, so as to create the energy-saving, ecologic, environment-friendly and low carbon architecture.

广州市扉越建筑设计有限公司

扉建筑在过去的6年中，与国内外多家知名企业（太古汇广州地产发展有限公司、岭南集团、保利地产、四海饮食集团、立信集团、中华民族文化促进会、例外服饰、流行美等）合作，创作了一批优秀的建筑及室内设计项目。在办公总部、度假酒店、文化艺术建筑、商业及三旧改造等几个专业领域积累了宝贵的经验，为客户提供从建筑策划、建筑设计、室内设计、景观概念至艺术筹划的整体解决方案。

建筑是艺术与技术综合素质的反映。公司在进行大量设计实践的同时于2007年开始运营广州一个最重要的民营当代艺术机构扉艺廊，并于2008年创办原创设计品店扉卖品。围绕“打开无限可能”的宗旨，与中外文化机构、艺术家、设计师一同举办了近百场当代艺术展览、活动、演出、沙龙和跨界合作，成为南中国一个重要的艺术文化交流平台。

扉建筑、扉艺廊、扉卖品涵盖了建筑及空间设计、艺术展览与文化交流以及原创设计与艺术衍生品三个方向的发展。扉建筑因跨界而创造了更多可能。

扫描查看更多信息

Functionality
机能
Elegance
诗性
Infinity
无限

通过设计企业总部，以独特的空间构成禅意的东方意象，将抽象的企业文化内涵转换为空间视觉语汇。

企业总部

亿达大厦

Estate Plaza

建筑面积：11 312.5 m²

Building Area: 11,312.5 m²

这幢大厦的功能不仅是为集团下属6个公司的500员工提供办公场所，更重要的是传递公司以人为本的企业精神和智取财富的营运能力。

首先要“舍得”。为获得纵深达15.5 m的无柱灵活分隔办公空间，增加近百万预算用于大跨度预应力超宽扁梁；为让出最具商业价值的首层，用手扶梯把人流带到宽敞的二楼大堂，这里有画廊、有咖啡，两层高的架空入口成为外观亮点，令不太高大的建筑亦显得气势不俗；因规划的限制标准层只有3.3 m层高，设计用“舍”的办法让内部空间“得”到更堂皇的效果——所有公司都设计成不同形式两层中空，虽然减少了几十平方米的使用面积，但换来了租用甲级写字楼都无法获得的空间魅力和丰富体验，体现了小中见大的设计特色。

其次，要成为员工的“家”。尽管寸土尺金，但设计师把消防前室稍稍加大，开放为两层一个的中空花园，是员工小憩的好去处，穿过花园是卫生间，有别于封闭式甲级写字楼。为鼓励步行，走火楼梯也设计了可以望向花园的圆窗，提倡节能环保之余亦可享受庭园景致。带有天窗的地下室则设有壁球室、淋浴间及红酒房，教同事如何能不爱“家”？

采用以人为本的设计手法，把低预算、高效、绿色、传递企业精神统一体现在这个小项目里。

广州岭南国际企业集团

Lingnan International Enterprise Group

建筑面积：4200 m²

Building Area: 4,200 m²

岭南集团是广州市政府于2006年重组的大型国有资产管理公司，负责管理广州花园酒店、中国大酒店、东方宾馆、爱群大厦等十数家四星及白金五星宾馆，同时拥有广之旅、广骏的士、皇上皇、致美斋等多个著名品牌商标。作为这样一个实力雄厚的本土企业，其总部既要展现岭南文化的特质，也要体现公司锐意改革的创新精神，其构思的源泉是“广州——海上丝绸之路的起点”。扬帆出海、不断探索、兼收并蓄，就是岭南文化的特质。海浪、帆船、渔网、沙滩、砾石都以崭新的材料，抽象的形态，融入了办公空间。

扉建筑 FEI ARCHITECTS 广州市海珠区南华东路草芳围2号自编151号财京商务公馆601-603 T:(020)37688779 F:(020)37688799 www.feiarchitects.com

广州鸣泉居环境综合整治工程
Guangzhou Oriental Resort

面积：3963 m^2

Area: 3,963 m^2

项目地处白云山风景区和大金钟水库，景色秀丽，为与园区现有建筑协调，建筑外观采取和谐策略，低调平和。建筑设计的重点是注重借景、造景手法的运用，达到窗窗有景；内部着重公共空间的序列组织构成不同风格的居住氛围。“欧式”以富丽堂皇的弧型楼梯中庭为核心展开对称的布局。“广东式”融入了岭南庭园的精华，拱桥、八角水厅、天井、小姐楼，景色层层展开。“现代”式则运用了自由流动的平面，螺旋梯，采光中庭等现代主义的经典。尽管施工周期短、经费不足，但由于建筑空间的比例尺度及组织方式已把各种风格表达得淋漓尽致，即便外观朴实平和，内装也相当简朴，但仍然创造了丰富而特色鲜明的空间体验和居住感受。

象泉山庄
Elephant Spring Inn

占地面积：12 000 m^2

Site Area: 12,000 m^2

象泉山庄位于距成都市区45min车程的石象湖风景区内，享有“冬到亚龙湾、夏至石象湖”的美誉。保利地产意在将其打造为进入成都市场5年后第一个标志性精品酒店。要精而不贵，也要简而不陋。项目充分利用现有地形的高差，形成由高低错落的庭园串联成的公共空间。原有的小山谷被改造成公共空间的主要景观。建筑与现有树林共生，屋中有树，树中有屋。坡顶平顶结合成小体量融于自然。以建筑空间的组织借景造景。减少室内装饰以便在潮湿的环境下便于维护并降低造价。把走火梯开放面向最美的景色，鼓励客人多用楼梯少用电梯，观念低碳而非技术低碳。以石、木等自然元素制作的艺术装置使建筑在质朴中透露灵气。

例外方所

Exception Fangsuo Commune

地点：广东 广州
建筑面积：1600 m^2

Location: Guangzhou, Guangdong
Building Area: 1,600 m^2

作为连续历史最长、最早登上国际舞台的中国女装品牌，例外Exception在商业运营和品牌传播上都获得了巨大成功。借品牌建立15周年之际，企业希望打造一间概念店，除了将业务由女装扩展至男装、家居服，更增加书店、咖啡店、生活设计用品、绿色植物及展览空间等。例外将不再仅是服装经营，而是另一种生活方式的倡导者，品牌的理念是用自然主义重新定义未来主义。空间设计从多个角度诠释了生活与艺术之间某种扑朔迷离的关系，植入例如书报亭、树屋等异型构造物，使得整个场地空间处处充满美学气息。

流行美首发空间

1st Stage Beauty in Fashion

面积：室内面积515 m^2

Area: 515 m^2 (indoor space)

流行美用10年时间完成了从一家小店铺到全国2000家连锁店中国发饰品第一领军品牌的蜕变。在获得联想基金的投资加盟后，企业希望通过打造一个概念空间以引领下一个10年的方向，同时达到对内对外宣传企业文化和塑造品牌形象的效果。

功能与空间组织：项目选址于旧纺织厂内一个8.5m×72m的车间，包括概念店、发型历史博物馆、图书阅览区、咖啡区、发型画廊、企业历史博物馆、展览及沙龙、工作坊、秀场、办公区等等十多项功能。设计巧妙地以弧线将窄长的管状空间分隔出相对独立而又自由流动的区域。

设计主题：编织梦想

头发——梦想长出来的翅膀，管状。

发型——编发——编织梦想。

全场以透明的管状物料配合LED灯光编织成立体空间网络包裹旧厂房的结构，在保存工业遗留的空间特色的同时，诗意地展现品牌外柔内刚、创造改变的内涵。

二沙岛文化艺术中心
Ersha Island Cultural Art Center

项目地点：广东 广州
建筑面积：18 000 m^2

Location: Guangzhou, Guangdong
Building Area: 18,000 m^2

项目位于二沙岛的南岸，是整个二沙岛南岸沿线文化展览建筑群的西侧端点。因而建筑的整体设计思路围绕如何塑造亲近、宜人的公共文化艺术建筑而展开。为此，建筑整体摒弃尖锐张扬的夸张造型，而以圆润柔和的弧线作为建筑轮廓的造型语言。建筑面向江面的首层整体以大草坡覆盖，草坡连同二层的平台全天向市民免费开放。平台以上的建筑体量利用出挑与退台等手法，既应对南方特殊的西晒气候，同时又塑造了多层的平台花园。

扉艺廊
Fei Gallery

项目地点：广东 广州
面　　积：室内面积300 m^2

Location: Guangzhou, Guangdong
Area: 300 m^2 (indoor space)

扉艺廊是亿达大厦免费向公众开放的当代艺术空间，是企业承担社会责任而发起的NGO组织。从大厦落成开始便在带有两侧天窗的12楼持续举办展览、沙龙，成为广州重要的当代艺术基地。2009年，因活动频繁，为方便观众迁至一层，完成了包括画廊、设计品、有机轻食的完整组合，把艺术、生活、交流共冶一炉，为各种艺术形式的展现，各类创意产品的面世，生活、设计、艺术态度的跨界传播搭建了平台。

入口处的旋转楼梯在油画家邓瑜的笔下慢慢幻化成海洋，让海底的生物自由畅快地游走着。整片海洋与扉艺廊的空中水池相得益彰，延伸出一个绚丽的虚拟海底世界。画廊内层层叠叠的台阶既是通道、舞台，又是坐席。复合的功能呼应了“扉”倡导的多元化理念。

扫描查看更多信息

SP® SHING & PARTNERS
漢森伯盛國際設計集團 SINCE 1993

汉森伯盛国际设计集团是一家久负盛名的大型专业国际设计集团公司，具备十分丰富的海外视野和国内实践经验，拥有大批专家团队，现有近300名国内外资深注册专业设计师，可为客户提供从规划概念、建筑方案、土建设计以及室内和景观设计等各领域的一站式服务。
汉森伯盛强调融入环境和尊重文化内涵，讲求现代设计创新，同时高度重视市场和客户利益，力求每一项目都维持高水平的服务，“令成功的客户更成功！”是集团的目标，公司也因此伴随众多成功的客户共同成长！
集团在海外和国内四十几个城市进行项目实践，具有十分丰富的工程设计经验。现代化的通信和管理确保全球各地的专业设计人员可以无缝协作，共同致力于同一项目中，创造出一个又一个的精品并获得多项荣誉。

广州
地址：广州市体育西路123号新创举大厦8楼（510620）
电话：+86-20-38218180
传真：+86-20-38288130
邮箱：SP@sp-arch.cn
网址：www.hsarch.com.cn

北京
地址：北京市朝阳区建国路93号万达广场3号1505（100022）
电话：+86-10-58205376
传真：+86-10-58205376转620

海南紫园
Hainan Purple Garden

项目地点：海南 海口
用地面积：230 000 m²
建筑面积：310 000 m²
容 积 率：1.2
建筑密度：13.5%
绿 化 率：60%

Location: Haikou, Hainan
Site Area: 230,000 m²
Building Area: 310,000 m²
Plot Ratio: 1.2
Building Density: 13.5%
Green Ratio: 60%

紫园项目位于海口风景优美的旅游休闲度假、高尚居住区的西海岸，距离市中心仅12min车程，离尘不离城，轻松“切换”都市与休闲度假生活节奏。
项目由“新派东南亚风格”小高层电梯公寓组成。创新设计的新派东南亚建筑，通过与东南亚自然人文亲密的感触和对国际时尚的现代简约风格的理解，营造出舒适、优雅的生活环境，让身心轻松和愉悦。不是完全的原始自然环境，但拥有最适合的自然悠闲生活尺度。东南亚热带风情水景庭园，蜿蜒流淌的水系，茂盛葱郁的热带植被园林，自然的坡地景观，以及散布在热带丛林中的温泉、瀑布跌水，宽阔水面的热带园林式花园泳池，让心情与自然亲密接触。

The project is located on the western coast of Haikou City, which is a picturesque region for traveling, relaxing, taking holidays and residing. It takes only 12-minute drive from the downtown. Far away from the civic life rather than from the city, here people can switch the life pace from working to relaxing easily.
The integral building consists of sub high-rise apartments of New Southeast Asian style. The innovative New Southeast Asian building, through the close relation with nature and humanity of southeast tropical architecture and the understanding of modern simplicity style of international fashion, creates a comfortable and elegant living atmosphere, making people feel relaxed and cheerful. The natural environment here is not completely primitive, but the life style here is suitably natural and relaxing. The southeast tropical waterscape garden, winding waters, verdant tropical garden, natural hillside land sights, and springs, fountains as well as swimming pools scattered in the tropical forest, all can make you get close to nature.

南京紫园
Nanjing Purple Garden

项目地点：江苏 南京
用地面积：104 836 m²
建筑面积：147 339.6 m²

Location: Nanjing, Jiangsu
Site Area: 104,836 m²
Building Area: 147,339.6 m²

项目位于南京市栖霞区环陵路99号，北侧为天泓山庄，东面与海军医学院相邻，南面与马群原有的商业街隔宁杭路相望，西侧为环陵路，正对全国著名风景区——南京历史人文气息最为浓厚的钟山风景区东南山麓。拥有得天独厚的生态、景观资源，是本项目最核心的优势。随着城市基础设施的进一步完善，周边环境景观的改善，项目有着极好的发展前景。地块呈"L"形，基本是正南正北走向，南北向约650 m，东西向约80～380 m。地块内东北高、西南低，大体沿"L"形的走向向北向和东向逐渐抬高，南北间高差约14～17 m，东西间高差约为8 m。

The project is located in No. 99 Huanling Road, Qixia District, Nanjing City. It is beside Tianhong Mountain Village in the north, adjacent to Navy Medical College in the east, facing the original Maqun Business Street across the Ninghang Road in the south, beside Huanling Road in the west, and opposite to the southeast foot of Bell Mountain Scenic Spot, famous all over the country for its strongest historical and humanistic atmosphere in Nanjing. The core landscape advantage of this project is the rich ecological environment resources endowed by nature. With the development of the city infrastructure and the surrounding environment, this project is very promising. The land is L-shaped, 650 m long from south to north and 80—380 m long from east to west. The land is high in northeast and low in southwest and gradually rises in the north and east direction along the turning point of "L". The altitude difference between south and north ranges from 14 to 17 m. The altitude difference between east and west is about 8 m.

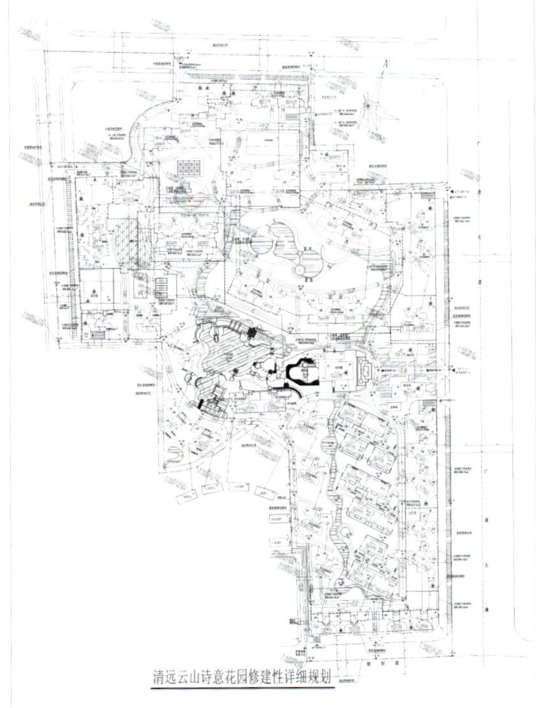

清远云山诗意
Qingyuan Poetic Cloud Mountain

项目地点：广东 清远　Location: Qingyuan, Guangdong
用地面积：200 000 m^2　Site Area: 200,000 m^2
建筑面积：500 000 m^2　Building Area: 500,000 m^2

建筑风格方面，项目继承了广州“云山诗意”系列的东方特色及优秀品质，同时又非常注重创新。在徽派建筑的设计风格中，通过现代技术手段、色彩技巧和装饰材料，赋予传统徽派建筑以现代且简约的气息，使之成为优雅、现代、尊贵的高档社区。
园林设计方面，8万m^{2}超大东方神韵环区山水园林。清远云山诗意整体以“水”为主题，通过跌级流水、涓涓小涧、人工湖泊等不同形态、不同类别的水体，结合精心雕凿的景观建筑、奇石、影壁等形成奇趣盎然的中式园林核心景观。项目将现代的技术和传统的建筑风格有机地结合，创造了现代、建筑、文化三位一体的和谐人居新理念。

The architecture inherits the oriental features and high quality of “Poetic Cloud Mountain” series in Guangzhou, and meanwhile quite innovative. Based on the Anhui architectural style, the project creates a modern and simple atmosphere in traditional Anhui buildings with modern approaches, color techniques and ornamental materials, making it an elegant, modern and noble residency.
The gardening design is an 80,000 m² huge garden with oriental features circling the residency. It is themed by “water”, including springs, streams, artificial lakes.With different kinds and forms of the water, combined with well-carved stones and screen walls, it creats the marvelous core landscape of a Chinese garden. The project integrates modern technologies with traditional architecture and creates a new residential philosophy of a harmonious trinity of modernity, architecture and culture.

广西阳朔河畔度假酒店
Yangshuo Riverside Holiday Hotel, Guangxi

项目地点：广西 桂林	Location: Guilin, Guangxi
用地面积：168 000 m^2	Site Area: 168,000 m^2
建筑面积：36 741 m^2	Building Area: 36,741 m^2

阳朔河畔度假酒店坐落在阳朔精华景区遇龙河畔，青山绿水，风景秀丽，地理位置独一无二，交通十分便利。

低调、自然——重归纯朴是阳朔河畔度假酒店的真实写照。酒店建筑风格以桂北侗族居民为蓝本，尊重地形、环境，遵循和谐共生的原则，严格控制建筑高度在10 m以内，采用传统连廊把建筑串联于原貌生态中。依地形设计出高低错落、生动有趣的建筑形式。部分建筑底层架空，结合庭院创造"生态潜入"空间。

每个客房面河观景，纵、横向设计无视线干扰，保证客房私密性，注重观赏和个性感受，真正体现了以人为本的理念。

The project is located on Yulong Riverbank, the best scenic spot in Yangshuo County, Guilin City, Guangxi Autonomous Region. The landscape is picturesque with green mountains and clear waters. The geographical location is unique and the traffic is very convenient.

Folksy, natural – returning to simplicity is the true portrait of Yangshuo Riverside Holiday Hotel. The architecture is based on the "Dong" ethnic residence in northern Guangxi, respecting the landform and environment, complying with the principle of harmonious coexistence. The height of the building is strictly limited within 10 m, and the building is integrated into the original ecosystem by traditional vestibules. Interesting architectural forms of different heights are designed according to the landform. The overhead ground floor of some buildings creates an "eco-intrusion" space with the courtyard.

People can see the scenery of Yulong River from every room without visual disturbance in vertical and lateral direction. The design guarantees the privacy of each room and emphasizes sightseeing and personal feelings, which truly represents the human-based attitude.

扫描查看更多信息

广东华方工程设计有限公司
GuangDong Huafang Architects & Engineers Co., Ltd.

广东华方工程设计有限公司团队最早成立于1992年4月，成立之初凭借技术服务优势迅速发展壮大，经过多年的发展，已成为一家综合性的设计服务机构。公司具备国家建设部颁发的建筑甲级（A144014879）、规划乙级（082037）、人防乙级（A244014876）、消防甲级、工程咨询丙级资质并获得GB/T19001−2008质量管理体系认证。公司成立10多年来，完成了大量优秀工程设计项目，业务范围涉及区域规划、住宅、商业、酒店、写字楼、医院、学校、室内外装修等设计业务，在同行中享有较高的知名度。公司在东莞、广州、中山、成都、南宁均设有分公司，拥有多名具有海外工作经验的建筑师，高、中级技术人员300多人，专业技术力量雄厚。公司拥有完善的质量管理体系，对设计产品前期策划至建成竣工验收阶段的全过程提供优质的设计咨询服务。

地址：广东省东莞市南城区元美路华凯广场A座19楼
邮编：523070
电话：+86-769-22820769
传真：+86-769-22820736
邮箱：winway866@163.net
网址：www.gdhuafang.com

Add: 19F, Tower A, Huakai Plaza, Yuanmei Road, Nancheng District, Dongguan, Guangdong Province
P.C.: 523070
Tel: +86-769-22820769
Fax: +86-769-22820736
E-mail: winway866@163.net
Wed: www.gdhuafang.com

碧桂园时代城
Country Garden Times Town

设 计 师：关大利
项目地点：广东 东莞
用地面积：24 086.80 m²

Designer: Dali Guan
Location: Dongguan, Guangdong
Site Area: 24,086.80 m²

本案位于东莞市塘厦镇原塘厦中学旧址， 周边为塘厦镇现有的商业中心区，人流密集，消费市场活跃。本项目提供的城市中心住宅和空中花园，为实现城市中心生活提供了可能，也能满足有老城区情结的人们回归老城中心的居住需求。
本地块只有在东北侧有约 80 m 左右的沿街面，在东侧有一个可供人行的出入口，其余三面均被现有建筑包围，无法形成机动车及人行出入口。项目用地周边的局限性给本地块的通达性带来很大的挑战，而人流和车流的便捷通达是本商业项目成功的关键。本设计力求通过塔楼的环绕式布局形成内聚的空间感受，商业通过横穿地块对角线形成主要动线，为内部引入人流和车流，从而盘活本地块，带动整个地区的商业和居住的升级。

The project is located in the old site of original Tangxia Middle School, Tangxia Town, Dongguan City. It is surrounded by existing CBD in Tangxia Town, with dense pedestrian flow and active consumer market. This project of city central residence and aerial garden provide a possibility of city center life and meanwhile meeting the living requirements of former habitants who return to old city center due to a nostalgic complex.
This Plot has only a 80 m wide side facing the street in the northeast, and one pedestrian entry/exit in the east side, and the other three sides are surrounded by existing buildings which cannot form entry or exit for motor vehicles or pedestrians. The peripheral limitation challenges the accessibility of this Plot, since pedestrian and vehicular flow passage is the key to the project. The design tries to form a cohesive space feeling by enclosing towers, where commercial buildings penetrate the Plot along the diagonal line to form the main generatrix, introducing pedestrian flow and vehicular flow, and thus activating the commercial and residential upgrading of the whole area.

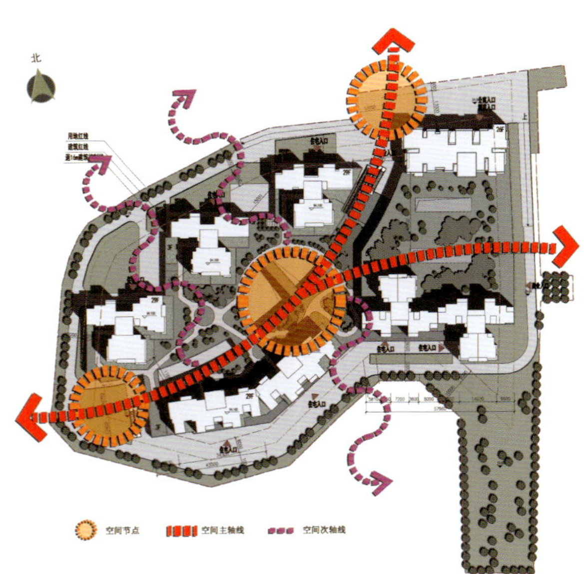

规划空间分析图

皇马郦宫
Royal Palace

设 计 师：伍志勇
项目地点：广东 东莞
用地面积：182 278.27 m²
建筑面积：511 400 m²

Designer: Zhiyong Wu
Location: Dongguan, Guangdong
Site Area: 182,278.27 m²
Building Area: 511,400 m²

项目位于东莞东城区八一路旁，背山面水的自然景观及人文资源，确立了其不可复制的价值。此区域必将成为东莞新的高尚国际化居住社区。规划地块以中间规划道路为界，划分为南北两个地块。
北地块内由一条景观水系分隔东北面联排别墅以及西面点式高层住宅部分。东面的联排别墅部分，地上 38 180 m²，107 户。别墅部分在地块东侧有社区入口，需要经过一段林荫道路进入，掩映在树影间的精致建筑更加彰显社区的尊贵品质。西面的点式高层住宅充分考虑对黄旗山脉山景和同沙水库水景的导入，并以人性化尺度营造居住的舒适感和温馨感。所有高层都依景循势地调整了方向，实现直面景观的视角，造就磅礴大气的开阔境界。
南地块规划为高层住宅及配套公建部分。

The project is located beside Bayi Road, Dongcheng District, Dongguan City. Its picturesque landscape and cultural resource have established its uncopiable value, and this area will certainly become a new noble international residential community in Dongguan. The land is divided into south lot and north lot by the central road.
In the north Plot, a landscaping water system separates the north & east townhouse section from the west point-style high-rise residential section. The east townhouse section, with aboveground floor area 38,180 m², includes 107 units. The townhouse section has a entrance in the east, and one can enter the section through a boulevard. Exquisite buildings seen from the boulevard will add more nobleness to the community. The west point-style high-rise residential buildings introduce Mt. Huangqi landscape and Tongsha Reservoir waterscape, and create habitable comfort and coziness to personalized extent. All high-rise buildings are changing their directions following the landform and landscape, realizing direct vision to landscape, and creating a generous and broad scenario.
The south Plot includes high-rise residential and supporting public buildings.

东莞市东城万达广场
Wanda Plaza, Dongcheng District, Dongguan

设 计 师：杨锦棠
项目地点：广东 东莞
用地面积：123 905.4 m²

Designer: Jintang Yang
Location: Dongguan, Guangdong
Site Area: 123,905.4 m²

东莞市东城万达广场位于东城区东纵大道，本项目由 A、B、C 三个地块组成，A 地块主要规划功能为综合百货、娱乐城、底商、步行街和商务酒店，B 地块主要规划功能为五星级酒店和写字楼，C 地块主要规划功能为 SOHO 办公和住宅。

The project is located on the Dongzong Avenue, Dongcheng District, Dongguan City. It includes three Plots: Plot A is planned to have department store, entertainment town, ground floor shops, pedestrian street, business hotel and other functions, Plot B is planned to have 5-star hotel and office building, and Lot C is planned to have SOHO office and residential building.

君御旗峰豪园
Junyu Qifeng Garden

设 计 师：李桁忠
项目地点：广东 东莞
用地面积：53 715.20 m²

Designer: Hangzhong Li
Location: Dongguan, Guangdong
Site Area: 53,715.20 m²

■消防车道：整个小区由周全的市政道路网架，小区车行道及区内的隐形消防车道共同形成消防车行系统。区内的园林式步行道作为隐形消防车道，在紧急时刻供消防车无障碍通行。

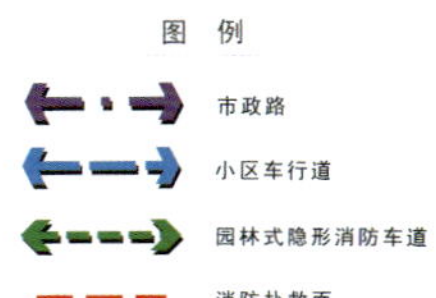

本项目位于东莞市莞城区，紧接中央商务区，临近东莞市政府和东莞市体育中心。就项目地块而言，基地北面为当地居民的私宅，多为3～4层。西北面是小山坡，西面是在建的商业银行，南接体育路，东面为建设中的大型商业建筑。
基地周边道路系统完善，交通十分便利，已建成的公交车站位于地块的南面，周边有餐饮、小超市、便利店等，基本配套设施比较齐全。基地内部为拆迁待建用地，与周边用地有3～4 m的高差。基地阳光充足，有良好的采光通风条件。

This project is located in urban area of Dongguan City. It is close to CBD, and near the office building of Dongguan People's Government and Dongguan Sports Center. In regard to the land Plot, the north is local residential houses, mostly 3 to 4-floor buildings.The northwest is a small hillside, the west is a commercial bank under construction, the south is adjoining Tiyu Road, and the east is a large commercial building under construction.
The Site is surrounded by improved road system, with very convenient traffic, and the established bus station is in the south of the Plot, There are also catering service, small supermarket, convenience stores and other facilities around this area, with complete infrastructure. Inside the Site is the land to be reconstructed, 3-4 m height difference with peripheral lands. The Site has sufficient sunshine and good ventilating condition.

芙蓉和苑
Lotus Peace Garden

设 计 师：伍志勇
项目地点：湖南 长沙
用地面积：374 804.95 m²
建筑面积：917 493.84 m²

Designer: Zhiyong Wu
Location: Changsha, Hunan
Site Area: 374,804.95 m²
Building Area: 917,493.84 m²

项目位于长沙市天心区芙蓉南路和韶山路交汇处北侧，绕城高速绿化带南侧，正处在长株潭城市群融城发展的最佳位置。
北区的主要出入口选择在芙蓉南路上，入口由一条横向步行景观轴线连接至会所，另外设计一条沿台地的公共景观带。其中中央的两个居住组团布置最大面积的情景洋房和多层洋房，其他组团综合景观因素布置阔景洋房和电梯洋房。
南区的主出口选择在韶山路上，同样有一条横向主景观轴线连接至会所。另设计一条横向的弧形步行道，结合纵、横轴线将南区划分为6个居住组团和一个学校组团。学校组团规划在地块的东北角，方便学校的对外开放。小区的配套商业布置在披塘路和韶山路。披塘路上的商业和北区的沿街商业，形成良好的商业氛围。

The project is located north of the juncture of Lotus South Road and Shaoshan Road, south of Ring Expressway Greening Belt, Tianxin District, Changsha City. It is in the best position of Changsha-Zhuzhou-Xiangtan trinity development area.
The main entrance/exit of north zone is on Lotus South Road, the entrance of which is connected to the club by a pedestrian landscaping axis, and an additional public landscaping belt is designed along existing terrace. In the central two groups are set the largest scenic houses and multi-floor houses, while in other groups are set broad view houses and elevator-equipped houses sharing landscaping elements.
The main exit of south zone is on Shaoshan Road, connected to the club by a transverse primary landscaping axis. Another transverse arc walking path is designed with the longitudinal and transverse axes to divide the south zone into six residential groups and one school group. The school group is planned on the northeast corner of the Plot, easy for opening to the outside. Commercial supports are set on Pitang Road and Shaoshan Road. Commercial buildings on Pitang Road and roadside commercial buildings in the north zone form a good commercial street climate.

御景湾 8 号
Royal Lagoon No. 8

设 计 师：杨锦棠
项目地点：广东 东莞
用地面积：159 895.83 m²

Designer: Jintang Yang
Location: Dongguan, Guangdong
Site Area: 159,895.83 m²

项目地块位于广东省东莞市峰景高尔夫球场东南端，基地环境资源极佳。背山面水的自然景观及人文资源，确立了其不可复制的价值，此区域必将成为东莞新的高尚国际化居住社区。
整个地块用地及周边地势变化较大，周边地貌形态多样。规划用地范围内基本上为缓坡及陡坡地。整体呈南高北低，西高东低的总走向，在180多m南北方向上总高差达20 m左右。用地最高点位于地块中间南端，高39 m左右；最低点位于基地中间西北端，高13 m。用地范围东面有水塘，区内景观水系设计以现有水塘为基础，结合排洪泄洪的需要进行了整合，以便取得最大化的景观覆盖面，同时也为水体自然净化提供良好的前提。

The Plot is located in the southeast end of Hill View Golf Course in Dongguan City, Guangdong Province. Based in excellent environmental resources, its picturesque natural landscape and cultural resources have established its uncopiable value, and this area will become a new noble international residential community in Dongguan.
The Plot has distinct topography with the periphery, and the peripheral landforms are diverse. The Plot mainly includes gentle slopes and steep slopes. The entire Plot is high in south and low in north, high in west and low in east, with approx. 20 m overall elevation difference in the over 180 m south-north direction. The highest elevation is in the central south end of the Plot, around 39 m; the lowest elevation is in the central southwest end of the lot, elevation around 13m.The east is a pond, based on which is designed a inner landscaping water system, to integrate with the requirements for flood release and discharge, so as to maximize landscaping coverage, while providing good premises for self-purification of the water.

扫描查看更多信息

天作國際
TEAMZERO

广州市天作建筑规划设计有限公司
Guangzhou TEAMZERO Architectural Design & Planning Co., Ltd.

广州市天作建筑规划设计有限公司(Guangzhou TEAMZERO Architectural Design & Planning Co.,Ltd.)成立于2002年，是广州市城市规划勘测设计研究院和美国TEAMZERO建筑规划设计公司联合成立的设计机构。公司充分发挥两大技术支撑平台的综合优势，本着专业综合联动的服务理念，在短短10年内已经在建筑设计、城市规划、城市设计和景观设计等领域取得令业界瞩目的优秀成绩，获得国家级、省部级、市级设计奖项30余项。公司拥有建设部颁发的建筑行业（建筑工程）甲级：A144016197、城乡规划编制资质证书：【建】城规编第（111239）甲级以及风景园林工程设计专项乙级：A244016194资质。
公司现有130多名专业技术人员，涵盖城市规划、建筑设计、景观设计、旅游规划、市政工程、工程咨询、城市交通、城市经济和生态环保等多个专业，是一个国际化、综合化、专业化且具有较高行业影响力的设计机构。

地址：广州市珠江新城华夏路28号富力盈信大厦6楼
电话：+86-20-83840927/83840847/83840875/83840583
传真：+86-20-83839072
邮箱：teamzero@teamzero.com.cn
网址：www.teamzero.com.cn

Add: Floor 6, R&F Yingxin Mansion, No.28 Huaxia Road, Pearl River New Town, Guangzhou City
Tel: +86-20-83840927/83840847/83840875/83840583
Fax: +86-20-83839072
E-mail: teamzero@teamzero.com.cn
Web: www.teamzero.com.cn

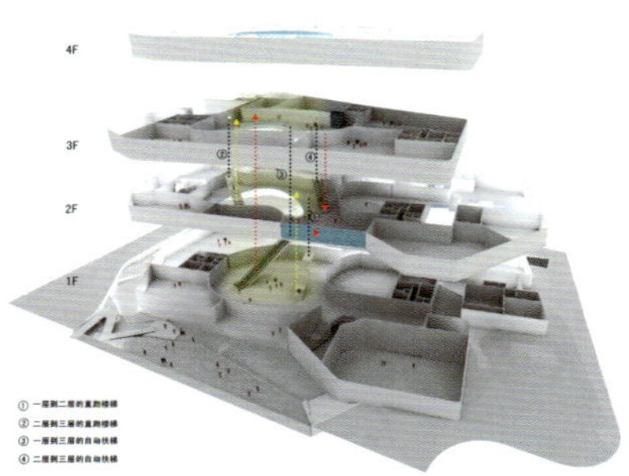

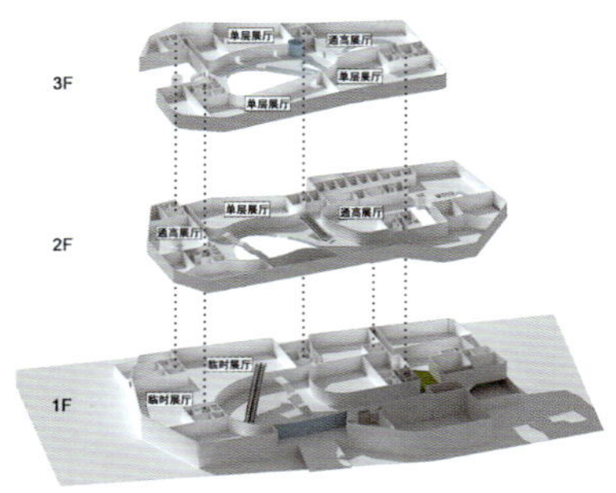

安徽省美术馆
Anhui Art Museum

设 计 师：李少云
项目地点：安徽 合肥
用地面积：16 667.5 m²
建筑面积：25 993 m²
地上建筑面积：19 180 m²
地下建筑面积：6813 m²

Designer: Shaoyun Li
Location: Hefei, Anhui
Site Area: 16,667.5 m²
Building Area: 25,993 m²
Building Area Aboveground: 19,180 m²
Building Area Underground: 6,813 m²

美术馆方案注重融合安徽文化尤其是书画文化的独特性。建筑造型通过现代建筑手法，对徽砚圆润、优美的形态进行抽象和升华，建筑主体犹如承托于基座上的歙石名砚，从不同方位观看都优美多变、古朴雅洁，既具有视觉的冲击张力，又不失浑实厚重。立面的处理强调虚实对比，开窗纹理借鉴了歙砚中“金晕”纹理的优美形态及浓淡变化，也体现了传统山水画的意蕴。观众不仅从展览中获得对美术作品的认知，更能通过建筑本身感受到传统文化及书画意境的熏染。

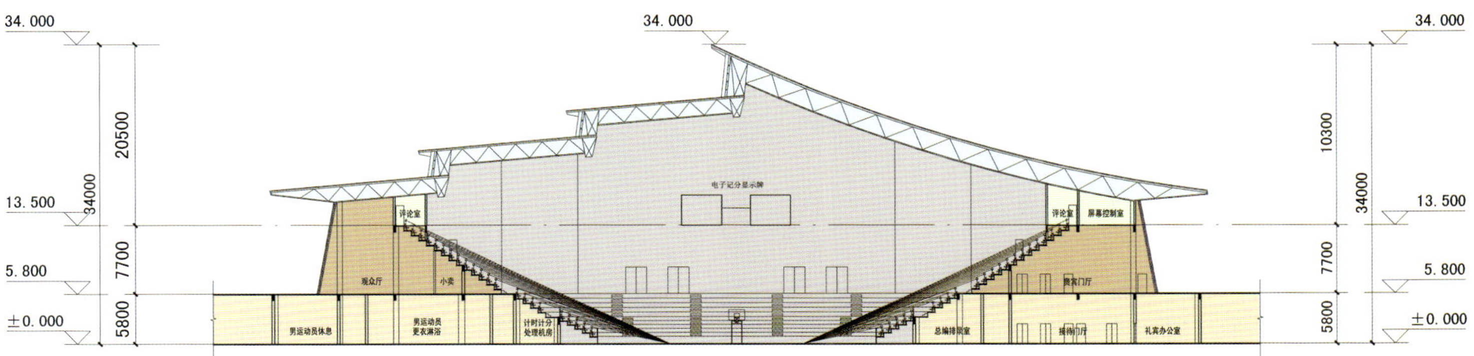

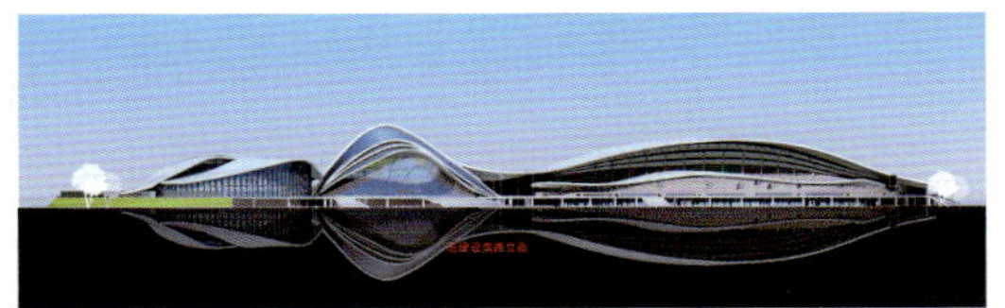

太湖县体育中心修建性详细规划及建筑方案设计

Detailed Planning and Architectural Design of Taihu County Sports Center

设 计 师：李少云	Designer: Shaoyun Li
项目地点：安徽 安庆	Location: Anqing, Anhui
用地面积：94 214 m^2	Planning Site Area: 94,214 m^2
建筑面积：21 748 m^2	Building Area: 21,784 m^2

项目位于历史悠久的太湖县，基于其深厚的水文化提出"湖之波"的基本概念，结合现代建筑特点形成流畅、优美的建筑形态。规划充分考虑了三个场馆的总体布局及竖向关系。主体育馆位于用地中心，成为统领全局的主体，建筑之间以交通平台连接，形成一气呵成的整体。体育馆建筑造型采用层叠的波浪形屋顶，与"湖之波"的设计概念契合。结构体系与建筑造型、体育场馆的室内空间很好地结合在一起，形成优美的形态，并达到表里如一的空间效果。

贵阳未来方舟G组团规划及建筑设计

Planning and Architectural Design of Zone G, Future Ark, Guiyang

设 计 师：李少云
项目地点：贵州 贵阳
用地面积：296 493 m²
建筑面积：874 727 m²

Designer: Shaoyun Li
Location: Guiyang, Guizhou
Site Area: 296,493 m²
Building Area: 874,727 m²

该项目位于未来方舟项目宗地的北部，南明河以西。地块坐落于山水之间，设计利用缓坡台地形优势，既缓解了高密度住区的空间压力，也为后排层次的建筑最大限度地争取日照朝向及景观资源。

整体的规划设计采用高层大组团的布局形式，建筑形式统一中强调细节的变化。居住环境注重山水景观的利用，体现绿色生态主题。

建筑整体注重营造滨水住区休闲氛围，建筑立面采用现代风格，强调住宅建筑公建化，同时通过阳台、露台富有韵律的变化实现与滨水区域氛围的匹配，立面轻盈而富有动感。

商业街通过错落有致的设计，注重商业街区的整体空间营造，与河道滨水景观及对岸的游乐城相呼应。设计强调商业立面的休闲感和风情感，营造独具特色的滨水商业氛围。

顺德美的君兰江山花园二期规划及建筑设计

Plan and Architectural Design of Midea Orchid Land Phase II, Shunde

设 计 师：李少云
项目地点：广东 顺德
用地面积：151 951 m^2
建筑面积：498 966.8 m^2

Designer: Shaoyun Li
Location: Shunde, Guangdong
Site Area: 151,951 m²
Building Area: 498,966.8 m²

整体建筑设计为现代风格，强调清晰的建筑体块关系，以及简约、现代的肌理逻辑，同时借用了古典建筑立面竖向三段式的处理手法，体现出建筑优雅、经典的品味。
项目规划围绕周边高尔夫球场景观资源做最大化的利用设计，规划以一梯两户板式设计，分独栋和双联两种形式，利用建筑群前后错位，实现每户均有一线高尔夫球场景观。
建筑外轮廓设计规整，提倡住宅建筑立面公建化，强调立面的竖向整体效果在视觉上的吸引力和冲击力，以及横向联系在体块造型上的穿插和交互关系。宽度过大的高层体量作前后错位处理，形成比例协调的"高、挺、竖"的体块。局部通过阳台的大小及错位变化，实现高层空中花园的空间体验，增强了空间感染力。

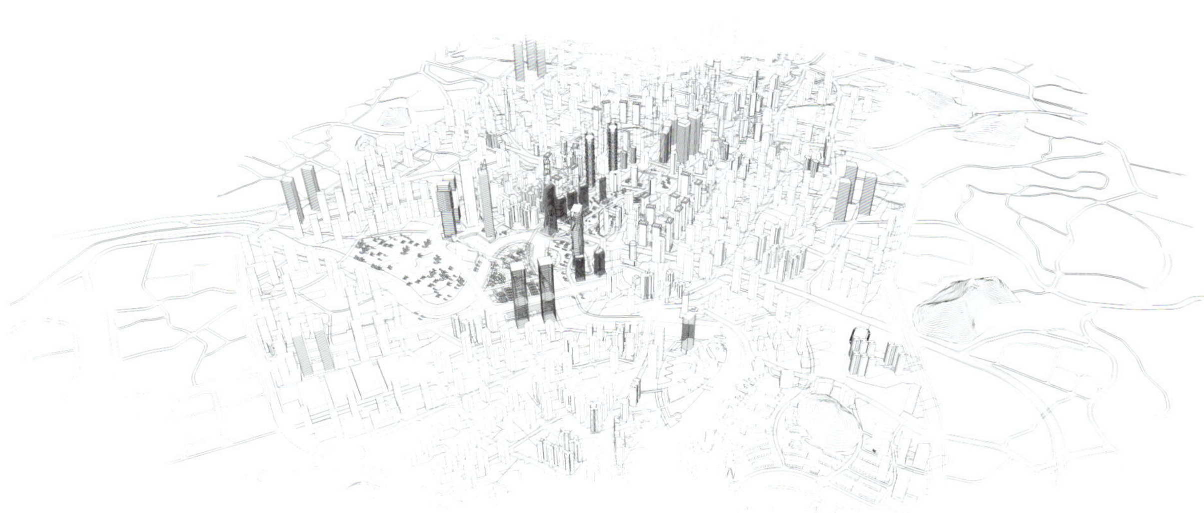

贵阳市旧城区更新概念规划方案（国际投标第一名）

Conceptual Plan of the Renovation of the Old Town of Guiyang City (First Prize in International Tendering)

设 计 师：李少云
项目地点：贵州 贵阳
用地面积：1020 hm^2
建筑面积：30 000 000 m^2

Designer: Shaoyun Li
Location: Guiyang, Guizhou
Site Area: 1,020ha
Building Area: 30,000,000 m^2

贵阳旧城区面临的重要问题是如何在建设容量提高的同时疏解交通压力，利用有限的土地提高基础设施的服务能力，在大规模旧城改造的同时保护历史文化，以及保障原居民的生存与发展权利。方案采用“公交都市”和“低碳城市”规划设计理念，提出结构策略、空间策略、交通策略和技术策略，将轨道交通规划建设与旧城更新改造紧密结合起来，通过围绕轨道交通站点进行TOD模式开发，引导高强度开发向车站地区集中，使在旧城中创造更多开放空间成为可能。

海南国际旅游岛先行试验区概念规划国际咨询方案（国际投标第一名）

Conceptual Plan of Hainan International Tourism Island First Pilot Area International Consulting Plan (First Prize in International Tendering)

设 计 师：李少云
项目地点：海南 陵水
用地面积：核心区6600 hm^2
建筑面积：16 800 000 m^2

Designer: Shaoyun Li
Location: Lingshui, Hainan
Site Area: Core Area 6600 ha
Building Area: 16,800,000 m^2

方案提出将先行试验区打造成联动港城、汇聚世界的生态湖海新城以及世界级的海洋文化旅游休闲胜地。通过引入多样化的旅游业，促进旅游产业健康持续发展并具备国际竞争力。方案采用组团式规划塑造可持续的景观生态格局。针对场地得天独厚的自然条件，设计方案利用一个综合交通换乘枢纽将两个泻湖与生态半岛联系起来，并借此在半岛中引入绿色交通系统，保证生态敏感区的景观资源不受过多干扰。巧妙的规划设计将其打造成为"国家领港、绿色海岸"。

扫描查看更多信息

TR Talent Co. ltd

广州市天任建筑设计顾问有限公司

Guangzhou Talent Architectural Design Co., Ltd.

广州市天任建筑设计顾问有限公司是专业从事规划与建筑设计、园林景观设计、城市广场与道路景观设计的公司。公司积累二十余年专业经验，以坚实的精英团队，完成了一批颇具影响力的优秀作品，涵盖行政办公、综合商业、文化教育、商品住宅等多个领域。

公司潜心研究传统地域文化和当下设计思潮，针对个案有独特设计概念。公司秉持苛求完美、精益求精的设计宗旨和诚信服务，以设计创造价值，以品质成就理想。

地址：广东省广州市天河区龙怡路91号省农机物资公司综合楼四楼

电话：+86-20-38483926/38483929

邮箱：tr3848@126.com

网址：www.tianren.8hy.cn

Add: F/4, General Building of Guangdong Provincial Agricultural Machinery & Materials Company. No.91 Longyi Road, Tianhe District, Guangzhou City, Guangdong Province

Tel: +86-20-38483926/38483929

E-mail: tr3848@126.com

Web: www.tianren.8hy.cn

1

2

1. 深圳香蜜湖商业中心

Honey Lake Business Center, Shenzhen

用地面积：50 000 m² Site Area: 50,000 m²

建筑面积：100 000 m² Building Area: 100,000 m²

2. 广州市帽峰山围屋酒店

Roundhouse Hotel, Maofeng Mountain, Guangzhou

用地面积：20 000 m² Site Area: 20,000 m²

建筑面积：18 000 m² Building Area: 18,000 m²

3. 鄂尔多斯东胜区文化中心

Cultural Center, Dongsheng District, Ordos

用地面积：30 000 m² Site Area: 30,000 m²

建筑面积：45 000 m² Building Area: 45,000 m²

3

4

5

4–5. 惠州东江城市综合体

Dongjiang Urban Complex, Huizhou

用地面积：60 000 m^2 Site Area: 60,000 m²
建筑面积：150 000 m^2 Building Area: 150,000 m²

6. 东莞民间艺术馆

Civil Arts Museum, Dongguan

用地面积：12 000 m^2 Site Area: 12,000 m²
建筑面积：10 000 m^2 Building Area: 10,000 m²

7. 广州市电信数据中心

China Telecom Data Center, Guangzhou

用地面积：10 000 m^2 Site Area: 10,000 m²
建筑面积：20 000 m^2 Building Area: 20,000 m²

8. 广州花都档案馆

Huadu Archives, Guangzhou

用地面积：1716 m^2 Site Area: 1716 m²
建筑面积：17 600 m^2 Building Area: 17,600 m²

6

7

8

9. 广州市科学城威创科研楼

VITRON Research Building, Science Town, Guangzhou

用地面积：20 000 m² Site Area: 20,000 m²
建筑面积：60 000 m² Building Area: 60,000 m²

10. 广州市民营科技园区科技大楼

Science & Technology Building, Private Business Science & Technology Park, Guangzhou

用地面积：12 000 m² Site Area: 12,000 m²
建筑面积：39 000 m² Building Area: 39,000 m²

11. 新塘全兴村国际公寓

Quanxing Village International Apartment, Xintang

用地面积：60 000 m² Site Area: 60,000 m²
建筑面积：120 000 m² Building Area: 120,000 m²

12–13. 华南农业大学跃进校区

Yuejin Campus,South China Agricultural University

用地面积：100 000 m² Site Area: 100,000 m²
建筑面积：150 000 m² Building Area: 150,000 m²

11

14

15

14—15. 广州恒福路汽配城
Hengfu Road Auto Accessory Market, Guangzhou

用地面积：48 500 m^2
建筑面积：62 000 m^2

Site Area: 48,500 m^2
Building Area: 62,000 m^2

16. 广州白天鹅流溪河水库度假中心
White Swan Liuxi River Reservoir Resort Center, Guangzhou

用地面积：60 000 m^2
建筑面积：50 000 m^2

Site Area: 60,000 m^2
Building Area: 50,000 m^2

17—18. 烟台市开发区F-14小区
Community F-14,Yantai Development Area

用地面积：90 830 m^2
建筑面积：231 500 m^2

Site Area: 90,830 m^2
Building Area: 231,500 m^2

16

17

18

深圳市东大建筑设计有限公司
Shenzhen Dongda Architectual Design Co.,Ltd.
东南大学建筑设计研究院深圳分院
Architectural Design & Research Institute of Southeast University, Shenzhen Branch

1992年，一群致力于推动建筑设计从教学走向市场的有识之士，在深圳特区创建了东南大学在深圳的综合甲级设计企业——东南大学建筑设计研究院深圳分院。2006年，分院经东南大学批准，进行改制，成立"深圳市东大建筑设计有限公司"。
经过几十年的发展，东南大学建筑设计研究院深圳分院，立足深圳，面向全国，从天涯海角到塞外雪林，从东海之滨到西域疆城，都留下了足迹，完成了一系列重大项目的规划和设计，获得了各界的赞誉。

地址：深圳市深南中路6031号杭钢富春商务大厦8楼
电话：+86-755-82996899
传真：+86-755-82995667
邮箱：ddfy_szb@21cn.net
网址：www.seusz.com

Add: 8th Floor, Fuchun Commercial Building, No.6031, Shennan Road, Shenzhen, China
Tel: +86-755-82996899
Fax: +86-755-82995667
E-mail: ddfy_szb@21cn.net
Web: www.seusz.com/

东莞人民医院（新院）
Dongguan People's Hospital (New)

项目地点：广东 东莞
用地面积：318 000 m^2
建筑面积：183 000 m^2
合作设计：美国CMC建筑与规划设计事务所；上海励翔建筑设计事务所

Location: Dongguan, Guangdong
Site Area: 318,000 m^2
Building Area: 183,000 m^2
Cooperators: CMC Architecture, Inc.; Shanghai Lixiang Architecture, Inc.

总体布局：根据现代医疗建筑设计理念、医技流程进行设计，平面布局灵活并有超前意识。项目采用最新脊骨式"医疗街"的设计理念，将医疗、科研、后勤以及预防保健等建筑部分，有疏有密地连接成为一体，由此展现了医院建筑如凝固音乐般的精彩华章。
立面及造型设计：整个建筑群完全融入自然绿化景观中，既具有系列的韵律感，又显得活泼、变化丰富，4幢形似风帆的病房楼以柔和的建筑体形，融入滨江环境中。

General layout: The project is designed with modern medical building concept and medical technique process. The layout plan is flexible and avant-garde. The project adopts the design concept of latest backbone type "Medical Street", and integrates medical building, scientific research building, logistic building and prevention & healthcare building properly to express the orchestral spirits of medical architecture.
Façade and shape design: The whole building cluster is completely integrated into natural green landscape, endowing the project with serial sense of rhythm and making it lively and variant. Four sail-shaped inpatient buildings are integrated with the riverside environment with their soft architectural figures.

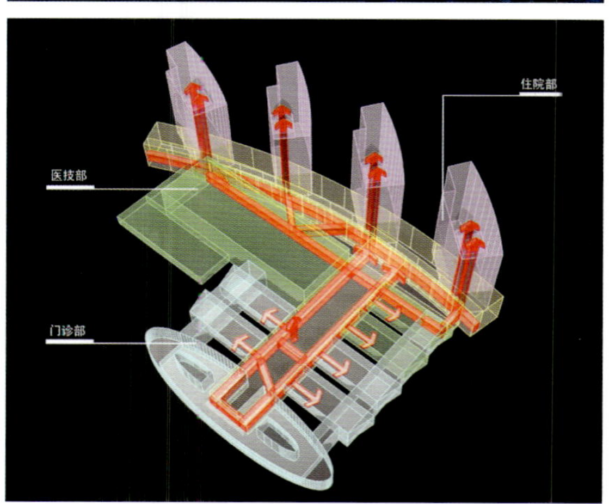

徐州市民活动中心

Citizen Activity Center, Xuzhou

项目地点：江苏 徐州
建筑面积：132 797 m²

Location: Xuzhou, Jiangsu
Building Area: 132,797 m²

项目作为云龙湖边最重要的城市标志之一，以整体、新颖、个性突出的形象，与音乐厅、体育馆和艺术中心形成呼应。建筑主体深厚凝重，博大雄浑，展示出动人的魅力。
功能和交通上，分成独立的三个区：中心主题区、影视娱乐区和商业开放区。它们各自独立，互不干扰。三个区域分区明确、独立，可分可合，按照活动要求，可单独开放或关闭个别区域，便于管理。
总之，活动中心为市民提供最大限度的服务，体现市民中心为人民服务的公共意识。

As one of the most important city landmark along Yunlong Lake, this project echoes with the Music Hall, Gymnasium and Art Center with its integral, fresh and characteristic image. The main body of the building is heavy and solemn, grand and magnificent, reflecting its touching charm .
According to the functions and traffic conditions, the project is divided into three independent sections, namely Central Theme Area, Film & Entertainment Area and Open Commercial Area. The three area are distinct, independent, separable and joinable. For specific requirements, the project may open or close a specific area for better management. Overall, the actiity center provides maximum service to citizens representing the public awareness of serving people.

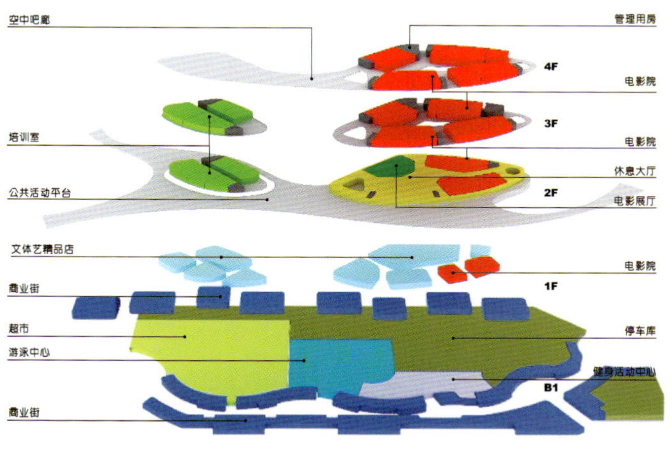

江西省龙虎山景区游客服务中心

Tourist Service Center of Dragon Tiger Mountain Scenic Spot, Jiangxi

项目地点：江西 鹰潭		Location: Yingtan, Jiangxi
用地面积：14 hm^2		Site Area: 14 ha
建筑面积：8290 m^2		Building Area: 8,290 m^2

游客中心由三大不同使用功能空间组成，三大不同功能融为一体，合中有分，分中有合。该项目既有强烈的现代建筑震撼力，又有深层的地域文化内涵，是龙虎山旅游区标志性建筑。项目2011年荣获教育部优秀设计三等奖。

The tourist center is composed of three spaces with different functions. The project integrates the three functions in the neatest design approach with the mode of separation and combination. It has strong power of modern architecture, as well as profound regional cultural meaning, therefore it becomes a landmark building in Dragon Tiger Mountain tourist area. The project has won the third prize for Excellent Design by Ministry of Education in 2011.

三亚金中海蓝钻
Jin Zhonghai Blue Diamond, Sanya

项目地点：海南 三亚
用地面积：46 346.52 m²
建筑面积：96 313.58 m²

Location: Sanya, Hainan
Site Area: 46,346.52 m²
Building Area: 96,313.58 m²

遵从地形，依照地势，该项目四周高、中心低，形成内敛的景观中心，为邻里交往营造出人性化空间，均匀分布的节点性公共交往场所为整个生活区创造了更加生动、鲜活的气氛。项目采用地下停车与地面停车相结合的设计，为居住区创建幽静、有序、安全的场所。项目2011年荣获教育部优秀设计一等奖。

Following the landform and topography, the project makes an introverted landscaping center fenced by high buildings. The project builds a human-oriented space in the neighborhood, and the evenly distributed sectional public facilities create moer Lively and fresher atmosphere for the whole living community. The design combines the underground parking lot and aboveground parking lot, providing a quiet, orderly and safe facility for the residential area. The project has been awarded first prize for Excellent Design by Ministry of Education in 2011.

美騰設計
M.T.DESIGN

香港美腾(M.T)设计工程有限公司
Hongkong M.T Design & Engineering Co.,Ltd.

香港美腾（M.T）设计工程有限公司始建于1989年，发展至今已成为国内及东南亚较大的国际化设计公司，主要业务包括建筑、规划、室内设计、景观设计。旗下拥有近100名经验丰富的专业设计人员和工程师，他们具有国际性视野，在建筑设计、室内设计、景观园林设计、总体规划和平面设计方面都颇有建树。M.T的设计师多数毕业于海外，如美国、英国、加拿大、法国和澳大利亚等，并已成为香港建筑设计师协会成员。

M.T把西方较为成熟的设计体系和理论运用到设计中，并加入中国本地的特色，使之具有独特的特征。毕竟，一种独特的设计将会吸引参观者的目光。

公司历经20多年的发展，现已凝聚了一支稳定的专业设计、施工及管理团队，建立了设计、施工、工程后续维护一体化的服务体系。公司总部注册员工近300名，有80多名优秀的室内设计师和具有丰富现场施工经验的项目管理人员。公司一直以优良的业绩取信于客户，业务范围拓展至全国20多个省、市、自治区，以优秀的设计作品、优良的施工工艺、优质的服务，为广大的客户创造出艺术与实用完美结合的高品质室内空间，得到了客户的一致好评。现公司重点致力于高档星级酒店与别墅豪宅样板房的设计与施工，并与中海地产、成都万华地产等知名客户建立了长期战略合作伙伴关系，已经在深圳、澳门、北京、上海、沈阳、成都、西安、哈尔滨、长春、青岛等多个城市完成了一大批精品工程。

身为中国香港著名的室内设计大师、香港美腾（M.T）设计工程有限公司创始人谭伟武先生，在过去的20多年里，承揽了数百项装饰工程项目，在装饰设计业界享有极高的盛誉，其设计作品遍及中国香港，澳门，东南亚，中国大陆各大城市。其主要代表作品有：澳门金莎二期（含香格里拉、喜来登等国际五星级酒店）、越南鸿运赌场酒店、沈阳万豪大酒店、沈阳国际皇冠假日酒店、北京名人大酒店、上海西郊大公馆（别墅群）、上海江南宴（豪宅小区）、杭州星光大道（酒店式公寓楼）等。

公司自成立以来一直秉承“以质量求生存，以信誉求发展”的经营理念，始终坚持以客户的需求和满意为核心追求，以“质量”当生命，以“诚信”为宗旨，坚持装饰工程施工规范，严格执行工程监督管理程序，严把装饰材料关，不断提高施工工艺水平，不断开发应用新技术、新工艺，在全国大多数城市创造了一大批别墅、豪宅、优质样板房工程，现已成为豪宅装饰行业中的典范企业。

地址：深圳市罗湖区人民南路深房广场B座11楼
电话：+86-755-82295888
传真：+86-755-25186832
邮箱：mt0755@163.com
网址：www.mtdesign.hk

Add: 11th Floor, Shenfang Building Tower B, Renmin South Road,Luohu district,Shenzhen
Tel: +86-755-82295888
Fax: +86-755-25186832
E-mail: mt0755@163.com
Web: www.mtdesign.hk

沈阳国际皇冠假日酒店

Crowne Plaza Shenyang Parkview

设 计 师：谭伟武、李娟、温梦琪
项目地点：辽宁 沈阳
建筑面积：70 000 m²

Designer: Weiwu Tan, Juan Li, Mengqi Wen
Location: Shenyang, Liaoning
Building Area: 70,000 m²

水晶，呈晶簇，在绚丽多彩中散发着神秘与奢华气质。
沈阳皇冠假日酒店外观造型为水晶的几何立方体结构旋转45°，并由主楼与裙楼组成，叠落的玻璃幕墙增加了建筑外观的线条节奏感。
工程位于沈阳市皇姑区黄河南大街88号，总用地面积9640 m²，酒店的建筑高度有18层，地下2层。拥有总统套房及各类豪华客房295间，是沈阳地区现阶段规模较大的一家国际五星级酒店。大堂中空门位置大面积采用七彩的水晶切面造型，美轮美奂，与主题相呼应。七彩水晶球环绕的天花吊灯奏出如风铃般清脆的乐章，令空间呈现一片和谐气象，突显出酒店高贵尔雅的气度。酒店的中餐部从中国传统文化中吸取养分，融合中国传统美学、哲学等元素，结合现代设计手法，活用清代园林厅堂空间概念，展现出清朝特有的文化内涵。

Crystal is colorful, exuding mystery and luxury in temperament.
Shenyang Crowne Plaza's appearance is shapped by the crystal cube rotated 45 degrees of the main building and the podium, overlapping curtain walls of the building increase the sense of rhythm of lines of the building appearance.
The project is located in the No.88 of south Huanghe Street of Huanggu District in Shenyang, with a total area of 9,640 m², the hotel have a height of 18 floors and 2 undergrond. It has 295 presidential suites and all kinds of luxurious guest rooms. It is the larger scale international five-star hotel in Shenyang. which is magnificent and echoes with the theme; The lobby uses large-scale colorful crystal facets colorful crystal chandelier plays a piece of music, such as wind chimes so that presents a harmonious space, and highlights the elegence of the hotel. The Chinese restaurarus of the hote absorbs nutrients from the Chinese traditional culture, blending Chinese traditional aesthetics, philosophy and other elements, combing with modern design techniques, utilizing the Qing Dynasty conceptin garden living room space, displaying a unique cultural charm of the Qing Dynasty.

扫描查看更多信息

深圳市武向兵建筑设计有限公司

Shenzhen Wu Xiangbing Architectural Design Co.,Ltd.

深圳市武向兵建筑设计有限公司于2006年成立。公司成立之初，正值中国房地产业方兴未艾之际。公司的业绩从一个侧面反映了这些年来房地产发展的轨迹，反映了建筑设计在市场竞争的环境下不断完善，自我更新的过程。

公司在成立之初就明确了以创新设计和全过程服务为追求目标，以高完成度设计和高质量的建成作品建立公司品牌，致力于在建筑设计领域创建自己的风格和口碑。设计注重从大局着手，关注用户需求，关注建筑细节，发现生活的真谛，创作具有领先于市场的作品。

近几年来，公司设计作品不断涌现，设计团队逐步壮大，正朝着精细设计、不断创新、团队协作、全面服务的方向发展，为建立下一个全新的设计平台而努力。

地址：中国深圳市华侨城东部工业区B-9栋302室
电话：+86-755-86106965
传真：+86-755-86106948
邮箱：project@wuarch.com
网址：www.wuarch.com

Add: Suite 302, Bldg.B-9, East Industrial Zone, OCT,Shenzhen,China
Tel: +86-755-86106965
Fax: +86-755-86106948
E-mail: project@wuarch.com
Web: www.wuarch.com

青岛 · 索菲亚国际大酒店

Qingdao · Soffia International Hotel

项目地点：山东 青岛
用地面积：24 844 m^2
建筑面积：73 400 m^2
容 积 率：2.95

Location: Qingdao, Shandong
Site Area: 24,844 m^2
Building Area: 73,400 m^2
Plot Ratio: 2.95

建筑设计充分挖掘滨海建筑的特点，以曲折的建筑形体表现海滨酒店的独特魅力，以丰富的线条和虚实对比反映现代建筑语汇的表现力。

The architecture is enriched by coastal features, for example, the curved volumes express special charm of seaside hotel, and the varying lines in contrast reflect the tension of modern architectural language.

东莞望牛墩镇市民中心

The New Civic Center of Wangniudun, Dongguan

项目地点：广东 东莞
用地面积：53 147 m²
建筑面积：100 000 m²

Location: Dongguan, Guangdong
Site Area: 53,147 m²
Building area: 100,000 m²

新市民中心反映了东莞望牛墩镇经济快速发展以后的崭新面貌。市民中心位于镇中心位置，由政府办公、市民办事处和图书展览等功能组成，建筑造型简洁、庄重，主楼、副楼和图书展览楼构成对称布局，并形成内外有别的院落 空间严谨又不失亲切。

The new Civic Center acts as a symbol for the fast growth of The local economy. Located in the town center, The Civec Center comprises of governmen offices, service facilities, library, exhibition hall and museum. The center is symmertrically laid out to reflect its governmental character. Gardens in different scales are stitched into the space to soften its rigidness.

长沙纳爱斯 · 秀山丽水
Nice · Xiushan Lishui in Changsha

项目地点：湖南 长沙
用地面积：232 500 m²
建筑面积：752 500 m²

Location: Changsha, Hunan
Site Area: 232,500 m²
Building Area: 752,500 m²

规划设计试图突破目前国内居住区规划的模式，以扭曲的点、线图形延展，形成富有层次、活而不乱、中心突出的全新规划模式。建筑风格以装饰艺术主义为蓝本，推陈出新，给人耳目一新之感。

The master plan was a brave breakaway from the usual residential plan patterns. The design played with dancing lines and dots, forming a layered and ever-changing series of spaces. The architecture inspired by Art-deco was nonetheless fresh creation for a brand new iconic impression.

合肥和平盛世
Hepingshengshi Residences in Hefei

项目地点：安徽 合肥
用地面积：71 262 m²
建筑面积：285 820 m²
合作单位：东南大学建筑设计院深圳分院

Location: Hefei, Anhui
Site Area: 71,262 m²
Building Area: 285,820 m²
Partners: Architects & Engineers Co.,Ltd. of Southeast University (Shenzhen Branch)

这是一个以高层为主、高密度的住宅项目，坐落在合肥市中心区。设计试图再现法式建筑的恢宏气质和浪漫情怀。成功在于细节，正是本项目的写照。

Located in the center of Hefei, this is a high-rise, high-density luxury residence development. The design tries to trace the glory and romance of the old French Classics. The trick is in the details.

大华 · 西野风韵
Dahua · Xiye Fengyun

项目地点：浙江 临安
用地面积：791 580 m^2
建筑面积：158 370 m^2
容 积 率：0.2

Location: Lin'an,Zhejiang
Site Area: 791,580 m^2
Building Area: 158,370 m^2
Plot Ratio: 0.2

在青山湖的青山绿水间进行商业开发，建筑师需要平衡建造的可行性和原生环境的保护。最终采取的策略是依山就势，建筑物适当集中，形成自然村落式布局，最大限度保留自然环境。

This real estate development is located in the beautiful mountains of Qingshan Lake. We grouped buildings into clusters of villages in responding to the natural contours and left as much forest areas untouched as possible.

大华 · 西溪风情别墅
Dahua · Xixi Fengqing Villa

项目地点：浙江 杭州
用地面积：666 666 m²
建筑面积：130 000 m²
合作单位：深圳华森建筑与工程设计顾问有限公司

Location: Hangzhou, Zhejiang
Site Area: 666,666 m²
Building area: 130,000 m²
Partners: Shenzhen Huasen Architectural & Engineering Designing Consultant Ltd.

杭州大华集团开发的西溪风情别墅，是位于杭州著名的西溪湿地边、约50万m²的联排别墅和洋房。在临河水处安排了17栋特别设计的经典大别墅，风格典雅，精致脱俗，亲水而不拘泥于水。

As part of the last wet land to be developed in the city of Hangzhou, this unique community of about 500,000M residences are mostly townhouses. Among them, about 17 luxury villas in various forms were built in the prime locations near the water edge. Special attentions were paid to create a classical design rich in detail while maintain a close relationship with the beautiful surroundings.

无极建筑设计有限公司（香港）
无极建筑设计（深圳）有限公司
WUA Architects Co., Ltd.(Hongkong)
WUA Architects (Shenzhen) Co., Ltd.

WUA香港公司于2010年成立WUA深圳公司，为日益增长的中国内地市场提供具有国际水准的规划、建筑和设计顾问工作。
公司擅长下列设计顾问工作：开发理念、规划地产、旅游地产规划、城市设计、城市综合体、酒店。
设计团队由国内外优秀的建筑和规划专业人士组成，富于进取的工作态度、娴熟的专业技能和丰富的实践经验，足以应对从小尺度建筑到大型规划的各类设计项目。把每个项目都当成独特的个案，综合考虑当地的气候特点、社会文化环境、项目的经济性和市场定位、可持续设计等因素，采取维度更广、程度更深的设计方法，为项目在给定的功能、时间、预算和技术条件下找到最适合的解决方案。

创意是公司的精髓，卓越是公司的品质。

地址：深圳市南山区蛇口南海意库2栋4楼408
电话：+86-755-26856312/26859375
传真：+86-755-26809387
邮箱：szwua@wua.com
网址：www.wua.hk

Add: Room 408, 4th Floor, Tower No.2, Nanhaiyiku Building, Shekou Nanshan District, Shenhen
Tel: +86-755-26856312/26859375
Fax: +86-755-26809387
E-mail: szwua@wua.com
Web: www.wua.hk

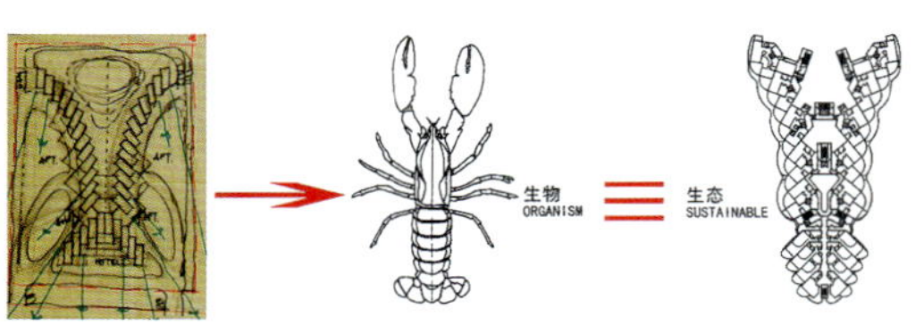

建筑的形态是源于场地100%的海景、日照、风向的需求发展而来；变成生态的体块，就像生物一样。

MASTERPLAN "SHAPED" BY NATURAL MIRCO CLIMATE ELEMENTS OF 100% SEA VIEWS,WIND DIRECTIONS AND SUN PATHS, AND EVOLVE INTO AN OPGANIC STRUCTURE OF A LIVING ORGANISM.

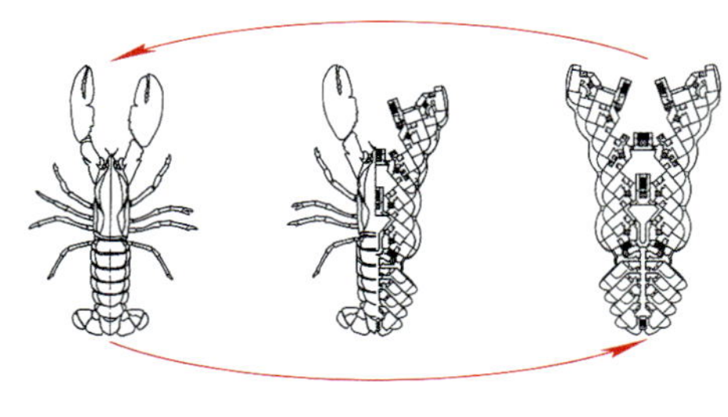

生物是生态的"机器"
生态的"机械"是一种生物

ORGANISM ISA LIVING SUSTAINABLE "MACHINE"
SUSTAINABLE "MACHINE" ISA LIVING ORGANISM.

北海市北苑中心
Beiyuan Center, Beihai

设 计 师：吴立志、谭龙、曾建新、江壬先、史建云、张岚、康鹏飞
项目地点：广西 北海
用地面积：13 341 m^2
建筑面积：129 482 m^2
容 积 率：8.0

Designer: Ng Lup Chee, Long Tan, Jianxin Zeng, Renxian Jiang, Jianyun Shi, Lan Zhang, Pengfei Kang
Location: Beihai, Guangxi
Site Area: 13,341 m^2
Building Area: 129,482 m^2
Plot Ratio: 8.0

项目要想获得100%的一线海景，只有采用非常规的设计方法。经过多次的尝试，发现在给定的地块形状内，唯一能实现100%海景的最佳方案是将酒店建筑设计成一个类似"V"字形曲线的布局。这种布局其实是与自然界微气候相适应的结果，对100%一线海景、最佳风向和太阳路径的考虑。曲线结构有助于提升气流带来的舒适性，垂直的风塔通过叶片将风能转化为电能。建筑立面的遮阳板则可以减少房间获得的热量，遮阳板上中空的太阳能集热管则为建筑提供热水。

Approach to achieve 100% front line sea views, The site was tested with numerous layout configuration within the parameters and the only optimum solution with 100% front line sea views is a curvilinear "V" shaped block masterplan. this curvilinear "V" shaped block masterplan is "shaped" by natural micro climate of 100% front line sea views, wind directions and sun paths provide the whole sustainable design with curvilinear porosity for air flow comfort, vertical wind turbine tower to harnessed wind energy for electricity, vacuum tube solar collector for hot water supply and sun shade to minimised heat gain.

成都锦江新天地

Jinjiang New World, Chengdu

设 计 师：吴立志、谭龙、江壬先、史建云、娄雨鹤、邬龙辉、张岚、康鹏飞、
曾建新、周毅、白鹤
项目地点：四川 成都
用地面积：127 936 m^2
建筑面积：941 048.7 m^2

Designer: Ng Lup Chee, Long Tan, Renxian Jiang, Jianyun Shi, Yuhe Lou, Longhui Wu,
Lan Zhang, Pengfei Kang, Jianxin Zeng, Yi Zhou, He Bai
Location: Chengdu, Sichuan
Site Area: 127,936 m^2
Building Area: 941,048.7 m^2

曲线型的建筑、流动和通透的空间设计，其灵感来自于四川独有的竹笋、黄龙池地质奇观、婀娜多姿的峡谷以及多彩的生活方式。建筑裙楼和塔楼的平面形式是由主要功能块的需求决定的，并遵循可持续设计的理念。

作为都市生活方式汇聚的枢纽，这里既有传统的文化、艺术，又融入现代的时尚风潮，东西方文化合二为一，让人们真正体验“生活是你所喜欢的，喜欢什么就是你要的生活”。

The curvilinear flow and fluid spatial porosity design inspiration soul comes from Sichuan pure iconic bamboo shoot, Huanglong magical geological pool wonders, curvaceous Canyon and Sichuan's colourful lifestyle.The Podium and Tower plans and forms are induce and derive from the function requirements of main Function groups and function requirements of sustainable design which epitome Function exceeds Form.

The urban Lifestyle hub where traditions of culture, arts are fused into the contemporary of fashion, trends, space, where East and West becomes One and a place to " Live what you Like, Like what you Live! "

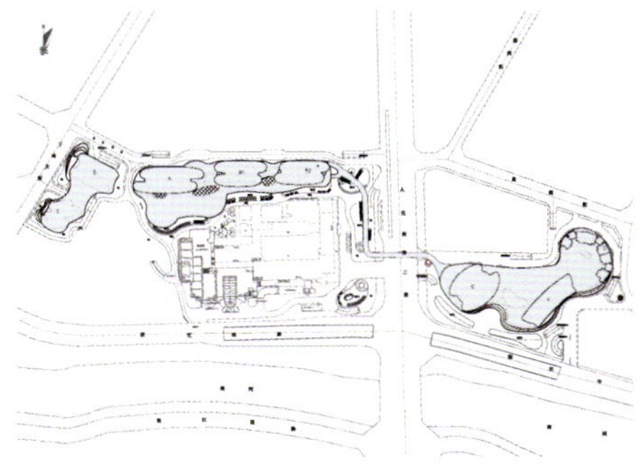

山水悠城 · 国际健康度假区

Shan Shui You Cheng · Health & Tourism International Resort

设 计 师：吴立志、史建云、康鹏飞、邬龙辉、谭龙、娄雨鹤、江壬先
项目地点：广东 佛山
用地面积：4 392 807 m^2
建筑面积：550 000 m^2

Designer: Ng Lup Chee, Jianyun Shi, Pengfei Kang, Longhui Wu, Long Tan,
Yuhe Lou, Renxian Jiang
Location: Foshan, Guangdong
Site Area: 4,392,807 m^2
Building Area: 550,000 m^2

本案最大的挑战是如何减少环境破坏、最大化利用自然资源和现有村落，创造一个国际化的健康旅游度假区，并提供多样化的产品、满足不同市场需求的可持续发展。项目一期启动区的目标是用最少的投资获得最高的市场曝光率，这就要求设计不仅是提供一张总规划图，而是做复合型的整体设计，包括环境利用策略、文化和社会肌理研究、旅游和经济解决方案等。

复合型的概念规划构建包括旅游、会展及酒店、康乐、住宅和企业会所、农业公园、运动和休闲六大功能区。

The primary challenge of this situation is how to minimised disruption and maximized utilisation of the natural green environment and existing farming villages to create an International Health & Tourism Resort with product diversity, market differentiation and economic sustainability, and to kick start project phase one with reasonable minimum investment and maximum market exposure. This situation require more than just a masterplan design but a compounded design, environment utilization, culture/social fabric, tourism and economic solution.

The compounded concept solution is a complimentary product mix of 6 components of Tourism, Convention Recreation Hotel, Well Being & Health, Residential & Enterprise Club, Agriculture park and Sports & Recreation.

广州亚泰建筑设计院有限公司
Guangzhou Asiatal Architecture Design Institute Co.,Ltd.

扫描查看更多信息

广州亚泰建筑设计院有限公司（原广州市荔湾区建筑设计院）始建于1981年，于2007年3月被华尚建筑设计有限公司收购，是国家建设部认定的建筑工程乙级设计单位，专业从事各类建筑工程设计及相关的景观设计、工程技术与城市规划咨询等业务。

华尚建筑设计有限公司成立于2002年，公司目前人才荟萃，拥有一批专业教育出身、工程实践经验丰富的优秀设计师。主要工程业绩有华南农业大学项目、广州天河软件园——IT商务中心（五星级酒店）、山西世茂大酒店、广州大学城二期工程、大宇宙信息创造（中国）有限公司办公大楼等。

地 址：广州市海珠区北约新街5号6楼
邮 编：510310
电 话：+86-20-84214629/84204936
传 真：+86-20-84214629-802/81908802-802
邮 箱：lwsjok88@163.com
网 址：www.gzatsj.com

Add: 6th Floor, Beiyue New Street, Haizhu District, Guangzhou
P.C.: 510310
Tel: +86-20-84214629/84204936
Fax: +86-20-84214629-802/81908802-802
E-mail: lwsjok88@163.com
Web: www.gzatsj.com

东新高速以东保障性住房项目
Indemnificatory Housing Project on East Side of Dongxin Expressway

项目地点：广东 广州
用地面积：133 742 m²
建筑面积：302 540 m²
建筑密度：0.28

Location: Guangzhou, Guangdong
Site Area: 133,742 m²
Building Area: 302,540 m²
Building Density: 0.28

本项目混搭中式文化的“神”和岭南风格的“意”来突破，创造一种全新的中式理念。在建筑设计上，一方面要顺应时代变化，采用适合现代生活方式和审美情趣的高层建筑形式和内部空间格局；另一方面，基于对中国岭南文化情结的深刻洞察，规划设计采用半围合、转折等手法创造出不同层次或开放或私密的休闲空间，营造出带有岭南印象的实用的休闲场所。结构布局上，充分满足通透、采光、通风的宜居舒适性，结合环保节能的新型建材，形成展现岭南居住文化之美且具有典型示范效果的新中式格局的现代化人居。设计中沿街商铺和架空层参考了岭南地域的建筑样式——骑楼，可避风雨、防日晒，特别适合岭南亚热带气候，利于营造商业氛围及增强组团内部的通风效果。

The project breaks through with the spirit of Chinese culture and the implication of Southernmost Chinese style, presenting a brand new Chinese style philosophy. On one hand, the architectural design follows the change of times by adopting high-rise building pattern of modern life style & esthetic sentiment and interior space layout; on the other hand, it creates open-or-private leisure space with semi-enclosure, turning and other approaches, based on the profound observation to Southernmost Chinese culture, and builds a practical leisure space with Southernmost Chinese impression. The structural layout creates habitable comfort with good permeability, lighting & ventilation, combining the environment-friendly & energy saving new material to present a modern residential building demonstrating the new Chinese style with the beauty of Southernmost Chinese inhabited culture. The design for the storefront and overhead layer refers to the building pattern in Southernmost China, namely sotto portico, to guard against wind & rain & sun, especially suitable for the subtropical climate in Southernmost China, better to create commercial atmosphere & ventilation effect inside the building cluster.

广州市黄埔区老人院二期工程
Rest Home Phase II in Huangpu District, Guangzhou

项目地点：广东 广州
用地面积：13 076 m²
建筑面积：28 662 m²
容 积 率：1.62

Location: Guangzhou, Guangdong
Site Area: 13,076 m²
Building Area: 28,662 m²
Plot Ratio: 1.62

本设计将“关怀与被关怀”作为核心价值观，有针对性地满足老年人的特殊需求，创造一个根植于社会和家庭的“关怀与被关怀”的温暖平台。建筑适当退让规划路减少噪声，通过园林植物降低噪声影响；建筑围合庭院，开口面向河涌，创造了一个宁静的内庭院；动静分区明确，形成三大分区：入口广场动区、内庭花园静区、沿街绿化隔离区。围合庭院与半围合庭院设计，结合了传统的园林景观设计手法，“虽由人作、宛自天开”，使景观更具诗情画意。本设计强调“小中见大”的设计手法，营造出多变的空间效果。地下一层，局部掏空档板，在地下一层种植树木、竹子，形成一个立体的景观效果，同时可以保证地下室活动空间有充足的光线及通风。

The core value of this design is “to care and to be cared”. This project aims at the special requirement of old people to create a caring platform with root in the society and family. The building retires properly from the planned road to reduce the noise, and reduce noise impacts though gardening plants. The enclosing courtyard faces to the river, forming a quiet inner yard, with clear distinction between active area and inactive area.While the three sections are formed, namely entry square (active area), inner inside garden (inactive area), and green isolation area along the street.

The design of closed and closing yards combines the traditional landscape design with the concept of “natural work though man-made”. The design emphasizes the approach of “multum in parvo”, creating more variable space effect. The ground floor adopts partially hollowed baffle and plants with forest trees in the underground, forming a 3D landscape effect while giving plenty lighting & ventilation to the activity spaces.

石丰路保障性住房项目

Shifeng Road Affordable Housing Project

项目地点：广东 广州
用地面积：87 104 m^2
建筑面积：244 538 m^2
建筑密度：2.8

Location: Guangzhou, Guangdong
Site Area: 87,104 m²
Building Area: 244,538 m^2
Building Density: 2.8

项目设计以中国民居建筑要素及岭南地域建筑文化为参考。规划上设计采用半围合、转折等不同层次的手法创造出或开放或私密的休闲空间，营造出带有岭南印象的实用休闲场所。布局上充分满足通透、采光、通风的宜居舒适性，结合环保节能的新型建材，形成展现岭南居住文化之美且具有典型示范效果的新中式格局的现代化人居。设计中沿街商铺和架空层参考了岭南地域的建筑样式——骑楼，可避风雨防日晒，特别适合岭南亚热带气候，利于营造商业氛围及增强组团内部的通风效果。通过提取中国传统的徽派民居建筑的一些基本元素，再结合现代简约自由的手法，同时又添加了岭南特色的建筑概念，形成了具有强烈中国风的立面风格，成为统一整个小区的总体印象符号。

The project design refers to Chinese residential building elements and peripheral texture elements of Southernmost Chinese architecture. It creates open-or-private leisure space with closing & turning etc. approaches based on the profound observation to Southernmost Chinese culture, and builds a practical leisure space with Southernmost Chinese impression. The structural layout creates habitable comfort with good permeability, lighting & ventilation, combining the environment-friendly & energy saving new material to present a modern residential building demonstrating the new Chinese style with the beauty of Southernmost Chinese residential culture. The design for the storefront and overhead layer refers to the building pattern in Southernmost China, namely sotto portico, to guard against wind & rain & sun, specifically for the subtropical climate in Southernmost China, benefiting the building of commercial atmosphere & ventilation effect of the group interior. Through abstracting some basic elements from the Chinese traditional Anhui-style residential buildings, combining the modern neat & free approaches, and adding the architectural concept with Southernmost Chinese features, the design creates a façade style with strong Chinese taste and builds a general impression symbol for the whole community.

西安祭台村城中村改造项目DK - 1

Urbanization Project DK-1 of Jitai Village, Xi'an

项目地点：陕西 西安
用地面积：50 800 m^2
建筑面积：398 496.33 m^2
容 积 率：5.87

Location: Xi'an, Shanxi
Site Area: 50,800 m²
Building Area: 398,496.33 m^2
Plot Ratio: 5.87

本项目位于西安市太乙路与南二环交汇处的东北角。该建筑群的建筑密度和容积率非常之高，住宅通过高低错落的排布，来弱化建筑群的高密度感。酒店采用简欧的设计风格，整体的色彩采用了土黄色，呼应了西安厚重的城市文化色彩。

The project is located on the northeast corner of the intersection of Taiyi Road and South 2nd Ring Road in Xi'an City. The building cluster has very high building density & plot ratio, and the residential buildings are scattered through framework, weakening the high density of the building cluster. The hotel adopts simple European design style, and the color is generally yellowish brown, responding to the heavy urban color of Xi'an.

华蓝集团
HUALAN GROUP

华蓝集团

华蓝集团是由广西华蓝设计（集团）有限公司作为母公司，与其14家控参股公司共同组建成立的集团企业，现有员工约3200余人。业务由工程设计与咨询、工程总承包、投资与房地产开发三大板块组成，持有咨询、勘察、设计、监理、施工、房地产等各类资质47项（甲级资质26项），涉及建筑、规划、市政、智能化、交通、岩土、能源、环保、房地产等领域。

广西华蓝设计（集团）有限公司

广西华蓝设计（集团）有限公司于2007年1月1日成立，是由广西建筑综合设计研究院（始建于1953年）改制而成的现代科技型企业，是目前广西规模最大的国家工程设计咨询综合性甲级企业之一。公司拥有建筑设计院、规划设计院、市政设计院等3个专业设计院，1个研究院和8个分公司，在北京、上海、广州、成都、深圳、福州、海口、昆明等地设立分支机构。

地址：广西壮族自治区南宁市华东路39号
邮编：530011
电话：+86-771-2438149
传真：+86-771-2435011
邮箱：hualan@gxhl.com.cn

Add: Huadong Road No.39, Nanning City, Guangxi Zhuang Autonomous Region
P.C.: 530011
Tel: +86-771-2438149
Fax: +86-771-2435011
E-mail: hualan@gxhl.com.cn

北海冠岭五星级酒店与会议中心
Guangling 5-star Hotel and Conference Center, Beihai

项目地点：广西 北海
建筑面积：88 533 m²

Location: Beihai, Guangxi
Building Area: 88,533 m²

五星级酒店与会议中心位于广西北海市冠岭滨海景区内，是一组和谐有机的整体。建筑形态似苍穹中展翅的凤凰、似海面滚动的浪花、似停泊在海滨的航船……这是一组美轮美奂、飘逸灵动、具有浓郁地域特色、让人浮想联翩的建筑，为北海市打造出一个国际级大型会议及高品质滨海休闲度假酒店，成为一张面向全国乃至世界的城市名片。

The project is located in Guanling seaside scenery in Beihai City, Guangxi Region, and it is a harmonious integrity. The building looks like a phoenix flying in the sky, a wave splashing on the sea and a ship anchoring in the harbor… it is a series of magnificent elegant buildings with strong regional features, which can inspire the people. The building cluster makes a huge international conference center and a high-grade seaside resort hotel, and it is a city card facing the whole country and even the whole world.

北海冠岭山庄
Guanling Mountain Villa, Beihai

项目地点：广西 北海
建筑面积：24 885 m²

Location: Beihai, Guangxi
Building Area: 24,885 m²

北海冠岭山庄位于广西北海市冠岭滨海森林公园景区内。建筑顺应地形、依山就势，以自然山脊为界进行散点式布局。建筑造型风格为地域式原生态风格和东南亚热带建筑风格。整组建筑具有强烈的广西山地建筑特点，舒缓的屋面和层叠的露台形成了高低错落的丰富的轮廓线，赋予了建筑独特的气质，营造出一种国宾级滨海休闲度假环境。

The project is located in Guanling Seastrand Forest Park in Beihai City, Guangxi Region. The building matches the landform and makes full use of the mountain, the layout of which is in scattered style and uses the natural ridge as the boundary. The architectural form is in regional ecological style and Southeast Asian tropical architectural style. The entire building has strong Guangxi mountain architecture features. Roofs and balconies of all heights form a diverse skyline in picturesque disorder, which gives the building a unique atmosphere, and creates a high-grade coastal resort environment.

南宁市综合档案馆（含南宁市方志馆）
Nanning Comprehensive Archives (Including Nanning Library of Local Records)

项目地点：广西 南宁
建筑面积：29 032 m²

Location: Nanning, Guangxi
Building Area: 29,032 m²

本项目包括南宁市综合档案馆和南宁市方志馆，两馆均是城市的记录者，是人类文化的积沉，折射城市的历程与未来。项目以具有民族性、地域性、科学性的建筑语言，强化城市机理，表达建筑身份，体现档案馆与方志馆的文化属性。项目将成为一个具有现代气息、文化品位、地方特色的南宁标志性文化建筑之一，成为“全国一流，广西第一”的市级综合档案馆和方志馆，成为具有“五位一体”功能的市级综合档案馆。

This project includes Nanning Comprehensive Archives and Nanning Library of Local Records, both of which are the recorders of the city, the deposition of human culture and a reflection of the course and future of the city. The design goal is to use national regional and scientific architectural languages to strengthen the city texture, show the identity of the buildings and present the cultural attributes of the two buildings as an archives and a library of local records respectively. The project aims to build them into one of the landmark cultural buildings in Nanning with modernism, cultural taste and local characteristics; "national first-rate and Guangxi's top" municipal comprehensive archives and library of local records; and a municipal comprehensive archives with "Five-in-one" functions.

百色市文化科技中心
Culture, Science and Technology Center, Baise

项目地点：广西 百色
建筑面积：59 105 m²

Location: Baise, Guangxi
Building Area: 59,105 m²

项目位于百色市龙景新区的中心位置，建筑内安排了民族艺术剧院、妇女儿童活动中心、科技馆、文化局、图书馆、综合档案馆和城建档案馆共7个单位。功能需求复杂，设计过程注重策划与设计相结合，注重与各方业主的沟通，解决了功能组织及空间分配等矛盾。建筑形象从美丽的传说《一副壮锦》中汲取灵感，以壮锦作为设计的核心元素，体现在建筑的形体、景观及装饰等各个方面。

The project is located in central Longjing New District in Baise City, and there are 7 parts in the building, namely Ethnic Art Theater, Woman&Children's Activity Center, Science and Technology Center, Library, Complex Archives and Urban Construction Archives. Due to complicated functional needs, the planning process emphasizes the combination of scheming and designing as well as the communications with all owners, to resolve the conflicts of functional organization and space allocation etc. The beautiful legend "Piece of Zhuang Ethnic Brocade" inspires the designers, so Zhuang ethnic brocade is the core element of the architectural image, displayed in all aspects like building form and structure, building landscape, decorations etc.

中国—东盟商贸物流中心B座
Tower B of China-ASEAN Trade & Logistics Center

项目地点：广西 南宁
建筑面积：180 807 m²

Location: Nanning, Guangxi
Building Area: 180,807 m²

项目寻求一种更自然、与生活更贴近的商业办公空间，空中庭院的错位设置，打破了高层办公空间的封闭空间感，使每一层办公空间都可以享受到自然的气息，降低了超高层建筑内部人们的恐惧感。空中庭院设置在电梯厅出入口处，实现了每一层办公空间都有自己的独立大堂和门厅，同时加入了休闲、交流等元素，增加了更多生活的情趣，提升了办公场所的品质。

The project is looking for a commercial office space that is more natural and closer to life. The dislocated arrangement of the hanging garden breaks the sense of enclosed space of the high-rise office buildings, so that the fragrance of the nature can be enjoyed in the office space on every floor and the fear felt by people inside the ultra high-rise building can be abated. The aerial courtyard is set at the entrance of the elevator lobby, which realizes the goal that the office space on every floor has its own hall and lobby. At the same time, the elements of recreation, communication etc. are also combined to add more enjoyment of life and promote the quality of the workplace.

广西壮族自治区林业综合楼
Complex Building of the Forestry Department of the Guangxi Zhuang Autonomous Region

项目地点：广西 南宁 Location: Nanning, Guangxi
建筑面积：34 995 m² Building Area: 34,995 m²

项目构思的主题是："生命之塔、常青绿洲"，整个建筑用地就像绿色的海洋。副楼斜屋面种植绿色草皮，由南向北渐行向上，形成绿色坡地，而主楼就是从森林中生长出来的水晶生命体，充满着无限的朝气。建筑风格简洁、大方，并从森林和树枝的特性出发，利用现代手法对建筑立面进行设计，象征着林业事业的蒸蒸日上。

The theme of the project design is "Tower of Life and Evergreen Oasis" and making the whole building lot look like a green sea. In the annex, the sloping roof is laid with a green turf, which goes up from south to north, forming a green sloping land; while the main building is a crystal creature growing from the forest, full of infinite vitality. The architectural style is simple and openhanded. In addition, with the characteristics of the forest and tree branches in mind, the project adopts modern methods to design the façade. The whole project symbolizes that forestry is on the rise, just like the upward sloping land.

广西建设职业技术学院
Guangxi Polytechnic of Construction

项目地点：广西 南宁 Location: Nanning, Guangxi
建筑面积：126 962 m² Building Area: 126,962 m²

学院的设计以砌筑的概念为出发点，体现出了建筑类院校的特征，建筑群强调了序列感，设计上运用了丰富建筑语言。教学区的空间变化具有趣味性，圆形与方形的体量在外形上形成呼应，同时具有强烈的雕塑感。共享空间的设计是序列中变化的语言，结合教学楼通风设计，给使用者创造了舒适的交流空间。红、白、灰为校园的主色调，营造出浓郁的学术氛围。

The project design centers on the masonry concept, typical of architectural school. The building cluster is design in a sequence, with rich architectural language. The spatial space variations in the teaching area are very interesting, with circular and cubic masses responding to each other, which, at the same time, shows a strong sense of sculptural beauty. The shared space is a changing language in the sequence, which, together with the ventilation design of the teaching building, creates a comfortable communication space for the users. The red, white and grey colors are the dominant tones of the campus, creating a strong academic atmosphere.

广西大学新闻出版人才培养基地大楼
The Building of Journalism and Press Talents Culturing Base in Guangxi University

项目地点：广西 百色 Location: Baise, Guangxi
建筑面积：25 145 m² Building Area: 25,145 m²

本项目校园的环境景观除了其物质功用外，还有为教职员工提供休闲、娱乐、学习、交往空间的作用，亦即"场所精神"。 项目功能分区明确合理，既相互独立又紧密联系，北面是培训住宿区，南面是教学办公区，中间是相对独立的大跨度的教学服务区，在建筑的中部形成一个开敞的中庭，为师生营造良好的教学环境和相互交流的氛围。

In this project, the campus landscape apart from material function can also provide the faculty and staff with leisure, entertaining, study and social spaces, that is, "venue spirits". The distinct functions are independent and closely related. The north is training and lodging zone, the south is teaching and office zone, the middle is large-span teaching service zone. In the middle of the buildings is an open atrium, to create a good teaching environment and communicating atmosphere for the faculty and students.

百色市龙景小学
Longjing Primary School, Baise

项目地点：广西　百色　　Location: Baise, Guangxi
建筑面积：25 145 m^2　　Building Area: 25,145 m²

项目位于百色市龙景新区北部，基地三边环路，紧邻右江。校园的平面打破了传统行列式布置方式，采用集中布置的方法，将全部建筑通过走廊连接起来，将建筑群连成一个整体。这种布置方式不仅可以避免风吹雨淋，还留出了大片供休憩娱乐的空地。建筑外墙局部采用红色涂料，以体现新校区的朝气和活力。

The project is located in the northern part of Longjing New District planned in Baise city. The base is close to Youjiang River with three sides facing the roads. The campus layout of concentration breaks the traditional parallel array. All the buildings are linked by the corridor, into a building cluster. With this method, the buildings can avoid exposure to wind and rain, and large area of open space is reserved for pupils' relaxation and entertainment. The exterior wall of buildings partly uses red painting, to represent the vitality and vigor of the new campus

蓝海银湾
Silver Bay on Blue Sea

项目地点：广西　南宁　　Location: Nanning, Guangxi
建筑面积：216 000 m^2　　Building Area: 216,000 m²

项目采用周边围合式布局模式，建筑南北向布置，通风采光景观视线良好。场地经过建筑单体的有机限定，划分出一个开放的大型中心景观庭院和四个周边小型院落。建筑立面以传统博古架窗格为设计元素，融合现代处理手法，凹凸有序，高低错落，使建筑造型既富有深厚的文化底蕴和清晰的历史文脉，又不失现代感。

The project adopts a peripheral enclosing layout. The sunward buildings receive a good ventilation, natural lighting and landscape view. Dynamically divided by individual buildings, the site forms a large open central landscape garden surrounded by four small courtyards. Antique-and-curio shelf panes are used in the façade design. Combined with modern treatment techniques, the panes are arranged in an orderly concave-convex manner, giving the building a sense of profound cultural background and clear historical context, and also a sense of modernism.

南宁东盟商务区印度尼西亚园区
Indonesian Park of Nanning ASEAN Business District

项目地点：广西　南宁　　Location: Nanning, Guangxi
建筑面积：71 380 m^2　　Building Area: 71,380 m²

项目是集住宅、商务、公寓、商业街区为一体的复合式旅游地产，独具印尼风情。规划以场地与现状为依据，以人为本，以整体社会效益、经济效益、环境效益三者结合为基准点，设计定位为"国际全开放式交流社区"，以高端住宅和特色化商业街为主体，结合印度尼西亚岛国风情的主题，为居民塑造开放宜人的居住空间。建筑布局上充分利用现状地形，因地制宜，尊重自然，突出生态观念，具有东南亚地区的居住建筑特征，造型的变化使建筑有个性和识别性。

The project is a complex tourism real estate that integrates residence, commerce, apartment and business district, characterized by Indonesian style. The planning is based on the site and its current conditions and centered on people and considers the overall social, economic and environmental benefits as a whole. The design positioning is to build an international total-open exchange community, take high-end houses and characteristic business streets as main body, combine them with Indonesian style, and create an open and pleasant living space for residents. In the architectural layout, the project makes full use of the existing landform and emphasizes ecology with a respect for the nature. In addition, the architectural features of Southeast Asia make the building unique and recognizable.

贵州省建筑设计研究院
Guizhou Provincial Architectural Design & Research Institute

扫描查看更多信息

贵州省建筑设计研究院建于1952年，是国内最早成立的全国综合甲级设计单位之一，具有国家建设部及国家计委颁发的工程设计、工程勘察、工程咨询、城镇规划、市政设计、工程监理、造价等多项资质，现有职工500余名，下设多个综合设计所，是全国优秀勘察设计单位。设计院将一如既往贯彻"坚持社会责任至上、客户价值至上、员工发展至上，树立行业风范，追求卓越品质，打造企业品牌"的宗旨，参与竞争，努力实践，以创贵州品牌企业为目标，竭诚为社会各界和广大客户提供最优质的服务。

地址：贵州省贵阳市遵义路48号
电话：+86-851-5573792
传真：+86-851-5572433
邮箱：gzsjy@vip.sina.com
网址：www.gadri.cn

Add: 48 Zunyi Road, Guiyang City, Guizhou Province
Tel: +86-851-5573792
Fax: +86-851-5572433
E-mail: gzsjy@vip.sina.com
Web: www.gadri.cn

遵义会展中心概念性规划
Concept Plan of Zunyi Exhibition Center

设 计 师：董明、许彤、王硕锋、孙雨杰
项目地点：贵州 遵义
用地面积：170.8 hm²

Designer: Ming Dong, Tong Xu, Shuofeng Wang, Yujie Sun
Location: Zunyi, Guizhou
Site Area: 170.8 ha

本案以湿地公园为圆心向四周发散，突出湿地公园在规划结构中的重要性。以湿地公园为中心的规划空间结构是整个规划范围的大形态，再辅以组团内的小空间核心，组成了整体一致、结构分明的规划体系。规划设计以湿地公园为中心，引伸出放射状的支路，根据不同性质地块的特异性，用一条区域环线分隔出滨水商业区与外圈住宅区，湿地公园内侧的慢行步道与建筑融为一体，勾勒出生态——建筑——人之间的有机联系。从景观设计的角度，湿地公园作为开放空间，视点由此放大，水体、建筑、景观，配合东南侧的山脉，整个景观结构呈由北向南的阶梯形景观架构。整个景观层次分明，山水结合必将成为城市一个新的亮点。

The significance of wetland park in the planning structure is highlighted by circling around the wetland park. The planned space structure circling around the wetland park is a large form of the overall planning area. Together with the small space core in the group, a planning system with distinct structure has been formed.Focusing on the wetland park, radial branches have been extended. According to the different specificities of the lands, a small district circles has been employed to separate waterfront commercial district and outer-race residential district, thus the crawl footpath inside the wetland park is fused with the whole building, mapping an organic connection among the ecosystem, building and human.From the landscape design, the wetland park is an open space, thus the viewpoint is enlarged. The water, building, landscape, as well as the southeast mountain ranges display a stepped landscape structure from north to south. With distinct levels, the combination of mountain and water will certainly become a new spot in the city.

遵义市规划展览馆
Planning Exhibition Hall of Zunyi

设 计 师：董明、许彤、刘敬业、王伟功
项目地点：贵州 遵义
建筑面积：25 000 m²

Designer: Ming Dong, Tong Xu, Jingye Liu, Weigong Wang
Location: Zunyi, Guizhou
Building Area: 25,000 m²

新区内连绵起伏的茶园、优美的自然景观使人不由地产生一种对遵义生态文化和茶文化的流连遐想。由优美的自然风光到茶文化再到园林式、生态型、可持续发展的现代化新城区，通过嫁接这种设计手法，使现实与文化之间完成良好的沟通。空间上使室内和室外"嫁接"。建筑的倾斜面作为参观者走向建筑屋面的便捷道路，由室外临水的步道沿等高坡道走向屋面，犹如置身于生态茶园一般；通过展厅连廊可以从建筑屋面走入规划展厅，体会到室内与室外交替变化带来的独特的空间感受。外部空间形态上使山、水、建筑进行"嫁接"，建筑以山为景、水为邻，为界面变化提供了良好的载体。架空连接的规划道路让建筑以连廊的方式穿过道路与西面水体产生联系，让大地景观向建筑延伸。

The rolling tea plantation and beautiful natural landscape will produce a reverie of the ecological culture and tea culture of Zunyi City. From the beautiful natural landscape to the tea culture then to the landscape-scale, ecotype and sustainable modern new town, grafting design method has been adopted to realize the communication between the reality and culture. In space: graft indoor space to the outdoor space. The inclined building turns to be a convenient path towards the building roofing, walking toward the roofing on the footpath over the water along the contour line. It seems as if they were in the ecological tea plantation. Walking into the planning exhibition hall through the building roofing along the corridor, interchange between the indoor and outdoor space will be experienced, enjoying a unique space perception. In outdoor space pattern: the grafting of mountain, water and building. Together with the mountain and water, the land has provide a good load for the interface change. The overhead planning path connects the path in the form corridor to the southwest water, greatly extending the landscape to the building.

中铁逸都国际
China Railway Leisure Town

设 计 师：任朝刚、张学源、黎涛
项目地点：贵州 贵阳
项目规模：210 hm²

Designer: Chaogang Ren, Xueyuan Zhang, Tao Li
Location: Guiyang, Guizhou
Site Area: 210 ha

中铁逸都国际在规划设计中摒弃了一般规划布局从总体到局部的设计流程和观念，而采用以研究人在室内空间的感受为切入点，从内到外进行规划设计，摒弃了一般总平面规划讲究"形"及"图面视觉"的误区，形成新型视域的规划理念。将不同居住结构及消费结构人群布置在不同院落，从居住心理上真正体现"以人为本"。单体设计打破了常见多层住宅中"标准层"的设计思路，以"非标准层"为切入点，打造居住感受均不相同的个性化住宅。

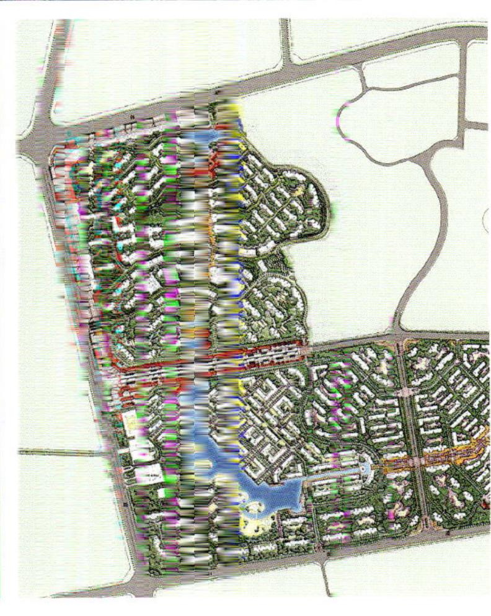

Desingers of ZhongTie Yidu International project abandoned the clue from whole to part of common design concept in this programme, but to use the study of feeling inside and outside rooms to make a plan from indoor to outdoor. The plan put away the idea of normal groud plan, which is based on "forms" and "drawing visual sense" to formed a new view field idea.

This project allocates diversified yards for people with different inhabitancy and consumption configuration to realize the idea of people oriented living. The theme of solo design breaks design clue of standard floors often seen in storied houses. The plan is based on concept of "a non-standard floors" to provide an individualized house which is built to suit local condition and bring different feelings of inhabitancy.

贵阳中医学院花溪新校区

Huaxi New Campus of Guiyang College of Traditional Chinese Medicine

设 计 师：程鹏、王凌志、李曦炜
项目地点：贵州 贵阳
建筑面积：550 000 m²

Designer: Peng Cheng, Lingzhi Wang, Xiwei Li
Location: GuiYang, Guizhou
Building Area: 550,000 m²

贵阳中医学院花溪新校区的规划建设充分契合贵州山水文化精神和中国传统文化特色，将中国传统哲学观念所追求的意境及表现手法融入规划建设中，积极吸收风水思想中有益的成分，在各个层面追寻山水城市的诗情画意。按照“天人合一，双翼环抱”的规划构思，采用“一轴两翼，向心发展”的总体规划布局原则，具有传统中医药文化和校园建筑的特征，同时具备贵州山地建筑特色和地域文化的特点。为强化学校整体特色，采用抽象建筑元素的提炼及重复运用，为校园的每一栋建筑打上统一烙印，使得整个校园既和谐统一又充满个性。

The new campus in Huaxi of Guiyang college of tradition Chinese medicine completely corresponds to the spirit of Guizhou landscape culture and the feature of Chinese tradition culture in its planning and construction. The artistic conception and expression technique what tradition philosophy in Chinese culture seek for have been involved in the design of planning and construction. Meanwhile, the campus plan absorbed actively the benefited idea of Fengshui thought and pursued a kind of poetic imagery of landscape city at all levels. According to the planning concept of "the unity of nature and humanity and surrounding from two wings", the principle of overall planning and layout of this project is "One axis and two wings, the development to the center area". The project has the feature of tradition medicine culture and campus building, along with the characteristics of Guizhou mountain building and regional culture. For strengthen characteristics of overall campus, the building make reduplicative use of abstract element. The "family" sign been marked on every campus building, making them particular and harmonious.

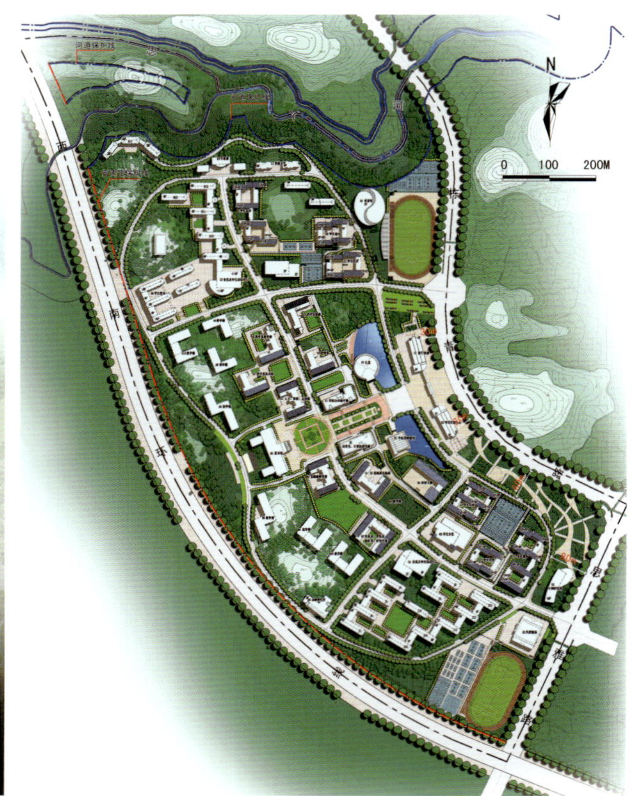

贵阳市供电局生产管理中心

Guiyang City Power Supply Bureau Production Management Center

设 计 师：程鹏、罗宇红、李曦炜
项目地点：贵州 贵阳
建筑面积：35 000 m²

Designer: Peng Cheng, Yuhong Luo, Xiwei Li
Location: Guiyang, Guizhou
Building Area: 35,000 m²

项目位于贵阳市金阳新区，由办公、电力客户服务中心、会议中心三部分组成，其形象具有一定的“地标”性质，在金阳新区商贸核心位置以三角形的平面布局诠释与众不同的道路交叉口处理模式。建筑造型简洁、明快，新颖、独特，采用复合金属玻璃幕墙，体现了贵州特有的岩层肌理和电力系统建筑的特殊性格。

The project is located in Jinyang New District, Guiyang City. It consists of office area, power customers' service center and conference center with an image of “landmark”. The triangular layout interpretes the different intersection mode at the commercial core location in Jinyang New District. The building shape is neat & bright, novel & unique, and adopts composite metal glass curtain walls to present the characteristic stratum texture in Guizhou and the special features of the power system architecture.

贵阳市公安局云岩分局业务及技术用房

Operation &Technology House, Guiyang Public Security Bureau Yunyan Branch

设 计 师：张翼、吴迪、胡振羽
项目地点：贵州 贵阳
建筑面积：36 000 m²

Designer: Yi Zhang, Di Wu, Zhenyu Hu
Location: GuiYang, Guizhou
Building Area: 36,000 m²

本项目设计为规划城市中金阳新区以及部分老城区（云岩区）的公安机关业务用房。建筑一方面体现公安机关勤政为民、廉洁奉公的精神，其风格又与当地建筑形式结合，并体现自己的独有特色；同时创造出舒适、宜人、美观的城市空间环境，并突出其公共性、开放性的一面。主体建筑向北京西路开口，平常民事流线、正常的公安干警流线以及干警办案流线互不交汇，内部核心交通便利。

The project is the business housing for public security organs of Jinyang New District in the planned city and partial old town (Yunyan District). The building represents the spirits of public security organs of diligent work & serving the people, honest & clean in official duties, and combines with the local architecture form in style, presenting its unique character; meanwhile, it creates comfortable, pleasant & beautiful urban space environment, standing out its characters of publicity & openness. The main building opens to the Beijing West Road, and has separate usual civil affairs streamline, normal police streamline and police case-handling streamline, therefore embracing convenient traffic at inner core.

贵阳职业技术学院
Career Technical College Guiyang

设 计 师：张翼、汪敬、彭贤迪
项目地点：贵州 贵阳
用地面积：20 hm²

Designer: Yi Zhang, Jing Wang, Xiandi Peng
Location: Guiyang, Guizhou
Site Area: 20 ha

贵阳职业技术学院由原贵阳职业学院及技术学院组建而成，具有26年高等职业教育办学历史，共设有高等职业教育专业10个，中等职业教育专业8个。设计中从基地独特的环境出发，构建一轴一核两中心组团式布局。通过合理的布局及设计塑造具有地域特色的整体校园环境，将人文氛围的营造和自然生态体系的培养作为着眼点，打造一所润泽师生、知行合一的生态校园。

Former Career Technical College and Guiyang University Guiyang occupation technical college to form, with 26 years of higher occupation education of higher occupation education in history, with a total of 10 major, medium occupation Education major 8. Design of the base from the unique environment, constructing a shaft two nuclear center group layout. Through the reasonable layout and design of campus building with regional characteristics of ring Exit, the humanities atmosphere of the natural ecological system and the culture as a starting point, to create a moisture of teachers and students, and the ecological campus

溪南高中（清华中学分校）
Xi'nan Senior Middle School Attaching Tsinghua Middle School

设 计 师：张翼、彭贤迪、霍子益
项目地点：贵州 贵阳
占地面积：53 000 m²

Designer: Yi Zhang, Xiandi Peng, Ziyi Huo
Location: Guiyang, Guizhou
Building Area: 53,000 m²

大方贵州宣慰府恢复重建工程

Defang Guizhou Xuan Wei Palace Reconstruction Project

设 计 师：阮志伟、张舜、刘超
项目地点：贵州 毕节
用地面积：7.5 hm²

Designer: Zhiwei Ruan, Shun Zhang, Chao Liu
Location: BiJie, Guizhou
Site Area: 7.5 ha

贵州宣慰府恢复重建的任务，选址在贵州传统的彝族聚居区大方县。彝族王宫的形制同《西南彝志选·论宏伟的九重宫殿》之中明确记载的九重宫，就是根据地形的高差层层向上的8个院落、9幢主殿组成的建筑群，这种院落高差变化的形式同西南多山的地理环境吻合。设计中综合彝族专家的意见，确定其形制应是“九层衙”，并应具备以下特点：九殿八院形制，两侧配辅房和辅院，并层层向上；背山面水，坐东朝西。

The project is located at dafang.The nine layer palace, which is recorded in the style if Yi palace "Journal of Southwest Yi selected, on the magnificent nine palaces", is designed as a eight courtyards and nine main palaces in order to form a architecture block, that is according to the height gap of t geography. And the style of the variation of the height correlated to south-east mountainous geography environment. In the design, we accompany the opinion of Yi community professors, and we decided that its style must be "Nine layers of Yamen". In addition, it should contain following characteristics:Nine palace and eight courtyards' style, and there should be accessory rooms and yards on both sides, and raise higher from the bottom.facing water and against mountain locate at east and facing west .

云岩区行政办公综合大楼

Administration Office Building Yunyan District

项目地点：贵州 贵阳
用的面积：2.6 hm²

Location: Guiyang ,Guizhou
Site Area: 2.6 ha

云岩区政府项目地处贵阳市北郊，大营坡地段，原黔灵乡政府旧址上。基地南低北高，地形复杂，最大高差有30 m。
充分考虑自然地形条件的制约，解决好地形与建筑功能完善交融的问题。
丰富空间层次，创造合理舒适的办公空间。

The project of Yunyan District government located in the Daying Slope section, northern suburb of Guiyang City and the former site of Qianling township government. The plot is lower topography in the South than that in the North, here also has complex terrain which can reach up to 30 meters.
Give a full consideration to the constraints of the natural terrain conditions, solve the problem of the perfect blend of the terrain and the function of the building.
Rich the space level in order to create a reasonably comfortable office space

扫描查看更多信息

贵州天海规划设计有限公司
Guizhou Tianhai Planning and Design

贵州天海规划设计有限公司成立于2002年7月，具有国家乙级旅游规划设计资质、国家乙级城市规划编制资质、国家乙级建筑工程设计资质，拥有各类工程技术人员60余人，是一支朝气蓬勃、勇于挑战、开拓创新、追求卓越的团队。

10年来，天海规划一直都在积极地探索规划设计之路，始终秉承“民族现代化，现代民族化”的理念，不断开拓创新，经过10年孜孜不倦的研究和实践，形成了自己的一套规划设计方法和模式：整合策划、规划、建筑、创意、效果表现、动画等与规划设计密切相关的各个行业于一体，使得规划设计从资源评估、背景分析到战略定位、规划布局、建筑创意、效果表现一气呵成，形成一站式全程服务的规划模式，这种模式最大限度地克服了传统规划设计模式中各个专业脱节的弊端。

多年来，凭借着对公司理念的执著，天海规划设计有限公司以独特的规划设计理念、颠覆性的操作模式以及极富想象力的创意，承担和完成了60多项大、中型规划设计项目，其中有贵阳市文化中心商业中心概念规划设计、云南省丘北县城市设计及普者黑景区规划设计、山东黄金主题公园规划设计、贵州凯里市城市设计、贵州铜仁市城市设计、贵州黔西县县城设计、华藏世界多国佛教园的规划设计等，这些设计或大气磅礴、匠心独运，或精美别致、风韵独具；既具备地域特质和民族文化底蕴，又充盈着浓郁的现代气息和时尚品味；既充满了深切的人文关爱和人性呵护，又洋溢着现代人的生活情调……引起了社会各界的极大反响，备受业界关注和青睐，更是深得开发商和项目开发地区民众的高度评价和认可。

地址：贵州省贵阳市中华南路181号华坤发展大厦16楼C座
电话：+86-851-5845337
传真：+86-851-5848386
邮箱：tianhai6699@163.com
网址：www.gzthgh.com

Add: 16th Floor,Tower C,Huakun Development Building,Zhonghua South Road No.181, Guiyang City,Guiyang Province
Tel: +86-851-5845337
Fax: +86-851-5848386
E-Mail: tianhai6699@163.com
Web: www.gzthgh.com

东方之梦
Oriental Dreams

设 计 师：伍新凤
项目地点：贵州 都匀
用地面积：58 000 m^2

Designer: Xinfeng Wu
Location: Dujun, Guizhou
Site Area: 58,000 m^2

“东方之梦”的设计来源于一位西方友人对东方梦境的追寻。在与这位西方友人的交谈中，设计师能深深地感受到他对于中国贵州这片土地的热爱和他浓浓的异国情怀，他的这份热忱深深地打动了设计师。为了他，为了他追寻的东方梦，于是设计师设计了“东方之梦”。

The original concept of the Oriental Dreams was inspired by a wester friend's chase for the oriental dreams. Communicating with this friend, we were acutely awre of his affection for Guizhou, China and his estraordinarily exotic feelings, which made designer touched a lot. Thus designer.

云南丘北火车站
Yunnan Qiubei Railway Station

设 计 师：伍新凤
项目地点：云南 文山
用地面积：11 684 m²

Designer: Xinfeng Wu
Location: Wenshan, Yunnan
Site Area: 11,684 m²

这是云南丘北火车站的创意设计，蜻蜓点水的形象直观地表达了丘北的秀美多姿。在设计的时候，设计师将蜻蜓的形象抽象化，在满足火车站功能要求的同时又与当地旅游文化资源紧密相扣。

This is the creative design of Qiubei Railway Station, the image of dragonfly playing water directly expresses the beauty of Qiubei. The disgner has modified a little abstractly when design the station in order to achieve the station funtion, and also link closely to the local tourism and cultural resources.

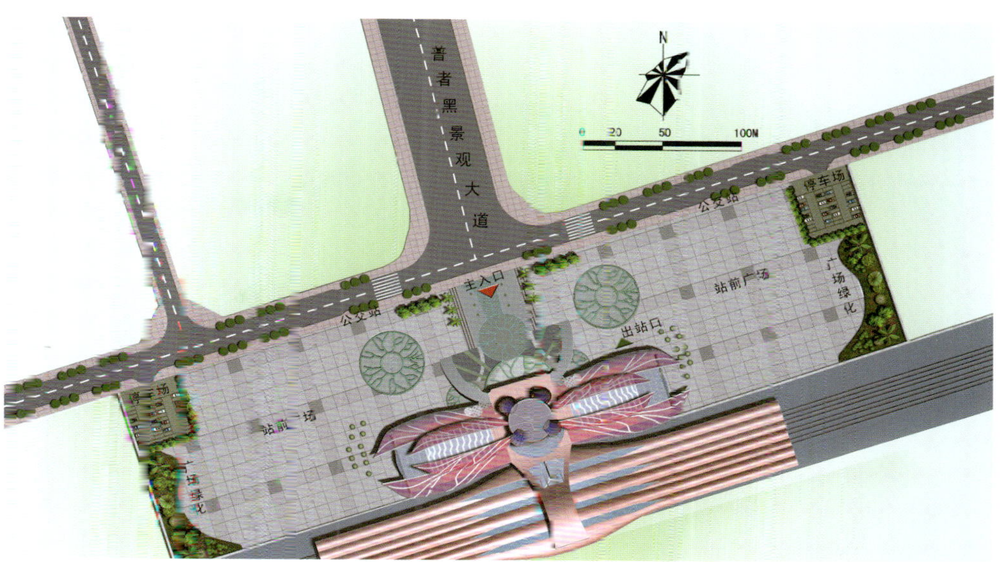

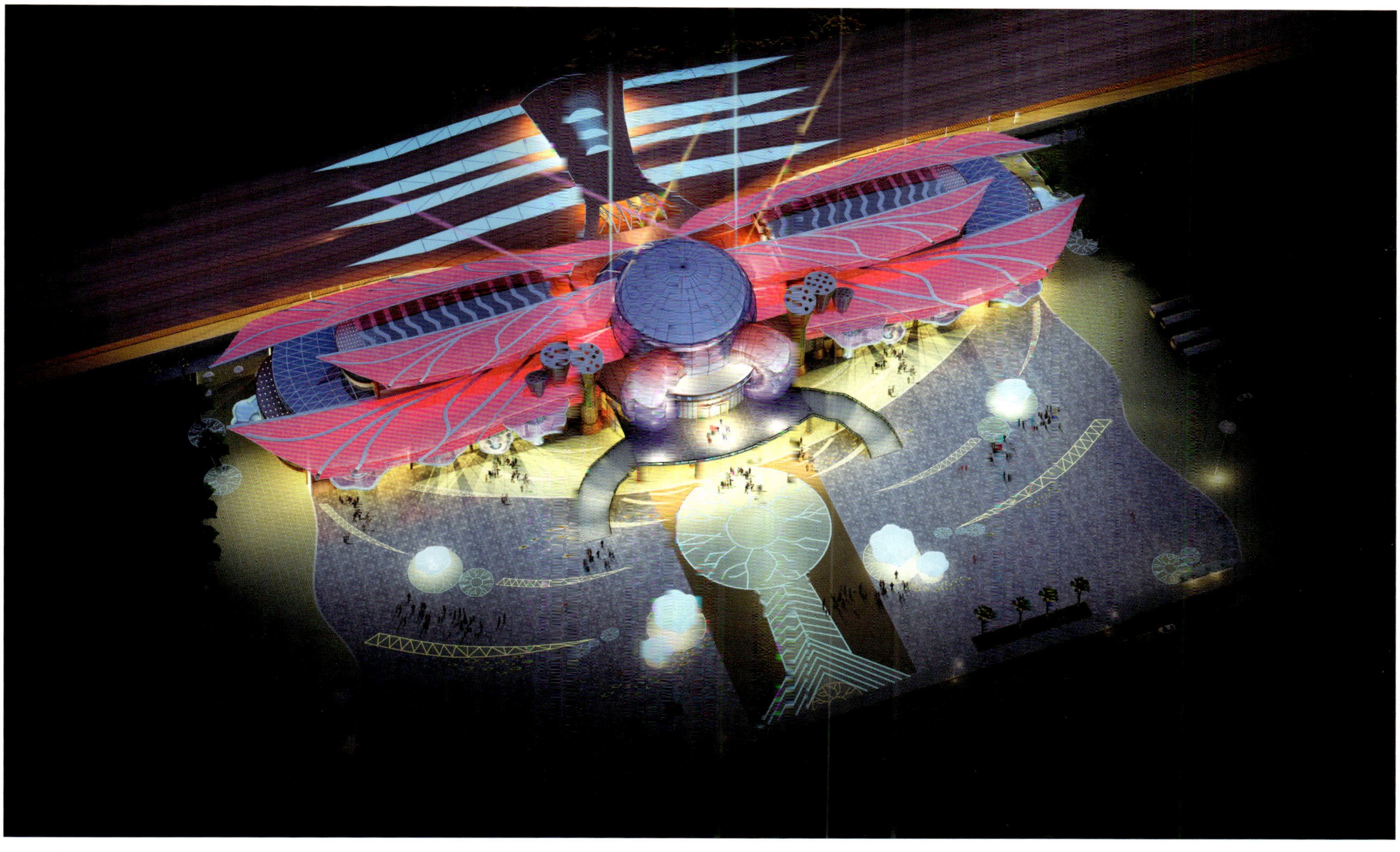

贵州民族文化展演中心
Guizhou Ethnic Culture Exhibition & Performance Center

设 计 师：伍新凤
项目地点：贵州 贵阳
建筑面积：38 889 m²

Designer: Xinfeng Wu
Location: Guiyang, Guizhou
Building Area: 38,889 m²

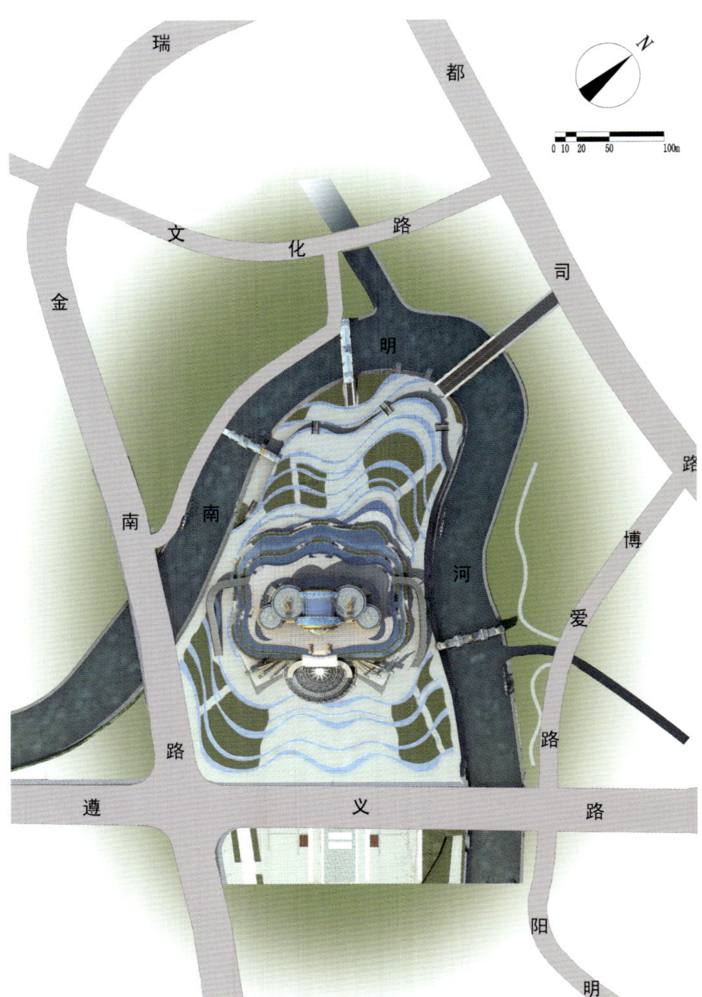

为省城贵阳构筑一处足令每个贵州人引以为荣、为之振奋的标志性建筑，为贵州打造一张交际四方、知会天下的名片，不独为天海规划设计公司的拳拳之心，更是社会各界有识之士和广大市民的热切呼吁。省城贵阳曾就贵阳一中原址的使用广泛征求市民意见，天海规划设计认为它无疑就是省城贵阳标志性建筑的最佳位置。
贵州民族文化展演中心规划设计曾在全国环境艺术设计大赛中荣膺最高奖项，实现了贵州省在此奖项上零的突破。

Guizhou Museum, magnificent and unique, designed deliberately by Tianhai Planning and Design Co., Ltd is beyond any doubt the best design proposal of symbol of Guiyang City to match with our cultural properties of Guizhou and be the spiritual pillar and soul of Guizhou.
Creating a symbol construction of which every Guizhou people is proud and heartstirring and a well-known card for Guizhou is not only the sincere heart of Guizhou Tianhai Planning and Design Company, but also the eagerness of the knowledgeable people of the society in all fields and the citizens. Guiyang, the capital of Guizhou had asked for citizens' opinions about the use of the original land of The First Guiyang High School, we Tianhai believed that it is the best location of a symbolic construction of Guiyang.
The planning and design work of "Guizhou Ethnic Culture Exhibition & Performance Center" had won the best praise in the National Environment and Art Design Compete in May 2006, which is the first of its kind in the annals of history in Guizhou.

江口县土家族博物馆
Tujia Ethnic Museum, Jiangkou County

设 计 师：伍新凤
项目地点：贵州 江口
用地面积：10 500 m²

Designer: Xinfeng Wu
Location: Jiangkou, Guizhou
Site Area: 10,500 m²

江口县土家族博物馆的设计灵感来源于当地的傩文化，傩鼓是傩文化中的重要道具，所以设计主体就以傩鼓为造型灵感，在入口处设有十分醒目的傩戏面具，大量的土家族装饰也被运用在设计中，并且还能用到吊脚楼的设计元素。总的来看，整个博物馆几乎就是江口文化的缩影。

The designing inspiration of the museum in Jiangkou County comes from local Nuo Culture. Nuo Drum is an important tool in Nuo Culture, so the main body of the design bears the image of Nuo Drum. The Nuo masks in the entrance are very conspicuous too. Many line decorations of Tujia Nationality have also been applied to the design. You can also see the designing elements of "wooden house projecting over the water". All in all, the whole museum is almost a miniature of the culture in Jiangkou County.

乳山黄金公园

Gold Park, Rushan

项目地点：山东
用地面积：[illegible] m²

Location: Shandong Province
Site Area: 80 ha

山东乳山黄金文化主题公园是以黄金矿区为依托，以黄金文化为主题的工业旅游园区。黄金迷宫作为园区的标志性建筑，最大限度地彰显着它的特色。金黄耀眼的色彩象征财富；哥特式建筑的尖顶及花窗玻璃美观大方；迷宫般的内部布局寓意要经过持久磨砺才能收获黄金般的灿烂人生。

This project is an industrial tourism park based on a gold mine. The gold labyrinth is a landmark building of the park, to maximize its features. The shining color is golden color, implying wealth. The Gothic pinnacle and stained glass have a beautiful appearance. The labyrinthine inside layout tells only long-term experience can result in shining gold.

扫描查看更多信息

西线建筑规划设计研究院是一支学术研究+实践的专业型设计研究机构，是贵阳建筑勘察设计有限公司（双甲级）最重要的设计分支之一。以研究通用技术体系的乡土建造与发展当代地域建筑的西线工作室以及专注于都市化状态下的设计方法有效贯彻的WB国际联合工作站（西线建筑与西班牙建筑师的联合体）为核心，以当代都市研究、当代地域实践为基本工作内容，关注乡土社会在现实商业社会与通用建造条件下的延续与再生，关注公共生活地的建设，关注新旧机制交替下的空间状态的研究等；高度重视以社会、经济与城市的维度探索都市及乡土建造的系统性控制方式与聚落的可持续经营；形成贯通现状分析→市场分析→运营研究→策划与传媒分析→原创理论支撑→具体解决方案→初步设计→施工图生产→可持续发展的系统化设计全流程，全面开展建筑设计、城市设计、景观设计领域的工作，在地域建筑实践、都市空间生产、公共生活空间创新、乡土材料的通用化建造、材料制造氛围与城市设计方面成果突出，是具有独立思考精神与系统工作方法的专业团队，是当代贵州建筑重要的领航者与探索者。

地址：贵州省贵阳市金阳新区金阳北路南段1号
电话：+86-851-8551870
传真：+86-851-8613501
邮箱：2864658963@qq.com

Add: 1 South Section of Jinyang North Road, Jinyang New District, Guiyang City, Guizhou Province
Tel: +86-851-8551870
Fax: +86-851-8613501
E-mail: 2864658963@qq.com

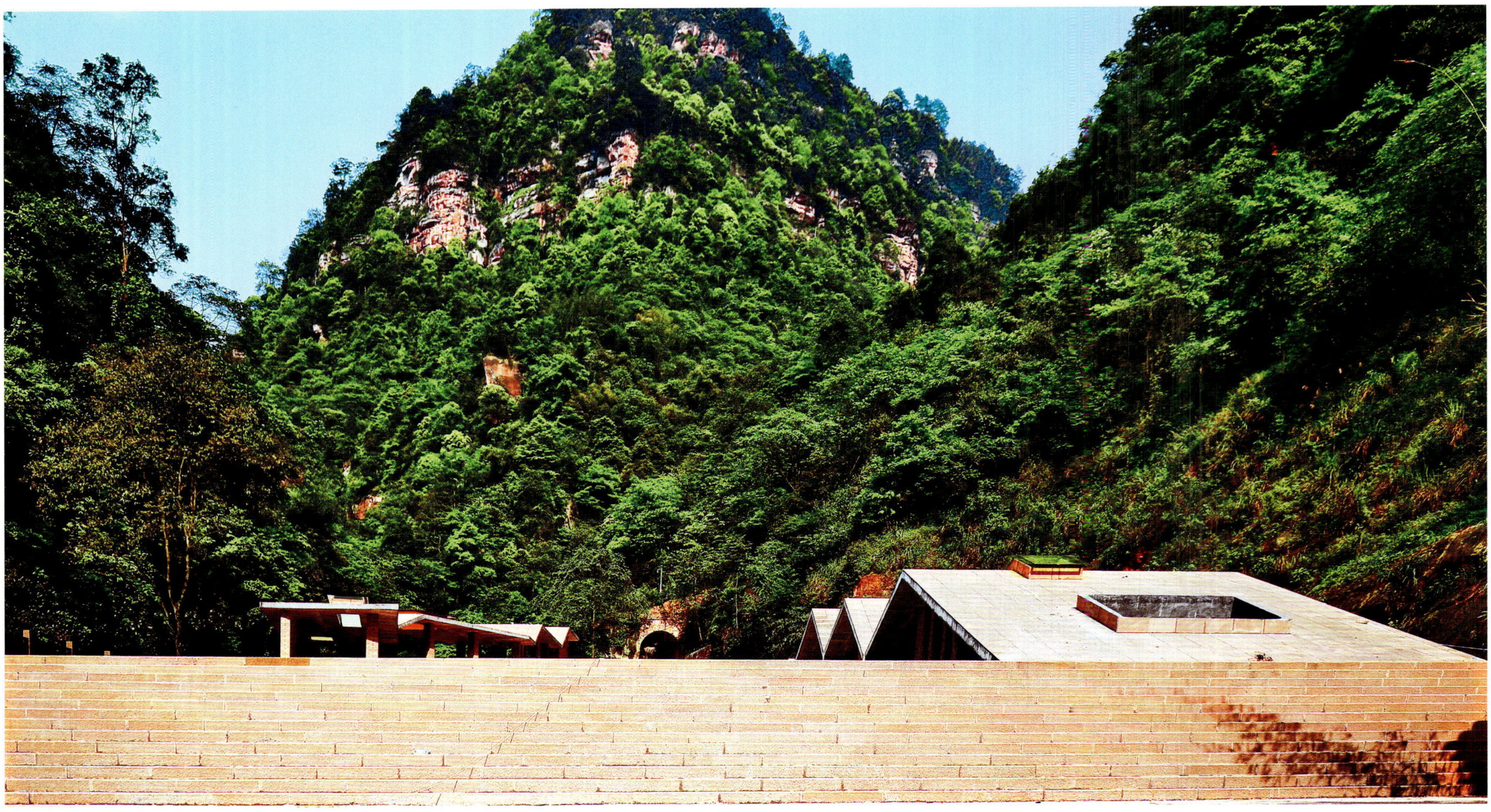

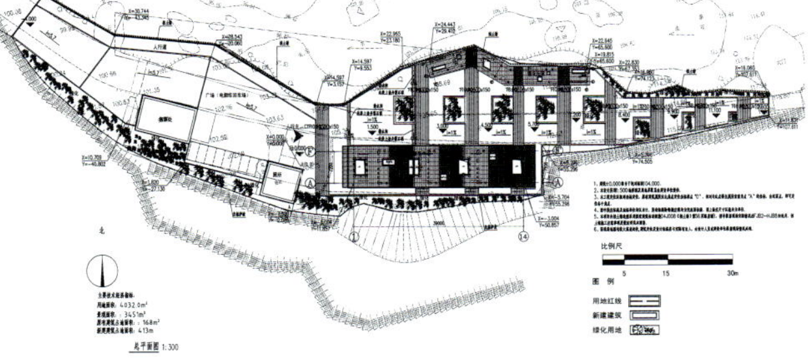

红场——中国丹霞世界遗产地赤水佛光岩游客接待中心与入口空间设计

Red Square—Chinese World Heritage Site Chishui Danxia Buddha Rock Tourist Reception Center and Entrance Space Design

一个巨大的丹霞条石铺砌的红场，显出静谧而神性的空旷感，红场与周围各界面的交接关系产生出三类不同属性的情景。
情景一："流动"是北侧观景廊与溪流奔腾照应的关键词。
情景二：起伏、石盒群、光的打击乐、井口、落雨。
南部的游客接待站受陡崖的影响，被蓄意处理成一组石头房，包含快餐、品茗与购物三种功能。系功能系统匹配光孔系统+天井系统匹配立体雨水系统的石盒群。

A huge Danxia stone paved Red Square, showing the transfer of the quiet and divinity "open" sense, Red Square and the surrounding interfaced with each other and its relationship scenarios produce three different types of property.
Scenario: flow: the "flow" of the north side of the viewing gallery and streams Pentium coordinate with the keywords.
Scenario two: downs, stone box group, light percussion, wellhead, rain.
Southern visitor's reception center has been influenced by cliffs, it is treated as a group of stone house, technically, it contains three functions of fast food, tea and SHOPPING. Functional system to match light hole system + patio system to match the stone box stereo rainwater system group.

潜移默化——全国红色旅游经典景区贵州赤水红军烈士陵园展陈馆

Subtle Effect—National Red Tourism Classic Scenic Spots of the Red Army Martyrs Cemetery Exhibition Pavilion in Chishui, Guizhou

项目地点：贵州 赤水

Location: Chishui, Guizhou

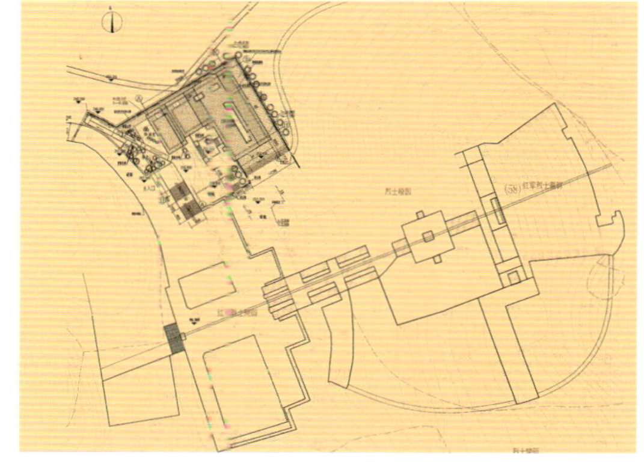

刻意组织一套多孔的天井在展陈馆中，目的是完成纪念空间的教化功能；三种性质的天井（仪式性的、诗性的、崇高的）被蓄意制造，或以咏物言志，或以应物象形，或以随类赋彩的方式与参观者建立某种心灵的默契，由此成就一种精神之悦，从而潜移默化地实现教化。

Deliberately organization set with edifying the porous model patio and make the system involved in showcasing Pavilion, the purpose is to enlightenment functions completion of the commemoration of space; courtyard of the three properties (ritual, poetic, lofty) deliberately established some kind of deep soul tacit understanding with visitors, thus the achievement of a spiritual Wyatt, thereby can achieve an function of enlightenment deliberately.

贵州赤水竹海国家森林公园景观

Landscape of the National Bamboo Forest Park in Chishui, Guizhou

项目地点：贵州 赤水

Location: Chishui, Guizhou

在充分理解竹群空间与其线条组织结构规律的基础上，有意识地将自然因素转换为空间控制要素，唤醒竹群空间的活性与身体感知。
以竞游走促成身体与竹群的互动；将置于顶面与垂直面的彩色透明玻璃体有效搭配，把自然光调配成竹海中常见的三种不同的光之氛围：白色——烟雨蒙蒙，一片苍茫；黄色——万道光芒，壮丽辉煌；红色——斜阳西下的辉煌……

The design of the national bamboo forest park is based on the full understanding of the lines between the bamboo and its rules of organization, deliberately change the natural factors into the spatial control elements, it is a presentation of the awakened bamboo and the body perceptiveness.
Our movement of cycling the bamboo gives an interaction of the body with the bamboo [illegible] with the movement of waking the bamboo, by placing the day surface and vertical color of the transparent vitreous together, we effectively create three kinds of natural light in the bamboo forest: white – the drizzly misty rain; the yellow – thousands of rays in magnificent glory; red - the brilliant sun sets ...

中国丹霞赤水世界遗产展示中心
The Danxia Chishui World Heritage Exhibition Center

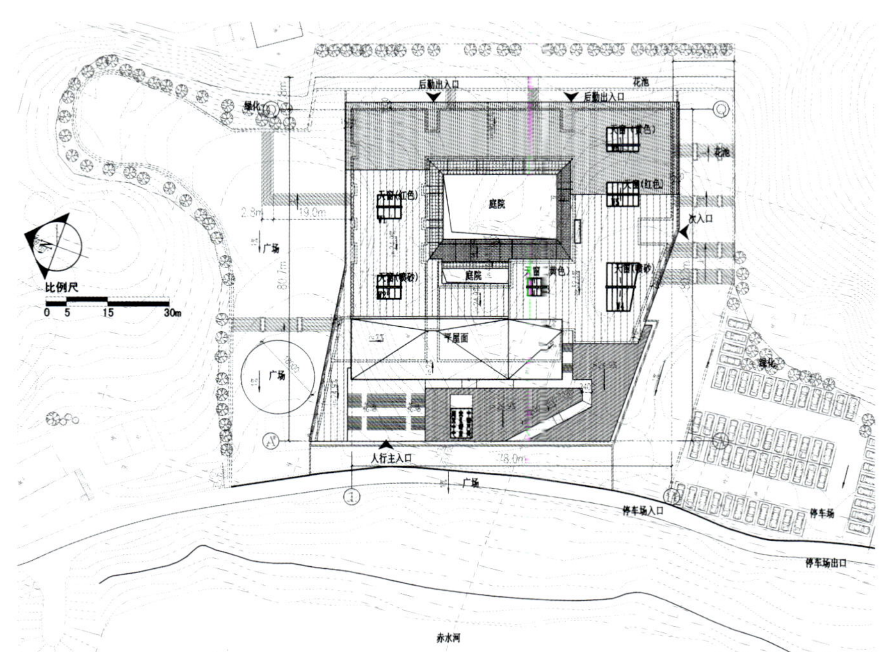

虽为人作，宛自天开。整个建筑犹如一块巨大的丹霞石镶嵌在山川间，以地景化的策略将庞大的建筑体量消匿于景观中，也表明对周边环境谦让的姿态。
将一组不同标高的石头群（其内设置不同的展厅）聚集组织，石体与石体间则是行云流水般的路径，并通过对不同石体运用不同的开口、拉槽、引光、扭转经营散点透视，激发人在石体与石体间移动探索。

Though it is designed by human, it looks like the work of nature.The whole building is like a huge landscape inserted in the mountains, by this way, the large building volume can be eliminated and hidden in the landscape, it also shows a gesture with humility surrounding environment.
The combination of a group of stones (different exhibition halls inside) with different elevation, routes between the stones go smoothly, and our body can move through the stones, with a scatter perspective of different opening, slot pull , light, the sensitiveness of the body can be activate.

独山大学城
Dushan University Town

项目地点：贵州 都匀　　　　Location: Duyun, Guizhou

贵州独山大学城充分贯彻“生态优先”与“整体优先”的绿色城市设计理念，以贯穿全城的如树枝状般布局的生态公园的方式，将绿色景观渗透到全城各区域，建立“绿心”“智心”“共心”三心合一的新型的城市营建理念；遵循以大学群为中心，以枝状生长的景观公园为骨架，以共享资源为纽带的城市生长框架，将大学城建设成为教育之城、创意之城和生态之城等多元复合的绿色山水公园城市。

Dushan university town in Guizhou province fully implement the concept of “ecology first” and the “overall priority”, the ecological parks throughout the city with a dendritic-like layout, penetrate into the green landscape of the city area, with the three hearts--“Green Heart” and “intellectual heart”, “common heart” it established a new city construction philosophy; take the University City University group as the center, the dendrite landscape park as the structure, the shared resources as a link Education city framework, we managed to make the university town an Education City, Creativity city and Ecological City, at the same time, a green park city.

扫描查看更多信息

海南华磊建筑设计咨询有限公司

Hainan Hualei Architectural Design Consulting Co., Ltd.

海南华磊建筑设计咨询有限公司成立于1988年，是海南建省最早、拥有中华人民共和国颁发的建设部甲级设计资质证书的民营企业之一，具有独立的企业法人资格，法定代表人于瑞为国家一级注册建筑师、国家注册公用设备工程师。现公司持有住房和城乡建设部颁发的甲级建筑勘察设计资质证书、乙级市政设计资质证书。主要承担建筑设计、规划、市政工程设计、总平面、竖向、配套专业站房、各类管网管线、园林绿化、室内外环境装修、动力、煤气、道路、消防、新型建筑技术开发、小区规划设计及项目前期策划，工程咨询、项目可行性研究、节能报告等业务。

地址：海南省海口市海甸岛32号金谷大厦202房
电话：+86-898-36365412
传真：+86-898-36365414
邮箱：hnhualei@sina.com/1091003047@qq.com
网址：www.hn-hualei.com

Add: Room 202, Jingu Building, 32 Haidian Island, Haikou City, Hainan Province
Tel: +86-898-36365412
Fax: +86-898-36365414
E-mail: hnhualei@sina.com/1091003047@qq.com
Web: www.hn-hualei.com

海口白沙门海上盛世

Sea Wonder, Baishamen Park, Haikou

项目地点：海南 海口
用地面积：36 078.59 m²
建筑面积：2012 m²
容 积 率：0.05
建筑密度：6.6%

Location: Haikou, Hainan
Site Area: 36,078.59 m²
Building Area: 2012 m²
Plot Ratio: 0.05
Building Density: 6.6%

白沙门公园北岸广场位于公园南北中线北端，由此向东西各伸展600 m为海滩旅游活动区整合范围，向北延伸约500 m至海峡内。为改变沙滩旅游区海水质量，自东、西两端点起建造沙坝，减轻海潮、旋涡对沙滩的影响。

The current Baishamen Park Northern Shore Square is located at the north end of south-north centerline of the park. The beach tourism zone extends 600m east and west from this square, and about 500m north from this square to the strait. In order to change seawater quality in the zone, the project builds a sandbar from the east and west ends, to reduce tide and eddy effects to the beach.

嘉源海韵修建性详细规划

Constructive Detailed Plan of Jiayuan Oceania Resort

独立别墅休闲屋注重景观与建筑内部空间的结合，每栋独立别墅休闲屋拥有独立的庭院，庭院内设有泳池和温泉泡池，坐在客厅或餐厅都可以享受到庭院的景观，顶层的卧室拥有观海的大露台。

酒店式公寓建筑布局是以中轴线对称的，大堂是开敞式的，大堂的南面与入口雨篷之间是绿化的庭院，大堂的北面是架空的景观平台下部，平台下部种植阴生的热带植物。酒店式公寓的标间有会客区、开敞厨房、景观阳台和生活阳台。酒店式公寓设有地下室，其功能为设备用房和地下车库。建筑室外采用花架，营造休闲空间氛围。

Independent leisure villas emphasize the combination of landscape and internal architectural space. Every independent leisure villa has an independent courtyard, with swimming pool and hot spring pool inside, where you can enjoy the courtyard views sitting in the living room or the dinning room and experience leisure of the swimming pool and hot spring pool. The bedroom on the highest floor has a sea-view balcony.

Apartment hotel layout is symmetric along the central axis. The lobby is open style, and between the south side of lobby and the entrance canopy is a greening courtyard. On the north side of lobby is the lower part of aerial landscape platform, on which grow tropic plants in shade. A standard room in the apartment hotel has a meeting area, an open kitchen, a landscape balcony and a living balcony. The apartment hotel has a basement, functioning as spare room or underground garage. The outside is flower rack to create a leisure space.

扫描查看更多信息

石家庄市建筑设计院
ShiJiazhuang Architectural Design Institute

石家庄市建筑设计院创建于1958年，现具有七项甲级资质，包括甲级工程设计、甲级建筑智能化专项工程设计、甲级工程监理、甲级工程总承包、甲级工程咨询、甲级施工图设计文件审查、勘察专业类（岩土工程（勘察、咨询、监理））甲级；三项乙级资质，包括乙级工程地质、岩土工程咨询，乙级市政公用行业（热力）工程设计及乙级城市规划等。

该院技术力量雄厚，综合性强，拥有各类专业技术人员175人，其中教授级高级工程师4人、高级工程师40人、工程师38人、国家一级注册建筑师和注册结构工程师26人、国家二级注册建筑师和结构工程师13人、国家注册监理工程师13人、国家注册岩土工程师4人，是知识技术密集的科技型单位。生产机构设置有5个综合建筑设计所、勘测所、石家庄汇通工程建设监理公司、石家庄徐志欣建筑工程设计咨询事务所，2003年已通过了ISO9001国际质量体系认证，2006年取得国际对外合作经营权。

地址：石家庄市光华路16号
电话：+86-311-87023414
传真：+86-311-87023414
邮箱：Reshine@heinfo.net
网址：www.sjzarc.cn

Add: 16 Guanghua Road, Shijiangzhuang, Hebei
Tel: +86-311-87023414
Fax: +86-311-87023414
E-mail: Reshine@heinfo.net
Web: www.sjzarc.cn

江南新城
Style Life

本项目位于石家庄正定县南城门地带，北靠古城墙，东北可遥望隆兴寺等古迹，紧邻107国道，文化久远，交通便利，地理位置优越。
生态城市理念——现代生态城市的观点是要重建“城市与自然的平衡”，从园区的角度来讲，必须要建立起完整的自然生态体系，并与城市的生态体系紧密联系，充分保护和利用城市的蓝带和绿带空间，如此才能形成和谐共生的城市生态环境。
人文和谐理念——充分体现“以人为本”的设计思想，在规划与设计中首先研究人的心理，营造空间尺度适宜，景观舒适怡人，构成优美和谐的居住休闲空间。
古城新景理念——充分考虑当地的地理特征、气候特点，利用现代和简洁的设计手法，正确处理建筑与城市的关系，使正定这座历史文化古城绽放新时代风采。
时尚前沿理念——通过独具匠心的前沿艺术设计，在园区规划中展现中西合璧、南北融汇、古今结合的和谐共生美景。将地方历史文化与国际时尚结合起来，形成集古今中外、经典时尚于一体的独特园区风格。
低密度理念——园区的密度是构建空间的基本要素，考虑疏密得当的群体和园区形态有助于形成亲切和谐的传统休闲气氛，适度的密度也体现了园区可持续发展的原则。

盛典苏州小区
Grand Suzhou Community

果岭湾

Green Bay

盛泽果岭湾位于石家庄市新华西路与西三环交汇处，北临307国道，交通便利，东临南水北调工程，西临市政府规划的西部山前生态新区，空气清新，环境怡人。周边临近电子54所、联合大学城、白求恩医院、傅山中医院、植物园、动物园、抱犊寨风景区等，科教、医疗、娱乐、休闲配套完善。小区总体规划从总体布局、景观绿化、小区内交通设计、建筑立面美学、户型设计等均有所创新，其中绿化景观借鉴高尔夫果岭空间形态，增加了社区空间层次感，打造出真正回归自然、亲近山水的生活体验。

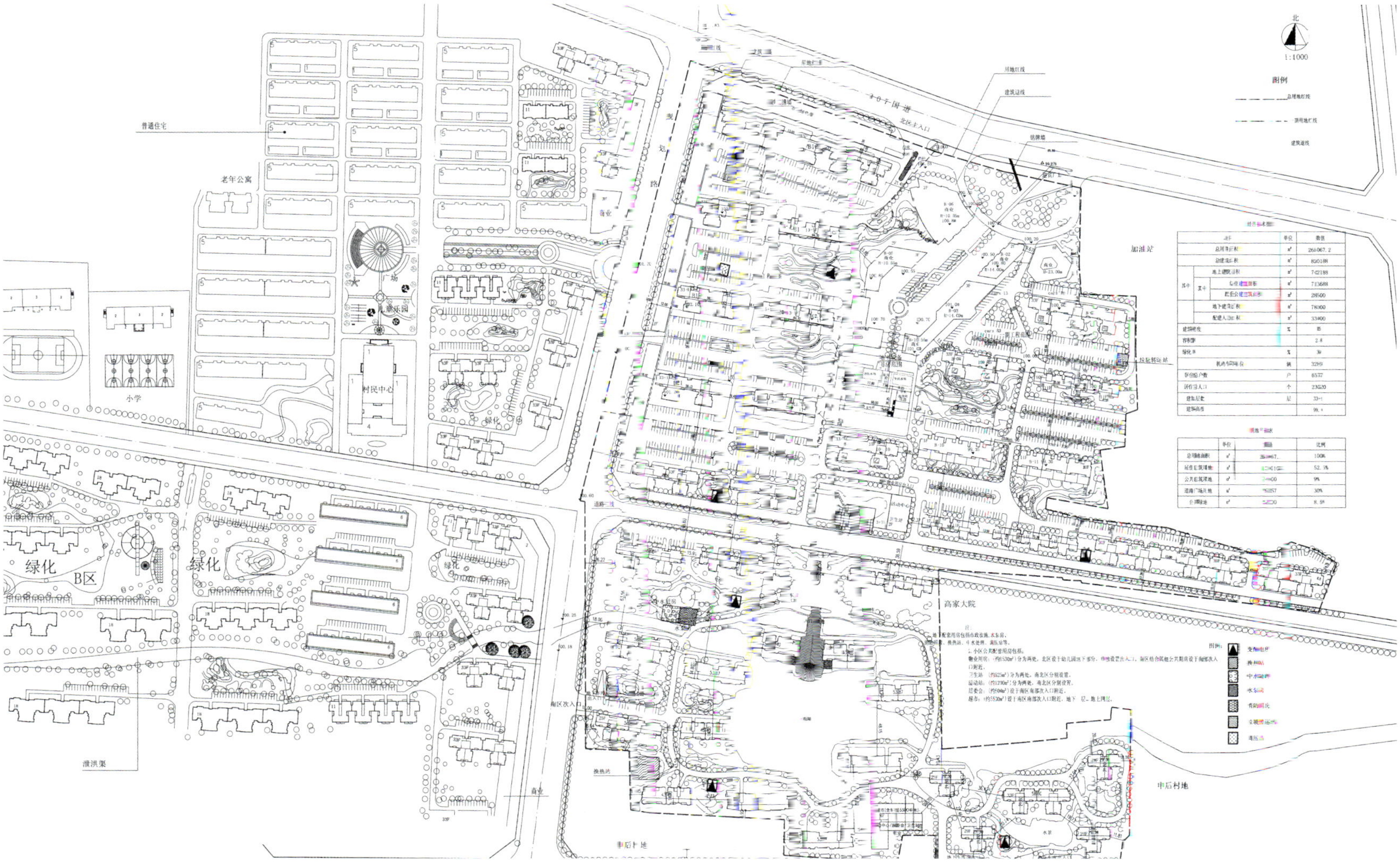

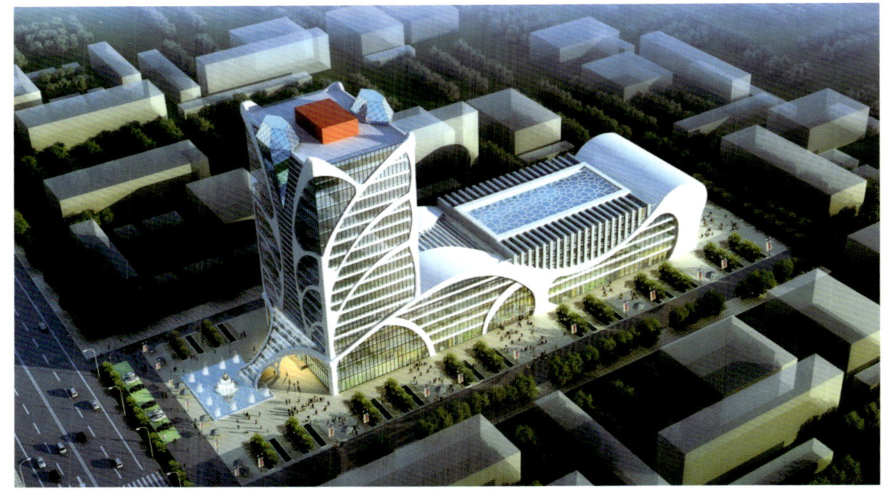

石家庄游跳中心
Shijiazhuang Swimming and Diving Center

石家庄游跳中心建筑面积5.5万 m^2，本项目地下一层为车库，地上建筑分主体建筑和裙楼两部分。建设内容包括全民健身馆、运动休闲馆、全民健身指导培训中心，包括乒乓球馆、羽毛球馆、游泳馆、跳水馆、体育用品商店、陆上训练房、更衣间、配套用房及工作办公用房、停车场等。

建筑方案主体以运动浪花形构成，外围的造型由钢构架支撑构成，整体呈现纤细而轻盈的形象，以钢结构屋盖、铝塑板墙面、特色玻璃幕墙的完美结合，传递出运动和力量的新理念，达到了造型与结构、力与美的高度统一。

昆仑盛阳门
Sunshine Arch of Kunlun

昆仑盛阳门位于石家庄南三条商贸城内，西临胜利大街，北临向阳路，建筑高度99.9 m，总建筑面积110 000 m^2，地上28层，地下3层，1～5层裙房为商业，6～28层主楼用途为办公，地下室为设备用房及车库。交通流线组织合理、顺畅，立面造型简洁、大方。

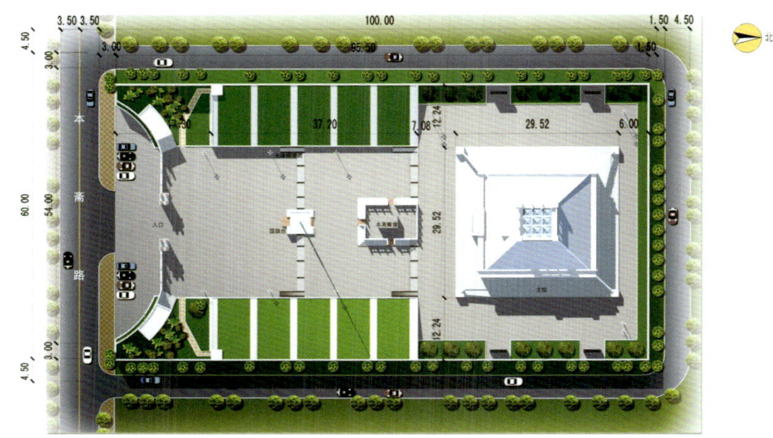

总平面图

马本斋英雄纪念馆
Ma Benzhai Hero Memorial Hall

项目地点：河北 沧州　　Location: Cangzhou, Hebei
用地面积：6000 m^2　　Site Area: 6,000 m^2

该纪念馆是为了纪念英雄马本斋而建，马本斋（1904–1944），河北献县东辛庄人，曾指挥回汉两族战士转战于冀中平原、渤海之滨，驰骋于冀鲁豫的广大敌后战场，成为威震敌胆的抗日名将、民族英雄。

该馆坐落在献县本斋回族乡本斋东村，马本斋母子陵对面。纪念馆由展馆、塑像、升旗台、矮墙、大门等组成，东西宽60 m，南北长100 m，占地6000 m^2。

纪念馆主体部分采用了伊斯兰风格，纪念馆入口设高大的穹窿，墙面上采用典型的伊斯兰窗及入口穹窿两侧的火焰券柱廊等。

石家庄市广播电视局采编播综合业务大楼

Comprehensive Business Building of Shijiazhuang City TV & Broadcasting Bureau

项目地点：河北 石家庄
建筑面积：45 246 m^2
主楼建筑高度：99.9 m

Location: Shijiazhuang, Hebei
Building Area: 45,246 m^2
The Main Building Height: 99.9 m

该项目主楼作为技术及行政管理中心，裙楼作为剧场公共空间。两者通过内部连廊相互连接在一起，从而使整个建筑内部空间及工艺流线顺畅，功能分区合理。

主楼主要包括演播制作区、技术区（网络传输播控区）、采编区、行政办公区、后勤服务区等。

裙楼设置大演播室空间及其配套用房、局部编辑制作用房，形成较独立的体块空间。剧场、演播厅结合影视娱乐、展览、观众互动、咖啡休息等其他对外开放的区域，形成内容丰富的视听文化大堂。

对主空间与次空间、工作人员与观众的内部交通组织进行了详细的分析和优化组合。观众座位及与观众互动的节目安排在裙楼以下区域，主楼主要为节目采编、制作、播出及工作人员工作区域，这样可以减少相互的交叉和干扰，动静分区明确且方便管理。

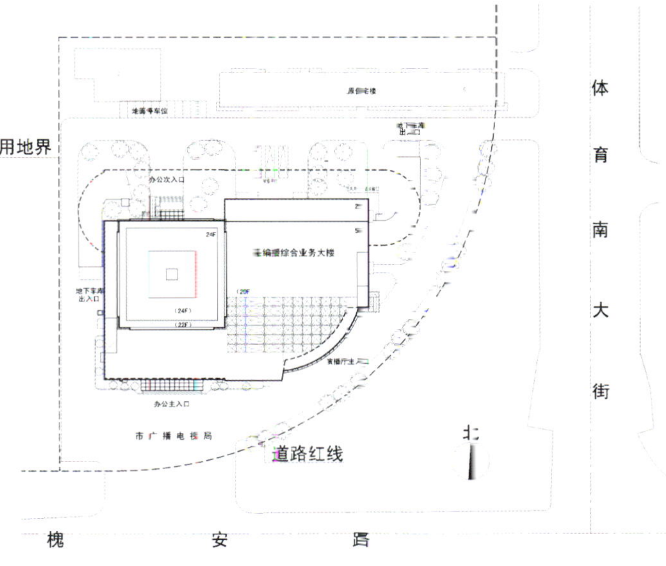

总平面图

河北医院

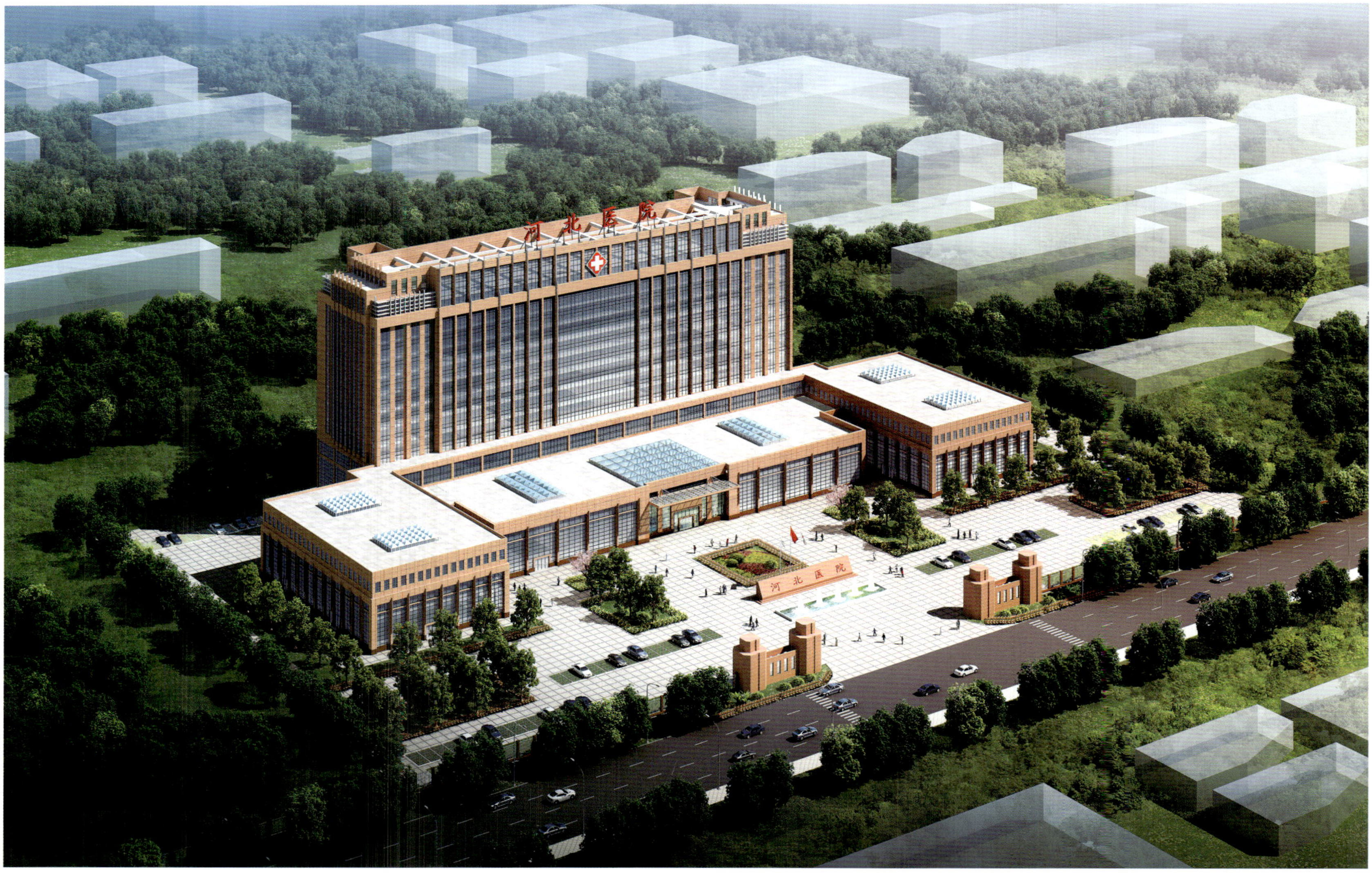

新疆巴州库尔勒市河北医院
Hebei Hospital of Korla City, Prefecture of Bayingolin, Xinjiang

项目地点：新疆 巴州
用地面积：66 621 m²
建筑面积：49 962 m²
容 积 率：0.7
建筑密度：16.4%
绿 化 率：45.2%

Location: Bazhou, Xinjiang
Site Area: 66,621 m²
Building Area: 49,962 m²
Plot Ratio: 0.7
Building Density: 16.4%
Greening Rate: 45.2%

新疆巴州库尔勒市河北医院（维吾尔医医院）用地位于库尔勒市中心56号小区，建国南路南侧，迎宾路西侧。周围市政公用设施齐全，交通便利，环境整洁，建设条件良好。
在满足建筑使用要求及其他专业功能的前提下，按照现行规范、标准及规程进行设计，做到技术先进，经济合理，安全适用。
本工程包括有采暖、通风、空调、防排烟、热能动力设计。采暖系统主要为地板辐射采暖系统；通风是自然通风与机械通风相接合；空调系统除MRECT设独立的恒温恒湿空调以外，其他部位均采用风机盘管加新风系统。

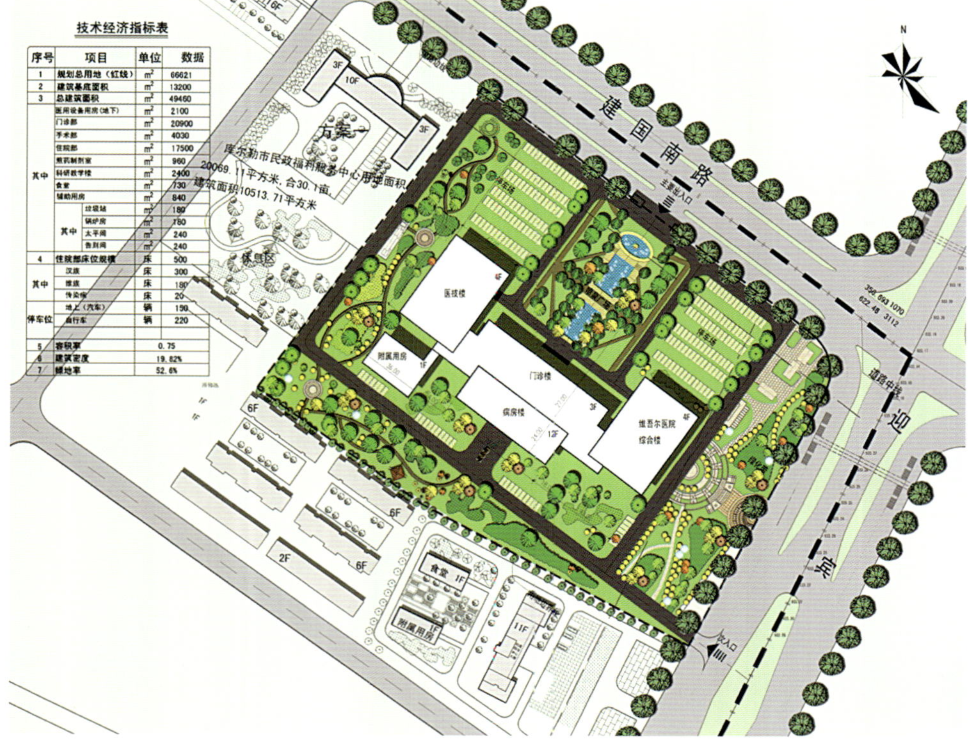

技术经济指标表

序号	项目		单位	数据
1	规划总用地（红线）		m²	66621
2	建筑基底面积		m²	13200
3	总建筑面积		m²	49460
其中	医用设备用房(地下)		m²	2100
	门诊部		m²	20900
	手术部		m²	4030
	住院部		m²	17500
	煎药制剂室		m²	960
	科研教学楼		m²	2400
	食堂		m²	730
	辅助用房		m²	840
	其中	垃圾站	m²	180
		锅炉房	m²	180
		太平间	m²	240
		告别间	m²	240
4	住院部床位规模		床	500
其中	汉族		床	300
	维族		床	180
	传染病		床	20
停车位	地上（汽车）		辆	150
	自行车		辆	220
5	容积率		0.75	
6	建筑密度		19.82%	
7	绿地率		52.6%	

建国南路
库尔勒市民政福利服务中心用地面积
20069.11平方米，合30.1亩，
建筑面积10513.71平方米
医技楼
门诊楼
病房楼
附属用房
维吾尔医院综合楼
主入口
次入口

河北医院

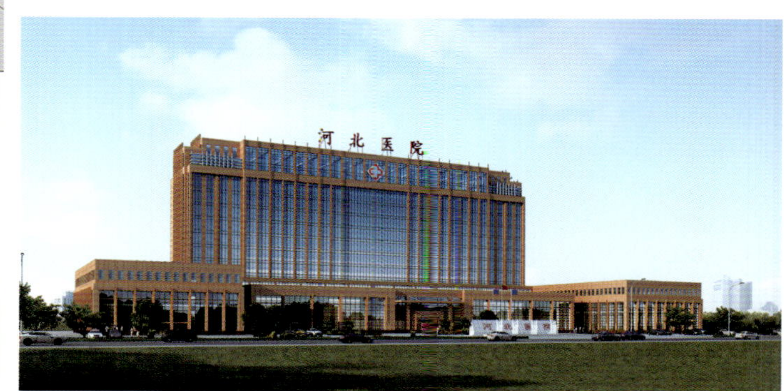

燕春花园酒店
Yanchun Garden Hotel

项目地点：河北 石家庄
建筑面积：55 000 m^2

Location: Shijiazhuang, Hebei
Building Area: 55,000 m^2

燕春花园酒店位于石家庄市建设大街与中山路交汇点西北角，属四星级标准酒店，总建筑面积5.5万m^2，地上26层，地下2层。采用框架剪力墙结构体系，功能分区明确，交通组织流畅，内设大堂，餐厅、娱乐室、棋牌室、游泳池、会议厅及客房。
该建筑造型新颖、简洁，采用流畅的S形，裙房柱廊的运用以及楼前开放性广场的设计使内外空间、城市与建筑空间有机结合，体现出动态空间的景观魅力。此项目荣获“河北省优秀设计一等奖”。

石家庄市政府办公楼
Government Office Building of Shijiazhuang City

项目地点：河北 石家庄
建筑面积：33 000 m^2

Location: Shijiazhuang, Hebei
Building Area: 33,000 m^2

石家庄市政府办公楼地上21层，地下2层，框架剪力墙结构，功能分区明确，内设大小会议室若干，造型简洁、大方，建筑性格体现充分，荣获“河北省优秀设计一等奖”。

保定市建筑设计院有限公司

BAODING INSTITUTE OF ARCHITECTURE DESIGN CO.,LTD.

保定市建筑设计院有限公司前身为保定市建筑设计院，始建于1952年，2011年11月完成深化体制改革。公司是河北省成立最早的综合甲级建筑设计单位之一，主要从事各种类型的大、中型民用与工业建筑设计、工程勘察、工程监理、施工图审查等。建院50多年来，完成了近万项工程项目的勘察、设计、监理等，项目遍及全国各省、市及自治区，积累了丰富的设计工作经验，拥有一支由高素质人才组成的设计队伍，在设计领域具有独特的地位和影响力。连续多年被评为"重合同守信用单位""建设系统信誉、信用AAA单位""优秀勘察设计院""河北省诚信示范单位"等。

公司具有建设部颁发的建筑行业（建筑工程）甲级资质证书、工程勘察专业类岩土工程（勘察、设计、咨询、监理）甲级资质证书、房屋建筑工程监理甲级资质证书、施工图审查一级资质证书和建筑行业人防设计乙级、市政公用行业（道路）设计乙级、城市规划乙级，咨询乙级、风景园林设计乙级、市政公用行业（热力、风景园区）丙级资质证书等。可承接各类大、中型民用工业建筑勘察、设计以及工程建设可行性研究，人防工程设计及相应的咨询与技术服务，市政公用行业给水、排水、热力、风景园林等工程设计及相应的咨询与技术服务，技术咨询，工程总承包，工程建设监理，施工图审查等业务。

公司现有职工240余人，各类专业技术人员218名，其中：教授级高级工程师9名、高级工程师52名、工程师66名、一级注册建筑师11名、一级注册结构工程师18名、注册公用设备工程师11名、注册电气工程师2名、注册城市规划师2名、注册咨询工程师（投资）6名、注册岩土工程师5名、注册监理工程师15名、注册造价工程师2名。

作为河北省最好的地市级建筑设计单位之一，公司以"精心勘察设计、公正科学监理、优质诚信服务、实现顾客满意"为己任，坚持以繁荣建筑创作为宗旨，不断完善、创新设计理念，力创建筑设计精品，在工程设计和科研方面屡获国家级、部级和省级优秀奖，在文化建筑、体育建筑、医疗建筑、教育建筑、居住建筑以及空间结构等设计领域具有独特的设计优势，而严格规范的ISO9001质量体系认证管理更使公司的设计质量为业界广泛认同。

公司注重推行和运用现代化设计手段，配备电脑网络管理系统、协同设计系统、办公自动化系统，并建立了一套完整严密的管理体系，形成了团结求实、严谨高效、进取奉献，严守职业道德的院风。

在未来，公司将不断开拓国内市场并争取打开国外市场，以"造就一流人才、创造一流设计、提供一流服务"为目标，竭诚为各界服务。

地址：河北省保定市五四中路951号（五四路地道桥东口北侧）
电话：+86-312-5997906/5997889
传真：+86-312-5997906/5997889
邮箱：bdad@vip.163.com
网址：www.bdad.com.cn

Add: No.951, Wu Si Zhong Road, Baoding City, Hebei Province
Tel: +86-312-5997906/5997889
Fax: +86-312-5997906/5997889
E-mail: bdad@vip.163.com
Web: www.bdad.com.cn

保定市人防指挥中心

Civil Air Defense Command Center, Baoding

项目地点：河北 保定
用地面积：8663 m^2
建筑面积：6500 m^2
容 积 率：0.98
建筑密度：24.50%
绿 化 率：35%

Location: Baoding, Hebei
Site Area: 8,663 m²
Building Area: 6,500 m²
Plot Ratio: 0.98
Building Density: 24.50%
Green Ratio: 35%

项目位于保定市北部，地处阳光北大街与隆兴路交叉口东北角，西侧为城市绿化带，东侧为保定市110指挥中心。该项目包括地面指挥中心、人防专业队用房、配套用房、地下车库及配套用房、指挥工程共五部分，其中指挥中心为地上七层，局部八层。

The project is located in northern Baoding City, at the northeastern corner of the junction of North Sunshine Street and Longxing Road, besides green belt in the west and the Baoding 110 Command Center in the east. It includes five parts, namely ground command center, rooms for civil air defense team, supporting rooms, underground garage and supporting rooms, and commanding works, among which, the command center has seven floors aboveground and eight floors in part.

电谷国际商务中心

Power Valley International Business Center

项目地点：河北 保定　　Location: Baoding, Hebei

建筑面积：61 674.86 m²　　Building Area: 61 674.86 m²

电谷国际商务中心位于保定市朝阳北大街以东、鲁岗路以南，为高层办公楼。建筑采用高性能围护结构保温系统和双银Low-E中空玻璃幕墙，实现节能50%；采用水源热泵中央空调系统，充分利用可再生能源；应用雨水回收处理二次利用技术，用于景观补水、绿化浇灌；采用节水卫生器具，减少水资源的消耗；通过采用高效的照明灯具、优化照明布置、选用声光控开关等方式减少照明用电，实现有效节能；充分利用雨水回渗所在各项措施，使雨水有效渗入地下，保持土壤湿度，补充地下水资源。

Power Valley International Business Center is a high-rise office building in Baoding City. It is beside North Chaoyang Street in the east and Lugang Road in the south. The Center is equipped with high performance building enclosure heat-preservation system. The curtain wall adopts double-silver Low-E insulting glass which can save 50% energy; the water source heat pump central air-conditioning system is employed for taking full advantage of renewable energy; rainwater recycling and reutilization technology is applied into landscape moisturizing and green watering; water-saving sanitary ware is used for reducing water consumption; lighting electricity reduction and effective energy conservation are realized through selecting efficient lighting fixtures, optimizing lighting arrangement and using sound and light-controlled energy-saving switch; and all measures of rainwater return percolation are fully exploited to make rainwater availably leak into ground and keep soil moisture supplementing ground water resources.

保定市儿童医院

Children's Hospital, Baoding

项目地点：河北 保定　　Location: Baoding, Hebei

用地面积：40 000 m²　　Site Area: 40,000 m²

容 积 率：2.5　　Plot Ratio: 2.5

建筑密度：30%　　Building Density: 30%

绿 化 率：35%　　Green Ratio: 35%

该项目位于保定市北二环北侧，西邻恒祥北大街，北临东君路，医院规划用地4万 m²，可建设用地3.275万 m²，总床位规模600床，门诊量1800人/天设计。新建医院设计内容为门诊医技病房综合楼，儿研所办公用房，辅助病房楼及地下车库，空调机房、高低压配电室，消防水池水处理间，餐厅、门卫、氧气站等辅助用房。项目建成后将形成7.969万 m²的建筑群体，是集门诊、急诊、病房、医技、手术、康复理疗科研于一体的综合性儿童医疗建筑。

医院设计的宗旨是："以人为本，以病人为中心"。

The project is located on the north side of North Second Ring Road of Baoding City. It is beside North Hengxiang Street in the west and Dongjun Road in the north. It covers a planned area of 40,000 m² and a land area of 32,750 m². It owns 600 beds in total and its outpatient service is designed with the scale of 1,800 patients per day. The design of the hospital includes a complex building combining outpatient, medical technology and inpatient, Children's Research Institute offices and auxiliary inpatient building, underground garage, air conditioner room, fire pool and water treatment with high/low voltage power distribution, restaurants, security room, oxygen station and other subsidiary rooms. After project completion, a building cluster of 79,690 m² will be formed, to be a comprehensive children's medical treatment building that integrates outpatient, emergency treatment, inpatient, medical technology, operation, rehabilitation and physical therapy and scientific research into one. To be human-based and patients-centered is the aim of the design of the Children's Hospital.

迈创新天地商厦

Maichuang Xintiandi Commercial Building

项目地点：河北 保定　　Location: Baoding, Hebei

建筑面积：50 000 m²　　Building Area: 50,000 m²

项目位于清苑县中心商业区，一大型综合性商业建筑。

本工程采用现代风格，简洁、庄重，通过建筑元素的重复与变化突出其韵律感；通过细部的点缀及变化，表现出其商业建筑的性格，通过体块的穿插，使如此大体量的建筑外形更加丰富，成为清苑县的亮点及地标性建筑。本工程主要外墙面采用石材板及玻璃幕墙为主要外墙材料，次要立面采用高级外墙涂料。设计充分考虑各种人群的使用方便性，每层均设置无障碍卫生间、无障碍电梯。整体建筑为人们提供集购物、休闲于一体的商住场所。

The project is a large-scale comprehensive commercial building located in the central business district of Qingyuan County, Baoding City.

The project takes modern style, which is simple but solemn; the repetition and change of architectural elements highlight its sense of rhyme, and the detailed ornaments and changes reveal its characters as a commercial building. With blocks inserting each other, the appearance of the large building shows much more variable and it becomes the shining point and landmark building of Qingyuan County. The main exterior wall selects granite slab and glass curtain wall as its material and the secondary façade selects senior outer wall paint. The project fully considers the easy use of facilities by different people, and therefore designs accessible toilet on each floor and barrier-free elevator, and provides people a commercial and residential places that integrates shopping and leisure into one.

扫描查看更多信息

唐山市规划建筑设计研究院
Tangshan City Planning and Architectural Design & Research Institute

唐山市规划建筑设计研究院于1953年建院，经历了唐山震后恢复重建的全过程实践，为新唐山的崛起及发展做出重要贡献。先后完成了第二届全国城运会主要场馆、省爱国主义教育基地、非典医院、援川、援疆等十余项国家、省、市重点工程项目。近几年又完成了万达广场等一批市重点工程项目。先后获得国家优秀设计银奖1项，新中国成立60周年国家建设创作优秀奖1项，建设部优秀设计奖7项，河北省优秀设计奖35项，感动河北工程设计成果奖及河北精品建筑奖各1项。为河北及唐山争得了荣誉，取得了良好社会、经济及环境效益。

地址：河北省唐山市路北区华岩路30号
电话：+86-315-2826806
传真：+86-315-2825904
邮箱：zgtsgiy@sina.com
网址：www.tsgjy.com

Add: 30 Huayan Road, Lubei District, Tangshan City, Hebei Province
Tel: +86-315-2826806
Fax: +86-315-2825904
E-mail: zgtsgiy@sina.com
Web: www.tsgjy.com

唐山市南湖西北片区震后危旧平房改造安置小区
Bungalow Reconstruction & Relocation Community after Earthquake, Northwest of South Lake, Tangshan

设 计 师：李双来、李丽梅、刘向东、富彬、张江、张浩、杨宇
项目地点：河北 唐山
用地面积：1 026 000 m²
建筑面积：3 260 000 m²

Designer: Shuanglai Li, Limei Li, Xiangdong Liu, Bin Fu, Jiang Zhang, Hao Zhang, Yu Yang
Location: Tangshan, Hebei
Site Area: 1,026,000 m²
Building Area: 3,260,000 m²

该设计是一回迁项目，在规划设计中希望每个组团都形成自己相对完整的空间结构和空间形态，以使其构成内向型的独立个体，使居民对其形成场所认同感和心理安全感。因本项目的特殊性，在设计过程中对建筑单体空间、外部形式采用标准化处理手法，使各个组团内以及村与村的建筑风格更易于协调，便于施工管理，提高施工速度，同时达到了绿色节能的目的。居住区内配置一所九年制学校，为该居住区内的重点工程。在用地较为紧张的条件下，学校的设计力图打破以往行列式的布局形式，采用了合院式空间组织方式，既做到了节约用地，又为孩子们创造出具有归属感的场所。

This is a relocation project. The design tries to make every group have its own complete space structure and space pattern, to form an inward independent unit, so residents will have local identity and security. Due to its specialty, the design standardizes single space and external pattern, to make the architecture within groups and between villages easier to coordinate, to facilitate construction management and increase construction speed, and meanwhile to achieve the purpose of green architecture & energy-saving architecture. In the residential community set up a 9-year school, serving as a key project in the community. With limited lands, the school design tries to break out previous linear layout, and organizes the space by an enclosing courtyard, to save the lands and also create a homelike venue for children.

唐山四中及联合村小学

Tangshan No.4 Middle School & United Village Primary School

设 计 师：陈合文、陈婧、宋维增、孙军成、刘海波
项目地点：河北 唐山
用地面积：58 365.94 m²
建筑面积：34 518.34 m²

Designer: Hewen Chen, Jing Chen, Weizeng Song,
Juncheng Sun, Haibo Liu
Location: Tangshan, Hebei
Site Area: 58,365.94 m²
Building Area: 34,518.34 m²

唐山四中及联合村小学分中学和小学两个部分，其中中学部分包括综合科技楼、中学教学楼、体育馆、综合艺术楼、中学男生宿舍楼、食堂及其附属建筑。小学部分包括小学教学楼、小学办公楼。本设计以南北向主路为中轴线把整个用地分为三个区域，包括教学区、食宿区和运动区，动静分区明确，以做到教学、活动、食宿互不影响。校区整体建筑造型以欧式风格为主。

Tangshan No. 4 Middle School & United Village Primary School is divided into middle school part and primary school part. Wherein, the middle school part includes a comprehensive technology building, a middle school teaching building, a gymnasium, a comprehensive art building, a middle school boy students dormitory building, a cafeteria and its subordinate buildings. The primary school part includes a primary school teaching building, a primary school office building, and a school area on the north side of the whole site. The design divides the whole site into three areas with south-north main road as the central axis, namely teaching area, dining & accommodation area and sports area, with clear division for action and inaction areas in order to separate teaching, activities and dining & living from interfering with each other. The overall architectural shape for the campus is mainly European style.

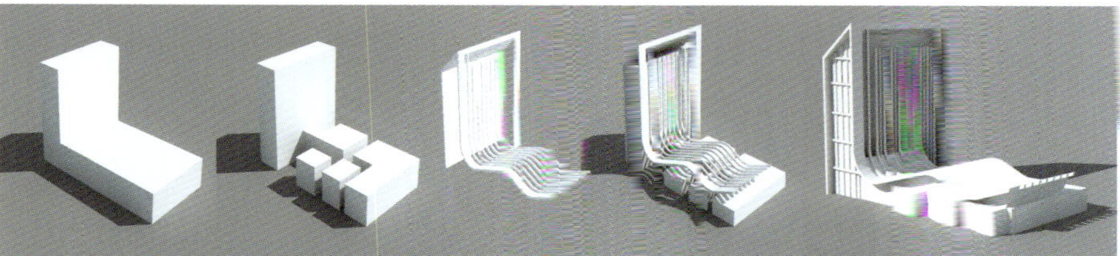

唐山市财会培训基地

Finance & Accounting Training Base, Tangshan

设 计 师：白晓航、赵彬、任颖莉
项目地点：河北 唐山
用地面积：20 000 m²

Designer: Xiaohang Bai, Bin Zhao, Tingli Ren
Location: Tangshan, Hebei
Site Area: 20,000 m²

培训基地项目所在地位于凤凰新城的东南区域，用地东侧是作为文化区的大学城，用地西侧是景色优美的凤凰湖，而西南2.4 km即为凤凰新城中心商务区。项目所处地点的特殊性赋予其地标建筑的属性。而建筑的地标性并不仅仅依靠建筑的绝对高度，更因其对城市空间的形象有所提升，即为凤凰新城提供了优美而性格鲜明的城市天际线。

The project is located in the southeast area of Phoenix New City, the east side of the site is University Town in Culture Area, the west side of the site is Phoenix Lake with beautiful views, and the Central Business Area of Phoenix New City is 2.4 km southwest of the site.
The particularity of the project position makes it a landmark building. The landmark not only relies on the absolute height of the building, but also its impact on the urban space. The final design result provides Phoenix New City with graceful yet characteristic city skyline.

唐山市公安局

Tangshan Public Security Bureau

设 计 师：李强、高建成、吴晓坤、任巍、高立、王雪原、张江、张鑫、聂婷、王越利、张凤霞、富彬
项目地点：河北 唐山
用地面积：27 000 m²
建筑面积：135 000 m²

Designer: Qiang Li, Jiancheng Gao, Xiaokun Wu, Wei Ren, Li Gao, Xueyuan Wang, Jiang Zhang, Xin Zhang, Ting Nie, Yueli Wang, Fengxia Zhang, Bin Fu
Location: Tangshan, Hebei
Site Area: 27,000 m²
Building Area: 135,000 m²

本方案在立面造型上吸取传统建筑的构图手法，端庄的形体、粗犷的轮廓、对称的立面、夸大的尺度充分体现出作为国家公安部门的庄严。

This plan follows the city texture, and absorbs the composing technique of traditional architectures on façade shape, so the decent structure, bold outline, symmetric façade and exaggerated size adequately reflect the dignity of the project as a state police.

唐山市古冶区民生商贸大厦与城市广场

Minsheng Commercial & Trade Building and City Square, Guye District, Tangshan

设 计 师：富松、高建成、富彬、苏海龙、刘丹、王亚生
项目地点：河北 唐山
用地面积：56 700 m²
建筑面积：47 600 m²

Designer: Song Fu, Jiancheng Gao, Bin Fu, Hailong Su, Dan Liu, Yasheng Wang
Location: Tangshan, Hebei
Site Area: 56,700 m²
Building Area: 47,600 m²

本楼是一栋现代化风格的标志性办公建筑，是古冶区的重点工程，整个项目分为南北两块，南侧是市民广场，北侧是主体办公楼及会议中心。建筑群和谐统一，强化行政区整体形象，体现出开放、公正、民主、亲民等特征。

现代、简约、高效是贯穿设计的主题，也是建筑意在表达的精神。通过虚实建筑材料的运用，水平质感与横向质感的对比统一，简洁而强烈地体现了建筑的性质。

This is a landmark office building in modern style, and also a key project in Guye District; the whole project is divided into south and north sections, of which, the south section is Citizen Square, the north section includes main office buildings and conference center. The building clusters are harmoniously integrated, to strengthen the overall image of the administration district, and represent open, righteous, democratic and people-oriented features.

The design is themed by modern, simple and high efficient features that are also the spirits to be expressed by the architecture. The application of void and solid building materials and the integration of horizontal sense of reality and lateral sense of reality simply and strongly present the quality of the building.

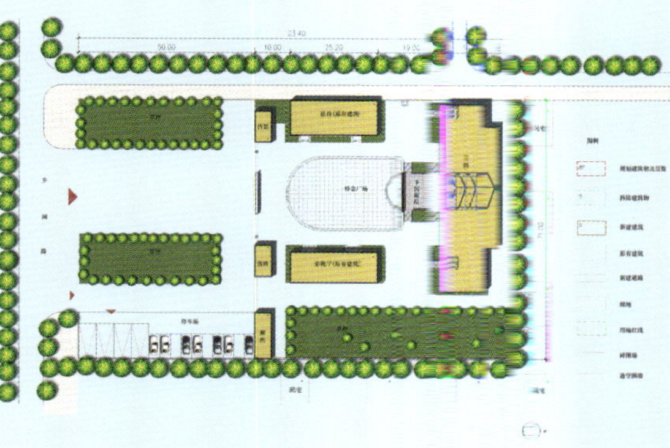
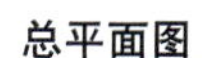
总平面图

滦南潘家戴庄惨案纪念馆

Panjiadai Village Massacre Memorial Museum, Luannan County

设 计 师：陈合文、马进中、富彬
项目地点：河北 唐山
用地面积：8733 m²
建筑面积：1580 m²

Designer: Hewen Chen, Jinzhong Ma, Bin Fu
Location: Tangshan, Hebei
Site Area: 8,733 m²
Building Area: 1,580 m²

该馆的创作不仅要具有纪念性建筑的陈展功能，还要具有永久保存的科学手段，能充分唤起人们对历史的回忆，表达正义的情感与呼唤，突出牢记历史、警钟长鸣这一主题。纪念馆采用现代设计语言，中心屋架及东侧墙上的孔洞象征着三天三夜大火烧毁民房1030间后的残垣断壁及斑斑弹孔的惨案。传统民居的灰瓦檐口、灰白相间的墙面、黑色铁钟及下沉空间及冤魂墙的处理、加之粗犷白石料，增强了建筑质感，再现了惨案的悲惨气氛。

This museum serves not only as a memorial display building, but also as an eternal storage of victims' skeletons by scientific means, to inspire Chinese people's memory of suppressed history, to express righteous emotions and calls, memorize this history, and alarm for this creative theme. The museum is expressed in modern design languages, for example the holes in central truss and east sidewall symbolize the relics and remnants of 1,030 civil rooms after burning for "3 days plus 3 nights" in that holocaust. Traditional residential grey tile eaves, grayish walls, black iron bell, sinking space and victim soul wall, as well as crude stone effects, all these strengthen the mass of architecture, reproducing the tragic scene of the massacre.

唐山丰润潘家峪惨案纪念馆

Panjiayu Massacre Memorial Museum, Fengrun County, Tangshan

设 计 师：沈谨、陈合文、许智梅、李志铮
项目地点：河北 唐山
用地面积：1152 m²
建筑面积：1000 m²

Designer: Jin Shen, Hewen Chen, Zhimei Xu, Zhizheng Li
Location: Tangshan, Hebei
Site Area: 1,152 m²
Building Area: 1,000 m²

项目与潘家峪惨案发生地"潘家大院"一河之隔，设计把作为历史文物加以完整保留的"潘家大院"组织在纪念馆的展示流线中，平面布局以"院落"为基本原型，用石头墙围合的院落式布局，形成沉重、封闭的院落空间。沿院落环线组织展线是建筑构成的基本骨架，把遗迹作为整体展示的重要的组成部分，呈现现场的真实感。钟声警示人们不要忘记发生在这里的悲惨一幕，警示人们不要忘记我们民族历史。"前事不忘，后事之师"引发人们对现实及未来的思考。

The project is near the historical massacre site "Pan Family Courtyard" on the other bank of the river. The design organizes this intact historical courtyard in the display flow of the museum. The layout plan takes the prototype of "court", enclosing a suffocating courtyard space by stone walls. The space sequence around the courtyard extends as the framework of architecture, integrating the relics as an important part of the display of historical reality. The bell warms people not go forget this tragedy, alarming Chinese people of this insulted history. This "precedence serving as a lesson" inspires people to think of the real world and the future world.

曙光大厦商住楼

Dawn Edifice Commercial and Residential Building

设 计 师：李志铮、安然
项目地点：河北 唐山
用地面积：19 964.07 m²
建筑面积：146 350.4 m²

Designer: Zhijing Li, Ran An
Location: Tangshan, Hebei
Site Area: 19,964.07 m²
Building Area: 146,350.4 m²

本项目为集公寓、住宅、商业为一体的综合建筑群，住宅和商业和谐共存是本项目的基本原则，大商业沿新华道布置，小商铺沿站前路布置，在有限的用地内高效地处理高层和商业的关系。

住宅类型有板式高层、点式高层等多种形式，辅以多样的户型；立面设计以现代欧式风格为主，整个建筑单体外部形式给人以平静、祥合的亲切感，体现家的美感和现代的气息。

This project is an inclusive building cluster integrating apartments, residential and commercial buildings. The basic principle is harmonious coexistence of residential and the commercial buildings. Large commercial buildings are set up along Xinhua Avenue, and small retail shops are set up along Zhanqian Road, and thus high-rise buildings and commercial buildings coexist within limited land.

There are various forms of residential buildings, such as plate high-rise buildings, point high-rise buildings, and the layouts vary too; the façade design is mainly modern European style, presenting the integral single building appearance with cordial feelings of peace and happiness, and giving better homeliness and modern sense.

唐山万达广场
Tangshan Wanda Plaza

设 计 师：李双来、王亚生、富松、王雪原、富彬、王丽艳
项目地点：河北 唐山
用地面积：212 700 m²
建筑面积：1 105 000 m²

Designer: Shuanglai Li, Yasheng Wang, Song Fu,
Xueyuan Wang, Bin Fu, Liyan Wang
Location: Tangshan, Hebei
Site Area: 212,700 m²
Building Area: 1,105,000 m²

唐山万达广场是唐山市第一个商业综合体工程，包括商业中心、住宅、办公、酒店，于2011年12月投入使用。通过对万达商业综合体的打造，将给予城市更完善的功能和更丰富的景观机理，带动经济和社会的发展，成为唐山的都市客厅。项目强调了它在城市空间体系和人们的物质生活、精神生活中不可取代的功能地位，提倡新城市主义和人文主义相融合的生活新概念。此次设计将着力打造时尚绚丽、精致热情且具有现代人文气息的景观空间，为未来的营销策划定下主基调。

The project is the first commercial complex in Tangshan City. It includes business center, residential buildings, office buildings, hotels, and was put into operation in December 2011. This creation will improve city functions and enrich landscaping texture, drive socioeconomic development and become local metropolitan hall, highlight its irreplaceable functional position in urban space system and resident material/spiritual life, and promote a new life concept that integrates neo-urbanism and humanism. Therefore, this design will make the best to create a fashionable, gorgeous, exquisite, enthusiastic and modern cultural landscaping space, keynoting the future marketing plan.

万达回迁办公楼
Wanda Relocation Office Building

设 计 师：高建成、富彬、任巍、王福军、谭媛媛、杨宇、陈猛、朱星彦、
孟卫坤、何嘉伟、马丽
项目地点：河北 唐山
用地面积：10 000 m²
建筑面积：65 000 m²

Designer: Jiancheng Gao, Bin Fu, Wei Ren, Fujun Wang, Yuanyuan Tan,
Yu Yang, Meng Chen, Xingyan Zhu, Weikun Meng,
Jiawei He, Li Ma
Location: Tangshan, Hebei
Site Area: 10,000 m²
Building Area: 65,000 m²

本方案力求简洁，采用现代手法着重处理细部，力争做到简约而不简单，庄重、大方。

The project tries to be concise, by modern means in details, to be simple yet not easy, to be solemn and generous.

扫描查看更多信息

[建築·規劃·景観·室內]
UDC 大成國際

河北大成建筑设计咨询有限公司
Hebei Dacheng Architectural Design Consulting Co., Ltd.

河北大成建筑设计咨询有限公司是一家多元化、全方位、拥有国家建筑工程甲级资质的设计公司，长期致力于城市规划、大型居住区、城市综合体、交通建筑、会展中心、体育场馆、博物馆、办公、公寓、酒店、商业、影剧院、学校、工业等项目专业的设计服务，在工程咨询、总体规划、景观设计、建筑设计等领域建立了广泛的合作客户群体。作品分布在河北、北京、上海、河南、山东、四川等地。大成公司一向务求达到尽善尽美，致力于寻找最新解决方法和施工技术。大成公司的作品博取众长，追求独特的风格和创意，以精益求精的工作态度，寻求最圆满的解决方案，重塑城市建筑的理想和高度。
大成公司现有专业技术人员60余人，其中一级注册建筑师7人、一级注册结构师8人，各类高级专业技术骨干50余人。公司技术装备配套齐全，办公场所适宜。目前公司设有建筑方案、建筑设计、结构设计、综合设计、市场和约、行政器材、档案人事共7个部门。完善的机构设置对高效管理提供了有效的保障。
大成公司主创人员有20余年的国内大型设计院工作经验及海外工作经历，多次获得国家鲁班奖、省优奖、部优奖等，具有丰富的设计实践经验、较强的创新能力及管理经验。公司成立近10年来，设计了众多项目，主要公建项目包括石家庄国际会展中心、石家庄民心广场、石家庄市赛格广场、河北名优特农产品城、石家庄财经职业学院、石家庄市一招广场综合体、壹公馆、北方药博园、九鼎搏击馆、河北数字印刷产业园、高碑店哇哈哈生产基地、保定五星级万杰广场酒店、邯郸鹿城国际、邢台红星美凯龙全球家居生活广场、三江家具城、威县巨腾商务中心等；民居项目包括石家庄的奥北公元、北城山水、柏林怡园、金蕾苑、四季福园、金海岸花园、中央时区、茗秀园、尚品雅居、金辉时代城、保定亢龙骏景、拉德芳斯新城、易县御景蓝湾、定州中山绿洲、晋州香江国际城、安新明珠丽景、山东夏津丽景国际城等。

地址：石家庄市裕华槐安路建华大街交口
万达广场写字楼A座12层
电话：+86-311-85899809
传真：+86-311-86212618
邮箱：dachenggs@163.com
网址：www.dachenggs.com

Add: 12th Floor,Tower A,Wan Da Plaza Office Building,
Yu Hua Huai An Road,Jianhua Street,Shijiazhuang City
Tel: +86-311-85899809
Fax: +86-311-86212618
E-mail: dachenggs@163.com
Web: www.dachenggs.com

石家庄市第一招待处改造项目
Shijiazhuang No.1 Reception Station Reconstruction Project

设 计 师：岳欣、刘建新、李倩
项目地点：河北 石家庄
用地面积：15 200 m²
建筑面积：158 800 m²
容 积 率：8.3

Designer: Xin Yue, Jianxin Liu, Qian Li
Location: Shijiazhuang, Hebei
Site Area: 15,200 m²
Building Area: 158,800 m²
Plot Ratio: 8.3

项目充分利用用地块特点及周边现有状况，对用地区域进行合理的功能分区和景观引入。集酒店和办公功能的主楼位于用地南侧，坐南朝北，减少了项目对中山路的压力。部分商务办公楼设在用地东侧，会议、餐饮等附属用房设在南侧，整体布局借鉴中国传统建筑围合式庭院风格，形成“围而不闭，通而不透”的空间形态；空间上形成景观广场→屋顶绿化→共享中庭景观的逐步渗透，以纵横的两条主轴线引领全局，形成多层次的空间关系。
项目在东、南、西、北均设置了商场出入口，形成四通八达之势，合理地解决了商场人流的疏散问题；办公主要入口设于北桥胡同与北人字街交叉口，以减少对中山路及平安大街的交通压力；酒店主要入口设于平安大街；南侧为疏散及货物通道。车行道沿建筑外围形成环路。

The project makes full use of landform features and peripheral existing conditions, to have proper function zoning and landscape introduction. Hotel and office building form the main building in the south of lot, facing northward, to reduce pressure on Zhongshan Road. Some commercial office buildings are in the east of lot, while conference, catering and other supporting houses are in the south, and the general layout learns from Chinese traditional enclosed courtyard to form a “enclosed yet not closed, ventilated yet not penetrated” space pattern. Gradual penetration from landscaping plaza to rooftop greening to atrium landscape sharing, together with the longitudinal and transverse main axes leading the general layout, forms multilevel space relations.;
The project sets up mall entrance and exit in all directions, to evacuate large pedestrian flow in the mall; office main entrance and exit are set in the crossroad of North Bridge Hutong and North Y-shaped Street, to reduce traffic pressure on Zhongshan Road and Ping’an Street; the hotel entrance is set on Ping’an Street; south side is evacuation and goods channel. Driveway forms a ring around the buildings.

五方中心
Wufang Center

设 计 师：岳欣、刘建新、栗少良
项目地点：河北 石家庄
用地面积：8500 m²
建筑面积：69 700 m²

Designer: Xin Yue, Jianxin Liu, Shaoliang Li
Location: Shijiazhuang, Hebei
Site Area: 8,500 m²
Building Area: 69,700 m²

项目包括写字楼、办公楼、公寓与市场四部分，各部分既相互独立，又在空间关系上相互依存。
项目分别在东、北、南三个方向设出入口，东侧为商务写字楼主入口，北侧为办公楼入口，南侧为公寓楼出入口，同时东侧和北侧设计商场的主出入口，在用地四周形成环形交通流线。通过合理的流线设计，达到办公人流、公寓人流及车流的分离。
项目外观通过大面积玻璃幕墙体现现代科技，整体造型稳重而不失灵巧，各部分之间互相辉映，建筑整体形成围合向心之势，通过空间穿插及体块的重组，达到共生共荣的效果。

The project includes office building, official building, apartment and marketplace which are independent and coexistent in space relationship.
The project designs entrance and exit in east, north and south directions, the east entrance is for commercial office building, the north is for official building; meanwhile, and the south entrance is for apartment building, the project designs main entrance and exit in the east and north for the marketplace, to form a traffic ring around the land lot smoothly. The streamline design separates office pedestrian flow, apartment pedestrian flow and vehicular flow.
The project displays modern technology by large glass curtain wall in stable yet not less agile shape, and all parts support each other to enclose a centripetal trend, while interlaces and block reorganization reach communication and co-prosperity effects.

石家庄国际会展中心
Shijiazhuang International Convention and Exhibition Center

设 计 师：岳欣、张涛
项目地点：河北 石家庄
用地面积：569 000 m²
建筑面积：259 500 m²
容 积 率：0.4
绿 地 率：30%

Designer: Xin Yue, Tao Zhang
Location: Shijiazhuang, Hebei
Site Area: 569,000 m²
Building Area: 259,500 m²
Plot Ratio: 0.4
Green Ratio: 30%

项目结合石家庄市历史背景，以纺织品褶皱般的建筑外观肌理作为基本造型元素。设计继承了当地传统文化并反映新城发展特色，贯彻“生态、人文、科技”的设计理念。
项目集展览、会议、商贸、科技、信息、休闲旅游、餐饮、娱乐等为一体，功能齐全，设施完备、结构合理，具有承办大型国际级商品交易会、大型贸易展览、大型国际会议的能力。设计遵循“以人为本”的原则，注重人流、物流、车流的交通组织。依据功能要求，处理好地上空间、地下空间及与邻近空间的联络关系。
方案设计加入新技术、新工艺、新材料、新能源的应用，并注重能源节约以及降低运行成本。

The project combines with historical background of Shijiazhuang City, and the architecture takes its basic shaping elements from textile creasing. The design follows local traditions and reflects new town development, carrying out the design ideas of “ecology, culture and technology”.
The project integrates exhibition, convention, commerce & trade, science & technology, information, leisure & tourism, catering, entertainment and other functions, with complete facilities and reasonable structure, capable of holding large international commodity trade fair, large trade expo, and large international conference. The design follows the “human-based” principle, paying attentions to the traffic diversion of pedestrian flow, logistic flow and vehicular flow, and make good liaison between aboveground, underground and neighboring spaces, according to functional requirements.
The design introduces new technology, new process, new materials and new energy resources, and pays attentions to energy saving and running cost reduction.

石家庄赛格广场

SEG IT Mall, Shijiazhuang

设 计 师：岳欣、刘建新、李倩、耿素军
项目地点：河北 石家庄
用地面积：325 500 m^2
建筑面积：2 495 900 m^2
容 积 率：5.78
绿 化 率：30%

Designer: Xin Yue, Jianxin Liu, Qian Li, Sujun Geng
Location: Shijiazhuang, Hebei
Site Area: 325,500 m^2
Building Area: 2,495,900 m^2
Plot Ratio: 5.78
Green Ratio: 30%

项目位于城市中心区域，无论是地理位置，还是传统氛围，都预示着此地将成为未来城市的发展中心，引领和带动周边区域的发展。项目整体布局恢宏大气，以一条商业景观大道贯穿始终，结合四条商业步行街，将商业区及住宅区划分开来，增加商业沿街面，并提升商业价值，使各功能区相对独立，便于管理，又相互融合，无明显界限。商业区整体考虑，大小商业相互融合，提升商业品质和价值，各商业间利用空中走廊相互连接，立体交纵，四通八达，使各种商业业态融为一体；居住区设计在地块东部，远离了城市主干道的喧嚣，自成一体，达到闹中取静的效果；绿地设计采用分散和集中相结合的方式，结合住宅设计大量宅间绿地、中心景观，再结合屋顶花园，形成立体的绿地系统；沿谈北路及谈南路引入公寓及商业，充分发挥原有街道的作用，同时结合日照分析，考虑对周边现状住宅的影响，适当增加商业面积。

The project is located in the center of Shijiazhuang City. Both geography and traditional climate predict that this land will become future city's development center, leading and driving peripheral development. The whole layout is grandiose, penetrated by a commercial landscaping avenue which, together with four commercial walking streets, divides the commercial community and residential community, to increase roadside commercial exposure and increase commercial value, to separate all functions for easy management, which are integrated without clear boundary; considering the whole commercial community, big or small commercial buildings are integrated to increase commercial quality and value, where all commercial spaces are interconnected by corridors in the air, which are accessible from all directions of the commercial integrity; the residential community is designed east of the lot, enclosed by commercial buildings, clear of noises from urban arterial roads, to form its own style gaining peace in noisy conditions; green land is designed in sparse and concentration manners, there are lots of green lands between houses, central landscapes, together with rooftop gardens, forming a 3D system of green lands; along Tan North Road and Tan South Road introduce apartments and commercial buildings, to make full use of original streets, in combination of sunlight analysis, to consider its impact on peripheral existing residential buildings, and properly increase commercial area.

邢台红星美凯龙全球家居生活广场

Redstar Macalline Global Household Plaza

设 计 师：庞海军、王小文、吕德芳、王新璞
项目地点：河北 邢台
用地面积：42.3 hm^2
建筑面积：168 600 m^2
容 积 率：3.2
绿 化 率：20%

Designer: Haijun Pang, Wang Xiaowen, Lv Defang, Wang Xinpu
Location: Xingtai, Hebei
Site Area: 42.3 ha
Building Area: 168,600 m^2
Plot Ratio: 3.2
Green Ratio: 20%

项目结合空间结构布局，从功能空间和建筑形式入手，根据建筑本身特色进行分区整合，从而形成既有各自特色又协调共存的整体组合，空间结构形态各自独立又相互统一，营造一种广场衬托商业、商业依托办公的多重空间紧密连接而又相互独立的空间层次。

项目延续红星美凯龙家居广场在全国各地的成功经验，尤其在购物空间的塑造上强调体验式的购物环境，使每一件商品都具有"家的文化、家的艺术"特质，创造一站式的家居购物中心，引领社会大众高品质的生活理念和生活方式。

The project considering space structure integrates all zones by functional spaces and building forms, according to the architectural features, so as to form a characteristic and coexistent integrity. The space structure is composed of independent parts which are united organically, to create a linking and independent multilevel space with a plaza supporting the commercial buildings which in turn support the office buildings.

The project follows the success of Redstar Macalline Household Plaza all over China, especially its shopping space is shaped in an experiencing environment, where every commodity is integrated into "home culture and home art", and the shopping space creates an one-stop household shopping center, leading the public to pursue high quality lifestyle and life philosophy.

邯郸鹿城国际

Deer City International, Handan

设 计 师：刘建新、吕德芳、张平平 Designer: Jianxin Liu, Defang Lv, Pingping Zhang
项目地点：河北 邯郸 Location: Handan, Hebei
用地面积：9100 m² Site Area: 9,100 m²
建筑面积：104 000 m² Building Area: 104,000 m²
容 积 率：8.5 Plot Ratio: 8.5
绿 化 率：30% Green Ratio: 30%

项目用地呈三角形，西临沁河，南侧为丛台路，东侧为陵西大街。设计充分利用地块特点及周边现有状况，对用地区域进行合理的功能分区和景观引入。在进行多方案比较之后，选择了三个主楼围绕中心布置的方案，使各部分使用空间均有良好的朝向、通风和日照。
立面设计采用现代风格，简洁、大方、时尚，具有较强的视觉冲击，对丰富城市景观起到一定的作用。整体造型寓意春天里的白桦林，通过凸出而富于变化的竖向线条、大面积玻璃幕墙和流畅有序的石材幕墙相组合，体现出现代办公的高效率和快节奏。

The project land is triangular, adjacent to Qin river in the west, beside Congtai Road in the south and Lingxi Street in the east. The design makes full use of lot features and surrounding conditions, to make reasonable function zoning and landscape introduction. After comparison with many schemes, the design has chosen the scheme of layout with three place buildings enclosing the center, allowing good orientation, ventilation and sunlight of all spaces.
The façade is designed in modern style, concise, generous and fashionable, with strong visual impact, enriching city landscapes. The integral shape implies birch wood in springtime, representing modern office efficiency and rapid pace through the protruding vertical lines rich of changes, large glass curtain wall and orderly stone curtain wall.

保定白沟新城万杰广场万杰酒店

Wanjie Hotel, Wanjie Plaza, Baigou New Town, Baoding

设 计 师：岳欣、庞海军、栗少良 Designers: Yue Xin, Pang Haijun, Li Shaoliang
项目地点：河北 保定 Location: Baoding, Hebei
用地面积：98 000 m² Site Area: 98,000 m²
建筑面积：150 000 m² Building Area: 150,000 m²

保定白沟新城因温泉度假而久负盛名，万杰广场坐落于新城中心地带，是联系老城与新城的纽带。
万杰酒店以五星级标准打造区域标杆，酒店集休闲、娱乐、餐饮等功能于一体，成为区域行业的领头羊。
万杰酒店建筑设计以地中海风情作为建筑设计风格，由内而外，从整体到细节，体现欧洲建筑特有的人文气息，致力于打造异域风情，同时借用当地特有的温泉资源，形成高贵、典雅的建筑风格。

Baigou New Town is famous for its hot spring resort. Wanjie Plaza sits in the center of the new town, linking with the old town.
Wanjie Hotel is built at 5-star standard into a regional landmark. The hotel integrates leisure, entertainment, catering and other functions to become a industry leader in the region.
Wanjie Hotel is designed in Mediterranean style, from inside to outside, from façace to detail, representing special cultural flavors of European architecture, committed to building exotic feelings, meanwhile loaning from local hot spring resource to form a noble elegant architecture.

九鼎搏击中心
Jiuding Kickboxing Center

设 计 师：庞海军、齐晓、耿素军
项目地点：河北 石家庄
用地面积：21.6 hm^2
建筑面积：14 000 m^2
容 积 率：0.65
绿 化 率：40%

Designer: Haijun Pang, Xiao Qi, Sujun Geng
Location: Shijiazhuang, Hebei
Site Area: 21.6 ha
Building Area: 14,000 m^2
Plot Ratio: 0.65
Green Ratio: 40%

九鼎搏击中心以巨大的鼎造型端庄地坐落在正定古城中，不同于古城中其他建筑现有的建筑形态，它以稳重、厚实又不乏张力的姿态与周边建筑形成强烈对比，巧妙地展现了与古城和谐共存的态势，并突出了自己张弛有度的特色。
在强调功能逻辑的前提下，将中国传统的重要元素之一"鼎"的概念赋予建筑本身，呼应了武术等体育竞技类项目同属中国传统文化一部分的特点，也暗含"问鼎"有胜利之意。

Jiuding Kickboxing Center is located solemnly in Zhengding ancient town, in form of a huge Tripod. Its architectural form is different from other existing buildings in the ancient town, whose stable, solid yet not less tensile posture contrasts with peripheral architecture, swiftly displaying the harmonious coexistence with the ancient town, while highlighting its features of controllable relaxation and tension.
Paying great attentions to functions, the design assigns the concept of "tripod", an important element of Chinese tradition on the architecture itself, corresponding to kickboxing and other sport programs as a part of Chinese traditional culture, while implying the winner's "challenging" for the first place.

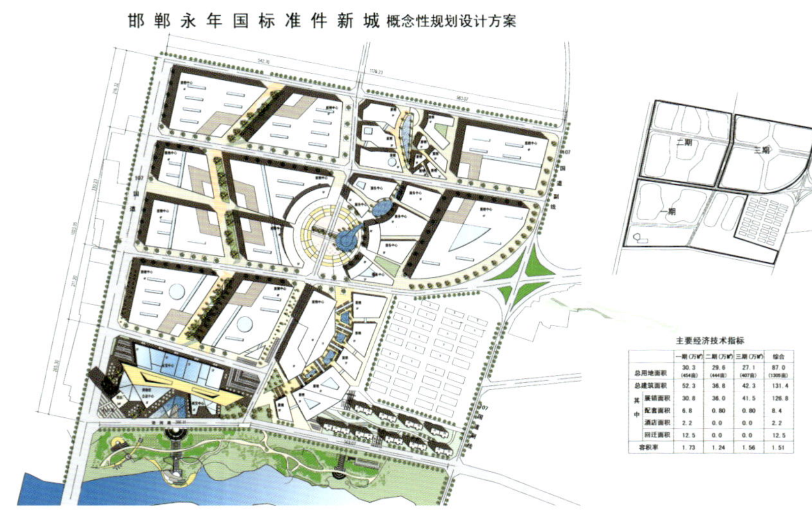

主要经济技术指标

		一期（万㎡）	二期（万㎡）	三期（万㎡）	综合
总用地面积		30.3（454亩）	29.6（444亩）	27.1（407亩）	87.0（1305亩）
总建筑面积		52.3	36.8	42.3	131.4
其中	展销面积	30.8	36.0	41.5	126.8
	配套面积	6.8	0.80	0.80	8.4
	酒店面积	2.2	0.0	0.0	2.2
	回迁面积	12.5	0.0	0.0	12.5
容积率		1.73	1.24	1.56	1.51

永年国际标准件新城
International Standard New Town, Yongnian

设 计 师：庞海军、刘建新
项目地点：河北 邯郸
用地面积：87 hm^2
建筑面积：1 310 000 m^2
容 积 率：1.51

Designer: Haijun Pang, Jianxin Liu
Location: Handan, Hebei
Site Area: 87 hm^2
Building Area: 1,310,000 m^2
Plot Ratio: 1.51

中国永年国际标准件新城是集商贸、酒店、会议、购物、展览、住宅于一体的综合商业中心。设计从城市地块环境出发，着眼整体规划，通过简约大气的外形的不断演变和发展，形成和谐统一、有震撼力的建筑集合体。建筑外立面采用高档印花玻璃，造型采用非线性造型，营造出丰富而动感的商业形象。
中国永年国际标准件新城人性化的空间布局、科学的客流引导系统，既保证了客流动线的均好性，又为消费者提供了完善、系统、人性化的消费空间。未来这里将成为整个地区商圈的核心。

This new town is a comprehensive commercial center integrating commerce, hotel, conference, shopping, exhibition and residence. The design starts from urban block conditions, develops concise and generous appearance based on general plan, to form a harmonious and astonishing building cluster. The façade is made of hi-grade printed glass, in nonlinear shape, to create a dynamic business image.
It has personalized space layout and scientific visitors flow guidance system, not only ensuring the evenness of visitors flow, but also offering a complete, systematic and personalized consumption space. In the future, here will become core of business circle across the region.

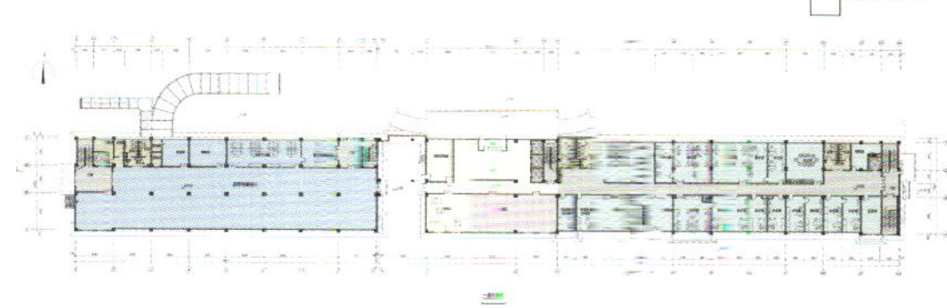

保证工艺流程顺畅、生产安全、物流合理是厂区建设的根本原则，也是本项目设计的出发点。現将数字印刷中心、综合楼及厂房以地块南北中线为轴线对称布置，最终形成“两区一轴”的规划结构。合理规划厂区出入口和厂区道路，将货物运输集中到厂区南侧，使得人流、物流分开，使生产运输流程顺畅，确保生产安全。以厂区物流最简洁为目标，将厂房设计成“U”字形，纸张进入纸库，经过印刷、装订等工艺，由成品库出场，形成一个连续、完整的物流体系，减少中间的运输环节。

Plant construction follows the radical principles of smooth process, safe production and reasonable logistics. Based on which, the project designs the digital printing center, comprehensive building and plant house in symmetry, which are divided into "two zones" by the axis of symmetry, the south-north centerline of the land lot. The design reasonably plans entrance and exit an plant roads, and concentrates goods transport in the south side of the plant area, so as to divide pedestrian flow and logistic flow, enabling smooth transporting process, and ensuring safe production. In order to make plant logistics the most concise, the plant house is designed in "U" shape, where paper sheets enter into paper warehouse, and after printing and binding, the sheets will come out from the finished products warehouse, forming a continual, integral logistic system while simplifying transport chain.

数字印刷产业园（石家庄基地）

Digital Printing Industry Park (Shijiazhuang Base)

设 计 师：齐晓、牛国安、耿素军
项目地点：河北 石家庄
用地面积：6 hm²
建筑面积：52 000 m²
容 积 率：0.81
绿 化 率：30%

Designer: Xiao Qi, Guoan Niu, Sujun Geng
Location: Shijiazhuang, Hebei
Site Area: 6 ha
Building Area: 52,000 m²
Plot Ratio: 0.81
Greening Ratio: 30%

香江国际城

Xiangjiang International Town

设 计 师：栗少良、信亮、王新璞、杨淑丽
项目地点：河北 晋州
用地面积：43 100 m²
建筑面积：295 500 m²
容 积 率：4.19
绿 化 率：35%

Designer: Shaoliang Li, Liang Xin, Xinpu Wang, Shuli Yang
Location: Jinzhou, Hebei
Site Area: 43,100 m²
Building Area: 295,500 m²
Plot Ratio: 4.19
Green Ratio: 35%

香江国际城位于晋州西南，是城市化进程和晋州新城区发展的重要节点，建筑设计在新都市主义理论的指导下以营造生态宜居社区为核心理念。
建筑采用ART-DECO建筑风格，在保留英伦格调的基础上，在现代建筑手法中融入中式居住理念，注入对新居住建筑风格的升华。
户型设计以实用经济为原则，南北通透。规划上，超大的楼间距，环抱的小区公园都为业主提供了惬意的休闲社交空间，项目自身规划有商业步行街和购物商场，为业主提供完善的服务。

Xiangjiang International Town is located southwest of Jinzhou City. It is an important point in the process of Jinzhou new district development and urbanization. The design is guided by neometropolitanism theory, to create an ecologic and habitable community as its core idea, with respect to the city and its history and culture.
It is designed in Art-Deco style, based on English pattern to integrate Chinese habitat idea by modern.
The unit design is practical and economical, ventilated south to north, large room and broad living visions, and in plan, there is over large floor space, and the circling community "park" provides an agreeable leisure and social space for owners. The project plans a commercial walking street and a shopping mall to provide better services to owners.

保定拉德芳斯
La Defense, Baoding

设 计 师：庞海军、李倩、刘建新	Designer: Haijun Pang, Qian Li, Jianxin Liu
项目地点：河北 保定	Location: Baoding, Hebei
用地面积：267 200 m^2	Site Area: 267,200 m^2
建筑面积：798 500 m^2	Building Area: 798,500 m^2
容 积 率：2.99	Plot Ratio: 2.99
绿 化 率：34%	Green Ratio: 34%

本项目所处位置具备较强的都市属性，设计引入“新城”概念，意在将项目打造成具有鲜明特色的城市名片，实现并提升项目的品质。
在一期入口处推出以“城”为核心的多元化公共复合体，成为项目本身乃至所属区域的活力源泉，加速城市节奏的脉动，旺盛人气和都市氛围直逼居住社区边缘。打造都市新城形象，突显项目重拳出击的恢弘气势。
在本项目的三个分区中，回迁区建筑密度较高，整体设计为住宅加沿街底商。一期地块位置优越，设计为商住空间，突出“城”的感觉，住宅设计采用几何式构图，开敞大气 体现“城”之气魄；二期地块相对安静，采用流线式布局，营造自然而灵活多变的社区环境。

This project is located in a parametropolitan area, with “new town” design, purporting to build a distinct name card for this city, and realize higher quality of the project.
At the entrance of phase 1 project launches a multifunctional public complex centered by “town”, to create the energetic source of the project or even the area, and speed up city pace. It leads large pedestrian flow and metropolitan climate to the marginal residential community. It builds a new town image, highlighting the splendor of this “heavy strike” project.
In the three zones of the project, relocation zone is densely designed into residential buildings and roadside ground floor stores. Phase 1 lot in advantageous location is designed into commercial and residential spaces, highlighting the “town” sense, and the residential design of which is geometric structure, spacious and generous, representing the spirits of town. Phase 2 lot is quiet, whose streamlined layout creates natural and flexible community conditions.

四季福园
Four Seasons Garden

设 计 师：庞海军、栗少良、魏蕊、吕德芳	Designer: Haijun Pang, Shaoliang Li, Rui Wei, Defang Lv
项目地点：河北 石家庄	Location: Shijiazhuang, Hebei
用地面积：24 400 m^2	Site Area: 24,400 m^2
建筑面积：52 900 m^2	Building Area: 52,900 m^2
容 积 率：1.16	Plot Ratio: 1.16
绿 化 率：30%	Green Ratio: 30%

四季福园由88套联排别墅及沿街商业组成，虽位于中心城区，却因得天独厚的地理位置而静谧天然。建筑设计传承中式传统文化，利用青、灰、白等清新、淡雅的色彩并结合中式园林文化创造一处舒适、安闲、惬意的家居环境。
建筑立面体现现代艺术与传统徽派风格的完美结合，通过体量、色彩、空间的处理，及对传统中式元素的提炼加工，使每栋建筑既古色古香又具有现代气息。
户型设计上，设计了宽敞的大露台，独享的小阳台，超大的阳光房，挑高的客厅、餐厅，从人居概念出发，将中国建筑的院落文化精髓贯穿始终。
景观设计以中式古典园林为蓝本，在规矩的平面用地上，通过水系、廊架、休闲亭及植被的变化与穿插，创造“庭院深深深几许”的诗意情境。

Four Seasons Garden is composed of 88 townhouses and roadside commercial buildings. Although located in central urban area, it is tranquil due to unique geography.
The design follows Chinese traditional culture, in green, grey, white and other light colors, in combination with Chinese gardening culture, to create a cozy, leisure and agreeable habitable environment.
The façade represents the perfect combination of modern art and traditional Anhui style, through volume, color and space processing, by extraction of Chinese traditional elements, to antiquate every building with modern sense.
The unit design features spacious terrace, exclusive balcony, overlarge sunshine room, living room, dinning room aloft, starting from habitat concept, and representing the essence of Chinese courtyard culture throughout the unit.
Landscaping design is based on Chinese classical gardening, with changes and interlays of water system, corridor, leisure pavilion and various vegetations, to crate a poetic scenario of “deep courtyard”.

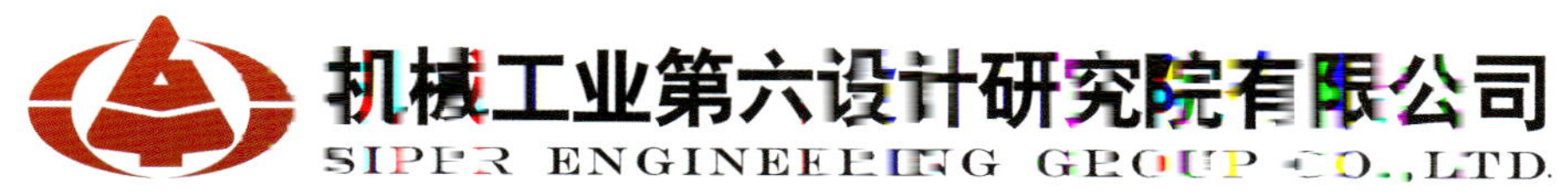

机械工业第六设计研究院有限公司（简称中机六院）创建于195[illegible]年，是拥有工程设计综合甲级资质的国家大型设计研究院，隶属中央大型企业集团——中国机械工业集团有限公司。

公司现有8个工程院、3个分院、4个子公司、3000多名员工，其中中国工程院院士[illegible]人、中国工程设计大师2人、英国皇家特许建筑设备注册工程师协会荣誉资深会员1人、享受政府特殊津贴专家28人、研究员级高级工程师[illegible]0人、高级工程师[illegible]人、各类国家注册工程师500多人。

地址：河南郑州市中原中路191号
电话：+86-371-67606087
传真：+86-371-67639571
邮箱：zjly@zz.sippr.cn
网址：www.sippr.cn

Add: No.191, Zhongyuan Road, Zhengzhou, He'nan Province
Tel: +86-371-67606087
Fax: +86-371-67639571
E-mail: zjly@zz.sippr.cn
Web: www.sippr.cn

1

2

3

1. 焦作市游泳馆、全民健身中心
Jiaozuo Natatorium & Fitness Center

项目地点：河南 焦作　Location: Jiaozuo, He'nan
建筑面积：40 000 m²　Building Area: 40,000 m²

2. 河南商会大厦
Office Building of He'nan Chamber of Commerce

项目地点：河南 郑州　Location: Zhengzhou, He'nan
建筑面积：90 000 m²　Building Area: 90,000 m²
建筑高度：120 m　Building Height: 120 m

3. 安阳东方明珠大酒店
Anyang Dongfangmingzhu Hotel

项目地点：河南 安阳　Location: Anyang, He'nan
建筑面积：100 000 m²　Building Area: 100,000 m²
建筑高度：150 m　Building Height: 150 m

4. 解放军第一五五中心医院
No.155 Central Hospital of the PLA

项目地点：河南 郑州
建筑面积：115 000 m²

Location: Zhengzhou, He'nan
Building Area: 115,000 m²

5. 郑州中原万达广场
Zhengzhou Zhongyuan Wanda Plaza

项目地点：河南 郑州
总建筑面积：531 000 m²

Location: Zhengzhou, He'nan
Total Building Area: 531,000 m²

6. 金成时代广场
Jincheng Times Plaza

项目地点：河南 郑州
建筑面积：2 000 000 m²

Location: Zhengzhou , He'nan
Building Area: 2,000,000 m²

4

5

6

7

7. 郑州国际会展中心
Zhengzhou International Conference & Exhibition Center

项目地点：河南 郑州
建筑面积：22.6万 m²

Location: Zhengzhou, He'nan
Building Area: 226,000 m²

郑州国际会展中心由中机六院和日本黑川纪章建筑都市事务所联合设计，是一个集展览、会议、交流、信息发布、观光等于一体的多功能、现代化、智能化的特大型公共建筑，总投资23亿元，建筑面积22.6 hm²，展览中心展厅跨度102 m，包含3560个国际标准展位。会议中心由一个5000人多功能厅、一个1200人国际会议厅、两个400人会议厅和十几个中小型会议室等组成；无线局域网覆盖会展室内外所有空间，30余个智能化子系统高度集成。

中机六院在设计中大胆突破、勇于创新，创下了国内会展中心八项设计之最：7 hm²的钛锌金属屋面，6 hm²原浆混凝土外墙饰面；360° 旋转电子大屏幕；大面积室外透水混凝土；3.4 hm²无柱展览大厅；10余米高的超重型活动隔断；单件重34t的铸钢件节点；30 m大跨度预应力梁板结构。

该项目工程设计先进、科技含量高，荣获中国建筑工程鲁班奖和第七届中国土木工程詹天佑奖。

Zhengzhou International Conference & Exhibition Center is an oversize multi-functional, modern and intelligence public building combined with the functions of exhibition, conference, communication, information and visiting. Its total investment is 2,300 million Yuan, building area 226,000 m², span of the exhibition hall is 102 m and international standard exhibition booths number is 3560. The Conference Center is consisted of a 5000-people multi-functional hall, 1200-people international conference hall, two 400-people meeting halls and more than 10 middle or small-sized meeting rooms; Besides, wireless LAN integrated by more than 30 subsystems covers all the space in and out of exhibition hall.

SIPPR creates eight "the first" in the design of Zhengzhou International Exhibition & Conference Center, that is: 70,000 m² titanium zincum metal roof, 60,000 m² virgin pulp concrete exterior wall finish, 360° rotary electric screen, large area exterior pervious concrete, 34,000 m² exhibition hall without column, over 10 m overweight movable partition, cast steel node with the single weight of 34t and 30 m long-span pre-stressing beam & slab structure. All these mentioned above reflect the comprehensive design ability of SIPPR in oversize public building design.

This project won Lu Ban Prize & 7th China Civil Engineering Zhan Tianyou Prize.

8—10. 现代（邯郸）汽贸城企业总部

Hyundai (Handan) Automobile Trade Headwuarter

总建筑面积：640 000 m²

Total Building Area: 640,000 m²

11. 沈阳机床（集团）有限责任公司发展数控机床及成套装备整体改造项目

Numerical Controlled Machine Tool and Outfit Development Transformation Project of Shenyang Machine Tool (Group) Co., Ltd.

项目地点：辽宁 沈阳
建筑面积：580 000 m²

Location: Shenyang, Liaoning
Building Area: 580,000 m²

该项目是中机六院承担“设计、监理、项目管理”业务的典型代表，总投资22亿，占地76万m²，建筑面积58万m²，被誉为“东北振兴的样板和典范”“中国最大的数控机床装备制造基地”。

SIPPR undertakes design, supervision & project management of this project. Its total investment is 2200 million, floor space 760,000 m² and building area 580 000 m². Its regarded as “the biggest intelligence and net factory” and model and sample factory in machine tool industry”

12. 中国北车集团大连机车车辆有限公司旅顺基地

CNR Group Dalian Rolling Stock Co., Ltd. Lushun Base

项目地点：辽宁 大连
建筑面积：735 000 m²

Location: Dalian, Liaoning
Building Area: 735,000 m²

13. 郑州煤矿机械集团有限责任公司高端液压支架生产基地

High-end Hydraulic Support Manufacturing Base of Zhengzhou Coal Mine Machinery Group Co., Ltd.

项目地点：河南 郑州
建筑面积：240 000 m²

Location: Zhengzhou, He'nan
Building Area: 240,000 m²

14

15

16

14—15. 上海烟草集团有限责任公司浦东创新科技园区

Shanghai Tobacco Group Co., Ltd. Pudong Innovative Technology Park

项目地点：上海

Location: Shanghai

16. 将军集团济南卷烟厂易地技术改造项目

Removing & Technical Innovation Project of Jinan Tobacco Company

建筑面积：474 000 m^2

Building Area: 474,000 m^2

该项目占地约7万m^2，建筑面积47.4万 m^2，建设投资20亿元，项目荣获机械工业科技进步二等奖。

This is the 1st technical innovation project of a tobacco company with the floor space about 70,000 m^2, building area 474,000 m^2, total investment about 2000 million Yuan.

17. 宇通重工新厂区建设项目

Yutong Heavy Industry New Plant Construction Project

用地面积：140 hm^2 Site Area: 140 hm^2

建筑面积：600 000 m^2 Building Area: 600,000 m^2

项目总征地面积140 hm^2，新建建筑面积约60 hm^2，总投资约30亿元，达产后年产各类工程机械设备30 000多台。项目建成后，将成为国内一流的工程机械及环保、专用车等产品研发制造基地。

Total land space of this project is 1,400,000 m^2, fresh construction area is 600,000 m^2, total investment is 3 billion RMB, annual production capacity of construction machine is more than 30 thousand once it's established. This project will be the leading domestic researching and manufacture base of construction machine, environmental friendly and special purpose vehicle.

17

河南圣地建筑景观设计有限公司
He'nan Shengdi Architecture & Landscape Design Co.,Ltd.

河南圣地建筑景观设计有限公司是华诚博远（北京）建筑规划设计有限公司在河南的分支机构，是一家集建筑、景观、规划设计、旅游地产及其他相关工程设计、策划为一体的设计单位。拥有十几年行业经验，在国内业界具有较高的知名度和权威性。对项目的整体规划、建筑设计、微环境设计、运营策划、技术服务等领域，具有熟练的专业操控水准和较高的前瞻性与行业带动性。公司自成立以来，先后承接了文化体育、办公、居住、医疗、工业、室内、商业综合体等多领域的工程设计及相关工程咨询服务，业务遍及全国。由资深设计师组成的高水平设计团队，以先进的设计理念、丰富的设计经验、高质量设计作品、良好的设计服务信誉，赢得了业主肯定和业内广泛好评。

地址：河南省郑州市经三路32号财富广场1号楼14层C室
电话：+86-371-65350762
+86-371-65350782
13592662087（郑州）/15010699369（北京）
传真：+86-371-65350763
邮箱：xsd808@163.com

Add: Room C, Floor 14th, Tower No.1, Fortune Building, Jingsan Road No.32, Zhengzhou City, He'nan Province
Tel: +86-371-65350762
+86-371-65350782
13592662087(Zheng Zhou)/15010699369(Bei Jing)
Fax: +86-371-65350763
E-mail: xsd808@163.com

老年公寓项目
Senior Apartment

设 计 师：刘世洛、杜刚成、周伟、王炳江、李维伟
项目地点：河南永城
用地面积：105 333.3 m²
建筑面积：107 440 m²
容 积 率：1.02
绿 化 率：43%

Designer: Shiluo Liu, Gangcheng Du, Wei Zhou, Bingjiang Wang, Weiwei Li
Location: Yongcheng, He'nan
Site Area: 105,333.3 m²
Building Area: 107,440 m²
Plot Ratio: 1.02
Green Ratio: 43%

标准型公寓
Standard Apartment

设 计 师：吴明友、王炳江、郑轲
建筑面积：24 512 m²

Designer: Mingyou Wu, Bingjiang Wang, Ke Zheng
Building Area: 24,512 m²

老年活动中心
Senior Center

设 计 师：刘世洛、杜刚成、李维伟
建筑面积：4687 m²

Designer: Shiluo Liu, Gangcheng Du, Weiwei Li
Building Area: 4,687 m²

医疗保健中心
Medicare Center

设 计 师：王炳江、周伟、郑轲
建筑面积：3111 m²

Designer: Bingjiang Wang, Wei Zhou, Ke Zheng
Building Area: 3,111 m²

老年大学
Senior University

设 计 师：刘世洛、吴明友、杜刚成
建筑面积：4644 m²

Designer: Shiluo Liu, Mingyou Wu, Gangcheng Du
Building Area: 4,644 m²

居家型公寓
Household Apartment

设 计 师：王炳江、李维伟、董希森
建筑面积：2020 m²

Designer: Bingjiang Wang, Weiwei Li, Xisen Dong
Building Area: 2,020 m²

老年公寓管理中心
Adminstration Centre of the Apartment for the Senior

设 计 师：董希森、杜刚成、尚瑞霞
建筑面积：5228 m²

Designer: Xisen Dong, Gangcheng Du, Ruixia Shang
Building Area: 5,228 m²

哈尔滨方舟建筑设计有限公司

Harbin Fangzhou Architectural Design Co., Ltd.

哈尔滨方舟建筑设计有限公司坐落于美丽的冰城哈尔滨，是我国建筑设计行业改革大潮中涌现出来的一支新军。多年来，方舟人凭借卓越的团队智慧及拼搏精神始终立于本地区建筑设计行业的不败之地。企业经历了由小到大、由弱到强的深刻变化过程，发展为拥有建筑、规划、景观、结构、给排水、暖通、空调、电气、工程咨询等专业设计资质的强大设计企业。目前公司拥有员工138人，其中包括一级注册建筑师、一级注册结构工程师、注册设备工程师与高级工程师48人，公司下设3个综合设计所，1个住宅研究所，1个景观设计所，5个工作室。公司成立10年来，为社会贡献了许许多多优秀的设计作品，得到了社会各界的广泛赞誉，多次被评为省优、部优，并在全国民营设计企业"华彩奖"评选中获奖。今天的方舟凭借着自己的实力与诚信，已经成为了黑龙江省内建筑设计行业顶尖机构，得到了社会各界的充分认可。

Harbin Fangzhou Architectural Design Co., Ltd. is located in the beautiful ice city of Harbin, is a new force emerged in the reform tide of domestic architectural design industry. Over the years, the Ark people have been holding an invincible position in the regional architectural design industry by outstanding virtues of team cooperation and fighting spirit, making the company experience a process of profound change from small to large and from weak to strong. Now Ark has become a powerful design company with professions including construction, planning, landscape, structure,water supply and drainage, heating/plumbing, air conditioning and engineering consulting, etc. Currently the company has 138 employees, 48 of whom are Class-A registered architects, Class-A registered structural engineers, registered equipment engineers and senior engineers. Over the past decade since its foundation, Ark has contributed lots of fine designs for the society, receiving wide praise from all works of life and winning many prizes in the province, ministry as well as the Huacai Award of National Private Design Enterprise. Today's Ark, by its own strength and credits, has become a top institution in the architectural design industry in Heilongjiang province, receiving full recognition from all social circles.

地址：黑龙江省哈尔滨市道外区红旗大街991号
电话：+86-451-87858515
传真：+86-451-87858504
邮箱：fz2001@vip.163.com
网址：www.fzjz.net

Add: No.991, Hongqi Street, Daowai District, Harbin, Heilongjiang Province
Tel: +86-451-87858515
Fax: +86-451-87858504
E-mail: fz2001@vip.163.com
Web: www.fzjz.net

黑龙江东方学院新校区

New Campus, East University of Heilongjiang

项目地点：黑龙江 哈尔滨
用地面积：593 054.84 m²
建筑面积：518 150 m²

Location: Harbin, Heilongjiang
Site Area: 593,054.84 m²
Building Area: 518,150 m²

哈尔滨工程大学61号楼
No.61 Building in Harbin Engineering University

项目地点：黑龙江 哈尔滨
用地面积：51 000 m²
建筑面积：100 436 m²

Location: Harbin, Heilongjiang
Site Area: 51,000 m²
Building Area: 100,436 m²

双鸭山高级中学
Senior High School, Shuangyashan

项目地点：黑龙江 双鸭山
用地面积：163 791 m²
建筑面积：102 823 m²

Location: Shuangyashan, Heilongjiang
Site Area: 163,791 m²
Building Area: 102,823 m²

绿地世纪城—牡丹江塞纳丽舍
Greenland Century City-Sennari House, Mudanjiang

项目地点　黑龙江 牡丹江
用地面积　148 103.87 m^2
建筑面积　283 440.72 m^2

Location: Mudanjiang, Heilongjiang
Site Area: 148,103.87 m^2
Building Area: 283,440.72 m^2

哈尔滨滨江国际
Binjiang International, Harbin

项目地点：黑龙江 哈尔滨
用地面积：42 706.2 m^2
建筑面积：144 066.86 m^2

Location: Harbin, Heilongjiang
Site Area: 42,706.2 m^2
Building Area: 144,066.86 m^2

哈尔滨纳帕英郡
Napayingjun Residences, Harbin

项目地点：黑龙江 哈尔滨
用地面积：138 901 m^2
建筑面积：447 000 m^2

Location: Harbin, Heilongjiang
Site Area: 138,901 m^2
Building Area: 447,000 m^2

齐齐哈尔锦湖雅居纯水岸
Jinhuyaju Water Bank Residence, Qiqihar

项目地点：黑龙江 齐齐哈尔
用地面积：53 540 m^2
建筑面积：244 131 m^2

Location: Qiqihar, Heilongjiang
Site Area: 53,540 m^2
Building Area: 244,131 m^2

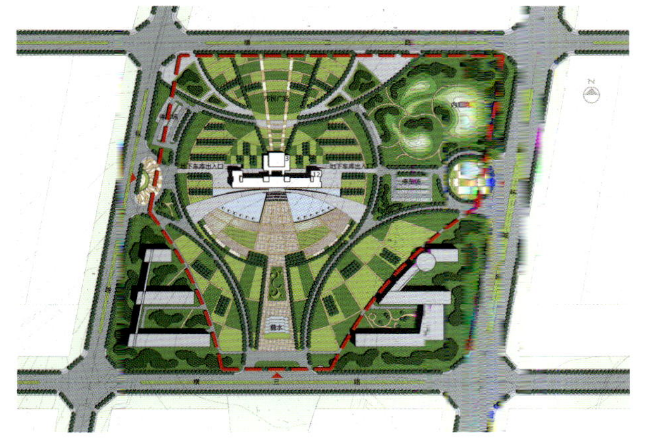

兰西行政办公大楼
Administrative Office Building, Lanxi County

项目地点：黑龙江 哈尔滨
用地面积：191 200 m^2
建筑面积：37 655.64 m^2

Location: Harbin, Heilongjiang
Site Area: 191,200 m^2
Building Area: 37,655.64 m^2

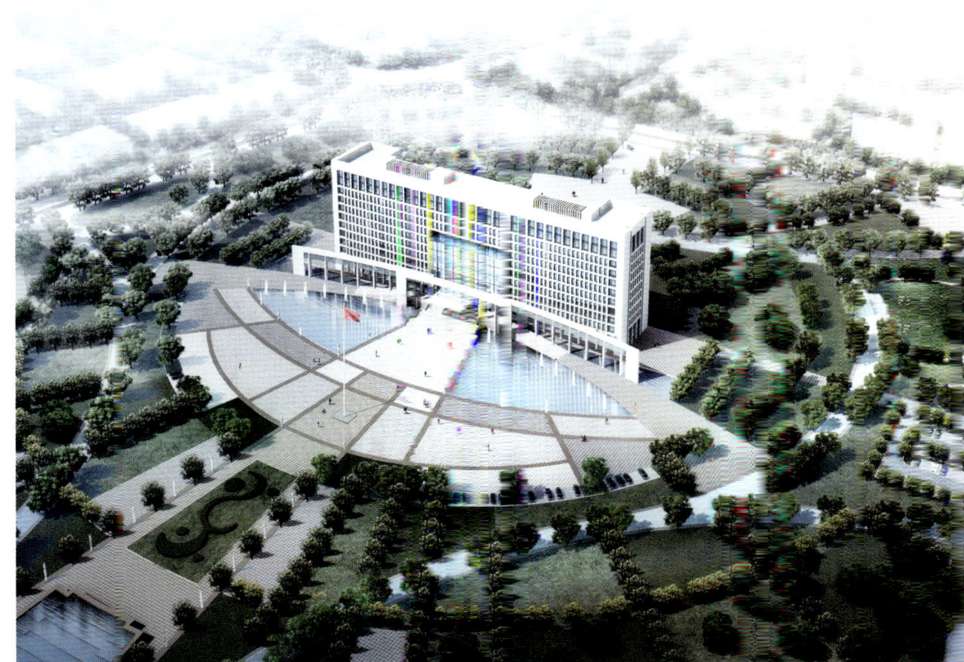

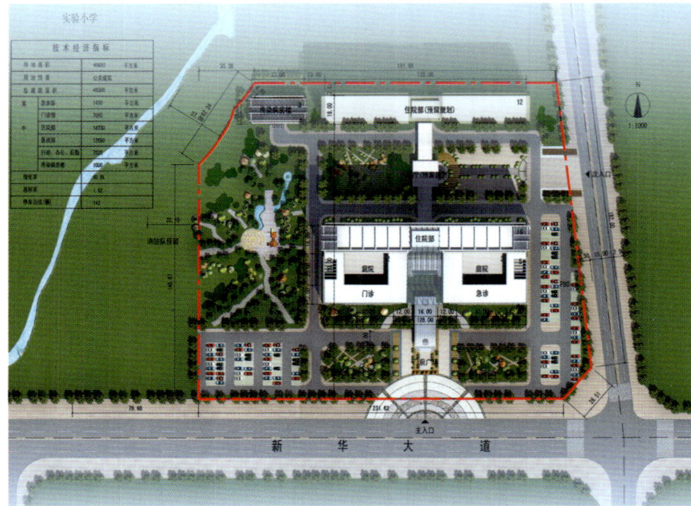

阿城区人民医院
Acheng People's Hospital

项目地点：黑龙江 哈尔滨
用地面积：46 689 m^2
建筑面积：39 590 m^2

Location: Harbin, Heilongjiang
Site Area: 46,689 m²
Building Area: 39,590 m²

双鸭山市人民医院
Shuangyashan People's Hospital

项目地点：黑龙江 双鸭山
用地面积：62 705 m^2
建筑面积：81 104 m^2

Location: Shuangyashan, Heilongjiang
Site Area: 62,705 m²
Building Area: 81,104 m²

省政府第一办公综合工作基地
The Provincial Government First Comprehensive Office Base

项目地点：黑龙江 哈尔滨
用地面积：4 hm^2
建筑面积：100 000 m^2

Location: Harbin, Heilongjiang
Site Area: 4 ha
Building Area: 100,000 m²

西城红场商业综合体项目
Commercial Complex, Xicheng Red Square

项目地点：黑龙江 哈尔滨
用地面积：118 899.45 m^2
建筑面积：320 000 m^2

Location: Harbin, Heilongjiang
Site Area: 118,899.45 m²
Building Area: 320,000 m²

扫描查看更多信息

HIAD 建筑 规划 景观 装饰 勘察 咨询 工程管理

黑龙江省建筑设计研究院
Heilongjiang Provincial Institute of Architectural Design and Research

黑龙江省建筑设计研究院（英文缩写HIAD），始建于1954年8月，是具有建筑工程设计、咨询、勘察甲级资质，建筑装饰、智能建筑、城市规划、园林、热力、测量等多项乙级资质的单位。现有职工352人，其中国家建筑设计大师1人，研究员级高级建筑（工程）师42人，取得国家各类专业注册资质人员80余人。HIAD设有建筑、规划、结构、给水排水、暖通、动力、电气、建筑经济、工程地质、工程测量等专业。同时还设有施工图审查公司、建筑装饰公司、工程监理公司、岩土工程公司等下属企业。

黑龙江省建筑设计研究院始终坚持“技术先进、质量可靠”的优化设计理念，发扬“严谨、扎实、传承、创新”的优良传统，以“服务优质，顾客满意”为经营宗旨。值得一提的是在严寒地区建筑设计领域，HIAD积累了丰富的设计经验。自建院以来，共有163项建筑工程设计、10项工程勘察成果和12项建筑标准设计规范、标准分别荣获建设部、东北地区和省级奖励；25项建筑科学研究成果分别荣获国际、国家和省级奖励；47个建筑设计方案分别在国家、部和省市建筑设计方案竞赛中荣获奖励。部分项目已达到国际先进、国内领先水平。

HIAD在工程设计领域较早通过了GB/T19001－ISO9001标准认证，使HIAD向社会提供优质、安全的设计产品有了更可靠的质量体系保障。HIAD多年来连续被省、市政府和主管部门评为建筑工程质量管理先进单位、贯彻城市规划法先进单位、省文明建设先进单位标兵；被中国资信评估学会和中国质量标准研究中心评为中国建设系统信誉、信用3A级企业；被中国勘察设计协会评为全国建筑设计行业诚信单位。

地址：黑龙江省哈尔滨市南岗区果戈里大街1号
邮编：150008
电话：+86-451-82694040
传真：+86-451-82694127/82694128
邮箱：HIAD@vip.163.com
网址：www.hljiad.com

Add: No. 1,Nangang District Gogol Street, Harbin, Heilongjiang province
P.C.:150008
Tel :+86-451-82694040
Fax :+86-451-82694127/82694128
E-Mail: HIAD@vip.163.com
Web: www.hljiad.com

哈尔滨麦凯乐休闲购物广场总店及公寓
Mykal Shopping Mall & Apartment, Harbin

项目地点：黑龙江 哈尔滨
用地面积：24 634.7 m²
建筑面积：197 182 m²
建筑高度：99.7 m

Location: Harbin, Heilongjiang
Site Area: 24,634.7 m²
Building Area: 197,182 m²
Building Height: 99.7 m

本项目是集休闲、购物、餐饮、酒店为一体的大型综合性休闲娱乐场所。项目位于哈尔滨市历史文化保护街区，周边建筑以传统欧式建筑风格为主，体现了浓厚的当地特色历史文化氛围，尤其是近邻的索菲亚教堂及建筑艺术广场，更是哈尔滨建筑艺术的精髓。该项目在建筑风格上充分考虑哈尔滨欧式建筑风格的历史文脉关系，重点在色彩、样式、线脚和比例上与早期经典建筑特征取得相应的逻辑关系，重点突出历史、文化特色。在建筑立面的处理上，以欧式古典建筑为蓝本，将其视为一种经典的语言形式，从中提取元素然后进行组织，创造了本建筑新古典主义的建筑形式。

建筑整体上强调横竖对比，产生丰富的空间层次与生动的形体关系。建筑层层叠落产生的开敞平台，为欣赏繁华街景、遥望索菲亚教堂提供了绝佳的观赏角度；高低错落的体量关系丰富了建筑的表现形体，更为城市天际线增添了丰富的变化。

五层商场内部采用通高中庭设计将日光最大限度地引入卖场，三部室内景观电梯贯穿其中。融合了巧妙灯光设计的店堂，结合了花园式布局的卖场空间，无论在严冬还是酷暑都为这里的人们营造出一派春意盎然的景象。

高层部分为具有商住性质的高级公寓，套内空间强调灵活的布局和超强的适应性，适用于SOHO式的工作、居住一体化的新型生活模式。其作为在哈尔滨商住一体化开发设计上新的有益尝试，取得了巨大成功。

香格里拉大酒店
Shangri-La Hotel

项目地点：黑龙江 哈尔滨
用地面积：15 209 m²
建筑面积：47 192.04 m²

Location: Harbin, Heilongjiang
Site Area: 15,209 m²
Building Area: 47,192.04 m²

黑龙江省电力勘察设计研究院新建设计办公楼工程

Newly-built Design Office Building of Heilongjiang Research Institute of Electric Power Exploration and Design

项目地点：黑龙江 哈尔滨
用地面积：17 380.75 m²
建筑面积：29 904.39 m²

Location: Harbin, Heilongjiang
Site Area: 17,380.75 m²
Building Area: 29,904.3[illegible] m²

黑龙江省电力调度信息中心

Electric Power Dispatching Information Center, Heilongjiang

项目地点：黑龙江 哈尔滨
用地面积：33 070.9 m²
建筑面积：64 688.6 m²

Location: Harbin, Heilongjiang
Site Area: 33,070.9 m²
Building Area: 64,688.6 m²

龙安大厦

Long'an Building

项目地点：黑龙江 哈尔滨
用地面积：5 975.25 m²
建筑面积：3[illegible]0.[illegible]3 m²

Location: Harbin, Heilongjiang
Site Area: 5,975.25 m²
Building Area: 38,1[illegible]0.63 m²

大连长兴岛临港工业区中心医院
Central Hospital, Lin'gang Industry Park, Changxing Island, Dalian

项目地点：辽宁 大连
用地面积：53 255 m²
建筑面积：100 887.58 m²

Location: Dalian, Liaoning
Site Area: 53,255 m^2
Building Area: 100,887.58 m^2

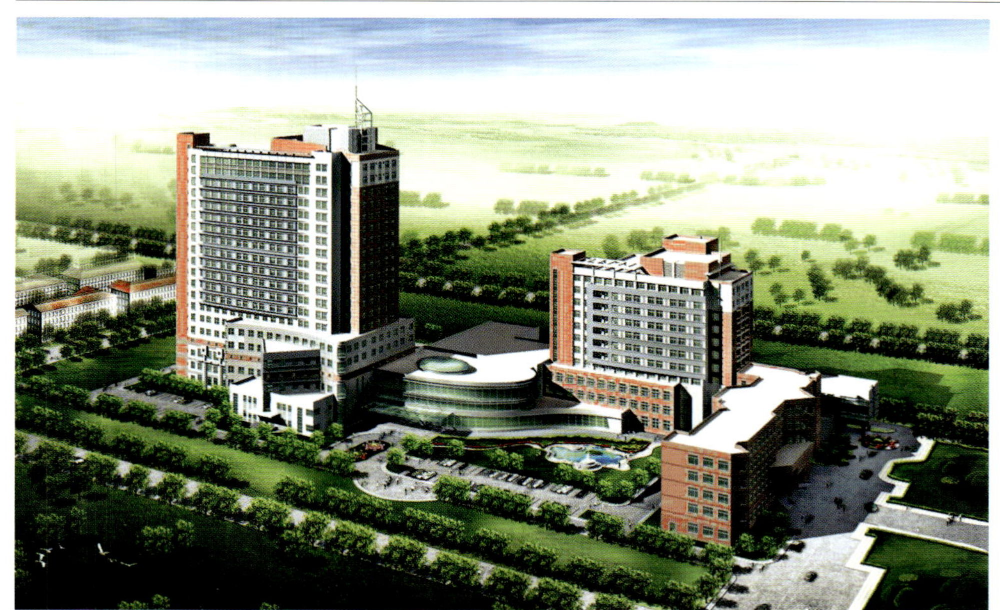

黑龙江省肿瘤医院
Tumor Hospital, Heilongjiang

项目地点：黑龙江 哈尔滨
用地面积：39 071 m^2
建筑面积：99 828.68 m^2

Location: Harbin, Heilongjiang
Site Area: 39,071 m^2
Building Area: 99,828.68 m^2

哈尔滨第十四中学高级示范校综合教学楼

General Teaching Building of Senior Demonstrative No. 14 Middle School, Harbin

项目地点：哈尔滨市
用地面积：64 300 m^2
建筑面积：6995 m^2

Location: Harbin, Heilongjiang
Site Area: 64,300 m^2
Building Area: 6995 m^2

哈尔滨一二二中

No. 122 Middle School, Harbin

项目地点：黑龙江 哈尔滨
用地面积：21 343 m^2
建筑面积：22 887 m^2

Location: Harbin, Heilongjiang
Site Area: 21,343 m^2
Building Area: 22,887 m^2

萧红纪念馆
Xiao Hong Memorial Museum

项目地点：黑龙江 哈尔滨
用地面积：2350 m²
建筑面积：4006.33 m²

Location: Harbin, Heilongjiang
Site Area: 2,350 m²
Building Area: 4,006.33 m²

纪念馆建筑方案创作于2010年，于2011年萧红诞辰百年之际竣工并投入使用。萧红纪念馆位于萧红故居“张家大院”一侧，基于对原故居的尊重及环境保护等要求，在南北狭长的用地范围内建筑形态采取化整为零、向地下拓展空间的方式，不仅满足了建筑的使用功能，同时降低了建筑的地面高度，以谦恭的姿态矗立在老故居旁边。建筑外立面采用质朴的灰色仿文化石水泥砖、玻璃幕墙和屋面采光黑色筒瓦。建筑风格朴实、隽永、富有张力，充分契合了萧红作为一名北方女作家坚强、质朴、向往自由的人格特征，达到了建筑与自然文化、人文文化的交融。建筑在室内空间处理上合理利用高低空间作为展览和办公附属用房，利用竖向采光井给有限的展厅和地下交通空间提供自然采光。纪念馆设计用地紧凑、资金节约、技术先进、环保节能，使建筑的艺术氛围达到了时间与空间的完美结合。

铁军书画院
Tiejun Gallery

项目地点：黑龙江 大庆
用地面积：4886.25 m^2
建筑面积：2762.48 m^2

Location: Daqing, Heilongjiang
Site Area: 4,886.25 m^2
Building Area: 2,762.48 m^2

友谊县知青博物馆
Educated Youth Museum, Friendship County

项目地点：黑龙江 双鸭山
用地面积：41 800 m^2
建筑面积：3995.69 m^2

Location: Shuangyashan, Heilongjiang
Site Area: 41,800 m^2
Building Area: 3,995.69 m^2

扎赉诺尔博物馆
Jalainur Museum

项目地点：内蒙古 满洲里
用地面积：24 818.96 m^2
建筑面积：8320.5 m^2

Location: Manchuria, Inner Mongolia
Site Area: 24,818.96 m^2
Building Area: 8,320.5 m^2

第二十四届冬季大学生运动会雪上竞技指挥中心

Snow Match Command Center of the 24th Winter University Games

项目地点：黑龙江 哈尔滨
用地面积：10 240 m^2
建筑面积：7423 m^2

Location: Harbin, Heilongjiang
Site Area: 10,240 m^2
Building Area: 7,423 m^2

哈尔滨国际会议展览体育中心

International Convention, Exhibition & Sports Center, Harbin City

项目地点：黑龙江 哈尔滨
用地面积：43 hm^2
建筑面积：400 000 m^2

Location: Harbin, Heilongjiang
Site Area: 43 ha
Building Area: 400,000 m^2

该项目规模宏大，功能复杂，占地43 hm²，建筑面积40多万m²，单体建筑最长618 m，横跨城市干道，功能集展览、体育、宾馆、新闻会议、商业为一体，是超高层大型综合建筑，为国内罕见、省内仅有。

一号工程为体育训练馆和万人体育馆，单体建筑面积20万 m²，平面长618 m，主跨宽128 m，采用索拱结构，最高点标高36 m。立面设计力求把现代技术与艺术完美地结合起来，在3 m高的大台基上，主体结构采用索拱屋架与人字“摇摆柱”配上点式玻璃幕墙的围护结构，透出隐含的水平韵律，玻璃幕墙长廊、弧形金属屋面构成了展览建筑的形体。整个建筑的室内外空间简洁、通透、明亮，营造出良好的空间形象，具有强烈的时代感，成为哈尔滨市标志性建筑之一。

二号工程为国际会议中心和宾馆，单体建筑面积10万 m²。宾馆高38层，高度170 m；国际会议中心为地上5层，高度40 m。立面及形体设计为“树叶”形，隐喻着建筑与自然和谐相处；裙房立面的大面积玻璃幕墙和条形铝合金装饰横带丰富了立面。高层部分以抽象的几何形体、材料的质感和色彩的变化，与裙房的水平划分形成强烈的对比。项目将成为黑龙江省政治、经济、文化的研讨中心。

三号工程为万人甲级体育场，是黑龙江省大型体育活动的中心。主体为钢筋混凝土结构，罩棚为钢结构。体育场除满足功能要求外，立面设计以外露壁柱形成连续的空间序列，240 m跨的钢拱双曲面罩棚形成体育场的第五立面。

项目采用技术复杂、先进并包括128 m跨预应力梁钢结构索拱体系，是目前国内同类结构中最大的。此外还有312 m跨钢结构罩棚体系，超大空间的空调系统，超大空间的消防灭火系统，大规模建筑的消防组织设计，建筑智能系统及智能灭火系统，大面积玻璃幕墙节能设计等。

伊春鹿鸣钼矿中心林场生活区

Luming Molybdenum Mine Central Forest Farm Living Area, Yichun

项目选址：黑龙江 伊春
用地面积：753 100 m^2
建筑面积：228 300 m^2

Location: Yichun, Heilongjiang
Site Area: 753,100 m^2
Building Area: 228,300 m^2

项目主要用地为三部分，一为东三场撤并合建于鹿鸣中心林场生活区、林场办公管理设施及林场生产配套仓储设施；二为鹿鸣钼矿办公生活区及相应服务配套设施；三为当地有关管理部门行政办公服务设施、市政公共设施。

本着以人居环境优化设计为出发点，在设计中主要体现了以下的设计理念。

设计中对原有规划用地内的次生林带尽量保留并加以妥善处理。通过与外部公路之间预留原有林带、镇区间保留大尺度原有林带等手段，将林区的环境特征引入和渗透到镇区内部，达到内外交融的效果，使自然的特征留存于生活之中。

对周边水系的引入和利用。通过对依吉密河支流河道增加分支和向镇区内部的引入，达到美化镇区中心与边界景观，增加微环境多样性的效果。同时在镇区与山之间形成一段天然的屏障，还能起到对山洪蓄防的作用。

利用原有地势和地质情况，优化竖向设计，保留自然地貌特征。在竖向设计和建筑单体规划时，尽量依托原有地势坡度和等高线，随形就势，对原有地形进行最小程度的改变。根据实际地质情况，明确地表径流的排水方式和途径。避免出现生硬的街区形象和过度的城市化特性，提高当地居民对居住环境的认同感。

强调对建筑朝向、采光和通风的规划设计。根据实际用地情况，适当加大居住建筑间距，保证所有居住建筑均为朝阳方向，满足大寒日日照均在6 h以上。各街区之间根据主导风向预留通道，保证镇区小气候能够形成良好的循环。

林场生活区整体采用欧式建筑风格，立面处理上以简洁的欧式线脚进行分割，体量和尺度的处理遵循传统欧式的构图原则，对相关建筑细部和符号的比例和形态则进行细致的推敲及灵活的处理，使各群体既相对独立又相互统一，融于林场生活区的整体环境中。

大正 · 莅江住宅小区

Dazheng · Riverside Residential Community

项目地点：黑龙江 哈尔滨
用地面积：149 151.3 m^2
建筑面积：470 869.25 m^2

Location: Harbin, Heilongjiang
Site Area: 149,151.3 m^2
Building Area: 470,869.25 m^2

燕郊"@"北京居住区设计

Design of "@" Beijing Residential Community, Yanjiao Town

项目地点：河北 三河
用地面积：577 452 m^2
建筑面积：240 829.09 m^2

Location: Sanhe, Hebei
Site Area: 577,452 m^2
Building Area: 240,829.09 m^2

哈尔滨学伟国际城住宅区

Xuewei International Town Residential Community, Harbin City

项目地点：黑龙江 哈尔滨
用地面积：100 000 m^2
建筑面积：200 000 m^2

Location: Harbin, Heilongjiang
Site Area: 100 000 m^2
Building Area: 200 000 m^2

01 黑龙江省北安市政府办公楼

项目地点：黑龙江 北安
建设功能：行政办公
建筑面积：20 006.45 m^2
建筑层数：16层
结构形式：框剪结构

02 黑龙江省农垦总局办公楼

项目地点：黑龙江 哈尔滨
建设功能：行政办公
建筑面积：39 604.5 m^2
建筑层数：21层
结构形式：框剪结构

03 黑龙江省农垦知青托管中心

项目地点：黑龙江 佳木斯
建设功能：康复医院
建筑面积：15 000 m^2
建筑层数：7层
结构形式：框架结构

04 黑龙江省农垦牡丹江分局培训中心

项目地点：黑龙江 密山
建设功能：宾馆
建筑面积：12 000 m^2
建筑层数：7层
结构形式：框架结构

05 哈尔滨国际会展名城

项目地点：黑龙江 哈尔滨
建设功能：商住楼
建筑面积：153 000 m^2
建筑层数：地上30层，地下一层
结构形式：框剪结构

07 黑龙江省密山市商业城

项目地点：黑龙江 密山
建设功能：商住楼
建筑面积：485 000 m²
建筑层数：地上18层
结构形式：框架结构及框剪结构

08 黑龙江省密山公安局办公楼

项目地点：黑龙江 密山
建设功能：办公楼
建筑面积：27 000 m²
建筑层数：地上6层
结构形式：框架结构

06 黑龙江省农垦科学院科研楼

项目地点：黑龙江 哈尔滨
建设功能：科研楼
建筑面积：[illegible]1 350 m²
建筑层数：地上9层
结构形式：框架结构

黑龙江省农垦建筑设计院

Heilongjiang Provincial Agricultural Reclamation Architectural Design Institute

建筑工程设计甲级资质 / 工程项目管理甲级资质
工程咨询乙级资质 / 城市规划编制乙级资质
公路设计丙级资质 / 装饰工程施工二级资质

黑龙江省农垦建筑设计院创建于1982年，是国家甲级建筑工程设计咨询机构，可提供大型公共建筑设计、民用建筑设计、城市规划设计、室内外装饰设计与施工、园林景观设计、建筑智能化系统工程设计、工程概预算编制、公路设计、工程项目管理等领域服务。

黑龙江省农垦建筑设计院具有建筑工程设计甲级资质、工程项目管理甲级资质；工程咨询乙级资质；城市规划编制乙级资质；公路设计丙级资质；装饰工程施工二级资质。总院坐落在黑龙江省哈尔滨市，现有职工80人，其中高级工程师39人，各类注册人员18人。该院专业人员配备齐全，技术力量雄厚，工程设计方法先进，工作管理规范。总院下设第一、第二、第三、第四设计所，即钢结构设计所，城市规划设计所，工程咨询所，汇雅装饰公司；在黑龙江省内还设有农垦建三江分院、即农垦红兴隆分院、农垦九三分院、农垦宝泉岭分院、农垦牡丹江分院；在省外设有黑龙江省农垦建筑设计院新疆分院。总院已通过ISO9001认证，具有健全的组织机构和先进的技术装备，完善的质量保证体系和以技术质量为中心的各级人员岗位责任制等规章制度，能确保工程设计的进度和质量。

在经营活动中，以严谨求实的工作作风，灵活经营的市场竞争意识，精心设计、力求精品的设计理念和对工程设计质量负责到底的承诺竭诚为用户服务，以求取得最佳的社会效益和经济效益。设计作品遍及省内外，主要作品有：黑龙江省农垦总局办公综合楼、天津中国轻工大厦、秦皇岛佳伦大厦、佳木斯同源大厦、国际会展名城、北安市政府办公楼、黑龙江省农垦科学院科研楼、牡丹江农垦培训中心、建三江金融大厦、农垦总局新农村规划、方虎公路至牡丹江分局新直道路工程设计等。其中“黑龙江省农垦总局综合办公楼”工程荣获2005年度国家优质工程银质奖，“北安市政府办公楼”工程荣获2009年度国家优质工程银质奖，中国轻工大厦、秦皇岛佳伦大厦、哈西国贸服装城、农垦干部管理学院等20余项工程被评为省优秀设计奖。

“守信重誉、勇于开拓、力求精品、竭诚服务”是黑龙江省农垦建筑设计院企业宗旨，该院决心以这一宗旨，与建筑业同仁交流合作，共同为建筑业的发展做出应有贡献。

地址：黑龙江省哈尔滨市香坊区公滨路441号
邮编：150036
电话：院长办公室：+86-451-55665[illegible]52
副院长办公室：+86-451-55664317
总工办：+86-451-55627151
办公室：+86-451-55630910
传真：+86-451-55662126
邮箱：hljsnkjzsjy@163.com
网址：www.nkjg.cn/fgs3/index.asp

扫描查看更多信息

扫描查看更多信息

中信建筑设计研究总院有限公司

CITIC General Institute of Architectural Design and Research Co., Ltd.

中信建筑设计研究总院有限公司（原武汉市建筑设计院）创建于1952年10月1日，是新中国成立之初诞生的大型国营设计机构之一，是持有甲级工程设计证书、甲级城市规划证书、甲级工程设计审查证书、甲级工程咨询证书、甲级智能建筑系统工程设计证书的大型综合性建筑设计院，是商务部认定的“骨干企业”之一。20世纪80年代，成为全国建设系统设计体制改革试点单位和全面质量管理（TQC）试点单位；1992年进入由建设部和国家统计局评定的中国勘察设计单位综合实力百强；1994年成为中国首家实现工程设计手段电脑化的设计单位；2000年通过ISO9001质量体系认证；2002年加入中国中信集团有限公司；2009年被授予国家高新技术企业称号；2011年被评为全国建筑设计行业诚信单位。

中信建筑设计研究总院有限公司人才荟萃，专业齐全，设备精良，技术先进。现有职工1315余人，其中专业技术人员1104余人，享受国务院特殊津贴专家12人，省、市专家33人。已有一级注册建筑师84人，一级注册结构工程师102人，注册城市规划师10人，注册公用设备工程师52人，注册电气工程师18人，注册造价工程师10人，注册咨询工程师11人，注册土木工程师（岩土）4人。6人取得内地与香港互认的建筑师和结构工程师资格，1人获得英国结构工程师学会会员资格。拥有中国绿建委员会委员1名，住房与城乡建设部绿色标志委员会专家5名。设有9个建筑设计院以及规划设计研究院、市政设计研究院、机电设计院和5个分支机构（北京、深圳、海南、上海、珠海），并投资兴办中信建筑工程顾问（武汉）有限公司和中外合资东艺设计公司。

60年来，中信建筑设计研究总院有限公司设计活动遍及全国，且远达亚洲、非洲、大洋洲、拉丁美洲的29个国家，并与美国、日本、德国、奥地利、英国、俄国、法国、意大利、新加坡等国家及中国香港、澳门地区开展了广泛的技术合作和人才交流。本院贯彻“改革兴院，技术立院，科学管院，人才强院”的治院方针，实施“发展高新技术，争创名优设计”的名牌战略，获得长足发展，获国家、部、省、市优秀设计奖200余项、科技进步奖100余项。

本院贯彻《北京宪章》，践行“诚信为根，创新为魂，以人为本，追求卓越”的企业核心价值观和“设计广厦千万间，人民安居尽欢颜”的企业精神，繁荣建筑文化，致力技术创新，严格质量管理，为社会贡献优秀设计和优质服务，为大众创造美好而公平的人居环境，创造可持续发展的现代城市与建筑。

CITIC General Institute of Architectural Design and Research Co., Ltd. (the former Wuhan Architectural Design Institute) was established on October 1st, 1952. As one of the largest state-owned design organizations when the new China was founded, it is now one of the major comprehensive architectural design institutes with Grade A Certificates for Engineering Design, Urban Planning, Engineering Design Examination, Engineering Consultation and Engineering Design of Intelligent Buildings. It is recognized as a “Key Enterprise” by the Ministry of Commerce. In 1980s, the Institute became the Pilot Unit for Design System Reform and Pilot Unit for Total Quality Control (TQC) in the national construction system. In 1992, it was found among the China's Top 100 Survey and Design Units in terms of comprehensive power assessed jointly by the Ministry of Construction and the National Bureau of Statistics. In 1994, it became the first design unit in China that adopted computer-aided engineering design methods. In 2000, it passed ISO9001 Quality Management System Certification. In 2002, it joined CITIC. In 2009, it was recognized as a National High-tech Enterprise. In 2011, it was awarded the “National Integrity Unit in Architectural Design Industry”.

The Institute is abundant in talented emplyees, and boasts full range of specialties, sophisticated equipment, advanced technology. It currently has 1315 staff, including 1104 professional technicians, 12 of whom are experts enjoying the State Council Special Allowance; 33 are provincial or municipal level experts. We now has 84 Class I Certified Architects, 102 Class I Certified Structural Engineers, 10 Certified City Planners, 52 Certified Facility Engineers, 18 Certified Electrical Engineers, 10 Certified Cost Engineers, and 11 Certified Consulting Engineers, and 4 Certified Civil Engineers. Among them, 6 have obtained architectural and structural engineer qualifications that are mutually recognized by China mainland and Hong Kong, 1 has obtained the chartered member of UK Institution of Structural Engineers, 1 is a member of China Green Building Council, and 5 are expert members of the Green Building Identification Committee of the Housing Ministry. It now owns 9 designing branches, planning and design research branches, municipal design branches, electromechanical design branches and 5 branch institutes (in Beijing, Shenzhen, Hainan, Shanghai, Zhuhai); also it has made investments in establishing the CITIC Constructional Engineering Consulting (Wuhan) Co., Ltd and the Sino-Foreign Joint Venue Dongyi Deign Company.

In recent 60 years, We have offered design services in almost every part of China and in more than 29 countries from Asia, Africa, Oceania and Latin America, and have made extensive technical cooperation and talent exchange with counterparts in USA, Japan, Germany, Austria, the UK, Russia, France, Italy, Singapore and the Hong Kong - Macau Region. The Institute marches forward by following the management guideline, namely, “push the development of the Institute with reforms, base the development of the Institute on technologies, manage the Institute in a scientific manner, and strengthen the Institute by giving full play to the talents”, and carrying out the brand name strategy of “developing high and new technologies, and striving to create brand-name and high-quality designs”. We have won more than 200 prizes for Outstanding Design and more than 100 prizes for Progress in Science and Technology of national, ministerial, provincial and municipal levels.

The Institute implements the Beijing Charter, and practices the enterprise core values of “taking credit as root, innovation as spirit and people as the foremost, as well as striving for excellence”. The Institute also carries forwards the enterprise spirit of “designing sufficient buildings to help people live suitably and cheerfully”. We make efforts for the prospering of the architectural culture, make commitment to the technological innovation, and enforce strict quality control measures, with the aim to contribute excellent design and services to the society, build a beautiful and fair habitat environment for the popular masses, and create modern cities and buildings that can develop in a sustainable way.

地址：武汉市汉口四唯路8号
邮编：430014
电话：+86-27-82722966
传真：+86-27-82726178
网址：www.whadi.com.cn

Add: No.8 Siwei RD, Hankou, Wuhan
P.C.: 430014
Tel: +86-27-82722966
Fax: +86-27-82726178
Web: www.whadi.com.cn

武汉辛亥革命博物馆（新馆）

Wuhan 1911 Revolution Museum (New Museum)

项目地点：湖北 武汉
建筑面积：22 000 m²

Location: Wuhan, Hubei
Building Area: 22,000 m²

武汉辛亥革命博物馆（新馆）位于武昌阅马场、首义南轴线上，于2009年12月动工新建，2011年10月竣工，是辛亥革命百年庆典纪念活动的重要场所。博物馆用地规模达14.6 ha，规划建筑面积22 142 m²，高23.3 m，地上三层设置7个展厅，地下一层为停车场。规划以"历史的峡谷""革命的闪电"为整体构思理念。建筑"V"字形外观具有很强标志性和象征性，与首义文化区和武昌老城区的整体景观相协调，体现出肃穆、凝重的纪念风格。建筑平面轮廓和外形为"正三角形"；建筑外墙采用粗糙的表面肌理，并利用自然雕琢、风化的纹理；建筑色彩以红色为基调，体现了楚文化的特色，又与原纪念馆红楼的色彩协调统一。

Wuhan 1911 Revolution Museum (New Museum) is located on the south axis of Yuemachang and Shouyi, Wuchang. Construction of the Museum commenced in December, 2009 and completed in October, 2011. It is an important place for centennial celebrations of the 1911 Revolution. The Museum, 23.3 m in height, has a plot area up to 14.6 ha and a planned floor area of 22,000 m². Seven exhibition halls are housed on the three floors aboveground, and one parking lot is provided underground. The general layout takes "historical canyons" and "revolutionary lightning" as its design conception. The V-shaped architecture, with identification and symbolization, coordinates with the overall landscape of Shouyi Cultural District and the old city town of Wuchang, which embodies a solemn and dignified memorial style. The plane profile and appearance of the building takes an "equilateral triangle" as its motif. The exterior wall adopts rough texture appearance and utilizes naturally weathered veins. The architecture adopts red as its keynote which not only embodies the features of Chu culture but also coordinates with the color of the original "Red Mansion".

湖北省博物馆总体规划和建筑设计

Master Plan and Architectural Design of Hubei Provincial Museum

项目地点：湖北 武汉
建筑面积：58 000 m²

Location: Wuhan, Hubei
Building Area:58,000 m²

湖北省博物馆位于风景秀丽的东湖之滨，与湖北省艺术馆、湖北省社会科学院、湖北日报社等文化设施为邻，人文气息浓郁。

省博物馆（含省文物考古研究所占地面积4667 m²）现有占地面积8.2 hm²，征用省文物总店占地面积约3000 m²，总体规划和建筑设计方案设计完成后，总建筑面积约为100 000 m²，其中拟保留原有建筑物的建筑面积42 000 m²，拟规划建设项目的建筑面积约58 000 m²。

设计立足于"植根于楚，寓意于博，凝神于藏，蕴型于钟"——"楚博藏钟"的核心构思，展现湖北省博物馆新馆"湖畔藏楚韵、编钟鸣盛世"的深远意境。

Hubei Provincial Museum is located at the scenic East Lake, and surrounded by several cultural institutions with strong humanistic atmosphere such as Hubei Museum of Art, Hubei Academy of Social Sciences, Hubei Daily Newspaper Group, etc.

The land occupied by Hubei Provincial Museum (among which 4667 m² possessed by Hubei Provincial Institute of Cultural Relics and Archaeology) amounts to 82,000 m² and the extra 3000 m² land is acquired from Hubei Provincial Cultural Relics Shop. After completion of overall planning and architectural design, the total floor area of Hubei Provincial Museum will reach 100,000 m², among which the floor area of the existing buildings will be 42,000 m², and the planned floor area will be about 58,000 m².

Based on the core conception of "taking root in Chu culture, representing the implication of the relics in the Museum, paying attention to Collection, and taking the shape of the Chime-bells (in short, Chu's Museum Collects Chime-bells)", the Museum will present the far-reaching concept of "collecting Chu rhythm on the lakefront, playing chime-bells to celebrate a flourishing age" of New Hubei Provincial Museum.

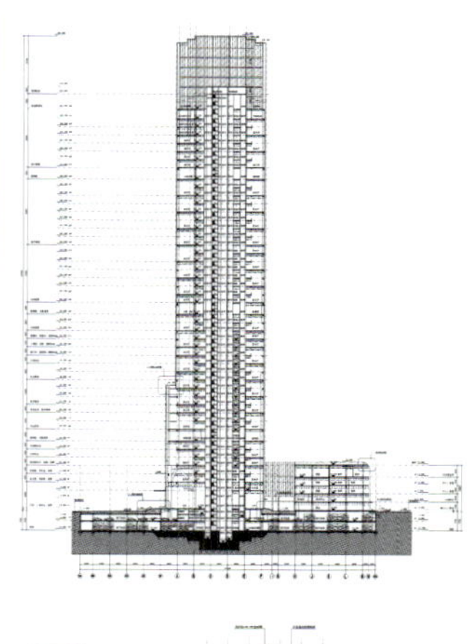

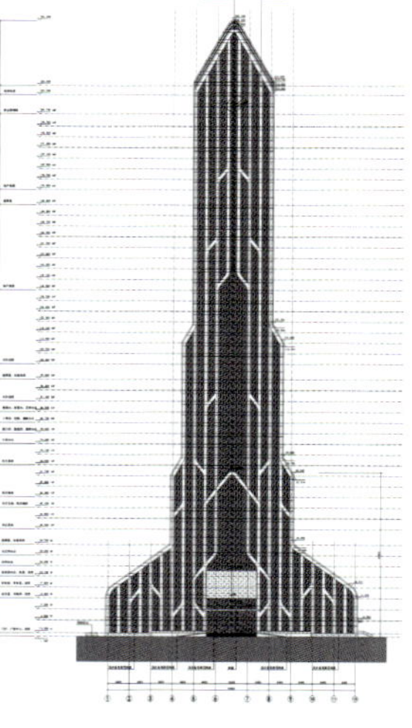

长江传媒大厦

Changjiang Media Mansion

项目地点：湖北 武汉
建筑面积：95 000 m²

Location: Wuhan, Hubei
Building Area: 95,000 m²

长江传媒大厦位于武汉市江岸区后湖街新村，交通便利。大厦地上建筑面积95 000 m²，总高度243 m。主立面正对金桥大道，前设主入口与入口广场，显示出建筑的恢弘气势。

建筑形体呈现出一种汇聚、向上、灵动的形态，造型自下而上依次收分，赋予了一种"直上云霄"的动势；外表以"人"字形为主题，着重体现传媒"社会瞭望者"的社会功能；竖向垂直的线条顺应建筑形态的走势，传达"以人为本"的建筑内涵。并通过节能技术的应用，体现"绿色、环保、舒适、和谐"的核心追求。

Changjiang Media Mansion is located in Xincun, Houhu Street, Jiang'an District, Wuhan, with convenient traffic conditions. The mansion is 243 m in height with a floor area aboveground of 95,000 m². The main facade of the mansion is right opposite to Jinqiao Avenue. A primary entrance and an entrance square are provided in front of the Mansion which set off its magnificence.

The architecture assumes a convergent, upward and flow form. The architectural form, gradually set back from bottom to top, endows the building a motion of "soaring straight up into the sky"; herringbone texture on the surface of the building is used as a motif, which mainly reflects the social function of media as "a masthead man for the society"; the straight vertical lines following the trend of architectural body, which embodied the architectural connotation of "human-orientation". The application of energy saving technology reflects the core pursuit of "greenness, environmental-friendliness, comfort and harmony".

新疆国际会展中心

Xinjiang International Exhibition Center

项目地点　新疆 乌鲁木齐
建筑面积　200 000 m²

Location: Urumqi, Xinjiang
Building Area: 200,000 m²

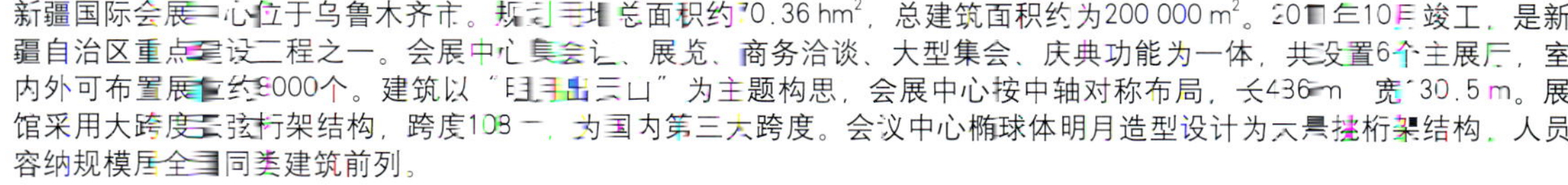

新疆国际会展中心位于乌鲁木齐市。规划用地总面积约70.36 hm²，总建筑面积约为200 000 m²。2011年10月竣工，是新疆自治区重点建设工程之一。会展中心集会议、展览、商务洽谈、大型集会、庆典功能为一体，共设置6个主展厅，室内外可布置展位约8000个。建筑以“日月出天山”为主题构思，会展中心按中轴对称布局，长436 m，宽130.5 m。展馆采用大跨度立体弦桁架结构，跨度108 m，为国内第三大跨度。会议中心椭球体明月造型设计为大悬挑桁架结构，人员容纳规模居全国同类建筑前列。

Xinjiang International Exhibition Center is located in Urumqi, with a total planned plot area of 70.36 hectares and a total floor area of 200,000 m². It was completed in October 2011, and was one of the key projects in Xinjiang Uygur Autonomous Region. It functions as a building for conferences, exhibitions, business negotiation, large assemblies and celebrations, etc. There are six main exhibition halls that can accommodate 8,000 exhibition booths. The building takes the notion of "a bright moon rising from the Mountain of Heaven" as its design conception. The exhibition center, 486 m long and 130.5 m wide, is arranged in symmetrical fashion along the axis. The exhibition halls adopt a 108 m-wide large span (the third largest in China) chord truss. The conference centre, in the shape of a spheroid (embodying a bright moon), is in the structure of a large cantilever truss, and its accommodating scale occupies a leading place in the whole country.

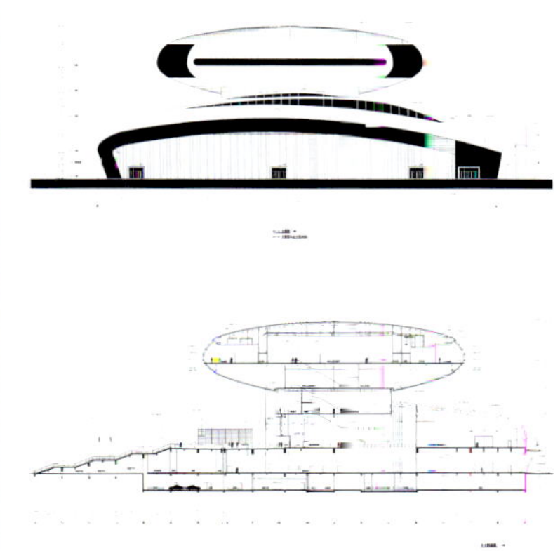

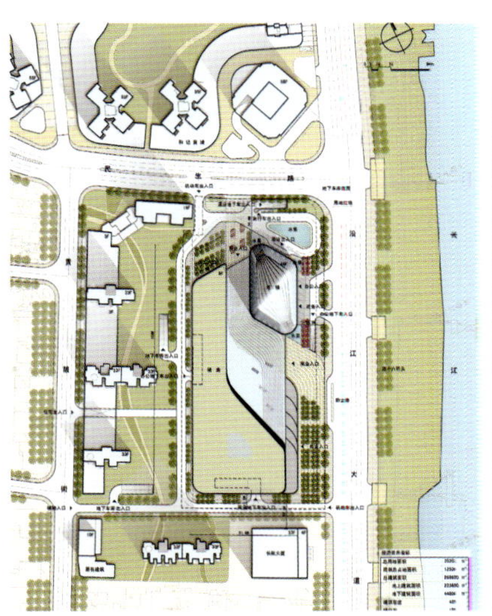

武汉航运中心大厦

Wuhan Yangtze River Shipping Center Tower

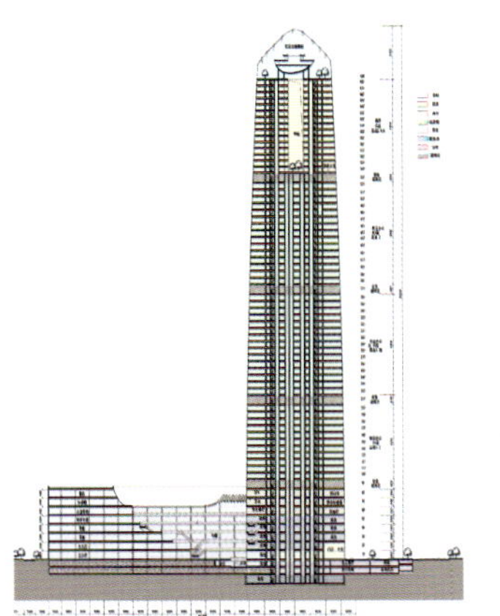

项目地点：湖北 武汉
建筑面积：230 000 m²

Location: Wuhan, Hubei
Building Area: 230,000 m²

武汉航运中心大厦位于武汉最繁华的江汉路核心商圈，地上总建筑面积约230 000 m²，塔楼高度330 m。

大厦设计根据城市环境总体规划，突出长江航运的地域文化，组合人性多元的功能模块。建筑造型生动和谐，层次丰富，表皮在夕阳与江水的相互辉映下，极尽柔美；顶部处理体现出“天河”神韵；底部设置水景景观，浪漫亲和。

Wuhan Yangtze River Shipping Center Tower is located in the most prosperous core business circle in Janghan Road. The floor area aboveground is around 230,000 m² and the Tower is 330 m in height. On the basis of the overall planning of the urban environment, the Tower design highlights regional cultures of Yangtze River shipping and integrates human-friendly and diverse functional modules. With abundant layers, the architectural form is vivid and harmonious. Texture on the surface of the building resembles gentleness of glistening light of waves in the sunset. A "Milky Way" verve is embodied at the top of the Tower; the waterscape at the bottom brings a sense of romance and affinity.

武汉市妇女儿童医疗保健中心综合业务楼

Comprehensive Business Building of Wuhan Medical Health Center for Women and Children

项目地点：湖北 武汉
建筑面积：200 000 m^2

Location: Wuhan, Hubei
Building Area: 200,000 m^2

武汉市妇女儿童医疗保健中心综合业务楼，位于香港路。保健中心功能一期包括门诊楼（含急诊）、医技楼、内科综合楼，二期包括妇产科、行政综合楼和两层地下室，总建筑面积200 000 m^2，设计日门诊量5000人，开放床位1800张，停车位1600个。

Comprehensive Business Building of Wuhan Medical Health Center for Women and Children is located in Hong Kong Road, including Out-patient and Emergency Building, Medical-technique Building and Comprehensive Building for Internal Medicine for the first phase, Administrative Complex Building of Obstetrics & Gynecology and two-layer basement for the second phase. Its total floor area is 200,000 m^2. As designed, it can provide service for 5,000 out-patients per day, with 1,800 beds available for in-patients, and parking space for 1,600 vehicles.

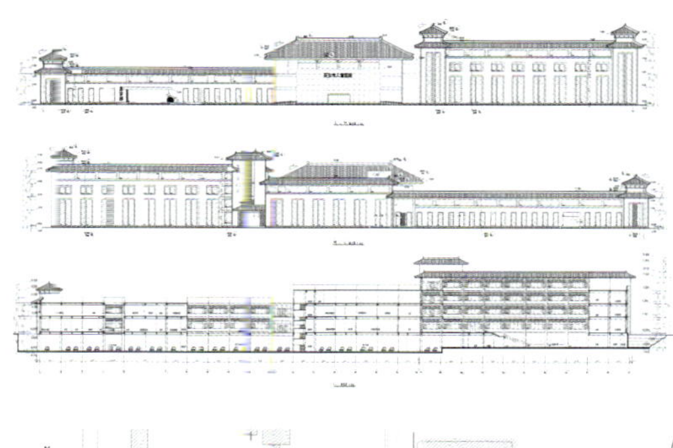

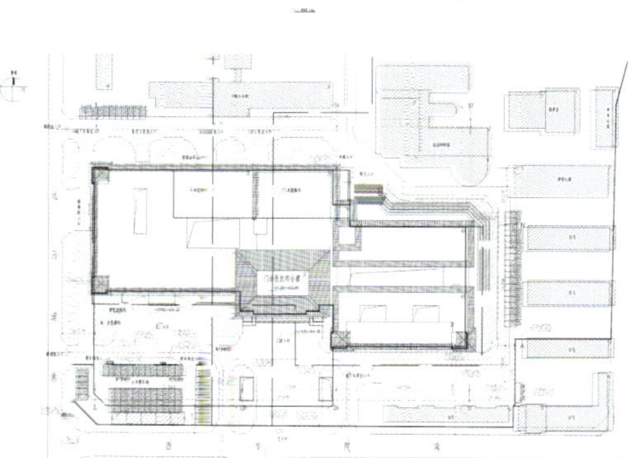

西安儿童医院门诊医技综合楼

Medical-Technique Comprehensive Building of Xi'an Children's Hospital

项目地点：陕西 西安
建筑面积：44 000 m^2

Location: Xi'an, Shaanxi
Building Area: 44,000 m^2

西安儿童医院门诊医技综合楼，位于西安市西大街。规划用地面积23 333 m^2，总建筑面积44 000 m^2。建筑总体规划布置呈"Z"字形，总平面为院落式组合布局。其急诊、医技、门诊三大功能分区相对独立，庭院和交通空间将三大功能区域有机组合。造型采用立体构成的手法将不同的体块整合为纵横有致、高低错落、富有雕塑感的空间形态。在细部处理上，选取极富古城风韵的建筑元素进行装饰。临西面城墙一侧，将建筑外墙微微倾斜，同时建筑上面提取西字城墙的部分片段以获得与环境融合，选用青灰色的外墙砖进行装饰，融入古城风貌。

Medical-Technique Comprehensive Building of Xi'an Children's Hospital, is located at the Western Avenue, Xi'an. As planned, its plot area is 23,333 m^2 and total floor area is 44,000 m^2.
The buildings take on "Z" appearance as per planning building control, and adopt a combined courtyard layout. The three functional areas of emergency, medical-technique and outpatient are organically combined by the courtyard and the traffic space. Based on figure integration, the building shape adopts three-dimensional composition technique to integrate blocks different in size and orientation into external space form looking like a sculpture. As for details, architectural elements, rich in the charm of ancient city, are selected for decoration. On the side adjacent to the Xi'an circumvallation, the external wall of the hospital is slightly inclined and a segment of the Xi'an circumvallation is incorporated so that the building is integrated into the environment. The caesious external tiles are selected to decorate the building external wall so as to go with the ancient city.

天津滨海火车站

Tianjin Coastal Railway Station

项目地点：天津
建筑面积：149 294 m²

Location: Tianjin
Building Area: 149,294 m²

天津滨海火车站，位于天津滨海新区，为滨海新区最大的地上综合交通枢纽。滨海站将环渤海城际铁路和津秦客运专线两个高速分别设置，共8站台18线。
建筑造型以"津风海韵"为建筑意象，简洁稳重、轻巧现代。立面采用横长线条，水平舒展。建筑流线型屋顶具有很强的现代感，体现火车站作为城市门户的建筑标志性。

Tianjin Coastal Railway Station, located in the Tianjin Binhai New Area, has a floor area of 149,194 m², and is the biggest comprehensive transport hub on the ground in Binhai New Area. The Binhai Stations are divided into two sections, namely, for the inter-city express railway around the Bohai Sea and the Tianjin-Qinhuangdao dedicated passenger line respectively, with 8 platforms and 18 lines in total.
The overall form takes the "Tianjin style and Binhai charm" as the architecture image, thus to form a building style which is simple and stable, light and modern. By adopting the horizontal long lines, the facade stretches horizontally. The streamlined roof with a sense of extremely dynamic and modern, demonstrates its architectural character for railway station as the city portal.

南宁东火车站

Nanning East Railway Station

项目地点：广西 南宁
建筑面积：195 945 m²

Location: Nanning, Guangxi
Building Area: 195,945 m²

南宁东火车站，位于南宁市凤岭分区北侧，道路围合范围约280 hm²，总建筑面积195 945 m²。远期规划在2030年旅客发送量达4779万人次，日均旅客到发总量15.7万人次。铁路车场中心里程轨顶设计标高为109.070 m，站台设计标高为110.320 m，站前广场设计标高为110.320 m，与站台平齐。
建筑造型结合当地气候特色，采用坡屋顶、多重檐，屋顶轮廓线向上收起汇聚，依次升起，寓意中国与东盟各国团结合作，共同发展。中间入口两侧设水平檐口立柱，表现出中国南大门的恢宏气势。屋面和柱廊的菱形纹饰来源于壮族织锦，抽象、简洁，具有民族特色和地域性。

Nanning East Railway Station, located on the north of the Fengling subregion, Nanning, occupies the area of 280 ha or so enclosed by the roads. The total floor area is 195,945 m². According to the long-term planning, by 2030, the dispatched passenger number will reach 47,700,000, the total number of passenger arriving-departing will reach 157,000 per day. The design elevation for the rail top in the center of the yard is 109.07 m, the elevation of the platform, 110.32 m, and the design elevation for the station square is 110.32 m, equal to that of the platform.
Considering local climate characteristics, the pitched roof and multi-tier eaves are adopted for the architectural form. The rooflines gather upward, rise successively, which implies the solidarity, cooperation and common development between China and ASEAN countries. The stand columns at the both sides of the middle entrance are powerful and grand, bearing mellow plump horizontal cornices, presenting the magnificence of the southern gate of China. The diamond-pattern on the roofing and colonnade are inspired by tapestry from the Zhuang nationality. The abstract and simple region expression makes the building more magnificent.

乌鲁木齐火车站
Urumchi Railway Station

项目地点：新疆 乌鲁木齐
建筑面积：100 000 m^2

Location: Urumqi, XinJiang
Building Area: 100,000 m^2

乌鲁木齐火车站，位于中国新疆维吾尔自治区乌鲁木齐市。站房总建筑面积100 000 m^2，东西面宽约220 m、南北进深约340 m、最高点标高40.000 m、檐口标高18.500 m，平面为工字形，南北面为圆弧形，呼应建筑与城市的关系。站房采用中间高，两侧低的形态，达到最合理的空间容积。

设计表达了开放包容、和谐共进的意象。结合场地地形，车站设有南北两个广场，南广场比北广场标高要高10 m。南广场与站台层平接，高架车道与城市快速路网形成多层次立体化衔接。建筑为体现"天山雪海，丝路明珠"的设计意象，以圆形为建筑形态，暗喻西域明珠。基座与屋顶形态相呼应，共同构成开放包容的建筑形象。

Urumchi Railway Station, located in the Uygur Autonomous Region, Urumqi, XinJiang Province. The total floor area of the station building is 100,000 m^2, with the width from the east to west side being about 220 m, the depth between the south and north side being about 340 m, the elevation of the highest point and the cornice being 40 m and 18.5 m respectively., the station building is in the shape of "H" viewed from the top and in the shape of arc viewed from north and south sides, which corresponds to the relationship between the architecture and city. The station building is high in the middle and low on both sides to obtain the most reasonable space volume.

贝宁共和国外交部办公楼

Office Building of Ministry of Foreign Affairs of the Republic of Benin

项目地点：贝宁共和国
建筑面积：11 503 m²

Location: The Republic of Benin
Building Area: 11,503 m²

贝宁共和国外交部办公楼位于贝宁外交部院内，总用地面积13 749 m²，总建筑面积11 503 m²，共5层。一层平面设各功能入口、新闻发布厅、签证厅、厨房、餐厅、图书室、档案室、办公室、计算机房、电话机房等；二、三、四层设办公室和会议室；五层设部长办公室、秘书长办公室和其他办公室。外交部办公楼作为国家对外的重要窗口，立面设计以庄重、新颖、实用为原则，不仅要充分体现国家政府机关稳重、庄严的气势，而且要展示国家鲜明的地域民族特色。因此根据当地美丽的热带风情，立面上设置水平遮阳板，舒展流畅；竖线条的连续柱廊托起出挑深远的大屋顶，轻盈飘逸；弧形坡道和大台阶气势非凡。

The office building is located in the courtyard of Ministry of Foreign Affairs, the Republic of Benin. It is a 5-floor building with total Site area 13,749 m² and total floor area 11,503 m². On the 1st floor are entrances to various functions, news hall, visa hall, kitchen, restaurant, library, archives, office, computer room and telephone room etc. On the 2nd to 4th floor are offices and conference rooms. On the 5th floor are minister's office, secretary general's office and other offices. As an important window of foreign affairs, the building has a solemn, novel and practical façade, not only representing its administrative authority, but also showing its strong national characteristics. Therefore, based on local beautiful tropic conditions, the façade is fixed with horizontal sun visor, extensive and smooth. Vertical continuous colonnade supports the profound roof, soft and swift. The curved ramps and large steps are magnificent.

莫桑比克共和国国家体育场

National Stadium of the Republic of Mozambique

项目地点：莫桑比克共和国
建筑面积：41 987 m²

Location: The Republic of Mozambique
Building Area: 41,987 m²

莫桑比克共和国国家体育场位于莫桑比克国家1号公路东侧，用地面积267 900 m²，总建筑面积41 987 m²，采用了现代先进的体育建筑结构类型。设计融合了当地人文、气候和自然环境特征，功能分区明确，布局紧凑，交通组织有序。立面造型庄重、简洁、明快，并以现代手法演绎当地传统民族文化。可满足国际足联和国际田联比赛要求，成为莫桑比克重要标志性建筑。

Stadium of the Republic of Mozambique located at the east side of the national No.1 highway. The plot area is 267,900m² and the total floor area is 41,987 m². The architecture adopts the modern advanced sports building structure. Incorporating the characteristics of cultural environment, climate and natural environment in this country, the architecture features clear functional zoning, compact layout and orderly traffic organization. The facade is solemn, simple and pleasant, incorporating the local traditional culture elements in a fashionable way. Satisfying the contest requirements of FIFA and IAAF the National Stadium is becoming a landmark of the Republic of Mozambique.

武汉积玉桥万达广场一期
Wanda Plaza Phase I, Jade Bridge, Wuhan

项目地点：湖北 武汉
建筑面积：139 101.98 m²

Location: Wuhan, Hubei
Building Area: 139,101.98 m²

武汉积玉桥万达广场位于武汉市武昌区积玉桥临江大道，武昌区中心地段，交通便利。A区由六星级酒店和5A级写字楼组成。在总体设计布局中，为缓解城市交通的压力，营造良好的城市空间，建筑后退临江大道35 m。两个汽车出入口分别设置在临江大道和前进路，较好地组织了交通，建筑内部设计考究，于精细处体现六星级标准。D区为超高层豪宅。结合用地基本特征，将长江的江景及绿地景观尽可能多地引入到小区内部来，使得更多的住户受益。品质居住区不等于一味的华丽，而应体现精益求精、尽善尽美的设计理念。底层商业部分，运用轴线对称布局，采用点，线，面成景的方式，根据所处位置及不同功能分割空间，营造一个多功能的、舒适的、令人愉悦的娱乐购物场所。

Wanda Plaza is located at Riverside Avenue, Jade Bridge, Wuchang District, Wuhan City. The location is in the middle of this district, with convenient traffic conditions. Zone A includes 6-star hotels and 5A-rating office buildings. The plaza is designed 35 m back from Riverside Avenue, to mitigate traffic pressure and building a good urban space. It has two automobile entrances separately at Riverside Avenue and Advance Road, organizing good traffic. The interior design is considerate, representing 6-star standard in details. Zone D includes overhigh-rise luxury houses, and river sights and greening views are introduced into this community as more as possible, to benefit more inhabitants. Quality residential community does not mean simple luxury, it must be perfect design. The ground floor commercial zone is clued by landscaping scenarios, symmetrically laid out by axis, and landscaped by point, line and plane, by reference to location and functional division, to create a multi-functional, comfortable and pleasant shopping and entertaining space.

武汉吉庆街民俗文化商业街
Wuhan Jiqing Folk Culture Commercial Street

项目地点：湖北 武汉
建筑面积：59 581 m²

Location: Wuhan, Hubei
Building Area: 59,581 m²

武汉吉庆街民俗文化商业街，位于武汉市汉口大智街区域。用地面积8.39 hm²。
在总体布局中引入风环境设计，利用武汉夏季东南向和南向的主导风向，留出“穿堂风”风道，延续老汉口建筑布局，遵循利用江河小环境和风向缓解夏季炎热气候的传统。建筑借鉴清末民初汉派商业街建筑风格，沿用各种古老的造型元素，如：红砖红瓦、骑楼、挑廊、圆窗、天井、露台等再现老汉口民俗文化商街风貌。

Wuhan Jiqing Folk Culture Commercial Street is located in Dazhi Street area, Hankou, Wuhan. With a plot area of 8.39 hectare.
Into overall layout wind environment design is introduced, which utilizes the prevailing wind direction of southeast and south in summer of Wuhan to create “draught” passage. In this way, a traditional practice in layout of buildings in old Hankou, namely, taking advantage of river microenvironment and wind directions to reduce the high temperature of summer, is preserved. As for the architectural style, the building style of Han Commercial Street in late Qing and early Republic is taken to reproduce the look and atmosphere of old Hankou Commercial Street. Moreover, various old elements are adopted, such as red bricks and tiles, arcade, overhanging corridor, round window, courtyard, and veranda, etc..

武汉中央文化旅游区（合作设计）
Wuhan Central Culture Tourist Area (Cooperative Design)

项目地点：湖北 武汉
建筑面积：300 000 m²

Location: Wuhan, Hubei
Building Area: 300,000 m²

武汉中央文化旅游区一期（J–2地块）为楚河汉街的一部分——水街，主要商业建筑为2～3层商业街，另外靠近东一路为一栋5层的大酒楼。J–2地块与K–3地块地下室连通，汉街部分为地下一层。
在建筑设计中注重城市空间的开敞，通过各种不同的水景与商业业态分布的结合方式，形成亲切趣味的具有水文化特色的商业模式。强调临水界面的景观渗透，体量变化丰富，并兼顾了街道界面的连续性。
建筑风格以中西合璧的建筑为基调，选取连续的骑楼、变化丰富的山墙面等传统建筑元素，打造传统建筑风格和尺度的现代商业街。

Wuhan Central Culture Tourist Area in the first phase (Plot J-2) is a part of Han Street, namely, Water Street. This street will be lined with 2 to 3 story shops; besides, a five-story restaurant will be located close to Dongyi Road. J-2 Han Street has one story underground which is connected to Plot K-3.
The architectural design emphasizes on the openness of urban space, incorporates various kinds of waterscape into the area where a variety of commercial buildings are distributed, forming a business mode characterized by cordial and interesting water culture. The design scheme stresses on landscape infusion into waterfront space; furthermore, the buildings vary greatly in form and volume; last but not least, the continuity of street interface is also guaranteed. The architectural style mixes Chinese and western elements. In particular, some traditional building elements like continuous arcades and gable wall with rich variations are adopted to build a modern commercial street with traditional building style and size.

万科 · 四季花城西半岛
Vanke · Wonderland West Peninsula

项目地点：湖北 武汉
建筑面积：262 336 m²

Location: Wuhan, Hubei
Building Area: 262,336 m²

该项目位于武汉市东西湖区，南临金银湖，西靠环湖路。用地面积约为20.18 hm²，总建筑面积为262 336 m²，建筑高度4~31 m，包括多层和小高层居住建筑及配套商业、会所、学校等用房。规划对绿化系统进行了“中心—组团—宅前”多层次的系统设计，使区绿地植物体系丰富多样。将金银湖主体水域延伸到地块中部，形成左右双半岛地形，建筑镶嵌在自然环境之中。设计采用新技术提高住宅的节能性。单体设计以白色基调为主，局部深色体块点缀，主要采用地中海元素，精心构筑丰富的外立面，创造了建筑与环境和谐之美。

The project is located in East-West Lake District of Wuhan City. It is south to Gold-Silver Lake, and west to Lakeside Ring Road. It has land area 201,800 m², total floor area 262,336 m², and it is built 4-31m high, consists of multi-floor or high-rise residential buildings and auxiliary commercial, club, school and other buildings. It has made systematic design on its greening system at all levels i.e. "center-group-house", to diversify greening system in the community. The water body of Gold-Silver Lake is extended to the center of this land lot, making a double-peninsula landform. The building is integrated with the Nature. The design adopts new technologies to increase energy-saving. Single design is keynoted by white color, and some details are in dark colors, mostly adopting Mediterranean elements. The outer façade is enriched elaborately to create the beautiful harmony between architecture and environment.

武汉万科金色家园B8东地块商住楼
Vanke Golden Home (Plot B8 East)

项目地点：湖北 武汉
建筑面积：58 242 m²

Location: Wuhan, Hubei
Building Area: 58,242 m²

武汉万科金色家园B8东地块商住楼为金色家园三期，位于汉口京汉大道和前进四路交汇处。主楼是高层内廊式住宅公寓，在裙楼和主楼之间，设立一个全架空的空中会所，富于田园气息。主楼中的平面布局的L形内廊式和公寓式主要由三个独立单元拼接而成。每层楼23户，户型面积从42~120 m²不等。裙楼由大小形状不一的小型商铺组成，造型迥然而异。立面造型上结合平面设计，利用出挑阳台、空调板，上下交错分布，创造出一种具有韵律节奏感的建筑语言。余辉之下，鳞次栉比，创造一种韵律建筑，与金色家园前两期住宅楼相映成辉。在色彩上，主色调采用浅黄色与深灰色，整体形象对比强烈，简约大方。

Vanke Golden Home (Plot B8 East) is a commercial & residential complex located at the crossroad of Jinghan Road and Qianjin 4th Road in Hankou District. The main building is a high-rise inner corridor style residential apartment and there is an elevated club in the air between the main building and skirt building, full of rural sense. In the main building, the L-shaped inner corridor style apartment comprises of three independent units. There are 23 units on every floor, and each unit has a floor area ranging 42-120 m². The skirt building comprises of small shops in different sizes and forms, with distinct shapes. The façade combines with flat surface, to create a building language with strong rhythm, by overlaying distribution of overhanging balcony and air-conditioning board. Under the sunlight, it creates an orderly rhythmic building, in comparison with previous two phases of residential buildings in Golden Home project. In color, light yellow and dark grey are main colors. The integral image has strong contrast, concise and generous.

中信金山湾
CITIC Jinshan Bay

项目地点：广东 佛山
建筑面积：445 356 m²

Location: Foshan, Guangdong
Building Area: 445,356 m²

中信金山湾项目选址于南海区里水镇和里丹江北路北侧地块。规划为商住区，北面约200 m处为海畔花园。项目规划总用地面积为119 667.3 m²，建筑面积为445 356.2 m²，共建51 29幢高层建筑，其中部分建筑首层、二层设计为商业、社区用房、物管用房及小区会所。该地块享有金溪河景观优势，位于美景大桥旁，具有丰富的山水资源，将被开发成为高端且富有特色的生活区。

CITIC Jinshan Bay project is located in North Side Plot, Heshunyanjiangbei Road, Lishui Town, Nanhai District. It is planned as commercial and residential area; about 200 meters to the north is Haipan Garden. The total planning plot area of the project is 119,667.3 m², the floor area is 445,356 m² with a total of 29 high-rise buildings constructed, some of which house spaces for commercial, community, property management, residential clubs on the first floor and second floor. The plot possesses the advantages of Jinxi River landscape next to the sceric bridge; and therefore with the rich resources of the landscape, it will develop into a unique high-level living area.

扫描查看更多信息

中国轻工业武汉设计工程有限责任公司

China National Light Industry Wuhan Design Engineering Co.,Ltd.

中国轻工业武汉设计工程有限责任公司（简称“中轻武设”）隶属中央企业中轻集团，为上市公司“中国海诚”的控股子公司，公司成立于1958年，是历史悠久、实力雄厚的知名大型工程、设计机构。

中轻武设拥有国家颁布的民用建筑、工程咨询、轻工、化工、医疗、环保设计、工程监理等七项甲级资质，同时还具备市政、规划、电子、机械等行业工程设计与咨询的乙级资质，已形成工程设计、咨询、监理和工程总承包四大主业。公司项目遍及全国并在海外10多个国家取得工程业绩。公司是国内最早通过ISO9001质量体系认证的单位之一，现有各类工程技术人员400余人，其中各类注册工程师160余人、教授级高工和高级工程师职称者共110余人、国家或省部级知名专家10余人。

公司奉行“尊重人、理解人、成就人”的企业文化，将企业打造成为一个温馨和睦的“家庭”、一所人才辈出的“学校”、一座美丽富饶的“矿场”。

地址：武汉市武昌区首义路176号
电话：+86-27-88075955
传真：+86-27-88041709
邮箱：rlzy@qgsj.com
网址：www.qgsj.com

Add: Shouyi Road 176, Wuchang, Wuhan, Bubei
Tel: +86-27-88075955
Fax: +86-27-88041709
Email: rlzy@qgsj.com
Web: www.qgsj.com

武汉巴登城温泉度假岛

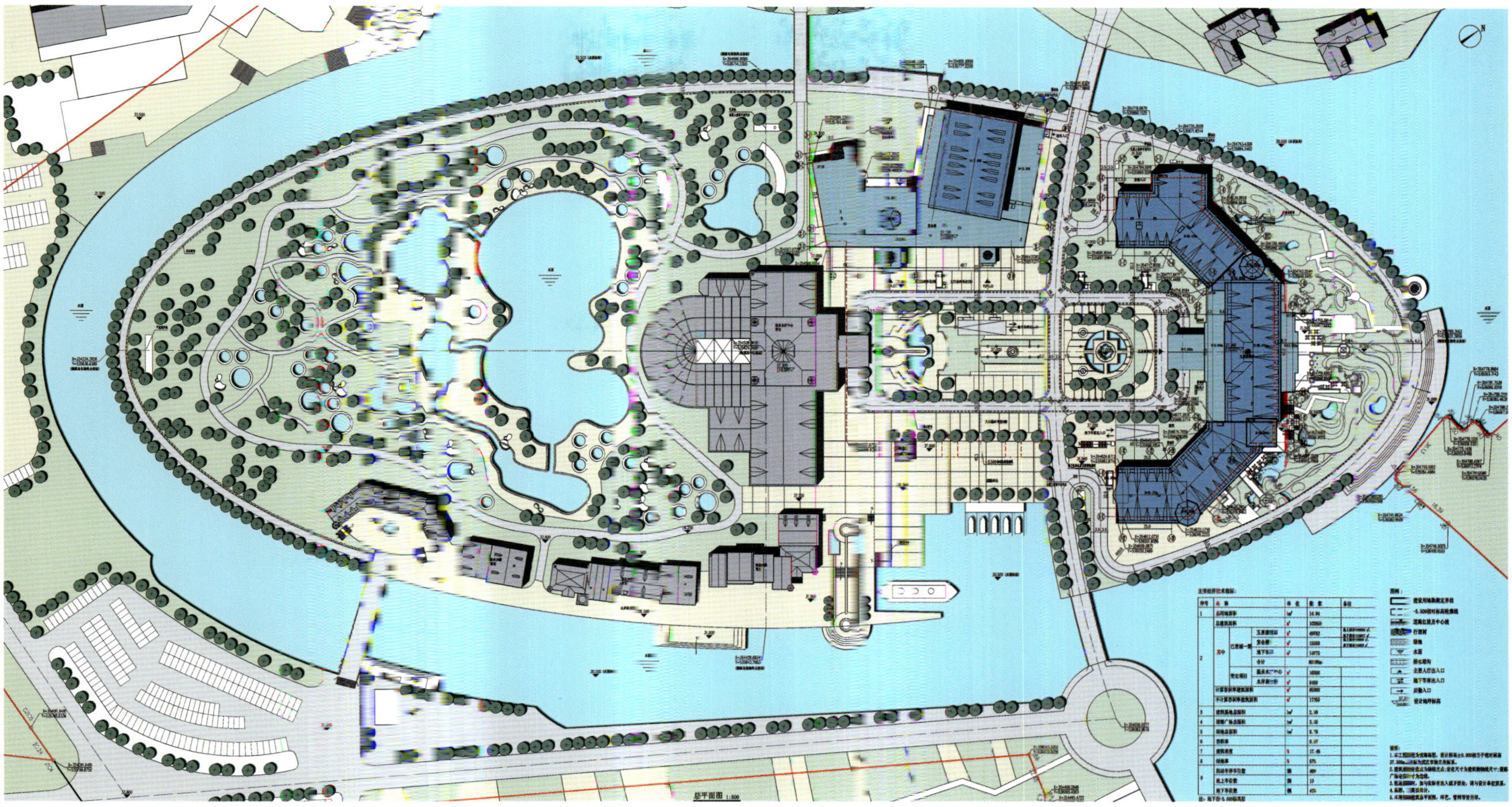

武汉巴登城温泉度假岛

Badeng Town Hot Spring Resort Island in Wuhan

设 计 师：王芳、韩飞燕、申俊芳、王[illegible]
项目地点：湖北 武汉
用地面积：14.84 hm^2
建筑面积：地上11万 m^2，地下2万 m^2
合作单位：北京清水爱派建筑设计有限公司

Designer: Fang Wang, Feiyan Han, Junfang Shen, Qinghui Wang, Rufeng Cheng
Location: Wuhan, Hubei
Site Area: 14.84 ha
Building Area: 110,000 m^2 aboveground, and 20,000 m^2 underground
Partners: Beijing Qingshui Aipai Architectural Design Co., Ltd.

巴登城项目包括五星级酒店、宴会楼、[illegible]、水岸商业街、配套公寓及地下工程等。项目总体定[illegible]地区时尚高端温泉休闲生活方式为核心，打造中国一流[illegible]、开放的、人文的、都市娱乐的，集温泉度假、人文居[illegible]为一体的中部地标型生态旅游开发项目。
巴登城的建筑风格呼应巴登城主题，引入[illegible]式风格与德国民居式建筑风格，强调高而窄，长而狭的[illegible]高拱、高耸的屋顶，高低起伏的老虎窗和又高又长的[illegible]，充分强调了结构与装饰的一致性、内部空间的几何[illegible]，形成独特的“光的空间”，成功打造了梦幻般异域风[illegible]。

The project includes five-star hotel, banquet building, hot spring SPA club, waterfront business street, matched with apartment and underground works. Overall project location is in Baden area, with high grade fashion spa leisure lifestyle as the core and builds up China's first class, ecological, open, human, city entertainment l, collecting hot spring resort, residential, ecological experience as one of the central landmark eco-tourism development project in Central China.The architecture style relates to the theme of Baden Town, introducing Gothic style and German residential building style. It emphasizes the high yet narrow, long yet confined proportion, such as continuous arc, high-rise rooftop, waving commant window, and broad glass window, which all emphasize the concordance of structure and decoration, the geometric integrity of interior spaces, to form a unique “space of lights”, and create a dreamlike exotic building cluster

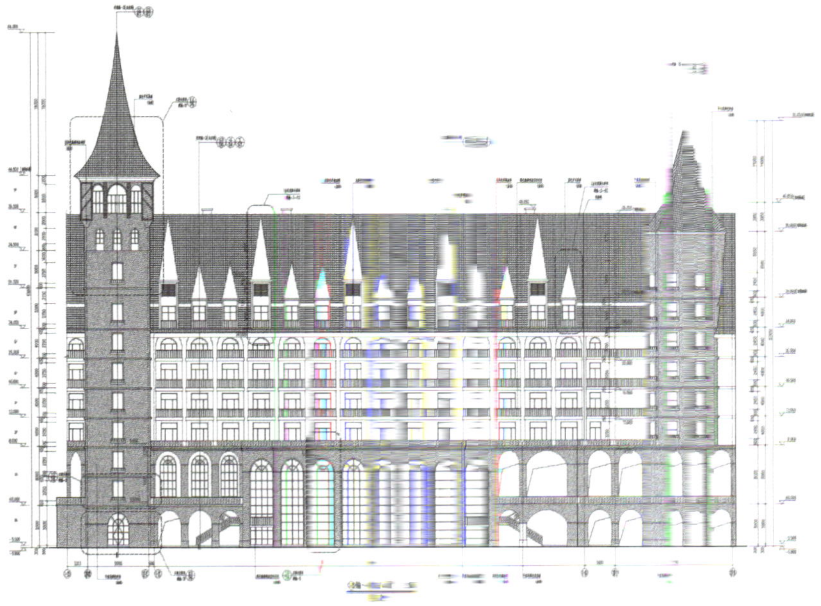

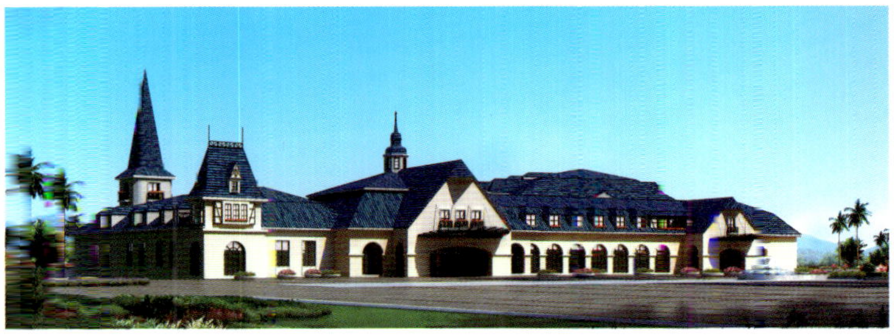

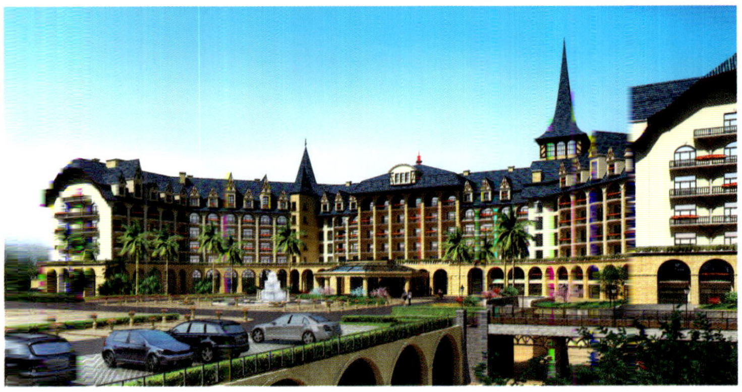

国家检察官学院湖北分院

National Prosecutors College Hubei Branch

设 计 师：第三设计所
项目地点：湖北 武汉
用地面积：221 698 m²
建筑面积：57 400 m²
容 积 率：0.26
绿 地 率：69%

Designer: No. 3 Design Institute
Location: Wuhan, Hubei
Site Area: 221,698 m²
Building Area: 57,400 m²
Plot Ratio: 0.26
Green Ratio: 69%

本项目位于武汉市江夏区梁子湖大道与中洲路交汇处，宽阔的汤逊湖水域以南。设计中秉承将规划、建筑、景观三者合为一体的整体思路，在规划设计之初即注重环境生态和培训基地功能性的研究，从整体规划入手，协调各功能区域建筑体量的平衡关系，营造出具有"山、林、湖、溪"特色的学院环境。

The project is located at the intersection of Liangzi Lake Avenue and Zhongzhou Road, on the south side of broad Tangxun Lake in Jiangxia District, Wuhan City. The design integrates planning, architecture and landscape three into a whole, paying attention on the research on environmental ecology and training base performance at the beginning of the design; starting from the overall planning, assorting with the balances relations in the building volumes in different function areas, to create a college environment featured by "Mountains, Woods, Lakes and Brooks".

文昌波溪丽亚湾项目

Bosnia Bay, Wenchang

设 计 师：金山、祝文波、喻晟、王芳、季如意、申俊芳、江执兵、程如风
项目地点：海南 文昌
用地面积：76 760 m²
建筑面积：115 160 m²

Designer: Shan Jin, Wenbo Zhu, Sheng Yu, Fang Wang, Ruyi Ji, Junfang Shen, Zhibing Jiang, Rufeng Cheng
Location: Wenchang, Hainan
Site Area: 76,760 m²
Building Area: 115,160 m²

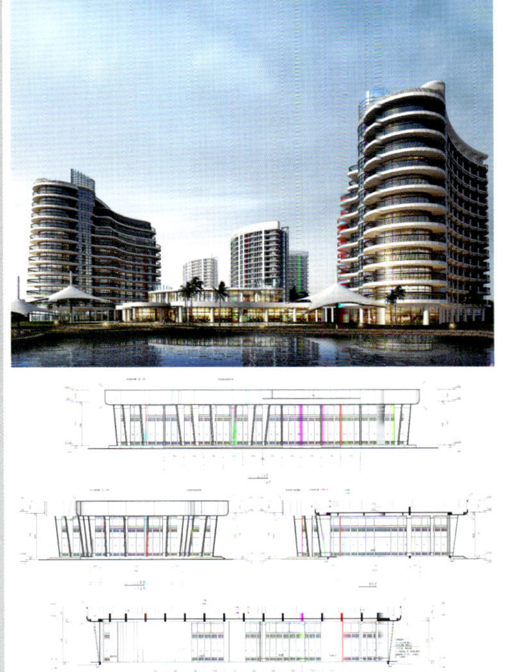

该项目为旅游地产项目，整体划分为星级酒店区和产权酒店区，包括星级酒店客房、普通客房、高层客房及独立式的豪华客房等多种物业形态。建筑采用分散式布局，以此化解狭长地块所带来的不利因素。设计中利用建筑间的间距和建筑层数的变化，保证每栋楼最大限度地观海。建筑采用现代风格，通过规整的柱网造就规整的客房平面，通过自由曲线的阳台来体现丰富的立面造型，从而塑造简洁、纯粹的现代滨海建筑。

The project is a tourism property item. It selects star-level hotel guest room, 11-floor common guest room, high-rise guest room, and independent luxury guest room and many property forms. The project divides into star- level hotel zone and proprietary hotel zone. The architecture adopts with scattered layout, solving all negative factors of the narrow strip land lot. The design makes use of inter-building distance and changing number of floors so as to ensure every building to enjoy the optimal sea views. The architecture uses modern style, where regular column network creates regular guest room layout, and the freely curved balcony expresses rich content facade shape in order to create the concise, pure, and modern seaside architecture.

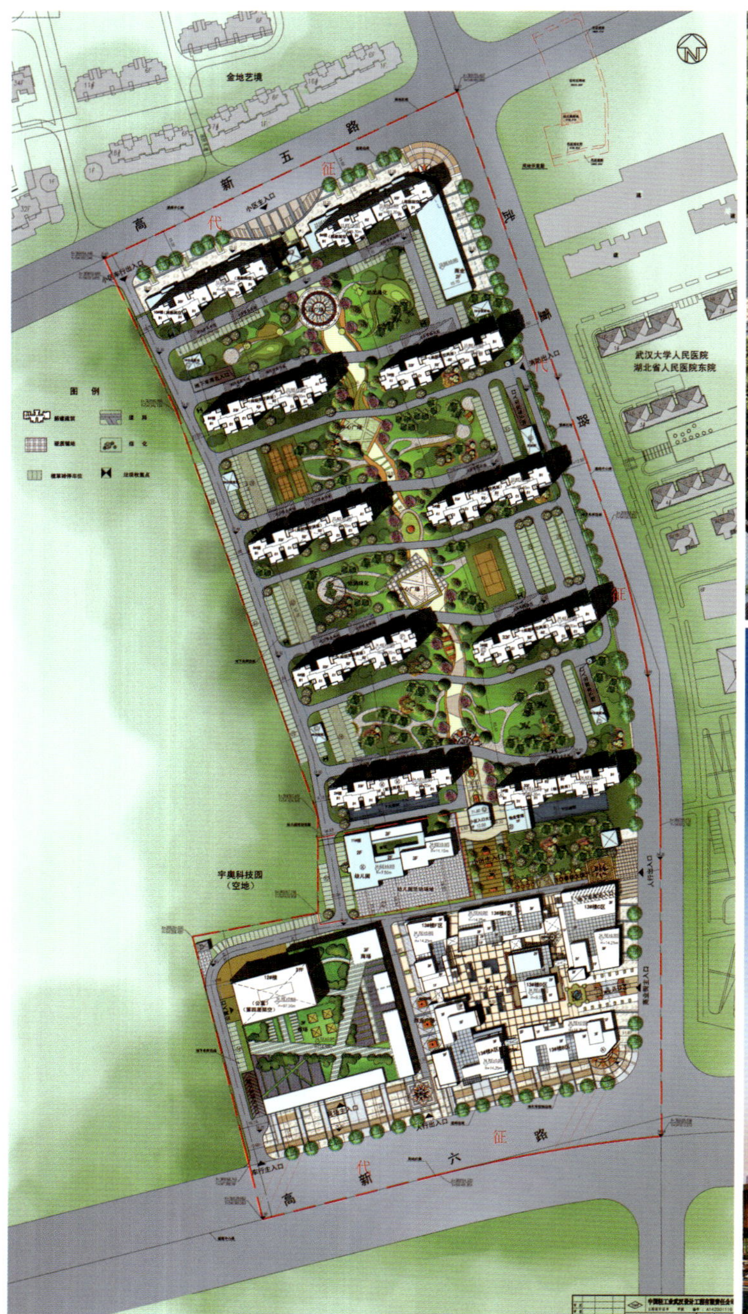

嘉欣园
Jiaxin Garden

设 计 师：金山、罗哲新、吴笃文、郭嘉琪、张忠兵、申莎、孙礼祥、张震云
项目地点：湖北 武汉
用地面积：118 297.37 m²
建筑面积：380 352.98 m²
容 积 率：3.10
绿 化 率：35.41%

Designer: Shan Jin, Zhexin Luo, Duwen Wu, Jiaqi Guo, Zhongbing Zhang, Sha Shen, Lixiang Sun, Zhenyun Zhang
Location: Wuhan, Hubei
Site Area: 118,297.37 m²
Building Area: 380,352.98 m²
Plot Ratio: 3.10
Green Ratio: 35.41%

该项目立足于"场所"的设计，在分析人的行为心理的基础上，通过空间的串联、流动、渗透，形成形态丰富的建筑群落。
根据基地形状将建筑物自然划分为南北两区，北区为住宅区，由三十余米的主景观轴串联各宅间绿化空间，住宅间距七十余米，空间尺度开阔；南区为商业区，由六栋低层独体商业围合成形态各异、互相渗透、自由流动的活动空间，既作为片区的商业功能节点，也作为住宅区居民的休闲、娱乐场所。整体建筑形态层次分明，秩序井然。

The project is established on the design of "Location", forming a variable building cluster through series connection, circulation and penetration of spaces based on the analysis to human behavioral psychology. The project is divided into south area and north area according to the landform. The north district is a residential area, greening spaces between buildings are connected by a main landscape axis over 30m long, and the distance between residential buildings is more than 70m, so the space is open. The south district is a commercial area, and six low-floor independent commercial buildings form active spaces in different shapes, integrating with each other and circulating freely; it is not only the commercial function joint of the area, but also a leisure and amusement facility for the residents in the residential area. The overall architecture is well arranged in order.

航天双城
Aerospace Twins City

设计师：第三设计所
项目地点：湖北 武汉
用地面积：65 047.52 m²
建筑面积：236 773 m²
容积率：3.64
绿地率：42.60%
合作单位：美国威尔考特建筑事务所；LEAP（杭州）

Designer: No.3 Design Institute
Location: Wuhan, Hubei
Site Area: 65,047.52 m²
Building Area: 236,773 m²
Plot Ratio: 3.64
Green Ratio: 42.60%
Partners: WTEOT; LEAP (Hangzhou)

项目位于武汉解放大道与二七路交汇处，建筑含住宅、商业及配套服务等，地下2层，地上26～32层。规划最大限度地利用地形条件，以“人本、自然、创新、融合”为原则，并从城市设计出发，建筑布置以点式、板式相结合，形成丰富的天际线，塑造出多变的社区景观。项目改变了区域形象，已成为区域地标项目。

The project is located in the junction of Liberation Avenue and Two Seven Road, Wuhan City. It includes residential buildings, commercial buildings and matched services, with 2 floors underground and 26-32 floors aboveground. The plan maximizes the use of landforms, follows the principle of "human-basis, nature, innovation and integration", and starts from urban design, in a combination of point type and plate type buildings to form a variable skyline, and shape a variable community landscape. This project has changed regional image, and has become a landmark in this region.

航天星都规划总平面图

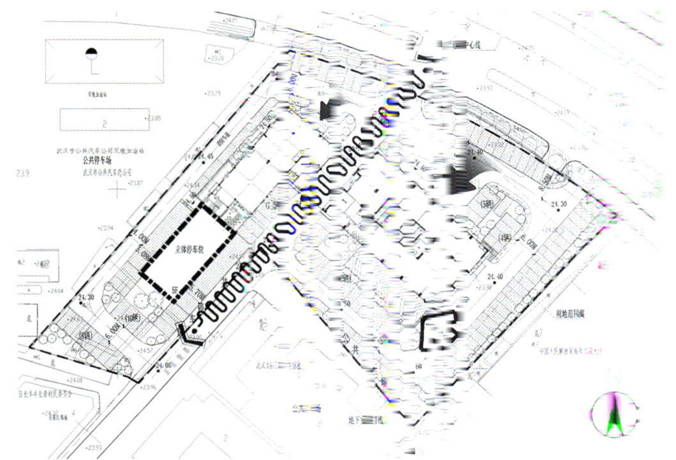

湘商大厦
Xiangshang Building

设计师：余磊、吴笃文、曾洁、陈谭鸿、李波、孙礼祥、张震云
项目地点：湖北 武汉
用地面积：7 994.83 m²
建筑面积：33 637.08 m²
绿化率：35.5%

Designer: Lei Yu, Duwen Wu, Jie Zeng, Tanhong Chen, Bo Li, Lixiang Sun, Zhenyun Zhang
Location: Wuhan, Hubei
Site Area: 7,994.83 m²
Building Area: 33,637.08 m²
Green Ratio: 35.5%

新建办公楼采用一字形布局，主立面朝向建设大道，呼应城市道路；南北朝向，兼顾采光和通风。临东侧塔楼为24层，西侧塔楼为16层，使天际线呈梯状上升。在基地西侧小块用地区域内设置立体停车库，最大地利用了土地资源，又解决了停车位紧缺的问题。
立面富有时代感，色彩淡雅、线条挺拔、简洁有力，突出地表现现代建筑清新、明快、时尚的特点。在形体上利用屋顶女儿墙的高度差别和布局走势，使建筑天际线高低错落、富于变化，有韵律感和层次感，其造型新颖、色彩靓丽，成为该区域的一道璀璨风景。

The new office building is in a linear layout, with the main façade facing toward Construction Avenue, corresponding to urban roads. The seat is facing south, with good skylight and ventilation. The east side tower building has 24 floors, and the west side 16 floors so as to increase the skyline in stair form. In the base on the west side, it is designed with a stereo garage with the regions of small land set in order to maximize to use land efficiency, and resolve parking difficulty.
The façade is epochal, in light and elegant colors, straight lines, concise and powerful, highlighting the clear, bright and fashionable modern architecture. The shape makes use of rooftop parapet height difference and layout trend, to make the building skyline variable in order, rhythmic and multilevel. It has novel shape, gorgeous colors, and become a charming landscape in the region.

武汉万科高尔夫城市花园四期，五期
Phase IV & V of Vanke Golf City Garden, Wuhan

设 计 师：第三设计所
项目地点：湖北 武汉
用地面积：135 620 m²
建筑面积：245 000 m²
容 积 率：1.81
绿 地 率：36.5%
合作单位：雅尚（加拿大AEL）

Designer: No. 3 Design Institute
Location: Wuhan, Hubei
Site Area: 135,620 m²
Building Area: 245,000 m²
Plot Ratio: 1.81
Green Ratio: 36.5%
Partners: AEL

项目位于武汉市金银湖畔，为武汉万科临湖高端住宅区，建筑含联排别墅、高层住宅、商业、小学等，地下1层，地上3～33层。本工程在空间结构上依托景观体系，力求达到对湖景最大限度的开放共享。

The project is located on Gold & Silver Lake shore in Wuhan City. It is a Vanke lakeside hi-end residential community, including platoon villas, hi-rise residential, commercial buildings, primary school, with 1 floor underground and 3-33 floors aboveground. The project relies on landscaping system in space structure, trying to open and share the lake views to the largest extent.

武汉经济技术开发区国际教育园区
International Education Park in Wuhan Economy and Technology Development Zone

设 计 师：第三设计所
项目地点：湖北 武汉
用地面积：29 687 m²
建筑面积：29 685 m²
容 积 率：1.0
绿 地 率：34%
合作单位：Perkins Eastman

Designer: No. 3 Design Institute
Location: Wuhan, Hubei
Site Area: 29,687 m²
Building Area: 29,685 m²
Plot Ratio: 1.0
Green Ratio: 34%
Partners: Perkins Eastman

项目位于武汉经济技术开发区，含一所为600名学生提供美国课程的英语学校和一所300人的法语学校，包括3座英、法语教学楼、1座剧院、1座体育馆等建筑，总建筑面积3万 m²。每一栋建筑在功能和形态上都各有其特征，但通过天然的材质色彩组合、现代建筑语汇、以及景观绿地系统等要素在校园内构建出整体、和谐的空间氛围。建成后将成为武汉两所纯外籍国际学校，为外籍专家解决子女教育的后顾之忧，加快高新技术产业园与国际接轨进程。

The project is located in Wuhan Economy and Technology Development Zone, including an English language school providing American Curriculum for 600 students and a French language school for 300 students; the building cluster includes three teaching buildings for English language and French language, one theater, one gymnasium, covering a total building area of 30,000m². Every building has characteristic function and outer appearance, yet an integral and harmonious space atmosphere is built on the campus through such space elements as natural material & color combination, modern architectural language and landscape green system. The project completion will provide two absolute foreign international schools in Wuhan, solving the school education problem for foreign experts' children and speeding up the integration and globalization of high & new technology industry.

曾益海工作室

Zeng YiHai Studio

中铝国际长沙有色冶金设计研究院有限公司
Changsha Engineering and Research Institute Ltd. of Nonferrous Metallurgy of CHINALCO

扫描查看更多信息

Tel: +86-731-84397140
Fax: +86-731-84458515
E-mail: zengyiha@yahoo.cn
Web: www.cinf.com.cn

湖南股权投资产业园概念性规划方案设计

Concept Plan of Hu'nan Equity Investment Industry Park

项目地点：湖南 长沙
用地面积：39 496.43 m²
建筑面积：284 123 m²
建筑高度：208 m

Location: Changsha, Hu'nan
Site Area: 39,496.43 m²
Building Area: 284,123 m²
Building Height: 208 m

双蜂县第一中学北校区建设项目工程设计

Engineering Design of North Campus of Shuangfeng County No. 1 Middle School

项目地点：湖南 娄底
用地面积：149 323.93 m²
建筑面积：87 363 m²

Location: Loudi, Hu'nan
Site Area: 149,323.93 m²
Building Area: 87,363 m²

中铝南方总部设计方案

Design Proposal of CHALCO South Headquarters

项目地点：上海
用地面积：9230 m²
建筑面积：67 125 m²
建筑高度：70 m

Location: Shanghai
Site Area: 9230 m²
Building Area: 67 125 m²
Building Height: 70 m

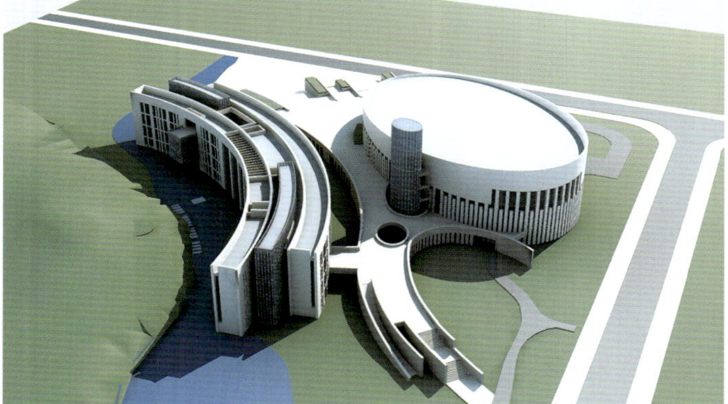

湖南工学院学生活动中心
Students Center of Hu'nan Institute of Technology

项目地点：湖南 衡阳
用地面积：7465 m²
建筑面积：1C 317 m²

Location: Hengyang, Hu'nan
Site Area: 7,465 m²
Building Area: 10,317 m²

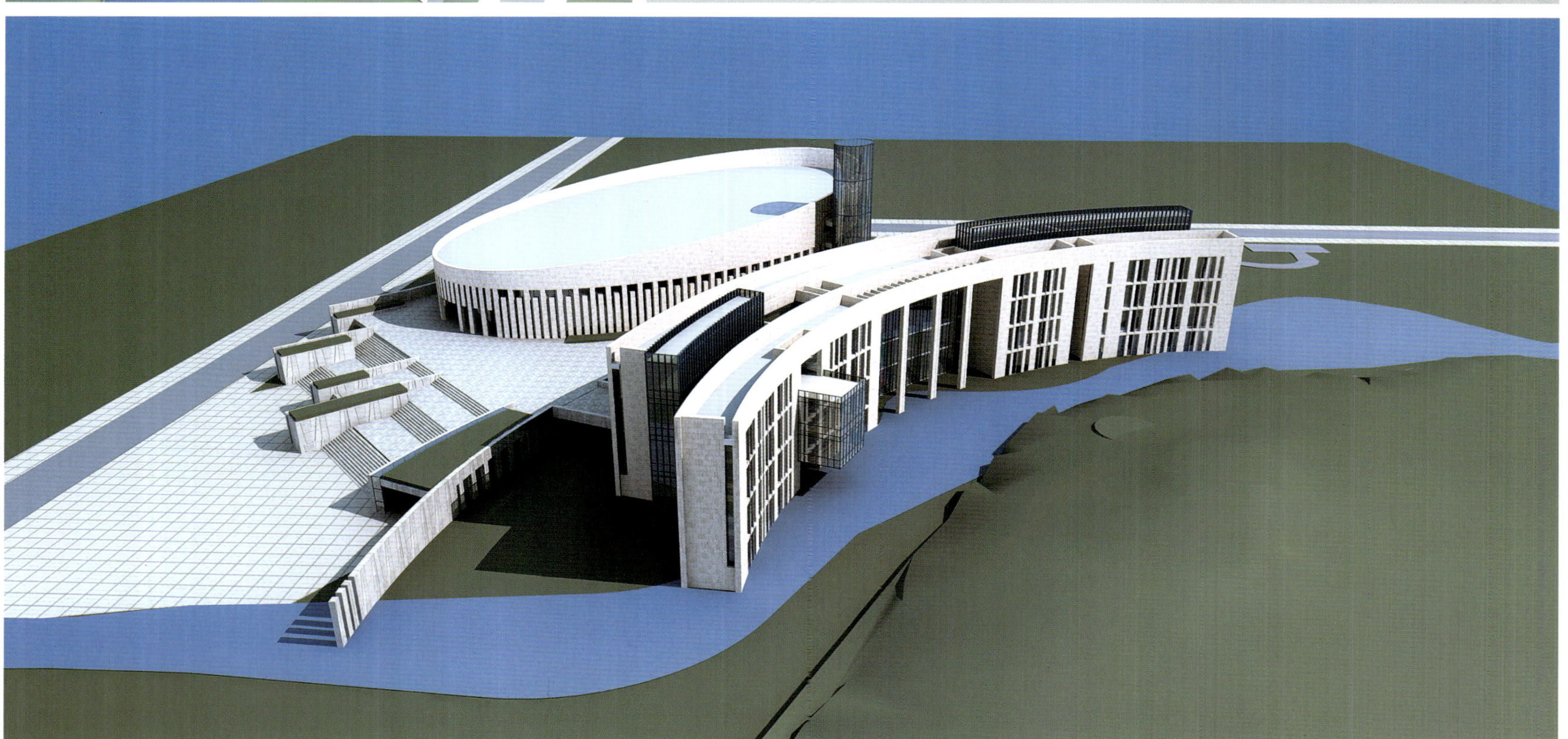

清溪川9号项目规划建筑设计

Architectural Design of Cheonggyecheon No. 9

项目地点	湖南 长沙	Location: Changsha, Hu'nan
用地面积	59 839 m²	Site Area: 59,839 m²
建筑面积	251 283 m²	Building Area: 251,288 m²

扫描查看更多信息

中国水电顾问集团中南勘测设计研究院
HydroChina Zhongnan Engineering Corporation

中南勘测设计研究院成立于 1949 年 5 月。在 60 多年的风雨历程中，中南院一直坚持走以水电、风电、建筑、交通、市政勘测设计为主业的发展之路，现已成为我国大型的骨干设计企业，是国资委下属的中国水电工程顾问集团公司的全资子公司。

中南院专业门类齐全、技术力量雄厚、工程经验丰富、设备配置精良。2008 年 3 月，获得建设部颁布的工程设计综合甲级资质，此外还拥有包括工程勘察综合、工程总承包、工程监理、工程咨询等在内的甲级资质证书 20 余项。中南院通过了 2000 版 ISO9001 标准质量体系认证和世界银行DACON 信息中心的资格认证，获得了质量管理体系、职业健康安全管理体系和环境管理体系认证。

企业核心战略：拓新致远。

企业的使命：心系工程，追求卓越，提升价值

企业的愿景：致力建设成为以技术和工程管理为核心竞争力的国际工程公司

企业的核心价值观：秉责、创新、卓越

地址：湖南省长沙市雨花区圭塘香樟路 9 号博远大厦 13 楼
电话：+86-731-85075241
传真：+86-731-85073221
邮箱：msdi2008@hotmail.com
网址：www.msdi.cn

Add: 13th Floor, Boyuan Building, Guitang Xiangzhang Road No.9, Yuhua District, Changsha City, Hu'nan Province
Tel: +86-731-85075241
Fax: +86-731-85073221
E-mail: msdi2008@hotmail.com
Web: www.msdi.cn

中南博远大厦
Zhongnan Boyuan Building

项目地点：湖南 长沙
用地面积：10 443.37 m²
建筑面积：61 256.69 m²

Location: Changsha, Hu'nan
Site Area: 10,443.37 m²
Building Area: 61,256.69 m²

中南博远大厦是中南勘测设计研究院新建的办公综合大楼，位于长沙市香樟路与沙湾路交汇的西南角。项目用地面积为 10 443.37 m²。整个工程由 A 座、B 座两栋高层建筑组成，工程总建筑面积为 61 256.69 m²，总高为 99.3 m。曾获得湖南省优秀勘测设计二等奖。

大厦是一栋集办公、住宅、商业、酒店为一体的综合性大楼。大厦建成后，不但极大地改善了中南院企业的办公环境和职工的居住条件，并以其独特、稳重、干练的建筑造型成为该地区的地标建筑。

The building is a new general office building built by HydroChina Zhongnan Engineering Corporation. It is located on the southwest corner of the crossroad between Xiangzhang Road and Shawan Road in Changsha City. The project, with land area 10,443.37 m², comprises of two high-rise buildings A & B, with total floor area 61,256.69 m², and total height 99.3m. The project won the 2nd prize of Hu'nan Excellent Survey & Design.

The building is a multipurpose building with office, residential, commercial and hotel functions. Upon completion, it not only greatly improves HydroChina Zhongnan's office conditions and workers' residential conditions, but also becomes a landmark in this region, for its unique, steady and concise shape.

长沙市湘江风光带

Xiangjiang River Scenery Zone, Changsha

项目地点：湖南 长沙	Location: Changsha, Hu'nan
用地面积：100 000 m^2	Site Area: 100,000 m^2

长沙市湘江风光带位于长沙老城区的湘江东岸橘子洲大桥与银盆岭大桥之间，全长1160 m，宽度 30 ~ 100 m 不等。

凭江临风，山水一体，景象因水而殊异，空间因水而生动，整体设计充分体现了“以人为本，天人合一”的设计理念；风光带的设计集城市景观工程及城市防洪于一体，较好地处理了两者之间的关系，是长沙市打造“山、水、洲、城”理念的一个代表性作品；湘江风光带滨临湘江，建成后是长沙市目前最大的市民休闲广场，深受广大长沙市民喜爱。

Xiang River scenic belt of Changsha is 1,160 m in length and ranging from 30-100 m in width, located in the old city between Juzizhou Bridge of Xiang River east bank and YinpenLing Bridge.

The project is near the river, facing the wind with integrated mountain and river, different scenes are shown because of the river, and the space looks vivid with the existence of the river; the overall design fully presents the concept of “human-based integration of human and nature”; the design of the scenery zone well combines the landscaping and the urban flood defense, properly deals with the connection of the two, hence is a representative work for demonstrating the concept of “mountains, rivers, islets, and the city” in Changsha; bordering along Xiangjiang River, the scenery zone will be the biggest leisure square and well-loved by the local citizens upon completion.

田汉大剧院

Tian Han Grand Theatre

项目地点：湖南 长沙	Location: Changsha, Hu'nan
用地面积：10 220 m^2	Site Area: 10,220 m^2
建筑面积：28 122 m^2	Building Area: 28,122 m^2

田汉大剧院坐落于长沙市侯家塘劳动路北侧，剧场容量 1204 座，音乐厅容量 883 座，属大型甲等剧场。大剧院外形设计宏伟壮观、浑然一体、气势轩昂。主立面以四根主柱为竖向支撑，加以横向连廊、钢网架结构和透明玻璃屋顶，构成建筑整体外轮廓，既具有中式建筑质朴稳重的风格，又富有西式建筑开敞现代的气息。主柱与连廊的立面构图及装饰的长城图案，既切合剧院命名，又巧妙地点明了纪念主题。此项目获得湖南省优秀勘察设计一等奖。

This large class-A theatre is located on the north side of Labor Road, Houjiatang Sub-district, Changsha City. Its theatre can hold 1,204 audients while its music hall 883.
The shape is magnificent, integral and spectacular. The main façade is supported vertically by four main columns which, together with the horizontal vestibules and transparent glass roof in steel grid structure, shape the integral contours, featuring not only pure steady Chinese architecture, but also open modern western architecture. The façade patterns of the columns and vestibules, as well as the great wall patterns of the decorations not only fit the name of the theatre, but also imply the memorial theme. This project has won the 1st prize of Hu'nan Excellent Survey & Design Award.

浏阳市体育馆

Liuyang Gymnasium

项目地点：湖南 浏阳	Location: Liuyang, Hu'nan
用地面积：28 372 m^2	Site Area: 28,372 m^2
建筑面积：25 835 m^2	Building Area: 25,835 m^2

体育馆位于浏阳市体育中心的体育场北侧，周围地形平坦，环境优美，位置显要。体育馆为举办地区性和全国单项比赛的体育建筑，体育建筑等级属乙级。体育馆内设两个场馆，由体育馆和游泳馆组成。体育馆的规模为 3226 人座，可满足篮球、排球、羽毛球、乒乓球、网球、体操等单项国际或国内比赛和训练使用要求。游泳馆的规模为 1116 人座，能进行各种游泳、水球等项目的正式比赛。

The gymnasium is located on the north side of the stadium in Liuyang Sports Center, with flat landform, beautiful environment and prominent geography. The gymnasium is a class-B sports building designed to hold local and national individual competitions.
The gymnasium has two venues: gym and natatorium. Scale of the gym: it has 3,226 seats and can be used for international or domestic individual competitions and trainings of basketball, volleyball, badminton, table tennis, tennis and gymnastics and so on. Scale of the natatorium: it has 1,116 seats and can hold a variety of official competitions like swimming and water polo events.

宜宾市酒都大剧院

Jiudu Theater, Yibin

项目地点：四川 宜宾
用地面积：18 086 m²
建筑面积：38 273 m²

Location: Yibin, Sichuan
Site Area: 18,086 m²
Building Area: 38,273 m²

项目位于宜宾市政府正对面西侧的F-5-1地块，与周边图书馆、体育场及市民广场共同构成城市的文化、体育、观演中心。
该项目剧场观众厅座位数为1485座，多功能厅座位数为512座。设计充分尊重城市的历史文脉，凝聚了宜宾市三江文化的灵魂，将建筑艺术与技术完美结合，创造出具有鲜明个性、极富文化内涵的公共建筑。

The project on the Lot F-5-1 is located at Shunan Avenue, Cuiping District, Yibin City. The land lot is west of Yibin People's Government Yibin City on the other side of the avenue. The project and the neighboring library, stadium and public square all together constitute a cultural, sports and performance center of the city.
The building covers a land area of 8,482 m², and its auditorium has 1,485 seats and multifunctional hall, 512. The design pays full respect to the historic content of the city and takes into consideration the soul of the "three rivers" culture of Yibin. With the perfect combination of architectural art and techniques, the project creates a public building with distinct characters and rich cultural contents.

常德防汛指挥调度中心
Changde Flood Control Command Center

项目地点：湖南 常德
用地面积：15 998 m²
建筑面积：21 830 m²

Location: Changde, Hu'nan
Site Area: 15,998 m²
Building Area: 21,830 m²

常德防汛指挥调度中心是一幢办公类高层建筑，主楼呈椭圆形，造型新颖，活泼生动，多个方向的流线型曲线似水波流动，令大楼的整体造型协调统一，充分体现了常德市水利局的行业特征。顶部通讯指挥塔也设计为椭圆的形式，通讯指挥塔顶端为钻石的形状，整个塔体与主楼协调统一，象一个璀灿的明珠镶嵌于沅水之滨。

Changde Flood Control Command Center is a high-rise office building, the main building is in elliptical shape, new and lively, and the streamlined curvy shape in multiple directions looks like the water wave flow, and makes the general shape of the building harmonious and integrated, adequately presenting the industrial characteristics of Changde Water Resources Bureau. The communication control tower on the top of the main building is also designed elliptic, and the top design is diamond-shaped, making the whole tower coordinated with the main building, like a bright pearl inlaid into the bank of Yuan River.

湖南中医药高等专科学校
Chinese Medicine College,Hu'nan

项目地点：湖南 株洲
用地面积：356 633.3 m²
建筑面积：168 034 m²

Location: Zhuzhou, Hu'nan
Site Area: 356,633.3 m²
Building Area: 168,034 m²

湖南中医药高等专科学校新校区学生规模为在校6000生，规划总人口数11 000人。学校用地位于株洲职业教育大学城智慧广场东南侧，占地总面积为356 633.3 m²。
规划布局强调“阴阳调和、天人合一”的中医理念，“因山势，就地理” 最大限度地保留和利用微地形和水系，减少土石方工程量，在总图规划中顺应地形机理，建筑与山水交融共生出一派和谐的景象。建筑造型将中国传统中医药文化与现代文化有机融合，体现山水交融的园林式校园精髓，展现中医药文化的特色魅力。

The new campus of Hu'nan Traditional Chinese Medical College can hold 6,000 students, and the total planned population is 11,000. The land is located on the south-east side of Wisdom Square in Zhuzhou Vocational Education University Town.
The layout emphasizes the TCM idea of “yin & yang balance and human & nature unity”, retains and uses the micro landform and water system as much as possible and reduces the earthwork/stonework volume. The general plan fits the landform well, and the architecture integrates the landscape, creating a harmonious climate. The architecture integrates TCM culture and modern culture, represents the essence of garden campus with waters and mountains, and displays the charming culture of traditional Chinese medicine (TCM).

竹山一中

Zhushan No. 1 Middle School

项目地点：湖北 十堰
用地面积：370 000 m^2
建筑面积：110 000 m^2

Location: Shiyan, Hubei
Site Area: 370,000 m^2
Building Area: 110,000 m^2

竹山一中新址位于湖北省竹山县潘口乡境内，距竹山县城约2 km，交通便利，闹中取静。新建内容包括教学楼、实验楼、教学综合楼、行政办公楼、图书馆、学生宿舍、学生食堂、体育馆、学术交流中心、体育场地等。
由于用地高差甚大，按典型的山地建筑规划格局，建筑形态层层递进，步步升高。在规划形态布局上形成一朵花苞，喻示“孩子是祖国未来的花朵，教育使花蕾绽放。”让方案展现出寓于“理性”的“感性”美。

The new site of Zhushan No. 1 Middle School is located in Pankou Township, Zhushan County, Shiyan City, Hubei Province. About 2 km from Zhushan county town, the school enjoys convenient traffic and quiet conditions amidst the noisy town. The new buildings include Teaching Building, Experiment Building, General Teaching Building, Administrative Office Building, Library, Students Dormitory Building, Students Dinning Hall, Gymnasium, Academic Exchange Center, Sports Ground and others.
The layout follows the hilly landforms, to extend the building forms and increase the building heights gradually. The layout looks like a bud, symbolizing “children are future buds of the country, while education will make them flower”. In this way, the plan shows its “sensational” beauty out of “rationality”.

远能·锦座综合体
Yuanneng · Golden Tower Complex

项目地点：湖南 长沙
建筑面积：17 702.5 m²

Location: Changsha, Hu'nan
Building Area: 17,702.5 m²

设计从城市空间形态出发，尊重原有城市肌理，尊重大楼的功能组成及灵活变通，创造气势恢弘又丰富多样的城市空间形态。设计采用现代建筑设计手法，运用材质对比及强调竖向构图，充分体现建筑的挺拔感、标志性。
设计将塔楼与裙楼在视觉上融为一体，功能上却相互分开，从而可以感受到主楼直接落地的磅礴气势，建筑形体完整简洁、雄伟稳重，很好体现其在这一区域内作为标志性建筑的要求。

The design starts from urban space forms, respects original city texture, respects building functions and flexibility, and creates magnificent and changing urban space forms. The design is a modern architectural design which uses material contrast and vertical structure to fully represent building straightness and landmark.
The design integrates the tower building and the skirt building visually, with separate functions, to give a grandiose experience of falling directly from the main building to the ground. The building shape is complete, concise, grand and steady, fully representing its necessity as a landmark in this region.

利比亚祖瓦拉海滩概念性规划
Concept Plan of Zuwarah Beach in Libya

项目地点：利比亚 祖瓦拉
Location: Zuwarah, Libya

利比亚城市祖瓦拉坐落于的黎波里地区的北部，这座城市被认为是组成利比亚首都的黎波里的五个分散地区之一。本方案设计运用大尺度、大手笔的线形构图和丰富自由的空间处理，形成与海岸平衡的景观序列，充分体现自然与人文的交融，形成“海洋—沙滩—软质铺装—景观建筑—绿化带—城市建筑背景”由海洋向城市过渡的景观层次，从而达到城市与海洋有机统一，形成设计在空间结构上的“一带三中心、七大功能区”。

Zuwarah is located in northern Tripoli, Libya. This city is considered as one of the five isolated districts of Tripoli, the capital of Libya. This project uses large scale linear patterns and free space planning, forms a landscape sequence in parallel with the coastline, and fully displays an integration of the nature and humanity. The project forms a landscape sequence from the sea to the city, namely “sea – beach soft pavement – landscape buildings – greenbelt – city building background”, to integrate the city and the sea, and form “one belt, three centers and seven functions” of the beach in the space structure.

梅州抽水蓄能电站办公生活区

Office & Living Areas of Meizhou Pumped-storage Power Station

项目地点：广东 梅州
用地面积：87 637 m²
建筑面积：22 430 m²

Location: Meizhou, Guangdong
Site Area: 87,637 m²
Building Area: 22,430 m²

广东省梅州抽水蓄能电站办公生活区位于梅州五华县杨梅隆下水库的西侧，规划包含办公区、餐饮区、体育运动区、宿舍区、酒店度假区、文化展示区。
总体规划强调基本的动静分区和使用功能的内在联系，建筑设计以现代风格为基调，吸取当地新岭南建筑特色，使建筑群具有强烈的地域特征。

The station owner's camp is located west of Yangmeilong Downstream Reservoir, Wuhua County Meizhou City. The office area, dining area, sports area, living quarters, hotel and resort area and cultural display area are included in the scope of planning.
The general plan emphasizes the basic static and dynamic zones, as well as the internal relations of different functions. The architectural design is keynoted by modern style. Absorbing the elements of local neo-southernmost architecture, the building cluster has strong regionalism.

TUMUFENG DESIGN

扫描查看更多信息

土木风设计是集规划设计、建筑设计、景观设计、装饰设计及景观施工为一体的集团化公司。持有国家住房和城乡建设部（简称“住建部”）颁发的建筑、装饰甲级资质及园林、规划乙级资质。公司通过设计实践，不断吸收、学习国内外优秀设计公司的成功经验和先进的经营理念，努力将公司建设成为省内一流、国内有影响的大型设计公司。

公司秉承“立足关东大地，营造寒地设计”的责任与使命，在总裁石铁军的带领下，虽然仅成立三年多，但是却取得了丰硕的成果，成为吉林省发展最迅猛的设计公司之一。被国家住建部评为“中国最具影响力的民营设计机构”。2011 年是土木风收获颇丰的一年，先后获得了国家住建部优秀建筑设计奖一项（是吉林省唯一获得国家级奖项的设计单位）；省优秀建筑设计一等奖两项、二等奖、三等奖各一项；规划设计优秀奖两项；保障性住房设计竞赛优秀奖两项。公司总裁石铁军也荣获首批吉林省青年设计大师的称号。2011 年末，公司成功登陆上海，成立了上海土木风设计有限公司。这预示着公司正向“立足东北，雄踞上海，幅射全国”的目标健康发展。

地址：吉林省长春市东民主大街 519 号
电话：+86-431-88591360
传真：+86-431-88591360
邮箱：tumufeng@126.com
网址：www.tumufeng.com

Add: No.519, Dong Min Zhu Street, Changchun City, Jilin Province
Tel: +86-431-88591360
Fax: +86-431-88591360
E-mail: tumufeng@126.com
Web: www.tumufeng.com

浩铭·乐府住宅小区
Haoming · Yuefu Residential Community

项目地点：吉林 长春
用地面积：24 700 m²

Location: Changchun, Jilin
Site Area: 24,700 m²

吉林省汽车小镇概念规划
Auto Town Conceptual Plan, Jilin

项目地点：吉林 长春
用地面积：920 800 m²

Location: Changchun, Jilin
Site Area: 920,800 m²

德惠市人民法院审判法庭
People's Court Trial, Dehui

项目地点：吉林 德惠
建筑面积：17 000 m²

Location: Dehui, Jilin
Building Area: 17,000 m²

桦甸市上海花园西区
Shanghai Garden West Zone, Huadian

项目地点：吉林 桦甸
建筑面积：121 000 m²

Location: Huadian, Jilin
Building Area: 121,000 m²

长春皇庭城市商业综合体一期
Phase I of Huangting City Commercial Complex, Changchun

项目地点：吉林 长春
建筑面积：249 000 m^2

Location: Changchun, Jilin
Building Area: 249,000 m^2

牛马行经贸广场
Cattle Exchange Commerce & Trade Square

项目地点：吉林 吉林
建筑面积：170 000 m^2

Location: Jilin, Jilin
Building Area: 170,000 m^2

汇才国际大厦
Huicai International Building

项目地点：吉林 长春
建筑面积：79 300 m^2

Location: Changchun, Jilin
Building Area: 79,300 m^2

文汇鼎文化商业中心
Wenhuiding Cultural & Commercial Center

项目地点：吉林 长春　　Location: Changchun, Jilin
建筑面积：102 891 m²　　Building Area: 102 891 m²

通化县城市展览馆
Urban Exhibition Museum, Tonghua

项目地点：吉林 通化　　Location: Tonghua, Jilin
建筑面积：9116 m²　　Building Area: 9116 m²

卓扬·中华城
Zhuoyang · China Town

项目地点：吉林 长春
建筑面积：103 000 m²

Location: Changchun, Jilin
Building Area: 103,000 m²

南航吉林分公司生产保障及训练综合楼
Production Security & Training General Building of CSAIR Jilin Branch

项目地点：吉林 长春
建筑面积：71 028 m²

Location: Changchun, Jilin
Building Area: 71,028 m²

中庆集团总部办公楼
Zhongqing Group HQ Office Building

项目地点：吉林 长春
建筑面积：25 000 m²

Location: Changchun, Jilin
Building Area: 25,000 m²

长春老年大学

College for the Elder People, Changchun

项目地点：吉林 长春
建筑面积：50 000 m^2

Location:Changchun, Jinlin
Building Area:50,000 m^2

扫描查看更多信息

吉林省光大建筑设计有限公司

Jilin Guangda Architetural Design Co., Ltd.

吉林省光大建筑设计有限公司是一家具有建筑行业工程设计甲级、城乡规划编制乙级、市政行业乙级、风景园林工程设计乙级资质的设计团队。公司成立于1995年，原为长春市光正建筑设计院。2005年晋升为建筑工程甲级设计资质单位，同时更名为吉林省光大建筑设计有限公司。

公司现有员工85人，其中吉林省设计大师2人、一级注册建筑师4人、一级注册结构师4人、教授级高工3人、高级工程师8人、工程师20人，构成了可承接各类建筑设计及城市规划、市政工程、风景园林等项设计的一支具有高度职业荣誉感和良好执业道德的设计团队，是吉林省内知名度较高、信誉度突出的设计企业。

公司秉承"以品质赢得信赖、以品牌赢得未来"的企业理念，追求设计作品从概念设计到最终投入使用全过程的高品质服务。苛求细节、创造技术与艺术的完美结合，不断创新、不断超越，力求打造卓越的经典建筑。

在居住建筑、教育建筑、酒店建筑、城市综合体建筑等类设计中都体现了本公司特有的风格和设计理念，建立了不断扩展的客户群体，为城市建设做出了应有的贡献。公司全体员工正以高昂的激情和扎实的步伐为企业的稳步发展进行着不懈的奋斗。

地址：吉林省长春市南关区人民大街207号
电话：+86-431-5355550
传真：+86-431-85355522
邮箱：JLGDJZ@163.com
网址：3229494.71ab.com

Add: 207 People Avenue, Nanguan District, Changchun City, Jilin Province
Tel: +86-431-5355550
Fax: +86-431-85355522
E-mial: JLGDJZ@163.com
Web: 3229494.71ab.com

珲春国际饭店

Hunchun International Hotel

设 计 师：于斌
项目地点：吉林 珲春
用地面积：37 500 m²
建筑面积：86 965 m²

Designer: Bin Yu
Location: Huichun, Jilin
Site Area: 37,500 m²
Building Area: 86,965 m²

珲春国际饭店以20世纪盛行的现代主义国际风格为主导设计思想，方案力求造型简洁、明快。

外立面造型：整栋建筑以暖色石材竖向线条搭配竖向通高玻璃幕墙为主体立面造型形式，营造韵律感强、光影变化丰富的立面效果。

内部空间组织：围绕建筑主体800 m²中庭组织餐饮、会议、办公、娱乐、客房等一系列功能空间，达到各主要使用人员能最大限度地共享中庭空间。二层回廊构成一个内与外、上与下的视线交织场所。

整栋建筑以或虚或实的墙面设计、或明或暗的光线组合、亦内亦外的景观融合、或高或低的空间变化，温和而清晰地表达出此次设计任务的主要目的。

The design is led by modernist international style prevailing in the 20th century. The project seeks for concise and bright shapes.

As to the façade, the whole building is shaped with warm color stone materials in vertical lines matching glass curtain walls in whole length, to create a rhythmic façade with rich changes in shadow and light.

As to interior spaces, the atrium (800 m²) of the main building integrates a range of functional spaces, including catering, conference, office, entertainment and guesthouse, where main pedestrian flow lines can optimally share the atrium spaces. The corridor on the second floor constitutes a dramatic visual crossroad of internal and external, upper and lower spaces.

The whole building varies or integrates with virtual or real walls, bright or dark lights, inside or outside landscapes, higher or lower spaces. The major purpose of this design is to express these changes in a mild and clear manner.

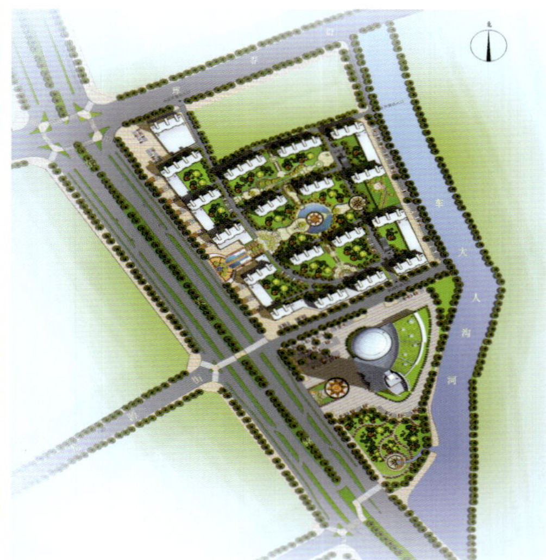

南湖假日

South Lake Holiday

设 计 师：刘冬华 何浩
项目地点：吉林 长春
建筑面积：39 535 m^2

Designer: Donghua Liu, Hao He
Location: Changchun, Jilin
Building Area: 39,535 m^2

前郭尔罗斯蒙古族大酒店
Front Gorlos Mongol Hotel

设 计 师：张成会
项目地点：吉林 松原
建筑面积：40 594.45 m²
合作单位：中国建筑设计研究院

Designer: Chenghui Zhang
Location: Songyuan, Jilin
Building Area: 40,594.45 m²
Partners: China Architecture Design & Research Group

延吉市香山国际社区
Xiangshan International Community, Yanji City

设 计 师：刘冬华 王晴
项目地点：吉林 延吉

Designer: Donghua Liu, Qing Wang
Location: Yanji, Jilin

规划二期用地

党校

梨

花

山

扫描查看更多信息

南京长江都市建筑设计股份有限公司前身为南京市民用建筑设计研究院，创建于 1976 年初。于 2003 年 9 月经南京市政府批准，由国家事业单位改制为民营股份制企业——南京市民用建筑设计研究院有限责任公司，2009 年底公司派生分立为南京长江都市建筑设计股份有限公司。公司从创建到现在经历了 30 年的发展，尤其是 20 世纪 90 年代后，公司进入了高速发展阶段。已初步形成以城市设计、办公建筑、商业建筑、科研教育建筑、居住建筑设计为特色和主要产品，成为具有一定实力的区域性、综合性、具有中等规模和鲜明特色的现代建筑设计企业。在江苏省乃至华东地区具较高声誉及较强竞争力。
公司目前具有国家建筑行业甲级等设计资质，现有在职员工 305 人，设计技术人员 269 人，占总人数 88%；高级职称共 64 人（研究员级 10 人），占专业技术人员的 21%。

公司始终坚持尊重客户、理解客户，持续提供超越客户期望的产品和服务的经营宗旨，精心设计、科学管理、客户至上、持续改进的质量方针。高度重视设计创新、技术质量、成本控制、设计服务等环节。公司于 2001 年已通过 ISO9001 质量体系认证，设计质量在省内外享有良好声誉。近年来，先后有几十项设计工程项目获得建设部、省、市级优秀设计，优秀工程和科技进步奖。
2009 年，长江都市为江苏省勘察设计企业综合实力第 5 名，并被授予江苏省勘察设计质量管理先进单位和诚信单位称号。2010 年获全国工程勘察设计行业优秀民营设计企业称号。

地址：南京市洪武路 328 号
电话：+86-25-84567205
传真：+86-25-84567205
邮箱：jihuake@nanjing-design.com
网址：www.nanjing-design.cn

Add: No.328 Hongwu Road, Nanjing City
Tel: +86-25-84567205
Fax: +86-25-84567205
E-mai: jihuake@nanjing-design.com
Web: www.nanjing-design.cn/

苏州玲珑湾三期
Suzhou Linglong Bay Phase III

设 计 师：汪杰、高亮、高华国、韦佳、刘顺
项目地点：江苏 苏州
用地面积：81 6847 m^2
建筑面积：13 915 414 m^2

Designer: Jie Wang, Liang Gao, Huaguo Gao, Jia Wei, Shun Liu
Location: Suzhou, Jiangsu
Site Area: 816,847 m^2
Building Area: 13,915,414 m^2

玲珑湾项目规划布局合理，具有得天独厚的环境优势，力求在全生命周期内减少对自然资源的消耗和对自然界的负面影响。在方案设计阶段就引入了电脑模拟对组团的日照、风环境、热岛效应等方面进行优化，并采用了雨水收集系统和透水地面设计，使玲珑湾三期成为舒适的绿色生态住宅小区。
建筑单体风格为优化过的 ART-DECO 风格，挺拔峻秀的建筑倒映于湖光波影中，丰富了万顷碧波尽头的天际线，为无边美景画上了一个完美的句号。

To achieve the goal of lessening the consumption of natural resources and the adverse impacts to the nature within the full lifecycle for the reasonable planning and layout and the unique environmental advantages of this area, the program introduces computer simulation, which simulates and optimizes the sunlight, wind environment and heat island effect and sets the rainwater collection system and the impervious surface design, to make it a comfortable, green and ecological residential community.
The style of single building is optimized Art-Deco style. Tall and great buildings are reflected on the surface of the lake, enriching the skyline beyond the end of waterline, all these end perfectly in the endless beautiful scenery.

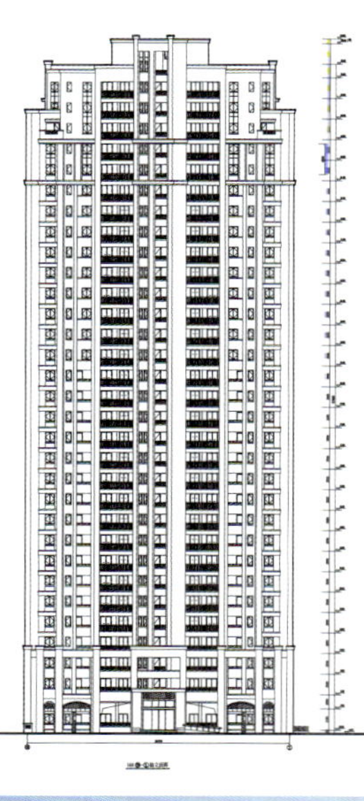

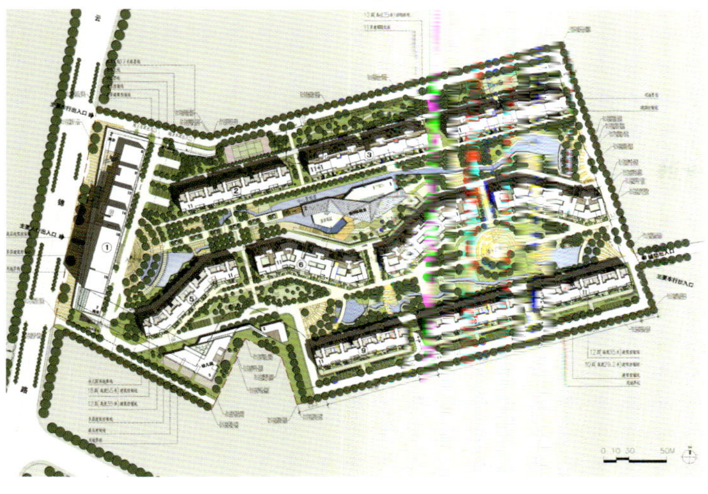

金地南京所街

Gemdale Nanjing Suo Street

项目地点：江苏 南京	Location: Nanjing, Jangsu
用地面积：90 760 m^2	Site Area: 90,760 m^2
建筑面积：203 475 m^2	Building Area: 203,475 m^2

项目位于南京河西新城，建邺区集庆门大街以南，云锦路以东。
“院落”是本项目最大的特点和亮点。住宅一层所有户型均设有私家庭院，形成半私密院落，构筑出了新友邻关系的公共院落空间。
住宅的入户花园重在营造入户院落新体验，06～08号楼顶层复式围合院落，突破传统院落形式，将院子搬到家中，带来只可与家人分享的私密院落空间。
所有一层住宅除有私家庭院外，还可赠送最大60 m^2的地下室；顶部则均为跃层户型，每户带有空中大露台。
多种形式空间的利用，增加了空间使用附加值，在别墅中才能体现的设计元素，在公寓中均得以体现，这种创新的空间设计在南京是从未有过的。
项目突出主入口形象设计，住宅主入口设在建筑物北侧，均设有双层入户大堂，入口大堂内装修典雅、宽敞明亮。车库采用下沉庭院与车库相结合，从而克服了以往车库必须采用人工照明且不开敞的弊端，使车库直接采光并自然通风，业主在后期使用上减少居住成本。
住宅在立面造型设计上，力图创造一个简洁、明快，具有较高文化内涵和强烈现代气息的建筑形象。建筑外观以砖红色面砖为基础，通过与浅色面砖相结合，达到明快、清晰的效果。通过浅色横线条穿插，在立面上形成一定的韵律感。

The project is located in Hexi New Town, south of Jiqingmen Street and east of Yunjin Road, Jianye District, Nanjing.
Courtyard is the most important feature and highlight. Every unit on the first floor has a private courtyard and the semi-private courtyard builds a new friendly public courtyard space.
Indoor garden gives the resident new experience in indoor courtyard. The duplex enclosed courtyards of floors 6-8 breaking through traditional ideas in courtyards to bring courtyard in door give you the personal spaces to share only with your family.
Besides private courtyards, every unit on the first floor will receive a basement of 60 m² at max. as an additional gift. The top is a loft with a big aerial terrace.
The application of various space forms adds the value of space usage. The design elements in villa are displayed in apartment. The innovative space design is unprecedented in Nanjing.
The project emphasizes the main entrance design. The main entrance is located on the north side of the building and is set with a double-layer indoor lobby. The lobby has elegant decoration and large bright space, reduce the cost of user.
The facade formation in the building design shows a concise, bright, contented and modern architectural image.
The exterior is shown in brick-red tile background combing with light-colored tile to achieve bright and clear visual effects. At the same time, the light-color transverse lines interlace in the facades form a kind of rhythm.

亚东 · 朴园一期

Yadong · Natural Garden Phase I

设 计 师：董文俊、耿川亮、沈璐、夏建阳
项目地点：江苏 镇江
用地面积：40 000 m²

Designer: Wenjun Dong, Chuanliang Geng, Lu Shen, Jianyang Xia
Location: Zhenjiang, Jiangsu
Site Area: 40,000 m²

本项目位于镇江市南徐板块和城区板块的交界处，紧邻城市中心大市口。西南面为南山国家森林公园，南面磨笈山近在咫尺，东面为规划中的黄山东路，并有一条规划中的城际铁路与之平行，北面是与城市道路相连的黄山中路，基地有得天独厚的自然资源和日臻完善的城市交通体系环绕。一期总用地面积 4 hm²，规划综合容积率为 0.72，居住户数为 152，居住人口为 486。

在规划中，将镇江的文化底蕴与中国传统居住价值相结合，把庭院作为核心的元素来考虑。住宅设计了前庭、天井和后院，将大量的植栽和环艺语汇融入宅院本身，形成独特的中国居住文化意境，重塑一种真正属于中国人的庭院生活。在空间的尺度上宜人、近人，通过“山、水、苑、林、人”来打造镇江独一无二的和谐生态社区。

The project is located in the junction of the Nanxu block and the urban block of Zhenjiang City, close to the Dashikou in city center. In its southwest, there is Nanshan National Forest Park; in its south, Mt. Moji is near at hand; in its east, there is the Huangshan East Road in planning, in parallel with an intercity rail line in planning; in its north, there is Huangshan Middle Road linking with the urban roads. The project has unique natural resources and improving city transport system. The total land area of the first phase is 40,000m², with plot ratio 0.72, totaling 152 residential units and 486 residents. In the plan, the courtyard is the core element after combining the culture and history of Zhenjiang with traditional Chinese residential value. The apartment designed with front-yard, patio and back-yard and a blend of plenty of planting and environmental arts into the house itself, forms a unique Chinese residential culture background and rebuilds a kind of courtyard life belonging to the Chinese people. It's also pleasant and close to people in space through the mountains, waters, gardens and woods to create a distinctive harmonious and ecological community in Zhenjiang.

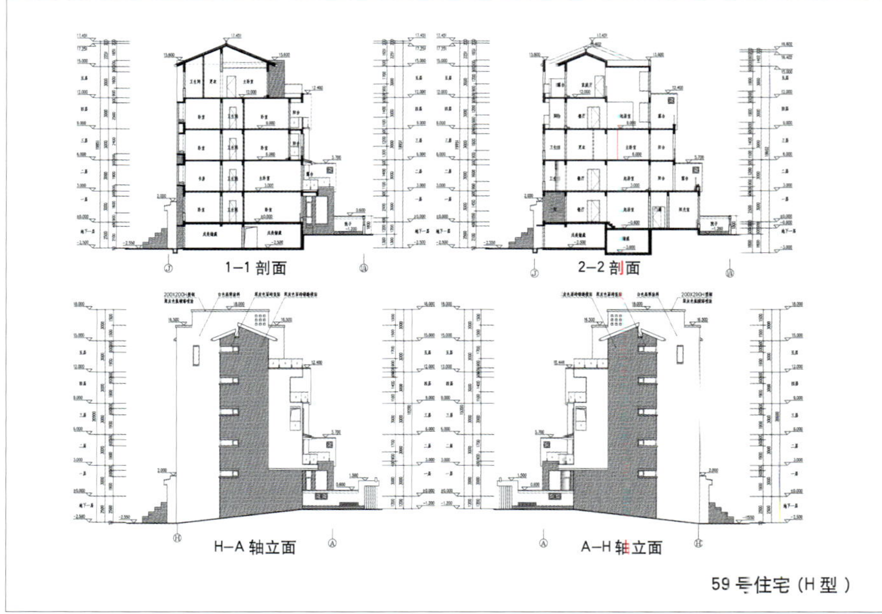

一期技术经济指标

项目		单位	数据
可建设用地面积		ha	4.00
地上总建筑面积		m²	29536
住宅总建筑面积		m²	26941
其中	联排住宅建筑面积	m²	10528
	叠加住宅建筑面积	m²	16413
公建总建筑面积		m²	2595
其中	商业及其他建筑面积	m²	1848
	公厕建筑面积	m²	40
	变电站建筑面积	m²	137
	监控中心建筑面积	m²	130
	社区服务中心	m²	370
	门卫建筑面积	m²	20
计容积率建筑面积		m²	23675
建筑总基底面积		m²	9491
公共绿地面积		m²	1960
容积率			0.72
建筑密度		%	23.73
绿地率		%	44.38
居住户数		户	152
居住人口		人	486
地下建筑总面积		m²	8538
机动车停车位		辆	192
其中	地面停车	辆	52
	地下停车	辆	140
非机动车停车位		辆	320

南航文科楼
Arts Building of NUAA

设 计 师：王畅、周璐、王亮、葛小林、胡旭明
项目地点：江苏 南京
建筑面积：13 600 m²

Designer: Chang Wang, Lou Zhou, Lang Wang, Xiaolin Ge, Xuming Hu
Location: Nanjing, Jiangsu
Building Area: 13,600 m²

项目根据周边建筑的肌理关系以及教学用房的采光需要，将建筑主体顺应校园肌理，以南北向为主。在总平面图布置上，将两个等大的矩形体块，以底层架空的“大空间”相联系，形成围合建筑布局，与北3号教学科研楼遥相呼应，使之与校园建筑整体布局协调，形成较好的空间关系。

文科楼在功能使用上分为两部分：外语学院、人文学院，主要入口在东侧开放空间，外语学院与人文学院可由各自独立门厅进入。小型国际会议中心和模拟法庭设置相对独立，在形体上将两学院体块联系起来，内部庭院则为院系间提供共享的趣味空间。建筑西侧底层架空，使文科楼内部景观与校园环境形成对话，使得空间连贯，景观层次丰富。

虽然建筑具有性质不同的功能空间，但其整体风格仍协调统一，同时在不同功能区又有形式上的变化。以大空间为主的部分，立面开窗上力求化零为整，并通过表皮的退让关系，使立面层次丰富；建筑色调以浅色和灰色为主，与校园内其他建筑形成整体风格的统一。

According to texture of the surrounding buildings and the daylighting of teaching function, the main building still follows the campus texture in north-south direction. In the general layout, the aerial ground floor combines two parallel rectangular blocks to form an enclosure layout. This arrangement well echoes the No. 3 Teaching and Research Building in north area to coordinate these buildings with the campus layout in a great space relationship.

In function, the Arts Building can be divided in two parts: the School of Foreign Language and the School of Humanities. The crowd mainly comes from the east side of the open space, while the School of Foreign Language and the School of Humanities can be entered respectively through separate entrances. The small international convention center and the moot court are set relatively independent, but they connect the two schools in shape, and the inner courtyard provides a common interesting area across the faculties. The aerial ground floor on the west side of the building communicates the inner courtyard of the Arts Building with the campus environment, to continue the spaces and enrich the landscape.

Although the buildings have different functional spaces, they are still coherent and unified with the overall style and different in form. In the part with large space, the windows in façade try to integrate and enrich the levels of façade by surface retiring relations. It is mainly in light-color and grey tune, which is unified with the style of other buildings on campus.

上海迪丰服饰发展有限公司新建厂房

New Factory Building of Shanghai Dofiny Costume Co., Ltd.

设 计 师：常宁、汪愫璟、王丹、王珏、赵振
项目地点：上海
用地面积：48 800 m^2
建筑面积：78 771.9 m^2

Designer: Ning Chang, Sujing Wang, Dan Wang, Jue Wang, Zhen Zhao
Location: Shanghai
Site Area: 48,800 m^2
Building Area: 78,771.9 m^2

项目位于上海市青浦区徐泾镇西郊工业园区，是一个集生产、办公、科研为一体的综合研发中心。总体规划中，南区以科研、办公、生产为主体，北区则以方便生活为宗旨，两区的规划设计灵活，既可分亦可合，成为有机整体。在条形地块的基础上，南北方向依次为综合办公区、科研物流区、设计科研中心和物流中心；生活服务区独立成区，位于地块北部，通过绿化休闲广场与办公区、科研物流区形成紧密联系。本项目造型大方，建筑形式现代、简洁，建筑细节设计精致且紧扣设计主题，建筑功能布局合理，较好地满足了业主对科研办公的要求，是科研办公类建筑的优秀典范。

This project is located in Xujing Town Western Suburb Industry Park, Qingpu District, Shanghai. In master planning, the south zone is composed of scientific research, office and manufacturing functions; the north zone is living area. The planning of the two zones is independent and integral. On the strip lot, from south to north stand Comprehensive Office Zone, Research Logistic Zone, Design Research Center and Logistics Zone; the Living Service Zone is an independent zone located in the northern part of lot, connecting with Office Zone and Research Logistic Zone by green leisure square. This is an elegant design of modern architectural style with exquisite detail and characteristic theme and reasonable function layout. It can satisfy client's demand for scientific research & office. It is an excellent example in research & office buildings.

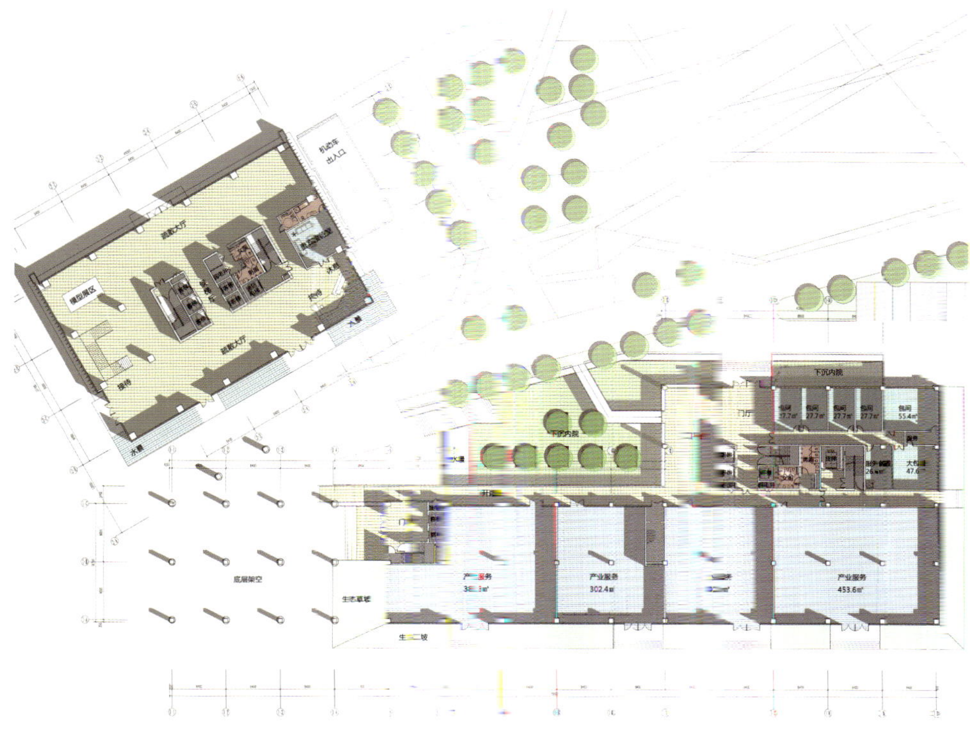

常州科教城国际创新基地科教园 1 号楼

Science Park No.1 building, CIIB in Changzhou Science and Education Town

设 计 师：常宁、王丹、汪愫璟、王珏
项目地点：江苏 常州
用地面积：58 573 m²
建筑面积：58 571.5 m²

Designer: Ning Chang, Dan Wang, Sujing Wang, Jue Wang
Location: Changzhou, Jiangsu
Site Area: 58,573 m²
Building Area: 58,571.5 m²

常州科教城国际创新基地科教园 1 号楼是科教城的启动项目及重点项目，是科教城区域的标志性建筑。项目以“科研、技术、文化、生态”为目标，在建筑简洁的体块上运用“减法”，通过一系列的切割与架空，形成丰富多变的形体，而且保留了基本体量的完整。建筑以竖向线条勾勒出高层建筑的力度感与挺拔感。

现代科教城主要功能是以研发办公为主，并结合一定的配套设施形成现代化的研究型城市空间。设计中将建筑的底层空间适当架空，将此虚空间与内院相互穿透，融为一体。建筑南北侧结合采光要求，采用“门”形构图，丰富了立面形象，并寓意“敞科教之门，迎四方宾客”。

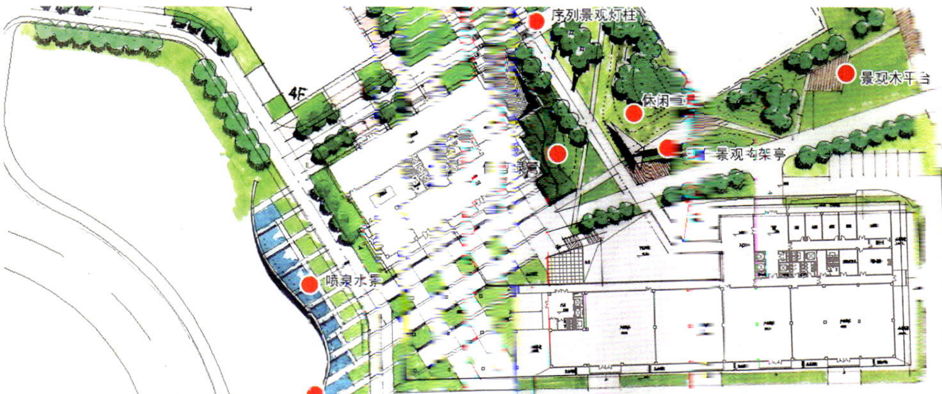

Science Park No.1 building of CIIB in Changzhou Science and Education Town, as a start-up and key project, is the landmark of the Science and Education Town. The project holds functions of “science, technology, culture and ecology” as target, applies “minus” in concise building, forms various forms and reserves the basic dimensions by a series of cutting and stilts. Vertical lines outline the strength and straightness of high-rise buildings.

Modern Science and Education Town is mainly used for research and development and work with corresponding facilities to form a modern research city space. The ground floor is appropriately aerial so that this virtual space integrates the inner courtyard. The door-shaped layout is applied considering the lighting in south and north of the building to enrich the elevation and to imply “opening the door of science and education to welcome guests from all over the world”.

徐庄软件基地贝尔实验室

Bell Laboratories in Xuzhuang Software Base

设 计 师：邱立刚、秦玲玲、董文俊、沈璐、屈亚芬、武锐、钱阳
项目地点：江苏 南京

Designer: Ligang Qiu, Lingling Qin, Wenjun Dong, Lu Shen, Yafen Qu, Rui Wu, Yang Qian
Location: Nanjing, Jiangsu

项目位于南京仙林新区西北部的徐庄软件产业基地，是国家级信息科技产业基地之一。
动态而又严谨的总体布局，提供了富于变化而又错落有致的建筑形体关系和外部空间格局，多层次的院落和中心公共空间，在大尺度的建筑群和私密的个人空间之间形成有机的转换和递进，既形成有着鲜明特色的整体建筑群，又为每栋办公空间提供良好视线。内外空间交融，能充分利用自然采光，又能将人工的办公环境和室外空间相连，创造生态且富有个性的办公环境。
建筑形体塑造采用“IT 车间”的概念，不规则斜面天窗的几何形体，仿佛是工业锯齿天窗的变异和抽象，经过当代视觉语言重新演绎的结果。
建筑立面上采用了多种现代的材质：Low-E 透明玻璃、U 型玻璃、钢材、波形穿孔金属板、防腐原木、素混凝土、欧文斯复合材料外墙板材等，结合建筑新颖而独特的形式与造型，还有建筑师严谨推敲后的搭配方式以及考究的比例选择，使得建筑充满了超强的视觉冲击和材质的韵律美感，符合软件园的整体视觉风格。

The project is located in Bell Laboratories in Xuzhuang Software Base, northwestern part of Xianlin District in Nanjing, which is one of the national information technology industry bases.
Dynamic and strict layout provide various and well-arranged form relationships and external spatial pattern. The multi-level courtyard and central public spaces organically convert and transfer from large building cluster to private spaces, forming a distinctive building cluster and providing great landscape views for each office building. The indoor spaces integrate the outdoor, to make full use of natural light, integrate the office environment with outdoor spaces, and create an ecological and personalized office environment for emerging industries.
The architectural form applies the concept of IT plant. The form of irregular cant skylight seems like the variation and abstract of industrial saw-toothed skylight.
Building elevation use various modern materials, such as Low-E transparent glass, U-shaped glass, steel, corrugated expended metal, anticorrosive log, plain concrete, Owens composite board, etc. It is the new and unique architectural form and style combining with the collocation and fine scale after architects' thorough study that gives the buildings avant-garde visual impact and beautiful cadence in material, which are all matched with the whole visual style of the software base.

狮子山风景区

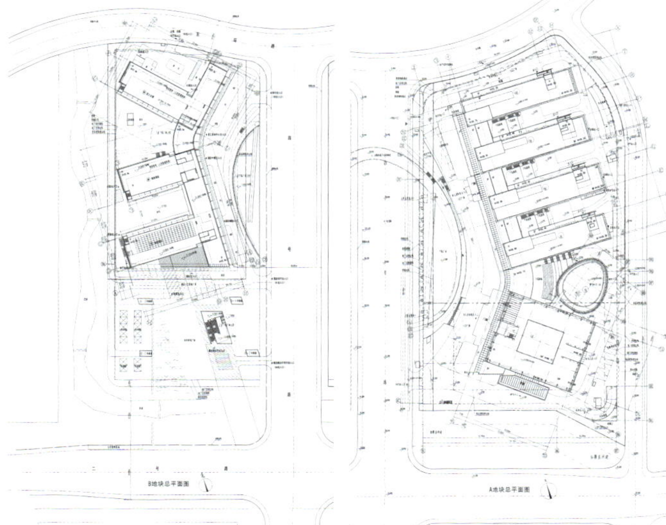

苏宁电器总部

Suning Appliance of China HQ

设 计 师：汪杰、傅世林、陈元俊、王丹
项目地点：江苏 南京
用地面积：A 地块 35 930 m²，B 地块 17 033 m²
建筑面积：A 地块 195 910 m²，B 地块 96 858 m²

Designer: Jie Wang, Shilin Fu, Yuanjun Chen, Dan Wang
Location: Nanjing, Jiangsu
Site Area: Lot A 35,930 m², Lot B 17,033 m²
Building Area: Lot A 195,910 m², Lot B 96,858 m²

苏宁电器集团总部办公基地，包括集团总部及房地产、酒店、百货等各产业办公区，是一个集办公、研发、展示、信息、接待、培训为一体的企业总部综合建筑群。该项目由美国恩比建建筑咨询（上海）有限公司完成方案设计。

项目特点是规划建筑群体与整个园区的空间尺度紧密结合，塑造良好和谐的园区整体空间环境，立面简洁，色彩明快，群体气势宏伟，充分展示国际一流企业的形象。

引入绿色、环保和可持续发展的现代建筑技术，创造舒适的办公环境和弹性的室内空间，充分凸显建筑的品质。

作为整个园区的主要交通空间，东西二区的 SPINE 共享连廊采用一气呵成的流线型形体。主立面向上略微后斜，更进一步地强调这一空间的独特作用。主立面的玻璃幕墙使室内外空间完全贯通，水平的遮阳板不仅为室内提供舒适的空间，同时强调了 SPINE 的水平动感。SPINE 内的各个低层建筑以独立形体插入，产生非常强烈的识别性。

建筑的细节通过石材、幕墙上丰富的细部及构件、不同透光度的玻璃、不同尺度的支撑框架及金属饰板、凸起或凹入的线条来强调，这些不同的细节刻画了建筑多个方向上的立面，使其更为丰富。

The office basement of Suning Headquarters consisting of headquarters of the group and other industries, such as real estate, hotel, department store, is a complex of corporate headquarters building cluster integrating office, research and development, display, information, reception and training. The program is designed by NBBJ Architectural Consulting (Shanghai) Co., Ltd.

The project features close combination between the building cluster and the whole park dimensions, to shape a good harmony of park conditions, with concise façade, bright color, magnificent buildings, to fully represent the image of a top international enterprise.

With green, environment friendly and sustainable modern building techniques, it creates comfortable offices and flexible indoor spaces to demonstrate the quality of the architecture.

As the main transport space of the whole park, the SPINE in the east and west areas share a corridor to form a streamline. The slightly backward main façade emphasizes the unique role of this space. The glass curtain wall of the main façade connects the inner and outer spaces smoothly. The horizontal louvers not only provide a comfortable indoor space, but also enhance the dynamics of the SPINE horizontally. All low-rise buildings are interlaid independently within the SPINE, to produce very strong identity.

The tile material, details on curtain wall, glass with different transparences, supporting frames and metal plates in different sizes convex or concave lines, all these emphasize the surfaces of the building and these details depict and enrich the different appearances of the building itself.

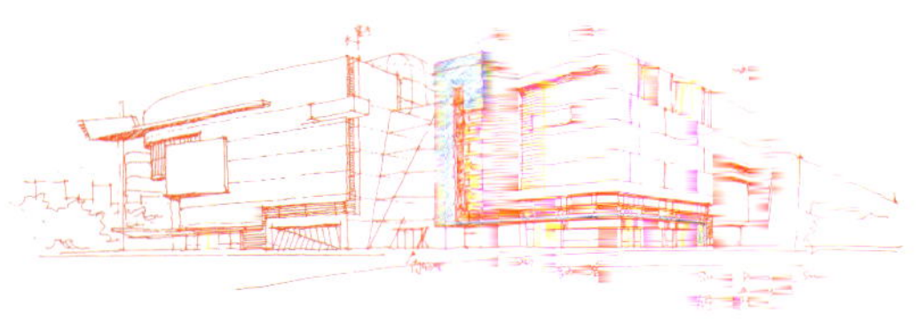

成都苏宁广场

Chengdu Suning Plaza

设计师：邱立岗、钟容、武悦、秦轶、王丹
项目地点：四川 成都
用地面积：267 36 m²
建筑面积：125 335 m²

Designer: Ligang Qiu, Rong Zhong, Yue Wu, Yi Qin, Dan Wang
Location: Chengdu, Sichuan
Site Area: 26,736 m²
Building Area: 125,335 m²

本项目位于成都地铁1号线上方，为了提高土地利用率，结构上采用33 m大跨度超大截面预应力转换梁跨越地铁隧道，将原本零碎的地块整合。

建筑功能整体上分为购物中心和苏宁电器店两大功能板块，根据这两部分的功能划分，建筑在形体关系上自然形成两个"L"形体量，围绕贯穿整个建筑的"L"形采光中庭空间展开。

作为苏宁电器集团开发的第一栋大型自建店，该项目在商业业态设计上把苏宁电器主力店和购物中心很好地结合在一起，两部分商业内容既相对独立，又互为补充，相辅相成，实现商业价值的最大化，是商业建筑设计的一次全新尝试。

The project is located overhead the Chengdu subway line one. To increase land utilization ratio, 33 m large span and large section prestress transition beam is applied in structure, which is crossing the subway tunnel and underpinning the six-story superstructure, in this way, the fragmentary land is used integratedly.

This building function can be divided into two parts, shopping mall and the Suning appliance store. Considering this, there are two L-shaped masses in architectural form, which are designed according to the atrium whose daylighting is through the whole L-shaped mass.

As the first self-built store by Suning Appliance Corp., the project well combined the key Suning appliance store and shopping mall in business store format. The commercial contents of the two parts are relatively independent but also complementary to each other to maximize the commercial value, which is a totally new attempt in business architectural design.

扫描查看更多信息

南京金宸建筑设计有限公司成立于1994年，是具有国家建设部工程设计甲级资质、人防乙级资质的综合性建筑设计公司，拥有一支由160多名优秀设计师组成的设计团队，公司现有第一建筑设计事业部、第二建筑设计事业部、第三建筑设计事业部、机电设计事业部、城市综合体设计研究院创作一部及创作二部等6个设计部门。本公司提供从策划、方案、初步设计到施工图设计的全过程建筑设计服务。

地址：南京市建邺区梦都大街150号建筑师工社4层
电话：+86-25-86500666
传真：+86-25-86501232
邮箱：kad@kingdomarch.com
网址：www.kingdomarch.com

Add: 4F Architects Commune, Mengdu Street No.150, Jianye District, Nanjing
Tel : +86-25-86500666
Fax : +86-25-86501232
E-Mail: kad@kingdomarch.com
Web: www.kingdomarch.com

玛斯兰德
Master Land

项目地点：江苏 南京
用地面积：482 000 m²
建筑面积：258 600 m²

Location: Nanjing, Jiangsu
Site Area: 482,000 m²
Building Area: 258,600 m²

玛斯兰德位于南京翠屏山风景区，由美国知名大师领衔规划，资深团队倾力打造，力图创造一种回归纯朴、自然的生活方式，希望人与人、人与社区、人与自然都处于一种相对和谐的状态。

Master Land is located in Green Screen Mountain Scenery of Nanjing. The project is planned by a famous US master, and built by a senior team with all efforts. It tries to create a pure lifestyle that returns to the nature, in the hope of a relative harmony between human and human, human and community, and human and nature.

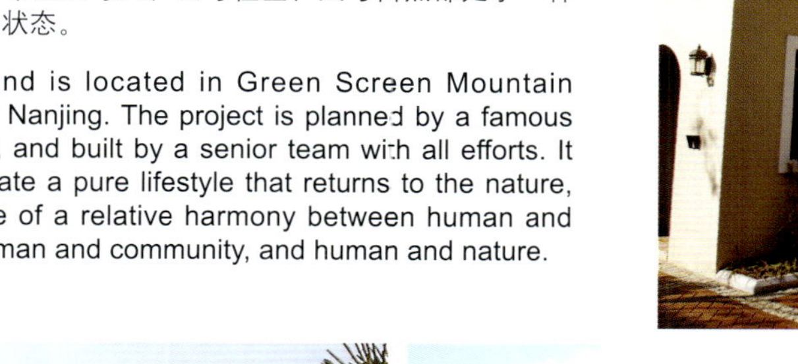

仁恒国际公寓

Yanlord International Apartment

项目地点：江苏 南京
用地面积：28 600 m²
建筑面积：135 700 m²

Location: Nanjing, Jiangsu
Site Area: 28,600 m²
Building Area: 135,700 m²

项目紧邻南京奥林匹克体育中心，是河西新城中央商务区最北端的重要标志性建筑。公寓借鉴美国都市高层豪宅的居住方式和空间构成，营造时尚开放的居家生活氛围；结合南京的地域特征和生活习惯，营造高度舒适、功能合理的居住空间；突破居住建筑固有的建筑外观，构筑挺拔优雅、卓尔不群的标志性建筑形象；注重节能环保和生态环境的塑造。仁恒国际公寓既具有豪华的高级酒店式日常生活服务，又极力营造高级公寓的生活情趣，注重朝向、景观等日常居住需求。

The project is close to Nanjing Olympic Sports Center, and is an important landmark in the northernmost end of Hexi New Town CBD. The apartment learns from US metropolitan hi-rise luxury residential style and space composition, to create a fashionable open household living condition. Considering Nanjing locality and lifestyle, it creates a residential space with comfortable height and reasonable functions. It breaks through the inherent appearance of residential buildings, with straight and elegant architecture, and outstanding landmark image. It pays attentions to energy-saving and ecology-shaping. Yanlord International Apartment not only offers luxury high-level hotel daily life services, but also tries to create the living interests in high-level apartment, highlighting the natural lighting, landscaping and other daily living experience.

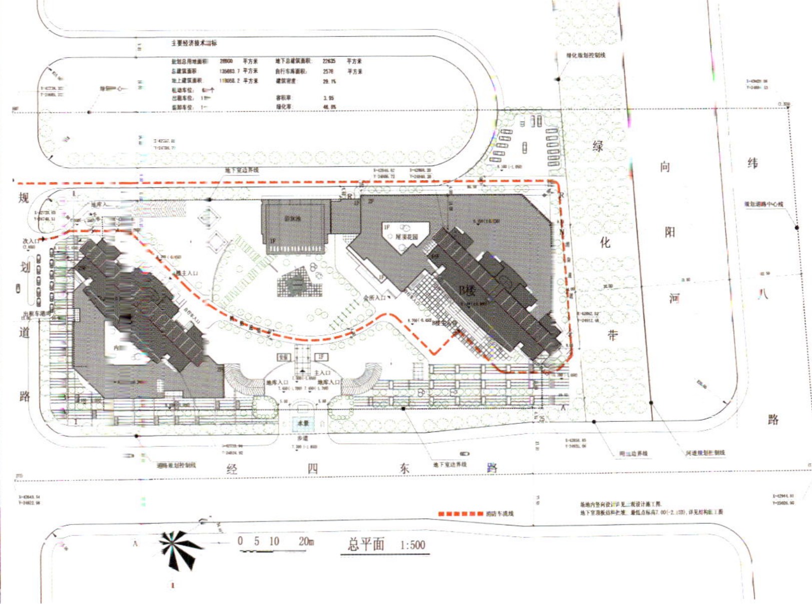

江苏省美术馆新馆（展览馆）
New House (Exhibition House) of Jiangsu Provincial Art Museum

项目地点：江苏 南京
用地面积：10 605 m²
建筑面积：27 827.06 m²

Location: Nanjing, Jiangsu
Site Area: 10,605 m²
Building Area: 27,827.06 m²

项目位于文化古迹轴线长江路的南侧，革命历史轴线中山东路的北侧。规划在基地的西北角形设计一个城市广场，延伸向原有的美术馆，并与原孙中山总统府从视觉上产生联系。在两个相互向对方旋转并咬合在一起的“C”形建筑体量之间形成了一个“峡谷”一样的中央大厅和共享交通空间。建筑的主要入口在建筑的西北侧，并朝向市民广场；次要入口位于东南侧，作为“峡谷”的结束。新馆以简洁富有雕塑感的造型给人留下深刻印象。

The project is located on the south side of Changjiang Road, the axis of cultural monuments, and on the north side of Zhongshan East Road, the axis of revolutionary history. The buildings form a city square on the northwest corner of the base, which extends to original art museum, and connects with Sun Yat-sen's Presidential Palace visually. Two C-shaped buildings are spiraled with each other, to form a valley-like central hall and common traffic space. The main entrance is on the northwest side of the buildings, and facing toward Citizens Square. The secondary entrance is on the southeast side of the buildings, as the end of "valley". The new house is impressing with its concise and sculptural shape.

南京水游城

Nanjing Aqua City

项目地点：江苏 南京
用地面积：26 700 m²
建筑面积：167 000 m²

Location: Nanjing, Jiangsu
Site Area: 26,700 m²
Building Area: 167,000 m²

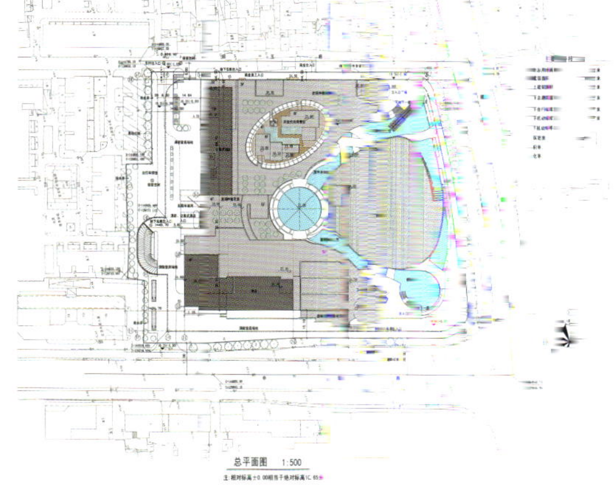

项目位于南京古城内城南地区，夫子庙商业圈核心地段。在设计中，水街空间是整个建筑内部空间的视觉焦点，在平面布置中构成了主要的室外步行空间，主要商业店铺均围绕室外步行空间而设置。因而在设计中力求使其舒展流畅并富于变化，弧线形的造型使室内空间更加丰富且有层次感。
营造独特的、具有浓郁商业气氛的高档购物中心，提高本地区城市品位是造型设计的指导思想。大面积玻璃幕墙和铝板幕墙的对比与凹凸处理，以及晚间的多处高度泛光照明和内部商业空间散发出的温暖光线，呈现出浓郁的商业氛围。

The project is located in southern old town of Nanjing, or the core of today's Confucius Shrine CBD. In the design, the aqua street, as the visual focus of the whole interior spaces, forms the main outdoor pedestrian space in the layout plan, around which set all main merchant stores. Therefore, the design tries to make it extensive, smooth and changing, and its curved shape enriches the indoor space and multilevel sense.
The guideline of the shaping design is to create a unique, densely commercial hi-end shopping center, so as to increase the city taste and grade of this area. The contrast and concave-convex processing between structural walls, large area glass curtain walls and aluminum curtain walls, plus floodlighting on the façade and warm lights from inside the commercial spaces in nighttime, give a full display of strong commercial climate.

辽宁省建筑设计研究院
LIAONING PROVINCIAL BUILDING DESIGN & RESEARCH INSTITUTE

辽宁省建筑设计研究院（LDI）是国家甲级建筑设计和甲级工程勘察单位，也是国家首批具有建筑智能化专项工程设计资质的单位之一。
LDI创建于1956年1月，至今已发展成为能够为工程建设提供全方位综合服务的专业设置齐全、设计手段先进、技术力量雄厚、社会信誉卓著的优秀勘察、设计和咨询单位。
LDI下设综合性建筑设计研究所、建筑方案研究所、规划设计研究所、市政设计研究所、环境艺术研究所、建筑经济所以及林立岩大师工作室、地源热泵应用技术研究所、技术研发中心（下设BIM工作室、特殊结构分析工作室）等部门，院属公司有岩土工程公司、项目管理咨询公司、施工图审查咨询有限公司等实体。并于2006年成立了大连分院。
LDI拥有职工540余人，具有国家级执业资格的有130余人，拥有中国工程设计大师1人，辽宁省创新领军人才1人，辽宁省优秀专家3人，辽宁省工程设计大师3人，辽宁省工程勘察大师1人，享受政府特殊津贴专家6人，入选辽宁省“百千万人才工程”百层次2人、千层次1人。
建院56年来，已经向社会提供了1.3万余项设计成果，工程项目遍布全国及海外部分国家。
20世纪80年代以来，共获国家、部、省、市优秀工程设计（勘察）奖228项和科技进步奖85项。

Liaoning Provincial Building Design & Research Institute (LDI) is a national top-level unit of architectural design and engineering reconnaissance, also a national first unit which has the special engineering design qualification of architectural intellectualization.
LDI was established in Jan. 1956, and has developed into an excellent reconnaissance, design and consultation unit which can provide omnibearing comprehensive service for engineering construction with complete professional equipments, advanced design method, abundant technical force and excellent social repute.
LDI's institute sets up four Comprehensive Building Design & Research studio, Architectural Plan Research studio, Conceptual Design Research studio, Municipal Design Research studio, Environmental Art Research studio, Building Economy studio, the Master Lin Liyan's studio, Applied Technology of Ground-source Heat Pump Research studio, Technical Research & Development Center (it sets up BIM studio and Special Construction Analysis studio), and so on. The companies which are subject to the institute include Geotechnical Engineering Company, Project Management Consultation Company, Working Drawing Examination & Consultation Co., Ltd., and so on. Dalian Branch was established in 2006.

There are more than 540 employees in LDI, the registered persons who have the national practicing qualification are more than 130 persons among them. Otherwise, there is a Chinese engineering design master, an innovation leading talent of Liaoning Province, 3 excellent experts of Liaoning Province, 3 engineering design masters of Liaoning Province, an engineering reconnaissance master of Liaoning Province, 6 experts who take governmental special allowance, 2 persons who are selected in the Hundred Level of “Liaoning Provincial Hundred Thousand and Ten Thousand Talent Engineering”, a person who is selected in the Thousand Level of “Liaoning Provincial Hundred Thousand and Ten Thousand Talent Engineering”.
For more than 56 years since the institute was established, it has provided more than 13,000 design achievements for the society, its engineering projects have covered the whole country and parts of overseas countries.
Since the 1980s, LDI has obtained 228 excellent engineering design (reconnaissance) prizes and 85 advancement prizes of science and technology of nation, ministry, province and city.

地址：辽宁省沈阳市和平区和平南大街84号
电话：+86-24-23392468
传真：+86-24-23389612
网址：www.ldi.com.cn

Add: No.84, Heping Nan Street, Heping District, Shenyang, Liaoning
Tel: +86-24-23392468
Fax: +86-24-23389612
Web: www.ldi.com.cn

辽宁省城市建设学校新校区
Liaoning Urban Construction Technical College New Campus

项目地点：辽宁 沈阳
建筑面积：81 873.76 m²

Location: Shenyang, Liaoning
Building Area: 81,873.76 m²

鲁迅美术学院大连校区

Luxun Academy of Fine Arts, Dalian Campus

项目地点：辽宁 大连
用地面积：382 868 m²

Location: Dalian, Liaoning
Site Area: 382,868 m²

东北传媒大厦
Northeast Media Building

项目地点：辽宁 沈阳
建筑面积：86 150 m^2

Location: Shenyang, Liaoning
Building Area: 86,150 m^2

本溪中国药都会展中心
Chinese Medicine Exhibition Center, Benxi

项目地点：辽宁 本溪
建筑面积：57 755 m^2

Location: Benxi, Liaoning
Building Area: 57,755 m^2

中海国际社区一期——中海龙湾
COPIC International Community Phase I — COPIC Dragon Bay

项目地点：辽宁 沈阳
建筑面积：64 000 m²

Location: Shenyang, Liaoning
Building Area: 64,000 m²

沈阳万宸建筑规划设计有限公司
Shenyang Wanchen Architecture & Plan Design Co.,Ltd.

沈阳万宸建筑规划设计有限公司成立于2005年9月，具有中华人民共和国建设部颁发的建筑工程甲级设计资质、规划乙级设计资质。
公司主要承揽建筑工程设计业务，自成立以来，坚持"以设计质量求生存，以诚信服务求信誉，以技术创新求振兴，以科学管理求发展"的宗旨，已先后完成了数百项、近千万平方米的工程设计项目，作品遍及辽宁省各主要城市以及吉林、黑龙江两省。万宸的每一项作品都凝聚着设计人员的聪明才智和创作激情以及对树立品牌的执著，正是因为不懈努力，使业主和社会认识了万宸、认可了万宸。

地址：沈阳市和平区胜利南大街92号胜利大厦14层
电话：+86-24-83507869-800/83507867
传真：+86-24-83507865
邮箱：wanchensj001@sina.com
网址：www.wcjzsj.com

Add: 14F, Victory Building, 92 Victory South Street, Peace District, Shenyang City
Tel: +86-24-83507869-800/83507867
Fax: +86-24-83507865
E-mail: wanchensj001@sina.com
Web: www.wcjzsj.com

阜新商业街水岸景观规划
Waterfront Landscape Plann of Commercial Street, Fuxin

项目地点：辽宁 阜新
建筑面积：7000 m²

Location: Fuxin, Liaoning
Building Area: 7,000 m²

鲅鱼圈商业居住区规划

Commercial & Residential Community Plan, Bayuquan District, Yingkou

项目地点：辽宁 营口
建筑面积：480 000 m²

Location: Yingkou, Liaoning
Building Area: 480,000 m²

盘锦广厦新城
Guangsha New Town, Panjin

项目地点：辽宁 盘锦
建筑面积：450 000 m^2

Locatior: Panjin, Liaoning
Building Area: 450,000 m^2

盘锦渤海路商业街入口

Entrance to Commercial Street of Bohai Road, Panjin

项目地点：辽宁 盘锦
建筑面积：12 000 m^2

Location: Panjin, Liaoning
Building Area: 12,000 m^2

抚顺雷锋体育场商业规划

Lei Feng Stadium Commercial Plan, Fushun

项目地点：辽宁 抚顺
建筑面积：900 000 m^2

Location: Fushun, Liaoning
Building Area: 900,000 m^2

盘锦辽滨接待中心
Liaobin Reception Center, Panjin

项目地点：辽宁 盘锦
建筑面积：40 000 m^2

Location: Panjin, Liaoning
Building Area: 40,000 m^2

沈阳怪坡办公楼
Office Building of Magic Slope, Shenyang

项目地点：辽宁 沈阳
建筑面积：2000 m^2

Location: Shenyang, Liaoning
Building Area: 2,000 m^2

沈阳财富中心售楼处
Sales Office, Fortune Center, Shenyang

大连市建筑设计研究院有限公司 C&Z建筑师工作室

Dalian Architectural Design&Research Institute Co.,Ltd. C & Z Architects Studio

山西剧院 1

C&Z建筑师工作室是大连市建筑设计研究院集团公司专业平台上的方案设计团队。由地域知名建筑师崔岩、赵涛领衔，与10多名有共同目标理念的职业建筑师组成。经过多年的设计实践，工作室逐步形成了一套成熟的方案创作方式，力争做到作品从前期研究至最终完成的每个工作环节中最大程度地体现艺术激情与专业理性的完美结合。 依托于大连市建筑设计研究院集团公司的综合技术平台，C&Z建筑师工作室在方案设计过程中提前引入专业配合的设计概念和方案优化信息，确保作品有条不紊地全方位推进直至投入使用。多年灵活严谨的设计经验及成功的建成作品，使C&Z建筑师工作室逐步成为地域建筑方案设计领域中一支有显著特色的设计力量。

C&Z Architecture Studio is the schematic design team among all the design teams of Dalian Architectural Design&Research Institute Co.,Ltd. Under the leadership of two locally distinguished architects,Yan Cui and Tao Zhao it consists of more than ten professional architects who share mutual goals and pursuits. After many years ofdesigning practice, it has developed a set of mature working protocols, aiming for a perfect combination of artistic passion and professional sense through each phase from preliminary analysis to the final construction. Thanks to the comprehensive technical support of Dalian Architectural Design& Research Institute Co., Ltd., C&Z Architecture Studio introduces the concepts and strategies of the cooperative fields to the schematic design,and advances every project with order and coordination. Because of the flexible and rigorous style and many successful cases, C&Z Architecture Studio has been upgrading its qualification remarkably fast and has been recognized as an esteemed star in the realm of local architectural schematic design.

生态科技城育明高中 5

地址：中国大连市西岗区胜利路102号308室
电话：+86-411-84313789
传真：+86-411-84303187
邮箱：ud_studio@126.com
网址： www.dlad.com.cn/

Add: No.102 SHENGEL RD.XIGANG DIST.DALIAN,CHINA
Tel: +86-411-84313789
Fax: +86-411-84303187
E-mail: ud_studio@126.com
Web: www.dlad.com.cn/

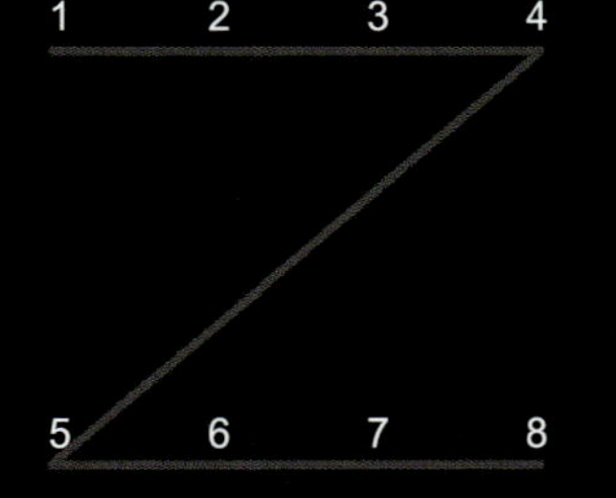

1. 山西剧院
2. 大连晟华科技大厦
3. 山西剧院
4. 大连国合国际公寓
5. 邢良坤陶艺馆
6. 远洋钻石湾运动中心
7. 大连国和东港
8. 大连国合国际公寓

DISCOVER CREATIVITY READY TO CHAMGE
LOOKING INTO THE FUTURE DREAMING
CAPABLEOF DREAMING DISCOVER POWERFUS
THINK MORECREAT MORE CREATIVITY
FREEING IDEAS COLLABORATIONS
INNOVATIONS ROCKS DREAMING
THINK MORECREAT MORE
CREATIVITY READY TO
CHAMGE THINK
MORECREAT
DISCOVER

DREAMING
CHAMGE THINK
CREATIVITY READY
TO CHAMGE IDEAS
THINK MORECREAT MORE
INNOVATIONS ROCKS DREAMING
FREEING IDEAS COLLABORATIONS
THINK MORECREAT MORE CREATIVITY
LOOKING INTO THE FUTURE DREAMI
DISCOVER CREATIVITY READY TO CHAMGE

扫描查看更多信息

沈阳都市建筑设计有限公司
Shenyang Urban Architecture Design Co., Ltd.

扫描查看更多信息

沈阳都市建筑设计有限公司成立于2001年4月，是中华人民共和国建设部批准的具有甲级设计资质并兼有地下人防工程乙级设计资质的一家民营股份制建筑设计及咨询机构。专门从事各类建筑设计、城市规划、区域规划、工程咨询、技术顾问、工程监理、装饰设计等多方位专业性服务业务，并通过了ISO9001：2008质量管理体系认证。

公司由多个专业设计及管理部门组成，共有职工186名。具有国家注册资格人员20名，其中拥有国家一级注册建筑师6名、国家一级注册结构工程师8名、注册公用设备工程师3名、注册电气工程师3名、42名高级工程师、52名工程师。任用具有多年工作经验的高级工程师负责本专业的技术组织和质量控制。

地址：辽宁省沈阳市沈河区西滨河路60-2号5门
电话：+86-24-23939337
传真：+86-24-23931370
邮箱：urban2001@126.com
网址：www.urban2001.cn

Add: Gate 5,West Binhe Road No.60-2,Shenhe District,Shenyang City,Liaoning Province
Tel: +86-24-23939337
Fax: +86-24-23931370
E-mail: urban2001@126.com
Web: www.urban2001.cn

北京宋庄公共服务平台
Songzhuang Public Service Platform, Beijing

建 筑 师：康慨
项目地点：北京
建筑面积：15 000 m²

Designer: Kai Kang
Location: Beijing
Building Area: 15,000 m²

方案在设计上力求建立建筑与街道、建筑与城市之间的良好关系。同时，这种临街封闭却内部中空的建筑方式与中国传统城市院落式的格局以及用简单形体构造丰富空间的传统建筑精髓也极为吻合。

建筑设计理念旨在创造一个空间，一个场所意味着设定或取消界限，概念设计中通过折叠从外部创造出界面变化，重新塑造了丰富多样的空间。松弛的建筑外表皮不在于叙述并解释内部功能下的结构，而是它表达了折叠自身运动中的统一性，给建筑空间带来了紧张与松弛、压缩与膨胀、连续与断裂、集中与消散等多种微妙感受。

The design tries to establish a good relationship between buildings and streets as well as buildings and the city. Meanwhile, the closed roadside and hollow inside architectural style is in accordance with the traditional Chinese urban courtyard layout and the traditional architectural essence of creating large space in simple forms.

The architectural design idea aims at creating a space, a venue, which means setting or abolishing the boundary. The conceptual design rebuilds a rich diversity of spaces by folding the exterior surface to create interface changes. The loose building surface no longer describes and explains the structure of interior functions. On the contrary, it explains the unity of folding motions, bringing subtle feelings of intensity and looseness, compression and expansion, continuity and rupture, concentration and evanishment.

盘锦水榭春城酒店

Waterside Spring Town Hotel, Panjin

设 计 师：康慨、顾全衡、黎晓黎、臧微
项目地点：辽宁 盘锦
建筑面积：110 000 m²

Designer: Kai Kang, Quanheng Gu, Xiaoli Li, Wei Zang
Location: Panjin, Liaoning
Building Area: 110,000 m²

项目选址的自然环境景观独一无二，如何最大限度地利用环境、融入环境，进而提升环境的整体品质是本次设计的重点。

The natural landscape of the project location is unique, so this design emphasizes on how to use the environment, integrate the buildings into the environment and in return enhance the integral quality of the environment as much as possible.

营口万隆广场
Wanlong Square, Yingkou

设 计 师：尹旭东、鄂丽明、邢征
项目地点：辽宁 营口
建筑面积：360 000 m²
合作单位：培特维建筑设计咨询（上海）有限公司

Designer: Xudong Yin, Liming E, Zheng Xing
Location: Yingkou, Liaoning
Building Area: 360,000 m²
Partners: PTW Architects Shanghai

项目位于营口鲅鱼圈，东临昆仑大街，地理位置优越。
建筑在合理设置平面功能及总体规划的基础上，立面采用“框架集装箱”的形式，以强烈、简洁的形体构成丰富的立面，创建一个多层次街景。
在满足使用功能及设备安装的条件下，尽可能降低层高、节省投资、增加经济效益，同时楼层高度按一定的规律设计，以保证立面效果，保证构件的统一化、模数化。

The project is loated in Bayuquan District of Yingkou City. It is beside Kunlun Street in the east, with advantageous geography.
Based on plane functions and layout planning, the building elevation adopts the “flat rack container” style, constructs a rich elevation in intense and simple forms and creates multilevel street sights.
This project tries to lower the height between floors, save investments and increase economic interests while meeting the requirements of functions and equipment installation. Meanwhile the height between floors is arranged according to certain rules, to ensure the façade effect and the structural unity and modulation.

瑞安 · 沈阳天地

Shui On Land · Shenyang World

设 计 师：尹旭东、鄂丽明
项目地点：辽宁 沈阳
建筑面积：290 000 m^2
合作单位：巴马丹拿建筑设计咨询（上海）有限公司

Designer: Xudong Yin, Liming E
Location: Shenyang, Liaoning
Building Area: 290,000 m^2
Partners: P&T Group Shanghai

项目位于沈阳市北陵大街中段，南邻巴山路，地理位置十分优越。
项目从城市设计入手，创造一个具有当今国际水平的现代建筑群体，使之溶于整个城市环境之中，丰富城市风貌，成为独特的地区性标志。建筑形式上借鉴现代欧式及现代简洁的建筑形式，展现出沈阳的商业文化特色。在建筑立面处理上，强化建筑的体量，以简洁的设计手法，效法古典的构图比例，实现城市设计与建筑设计的完美组合。

Located in the middle of Beiling Street of Shenyang City, beside Bashan Road in the south, this project has great geographical advantage.
This project creates a modern building cluster to the international standards by starting from city design. The building cluster integrates with the whole city environment, enriches the city landscape and becomes a unique regional landmark. The architectural style fully borrows the idea from modern European style and Shanghai style of simplicity and represents the special themes of Shenyang commercial culture. The façade emphasizes the building volume, with simple design method and classical composition proportion, to achieve a perfect combination of city design and architectural design.

内蒙古工大建筑设计有限责任公司

Inner Mongolia Grand Architecture Design Co.,Ltd.

10年，不长亦不短，对内蒙古工大设计来说，既是起步的10年，也是积累的10年，更是价值观形成的10年。在公司改制10周年之际，从每年完成的项目中选择一个作品回溯10年的创作历程。

地址：中国内蒙古呼和浩特市哲里木路49号
邮编：010051
电话：+86-471-6576005/575347/6577170
传真：+86-471-6576322
邮箱：Gdsjy@yahoo.com.cn

10年，10个建筑

内蒙古工大设计（内蒙古工大建筑设计有限责任公司），其前身是成立于1979年的内蒙古工业大学建筑勘察设计研究院。2002年，响应建设部号召，设计研究院改制为股份有限责任公司，建立起符合市场规律的现代企业制度。同时，公司建筑设计资质由乙级升为甲级，并于2004年通过了ISO9001：2000质量体系认证，步入了快速发展的轨道。

公司现有职工近百人，其中教授4人、副教授6人、高级工程师18人、工程师31人、一级注册建筑师6人、一级注册结构工程师5人、注册岩土勘察师5人、注册设备师3人、注册电气师3人。同时公司以高校学术力量为依托，充分发挥广大专业教师的设计潜能，在文化、教育建筑设计、本土建筑研究、绿色建筑策略及技术等方面成绩斐然，在自治区树立了自身独特的行业地位。

经过10年的发展壮大，公司管理制度日臻完善，综合实力逐步提高。目前，公司业务范围涵盖建筑、规划、室内、环境景观、地质勘查、工程咨询等。

经过10年的潜心创作，默默耕耘，公司完成了近百项大中型工程设计项目。通过"立足内蒙古，面向全国"的经营思路，和实施"传建筑文化，争创名优设计"的战略，创作出了大量的优秀建筑精品，获建设部、勘察设计协会、世界华人建筑师协会及自治区优秀设计奖十余项，受到社会广泛好评。

1. 内蒙古师范大学逸夫艺术楼
2. 鄂尔多斯影剧院改造
3. 内蒙古文化大厦
4. 内蒙古家具博物馆
5. 内蒙古盛乐博物馆
6. 呼和浩特市第二中学新校区
7. 乌兰察布市博物馆、图书馆
8. 内蒙古工业大学建筑馆
9. 恩格贝沙漠科学馆
10. 斯琴博物馆

Inner Mongolia GDSJ (Inner Mongolia UT Architectural Design Co., Ltd.), preceded by the Architectural Survey and Design Institute of Inner Mongolia University of Technology, was founded in 1979. In 2002, in response to the call of the MOHURD, the institute was transformed into a limited liability company with a modern market-oriented corporate system. Meanwhile, the architectural design qualification of the company ascended from Class B to Class A. In 2004, it adopted the ISO9001 Quality System (2000), entering a rapid development stage.

GDSJ currently has about 100 employees, including 4 professors, 6 associate professors, 18 senior engineers, 31 engineers, 6 first-class registered architects, 5 first-class registered structural engineers, 5 registered geotechnical investigation engineers, 3 registered equipment engineers and 3 registered electrical engineers. Meanwhile, with the academic strength of Inner Mongolia University of Technology, GDSJ makes full use of the design potentials of specialized teachers and has made great achievements in the design of cultural and school buildings, the study of local buildings and the strategy and technology of green buildings, establishing a foothold in the industry in the Inner Mongolia Autonomous Region with its own characteristics.

Through ten years' development and expansion, GDSJ has become an enterprise with a much improved management system and gradually enhanced comprehensive strength. GDSJ is currently engaged in building, planning, interior design, environmental landscape, geological survey and engineering consulting and so on.

In the past decade, GDSJ, with painstaking work, has completed nearly 100 large and medium-sized engineering design projects across the Inner Mongolia Autonomous Region and the People's Republic of Mongolia. Following the management concept "Based in Inner Mongolia and Facing the Whole Country" and the strategy of "Inheriting Traditional Architectural Culture and Creating Famous Brand and High-quality Design", GDSJ has created a great number of excellent architectural works and is well received by the society, with over ten awards from the MOHURD, China Exploration & Design Association, World Association of Chinese Architects and Inner Mongolia Autonomous Region People's Government.

扫描查看更多信息

山东建大建筑规划设计研究院
Shandong Jianzhu University Architecture & Urban Planning Design Institute

山东建大建筑规划设计研究院（原山东建筑大学设计研究院）成立于1960年，目前设计研究院已具有建筑工程设计、城市规划、风景园林、工程咨询四种甲级资质以及施工图设计文件审查机构认定书（建筑工程一类——钢结构），同时具有市政工程设计乙级设计资质。主要从事民用与工业建筑设计、城市规划、风景园林以及相关专业的工程设计与工程咨询业务。

设计研究院下辖行政与人力资源部、生产经营部、技术与质量管理部、财务部4个行政管理部门，以及建筑设计一分院、建筑设计二分院、规划设计一分院、规划设计二分院、建筑创作研究中心、咨询审图中心6个生产部门。全院共有员工160余人，设计人员中具有中高级技术职称人员约占60%以上，其中国家一级注册建筑师、注册结构工程师共27人，国家注册规划师、注册公用设备工程师、注册电气工程师、注册咨询工程师、注册造价工程师共44人，香港注册建筑师、结构工程师学会会员共5人，经过50余年的努力，山东建大建筑规划设计研究院已经成为一所集设计、教学与科研相结合，技术实力雄厚，管理先进，在省内外有良好影响和较高知名度的设计研究单位。

地址：山东省济南市历下区历山路96号
电话：+86-531-86367169
传真：+86-531-86956156
邮箱：sdjdsjy@163.com
网址：www.jdsjy.com

Add: 96 Lishan Road, Lixia District, Ji'nan City, Shandong Province
Tel.: +86-531-86367169
Fax: +86-531-86956156
Email: sdjdsjy@163.com
Web: www.jdsjy.com

燕子山庄扩建改造工程方案设计
Concept Design of Swallow Resort Extension and Reconstruction

设 计 师：赵学义、王润政、安俊贤、孙斌
项目地点：山东 济南
建筑面积：15 473 m²

Designer: Xueyi Zhao, Runzheng Wang, Junxian An, Bing Sun
Location: Ji'nan Shandong
Building Area: 15,473 m²

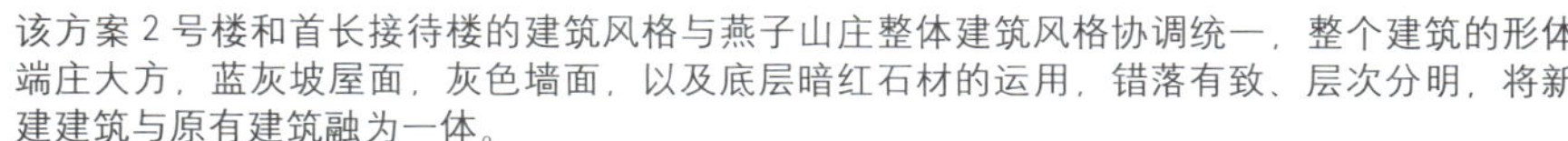

该方案 2 号楼和首长接待楼的建筑风格与燕子山庄整体建筑风格协调统一，整个建筑的形体端庄大方，蓝灰坡屋面，灰色墙面，以及底层暗红石材的运用，错落有致、层次分明，将新建建筑与原有建筑融为一体。

首长接待楼位于 2 号楼的东侧，1 号楼的南侧，共 3 层。东西长 44.6 m，南北长 21.8 m。用地面积 935 m²，建筑面积 2668 m²，每户 1334 m²，两个首长楼，分开布置，通过电梯解决上下交通问题。

2 号楼位于燕子山庄南部，借助地形呈高低错落状，东西向布局于山庄绿地南侧。主体 5 层，局部 6 层，用地面积 2194.88 m²。主要功能是客房、会议以及餐饮。2 号楼共包含客房 134 间，其中小套间 30 个、单人间 70 个、标准间 34 个，共计容纳 168 人。小套间布置在 1 ~ 5 楼的北向，5 楼的布置在南向。

The style of Building No. 2 and Principal Reception Building is concerted with the integral style of Swallow Hill, the whole form is elegant and generous, blue grey sloping roof, in order and distinct levels, grey wall and bottom dark red stone materials, enabling the new architecture to integrate with existing architecture.

Principal Reception Building is east of Building No. 2, south of Building No. 1, with 3 floors. It is 44.6 m long in east-west direction, and 21.8 m wide in south-north direction. With land area 935 m², total floor area 2,668 m², each unit 1,334 m², the two Principal Reception Buildings are separately, an elevator communicating the two.

Building No. 2 is in the south of Swallow Hill, following the landform, in different elevations, with its east-west layout in the south side of green land. The main building has 5 floors, some 6 floors in some part. With land area 2,194.88 m², floor area 12,805.56 m², it has main functions including guesthouse, conference, and catering. It has 134 guestrooms, including 30 small suites, 70 single rooms, 34 standard rooms, with total capacity of 168 guests. The small suites are set in north direction from the 1st to 5th floor, and in south direction on the fifth floor.

临沂市公安局业务技术用房方案设计

Concept Design of Linyi Public Security Bureau Technical House

设 计 师：王润政、赵学义、安俊贤、辛晶、孙斌
项目地点：山东 临沂
建筑面积：86 887 m²

Designer: Runzheng Wang, Xueyi Zhao, Junxian An, Jing Xin, Bin Sun
Location: Linyi, Shandong
Building Area: 86,887 m²

建筑形体充分考虑与周边建筑环境协调统一。场地周边现有建筑大都为高层建筑，因此为了体现建筑的气势与挺拔，设计师考虑把建筑主体做成22层，成长条形布置在中轴线上，从而与周边高层建筑协调统一又不失特色。同时为了使新建建筑在较大地块中不至于显得过于单薄，在主体南侧设计了5层裙房，尽可能拉大裙房面宽，增加建筑沿街面长度，使建筑显得更加宏伟大气。厚重裙房与挺拔高层的穿插象征着在稳固的基础上不断提升、勇攀高峰。

竖向线条突显挺拔气势。

建筑主体立面采用竖向线条的设计，增加质感的同时也突出了建筑的挺拔感。充满韵律的线条、强烈的虚实对比与体块的穿插给建筑注入了现代气息。建筑一层外部设有大面积浮雕墙，带给了建筑另一种文化韵味。色彩上，整个建筑以浅色石材为主，配以深色横线条，纵横交错增加了建筑的细节。远远望去整个建筑简洁明快、大气稳重。

The architectural shape considers the harmony with surrounding architecture.

The site surrounded by high-rise buildings mostly, so we consider to lay the 22-floor main building on the central axis, representing the strength and straightness, so the building with be in harmony with surrounding hi-rise buildings; meanwhile, in order to strengthen the new building in the large lot context, we consider to design a 5-floor skirt south of the main building to maximize the breadth of the skirt and increase the roadside length to make the building even grander. Massive skirt and straight hi-rise building symbolize continuous increase on solid basis and climbing for the peak.

Vertical lines highlight straight vigor.

The façade of main building is designed in vertical lines, to increasing the sense of reality while highlighting the straightness of building. The rhythmic lines, strong contrast and interlaying blocks infuse modern flavors into the building. The external ground floor is set with large relieve walls, bringing another cultural flavor to the building. In color, the whole building is keynoted by light color stone materials, with dark color horizontal lines, to make crossings increasing the details of building. Looking afar, the whole building is concise, clear, generous and stable.

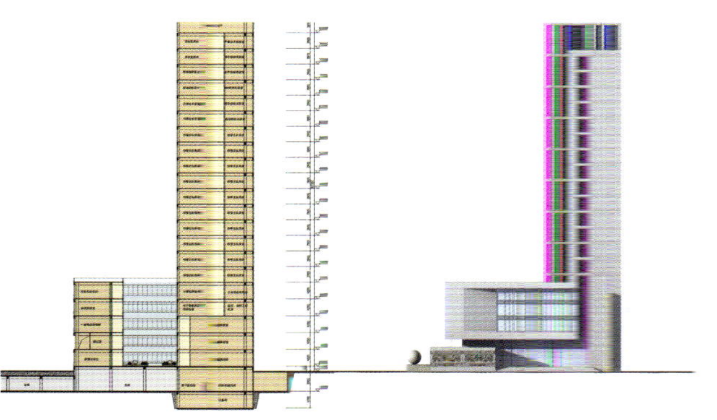

剖面图　　东立面图

一层平面图

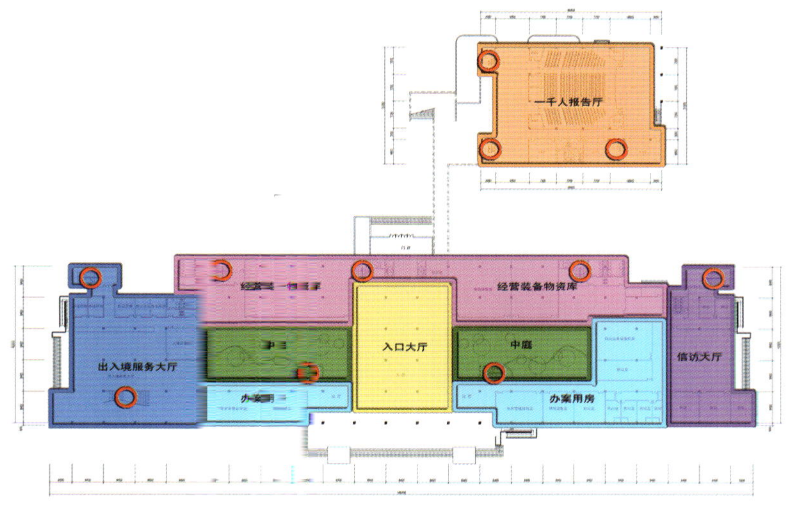

山东省人力资源市场方案设计

Concept Design of Shandong Human Resources Market

设 计 师：王润政、王乾、杜浩、刘杰民
项目地点：山东 济南
建筑面积：56 830 m²

Designer: Runzheng Wang, Qian Wang, Hao Du, Jiemin Liu
Location: Ji'nan Shandong
Building Area: 56,830 m²

打造以人为本、合理高效的功能空间
本方案功能空间比较多样化，包括对外服务设施、行政办公、会议、餐饮、活动室、档案存放、停车库等功能用房。如何对它们进行合理的布局，使其既便于联系，又互不干扰，成为了本方案需要重点解决的问题。

回归建筑的绿色使用方式
在项目的设计中引入了生态环境的设计理念，利用几处多层部分的顶部平台作为屋顶绿化，形成不同高度、不同层面的生态景观场所，不仅改善了办公环境，创造了舒适、宜人的休闲空间，也取得了节能环保的良好效果。

创造独特而又具有时代特性的建筑形象
本项目为一个集办公与服务用房为一体的综合类建筑，每个功能使用空间所要求的高度也有差异，但是力求做到形式与功能相协调且具有一定的统一性和整体性。

Build human-based, reasonable and efficient functional spaces.
The plan has diverse functional spaces, including external service facilities, administrative office, conference, catering, sports room, archive storage, garage and other functional houses. It is a priority to make reasonable layout, easy to contact without interference.
Return to green use of buildings.
The design introduces eco-environmental ideas, rooftop greening is made on top terrace in some multi-floor sections, to form eco-landscapes in different heights and levels, not only improving office conditions, and creating agreeable leisure space, but also achieving good effects of energy saving and environmental protection.
Create unique and epochal architectural image.
The project is a comprehensive architecture integrating office and service houses, where every functional space requires different height, but we are trying to ensure some concord and integrity with concerted forms and functions.

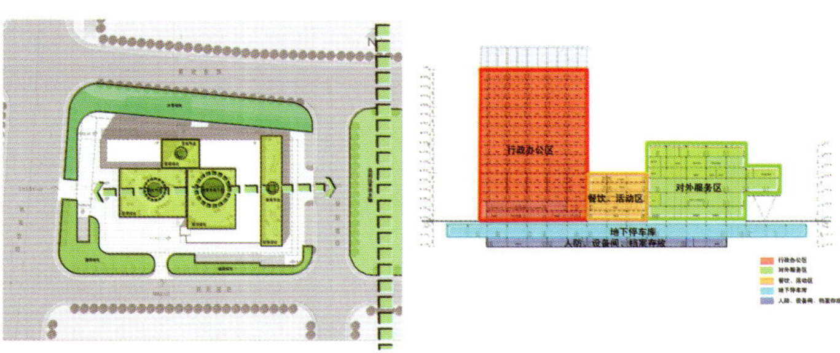

场地绿化景观分析图　　竖向功能分区分析图

孔繁森同志纪念馆改建规划及建筑方案

Mr. Kong Fansen Memorial Museum Rebuilding Plan &Architectural Concept

设计师：王润政、赵学义、孙斌、李冬、辛晶
项目地点：山东 聊城
建筑面积：7156 m²

Designer: Runzheng Wang, Xueyi Zhao, Bin Sun, Dong Li, Jing Xin
Location: Liaocheng, Shandong
Building Area: 7,156 m²

建筑及环境设计体现孔繁森同志工作环境特色。
建筑形体以简洁而现代的体块组合而成，通过白色与红色的穿插处理，并且在檐口、窗户等建筑细部借鉴了西藏当地的建筑符号，使建筑具有典型的西藏建筑特色。广场的设计借鉴了布达拉宫广场的设计手法又融入藏族哈达的文化特色，建筑的周边设计了碎石与低矮的植物来营造高原戈壁的氛围。
展览流线及空间设计结合孔繁森同志事迹展开。
从入口至瞻仰大厅，设计了一条坡道，尽端是一幅西藏雪山的壁画，而孔繁森同志的雕像也陈列其中，与屋顶采光窗所洒下的光线营造出一种充满地域特色而又庄严肃穆的场景。沿右侧进入孔繁森同志纪念馆的主题参观空间，事迹陈列以孔繁森同志不同生活时期为主线依次展开，整个陈列厅围绕大厅布置，空间开阔、流线清晰。

The architectural and environmental design represents the working conditions of Mr. Kong Fansen.
The shape is made from concise and modern blocks, interlaid by white and red colors, while borrowing Tibetan architectural symbols in cornice, window and other details, so the architecture is vested in typical Tibetan features. The plaza is designed in manner of Potala Palace Square, integrating with Tibetan Hada culture, and the periphery is designed with gravel and low plants to create a tableland gobi climate supporting the integral building atmosphere.
Display streamline and space design combine with Mr. Kong Fansen's stories.
From the entrance to the Reverence Hall, we design a ramp, the end of which is a mural painting of Tibetan snow mountain, and Mr. Kong Fansen's scripture is also displayed there. The sunlight from rooftop skylight creates a solemn and local scene. On the left hand side is the themed visiting space of Mr. Kong Fansen's Memorial Museum, with stories displaying Mr. Kong Fansen in different historic times. The whole display hall is arranged around the great hall, with broad space and clear streamline.

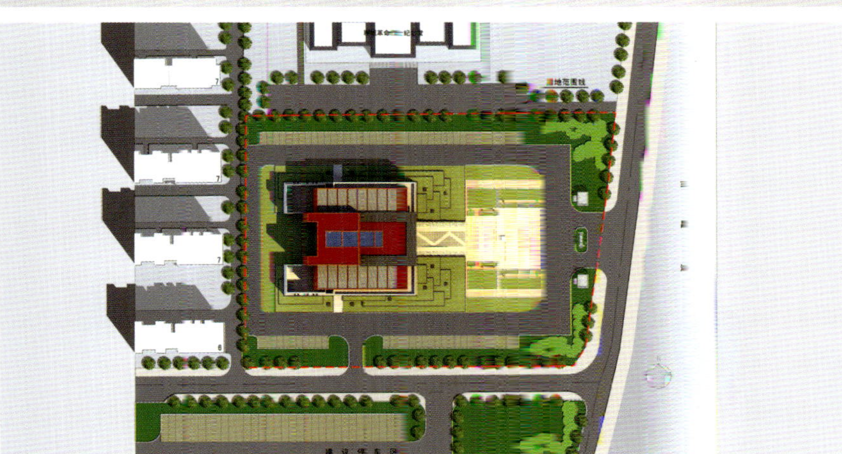

山东省华都建筑设计院有限公司
Huadu Architectural Design Institute Co.,Ltd.

山东省华都建筑设计院有限公司由山东省金城建筑设计院有限公司和山东省建设监理咨询有限公司于2009年11月联合组建，是山东省住房和城乡建设厅直属企业。具有建筑设计甲级资质与风景园林工程设计乙级资质。

山东省华都建筑设计院有限公司专业配备齐全，技术力量雄厚，综合实力强大。公司将秉承一贯的经营理念、服务宗旨，坚持科学公正、诚信团结、敬业奉献、务实创新的精神，为社会和城市的发展设计出更多更好的建筑精品。

地址：山东省济南市历下区历山路173号历山名郡B座5层
电话：+86-531-68600068
传真：+86-531-68600069
邮箱：sdhdsj@163.com

Add: 5th Floor, Tower B, Lishan Residence Building, Lishan Road No.173, Lixia District, Ji'nan City, Shandong Province
Tel: +86-531-68600068
Fax: +86-531-68600069
E-mail: sdhdsj@163.com

济南历山名郡
Ji'nan Lishan Mountain Quarter

项目地点：山东 济南
建筑面积：300 000 m²

Location: Ji'nan, Shandong
Building Area: 300,000 m²

本项目为济南最高档小区之一，位于美丽的千佛山脚下。小区规划合理，环境优美，充分利用了不规则地形和较大的高差，置身其中仿佛畅游于花园之内。

Located at the foot of the beautiful Mil-Buddha Mountain, this project is one of the highest-grade residential communities in Ji'nan. The residential quarter is reasonably arranged and makes full use its irregular landform and relatively large altitude difference. Combining with a beautiful environment, it seems to travel leisurely in a garden in this community.

泰安新湖绿园小区

Taian New Lake and Green Garden Residential Community

项目地点：山东 泰安　　Location: Taian, Shandong
建筑面积：140 000 m²　　Building Area: 140,000 m²

小区由高层、多层、联排别墅三种建筑类型构成，并吸收中国古代哲学思想，融入泰山精华，将住宅与景观相结合，构建山水社区，将泰山的精髓文化融入人们的生活空间。
小区由“一轴、一带、三中心、四组团”组成，包括叠石瀑布和踏步跌水构成的水景轴；青山、绿树组成绿色景观带；三个中心活动广场和“秀山、乐山、泰然、安然”四个组团，让人体会到人与自然和谐共生的生活理念。
项目曾获得了国家级最高奖项——“全国人居经典建筑规划设计方案综合大奖”。

The residential community is composed by three architectural types including high-rise multistoried building and townhouse. Absorbing the thoughts of ancient Chinese philosophy and blending the essence of Tai Mountain, it combines the houses with the landscape into a landscape community and infuses the essential culture of Tai Mountain into people's living space.
The residential quarter is constituted by a structure of “one axis, one belt, three centers and four groups”: The waterscape constituted by the waterfall with laying stones and the stepping stones and water drop channels; the green landscape belt constituted by the green mountain and trees; the three centers refer to the three center activity squares and the four groups to the “Xiushan (beautiful mountain), Leshan (happy mountain), Tairan (composedness) and Anran (peace)” residential house groups, allowing to experience the ideal of life that man and nature live in harmony.
This project have won the highest national prize – "Comprehensive Award for National Habitat Classical Architectural Planning and Design".

新疆五家渠煤电公司住宅小区

Residential Community of Xinjiang Wujiaqu Coal and Electricity Company

项目地点：新疆 五家渠　　Location: Wujiaqu, Xinjiang
建筑面积：340 000 m²　　Building Area: 340,000 m²

通过一条商业步行街将用地分为东西两部分，东地块为住宅区，西地块为科技研发中心及商业会所。东侧居住区规划形成“两心、一带、多节点”的布局。住宅单体设计采用了新古典主义风格，运用经典的三段式，强调建筑的细节和品质。建筑外墙采用米黄色面砖，线脚和窗下墙颜色略深，从而强调了竖向线条和向上的动感；顶层退台局部采用弧线处理，和当地民族风格相呼应。
西地块综合体采用极具动感的弧形外观以及奢华的室内装修，和南侧酒店、超市、会所等形成了一个新的城市商业、休闲中心。为市民提供餐饮、会议、娱乐、购物等一站式服务，同时也成为了五家渠市的新地标。

The land is divided into two parts by a commercial walking street, with the east lot as the residential area and the west lot as science and technology research and development center and business chambers. The east residential area is planned into a layout of “two centers, one belt and multiple nodes”. The single residential houses adopt the Neoclassicism style and the classic three-section type, which emphasizes the detail and quality of the buildings. Steady and lavish warm color system is used. The outer wall is paved with beige face bricks, with architrave and breast paved with darker colored bricks, which thus emphasizes the vertical lines and the upward movement. On the top there is terrace with arc design in some parts, responding to the local national style.
The complex in the west lot adopts a dynamic cambered appearance and deluxe interior decoration, forming a new city commercial and leisure center together with the hotel, supermarket and clubs in the south side. It provides local people with a one-stop service including catering, meeting, entertainment and shopping and meanwhile it has also become a new landmark of Wujiaqu City.

南郊宾馆餐厅翻扩建工程

Renovation and Extension Project of Canteen of Ji'nan South Suburban Hotel

项目地点：山东 济南
建筑面积：13 000 m^2

Location: Ji'nan, Shandong
Building Area: 13,000 m^2

南郊宾馆位于泉城公园南侧，为山东省国宾馆。本项目一层为千人"蓝色大厅"，可容纳110桌就餐；二层为政务接待。

Located in the south of Quancheng Park, Jinan South Suburban Hotel is a state guest house of Shandong Province. The first floor of the project is a one-thousand-people "Blue Hall", which can hold 110 tables; and the second floor is used for reception of political affairs.

北京阜外医院齐鲁分院综合楼

Beijing Fuwai Hospital Qilu Branch Complex

项目地点：山东 邹城
建筑面积：113 800 m^2

Location: Zoucheng, Shandong
Building Area: 113,800 m^2

功能上一个塔楼为医院，一个塔楼为宾馆，裙房为商业。立面采用欧式简约风格，立面上通过门、窗、柱的组合，形成了富有韵律感的竖向线条，同时通过局部的收分等手法解决了主楼高宽比过小的问题，从而加强了建筑的挺拔感和标志性。外墙面采用富有质感的暖色石材，体现建筑的古典气质和高贵形象，展示个性化的时代特征。

In function, one tower building is used as hospital and one as hotel and the skirt buildings are for business use. The facades, designed in the simple European style, will have rhythmical vertical lines because of the arrangement of doors and windows and pillars on them. Meanwhile, the problem that the height-width ratio of the main building is too small is solved through local entasis, which thus enhances the tallness and straightness of the building as well as its prominence as a landmark. Warm-colored stones with rich texture are adopted in the outer wall surfaces, presenting the classical quality and noble image of the buildings. Our goals are to design buildings that can highlight their character and place.

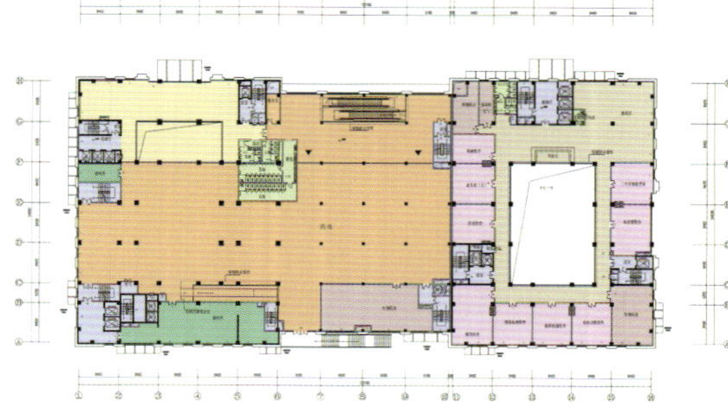

二层平面图

三层平面图

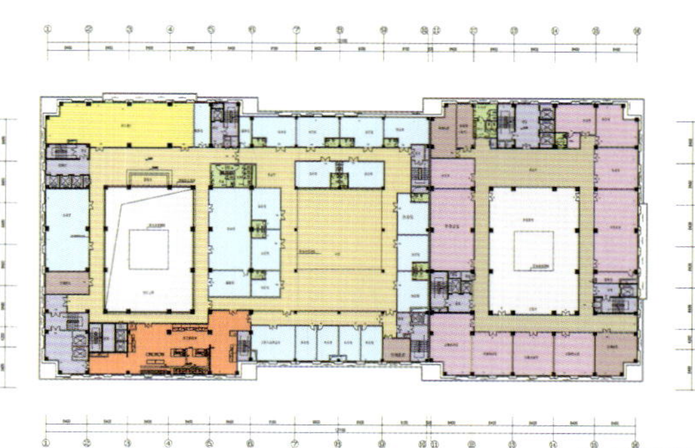

四层平面图

济南皇亭体育馆改造项目

Renovation Project of Jinan Huangting Stadium

项目地点：山东 济南

Location: Ji'nan, Shandong

项目为第十一届全运会篮球和举重的比赛场馆，在原有老体育馆基础上增加了二层的回廊，彻底解决了人员疏散问题，重新进行了立面、内装、管线、设备设计。项目投入使用后取得了较高的评价。

The stadium was the revenue of the basketball and weight lifting competition of the 11th National Games. On the basis of the original stadium, a two-storey ambulatory was added, which completely solve the personnel evacuation problem. The facades, interior decoration, pipelines and devices were redesigned. It has been spoken highly of after it was put into use.

邹平金融中心及创新家园设计

Zouping Financial Center and Innovation Home Design

项目地点：山东 邹平
建筑面积：250 000 m^2

Location: Zouping, Shandong
Building Area: 250,000 m^2

项目位于邹平鹤伴二路南侧，分为两大板块金融中心和创新家园。住宅地块建筑设计为22～28层，建筑面积为14万 m^2；金融中心地块建筑设计为27层，建筑面积为11万 m^2。设计上综合分析东侧鲁班大厦和西侧南水北调大厦的特点，风格上与其形成呼应，从而打造一组现代化高档高层办公组群；在与周边建筑协调的基础上形成自己独立的风格，打造地标性建筑。

住宅区的建筑采用了稳重、大方的灰色系现代风格，裙房采用干挂深色石材，稳重大气，和北侧创新金融中心协调。

The project, located in the south side of Hebam Second Road, Zouping, is divided into two blocks: Financial Center and Innovation home. The residential lot has 22 to 28 floors with floor area of 140,000 m^2, while the Financial Center lot has 27 floors with floor area of 110,000 m^2.

In the design, the Luban Building on the east side and the South Water to North Building in the west were analyzed comprehensively so the project can respond to them in style, so as to form a modern top-grade high-rise office complex. The design is also to build a project of its own style while in harmony with its neighboring buildings and build it into a landmark building.

The steady and openhanded grey system is used for buildings in the residential area: modern style. Cladding dark-colored stones, which look also steady and openhanded, are used in skirt buildings in response to the Innovation Financial Center in the north.

青岛维多利亚广场
Qingdao Victoria Square

项目地点：山东 青岛　Location: Qingdao, Shandong
建筑面积：53 000 m²　Building Area: 53,000 m²

项目位于青岛火车站广场，为高档酒店式公寓，1～3层为商业，4层为配套餐饮，5层为酒店大堂。外形为不规则椭圆形，流畅的曲线如海水的波浪一般，内部的中庭设计给人带来丰富的空间感受，为青岛滨海景观又添一景。

This project is located in the square of Qingdao Railway Station. It is a top-grade apartment with hotel-styled services, with the first to third floors for business use, the fourth floor as supporting restaurants and the fifth floor as the hotel hall. On its irregular ellipse appearance, the fluent curves are like the sea waves. The atrium design inside brings people rich spatial feelings. The project, standing facing the sea, adds another view to the coastal landscape of Qingdao.

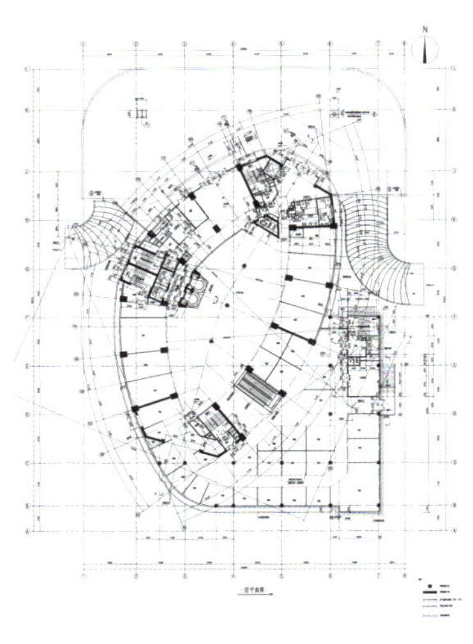

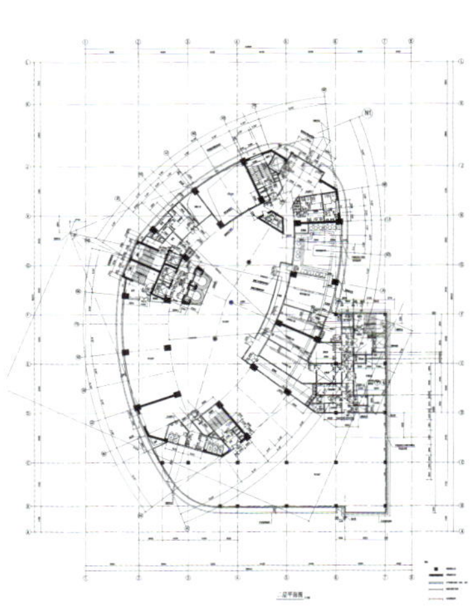

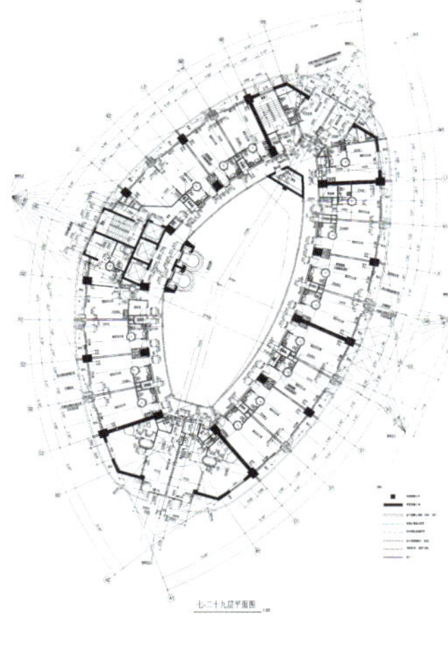

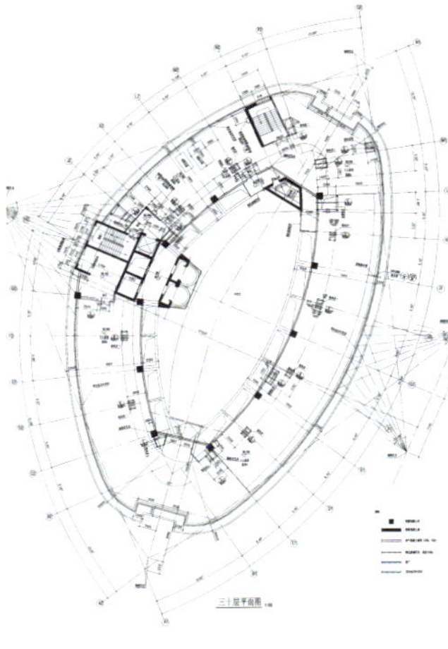

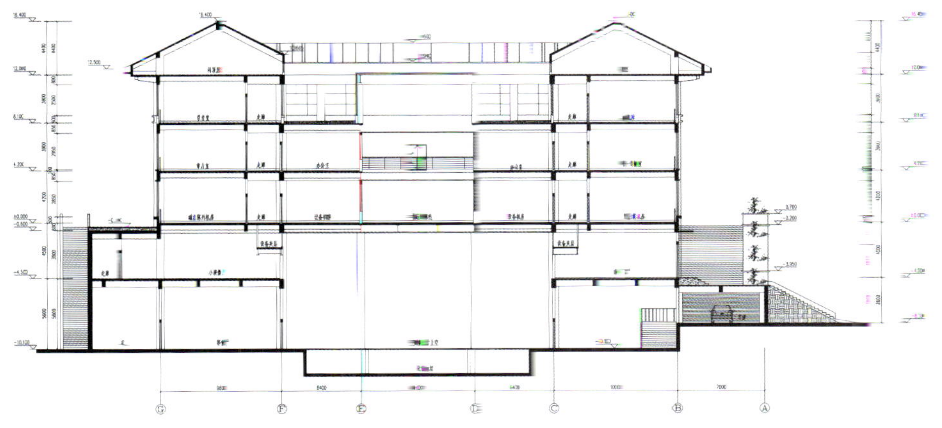

剖面图

山东教育电视台节目制作中心

Shandong Education Television Programme Production Center

项目地点：山东 济南
建筑面积：10 736.59 m²

Location: Ji'nan, Shandong
Building Area: 10,736.59 m²

位于济南市旅游路南侧，环抱于青山之中，现主体建筑由东西两个新旧单体建筑围合而成，通过改造将形成一个整体。采用现代中式风格，既和周边现存民族风格建筑、葱郁山景融为一体，又能体现电视台的现代气息。

Shandong Education Television Programme Production Center is located in the south of Tourism Road, Ji'nan City and embraced by green mountains. The main building, transformed into a whole, is formed with the new and old single buildings in the east and west enclosed together. With a modern Chinese style, it is not only integrated with the neighboring buildings of ethnic style and the beautiful view of green mountains, but also presents the modern atmosphere of Shandong Education TV Station.

扫描查看更多信息

青岛中景建筑设计有限公司
Qingdao ZhongJing Architectural Design Co.,Ltd.

青岛中景建筑设计有限公司自2004年成立以来，努力奉行“诚信至上、质量为本、绩效卓著、服务一流”的经营方针，积极倡导“砺炼品质、熔铸精华”的企业精神；以青岛及周边地区为依托，业务辐射全省，并在西北、东北地区拥有良好的市场；遵纪守法、依法经营，重合同、守信誉，不断提高顾客的满意度；竭诚为社会各界提供更好的设计产品和更优质的服务。

公司分别于2007年获得建筑行业建筑工程甲级资质、2008年取得城市规划编制单位丙级资质、2010年取得风景园林工程乙级资质，并以此为契机促进全面发展。近年来公司在同类建筑甲级设计单位中发展速度较快，业务成果显著，企业效益及年产值逐年上升；技术水平、管理水平等稳步提高。

公司扎实做好各项基础管理工作，注重质量管理，并于2007年通过ISO9001质量管理体系认证。

公司自成立以来所完成的各类项目设计均通过各级施工图审查机构审查。

公司坚持制度创新和管理创新，重视人才培训，积极引进和采用新技术成果；逐步深化企业标准化、信息化建设，并与北京理正软件设计研究院有限公司共同开发了《青岛中景建筑设计有限公司管理信息系统》。公司近三年来积极组织设计人员参加各项建筑设计竞赛，并荣获国家级金奖4项、国家级银奖1项、省级二等奖1项、省级三等奖11项、市级金奖1项、市级银奖3项、市级铜奖2项。

公司已连续多年荣获“山东省优秀勘察设计单位”“青岛市勘察设计咨询业先进单位”和“青岛市AAA级诚信勘察、设计单位”等荣誉称号。

公司目前是首批进入山东省勘察设计行业管理“绿色通道”单位、青岛市勘察设计协会理事单位之一。

公司于2009年在青岛市抚顺路竞拍土地成功，投资4000万元建设“中景设计创意中心”，总建筑面积近6000 m²。公司引入全新的设计企业办公理念，精心设计，并按照高标准建设该办公楼。新办公楼已于2010年底建成投入使用。一流的办公环境堪称国内先进水平，企业人员及规模也将继续扩大。中景设计创意中心自建成后已荣获2011年度全国人居经典建筑规划设计方案竞赛——建筑科技双金奖、2011全国环境艺术大展银奖、山东省优秀工程勘察设计评选三等奖、青岛优秀建设工程勘察设计评选一等奖、山东省“同圆杯”绿色建筑设计方案竞赛三等奖等6项大奖。

地址：中国山东省青岛市抚顺路15号甲
电话：+86-532-68876887
传真：+86-532-68876886
邮箱：hr@zjdchina.com
网址：www.zjdchina.com

Add: 15-A Fushun Road, Qingdao City, Shandong Province, China
Tel: +86-532-68876887
Fax: +86-532-68876886
E-mail: hr@zjdchina.com
Web: www.zjdchina.com

中景设计创意中心
Zhongjing Design Creation Center

项目地点：山东 青岛 Location: Qingdao, Shandong
建筑面积：6000 m² Building Area: 6,000 m²

中景设计创意中心是青岛建筑设计创意产业园的首个启动项目，也是青岛中景建筑设计有限公司的日常办公经营场所，位于青岛市四方区抚顺路青岛理工大学旁，其设计既极具现代感，又与周边的城市色彩、肌理相协调。项目采取了一系列先进的设计建设理念，是一座采用了地源热泵空调、外墙干挂中空陶板、屋顶花园等多项节能环保新技术、新材料的现代化、智能化办公楼。室内设计也依据其特有的运营模式及鲜明的企业文化进行创作。项目自建成后已荣获2011年度全国人居经典建筑规划设计方案竞赛——建筑科技双金奖、2011全国环境艺术大展银奖、山东省优秀工程勘察设计评选三等奖、青岛优秀建设工程勘察设计评选一等奖、山东省“同圆杯”绿色建筑设计方案竞赛三等奖等6项大奖。

青岛南京路1号酒店
Nanjing Road No.1 Hotel, Qingdao

项目地点：山东 青岛　　Location: Qingdao, Shandong
用地面积：10 016 m²　　Site Area: 10,016 m²
建筑面积：87 075 m²　　Building Area: 87,075 m²

青岛南京路1号五星级酒店位于青岛市中心黄金地段，南京路南端，是一栋集办公、五星级酒店于一体的42层超高层综合体，建筑高度为159.6 m。塔楼的立面采用Low-E低辐射玻璃结合花岗石材，体现出简洁、高雅的格调。在塔楼顶端，以灯塔为意象设计了玻璃盒子，为青岛滨海一带的天际线带来崭新的变化，将形成强有力的地标形象。

乌兰巴托天地泉国际温泉大酒店
Ulan Bator Tiandi International Hot Spring Grand Hotel

项目地点：蒙古 乌兰巴托　　Location: Ulan bator, Mongolia
用地面积：24 000 m²　　Site Area: 24,000 m²
建筑面积：86 500 m²　　Building Area: 86,500 m²

乌兰巴托天地泉国际温泉大酒店地处蒙古共和国首都乌兰巴托市南部，邻近著名的温泉水域，是蒙古共和国首家五星级酒店综合体。地上17层主要用于五星级酒店、写字楼、酒店公寓、综合娱乐场所、会议厅等，地下层设置停车场和温泉洗浴中心。

青岛新兴体育馆及文化体育大厦

Xinxing Gymnasium and Cultural & Sports Building, Qingdao

项目地点：山东 青岛
用地面积：47 000 m^2
建筑面积：157 500 m^2

Location: Qingdao, Shandong
Site Area: 47,000 m^2
Building Area: 157,500 m^2

青岛新兴体育馆及文化体育大厦位于青岛市北区南端延吉路，东靠徐州路。功能上包括体育馆、商业娱乐、文化体育大厦三部分，三个区域自南向北独立排列。南部体育馆建筑面积约7000 m^2；中部商业娱乐利用下沉广场结合地下空间，扩大了商业服务面积；北部为文化体育大厦，高45层的双塔建筑是以文化体育产业为主的综合写字楼。该项目在2011年全国人居经典建筑规划设计方案竞赛活动中荣获建筑金奖，在2011年全省优秀建筑设计评比中荣获三等奖。

即墨市实验高级中学

Jimo Experimental Senior Middle School

项目地点：山东 即墨	Location: Jimo, Shandong
用地面积：200 000 m²	Site Area: 200,000 m²
建筑面积：80 727 m²	Building Area: 80,727 m²

即墨市实验高级中学位于即墨市经济开发区辽河一路以南，埠西路以西。主要分为教学楼、科技楼、行政楼、宿舍餐厅、艺体楼五大功能区。
项目荣获2011年6月青岛市建设系统建设工程质量奖，2011年山东省优秀工程勘察设计三等奖，2011年青岛优秀建设工程勘察设计二等奖。

首创空港国际中心

Shouchuang Airport International Center

项目地点：山东 青岛　　Location: Qingdao, Shandong
用地面积：92 455 m²　　Site Area: 92,455 m²
建筑面积：235 009 m²　　Building Area: 235,009 m²

首创空港国际中心位于山东省青岛市城阳区流亭机场东侧，紧邻机场。体量不大，业态丰富。主要建筑功能包括9层SOHO、3–4层独栋办公楼、商业建筑、露天地下停车场等。

交运集团莱西市客运总站枢纽工程

Passenger Terminal Hub of Communication and Transportation Group, Laixi

项目地点：山东 青岛　　Location: Qingdao, Shandong
用地面积：128 530 m²　　Site Area: 128,530 m²
建筑面积：82 000 m²　　Building Area: 82,000 m²

交运集团莱西市客运总站枢纽工程，为投标中标方案，该方案立意新颖、造型现代，建成后将成为青岛以及胶东半岛地区的地标性建筑。

青岛财宝山庄度假村
Qingdao Treasure Resort

项目地点：山东 青岛　　Location: Qingdao, Shandong
用地面积：68 700 m²　　Site Area: 68,700 m²
建筑面积：38 168 m²　　Building Area: 38,168 m²

青岛财宝山庄度假村位于青岛城阳区惜福镇别墅区，以连体式低层住宅为主。度假村北靠空港路，与青岛理工大学分校隔路相望；东依群山，地势错落，环境优美。

石嘴山市大武口亘元地产项目
Dawukou Genyuan Real Estate Project, Shizuishan

项目地点：宁夏 石嘴山　　Location: Shizuishan, Ningxia

宁夏石嘴山市大武口亘元地产项目位于石嘴山市大武口区，项目规划总用地为50 950 m²，规划建筑面积为182 900 m²，其中地上建筑面积152 900 m²，地下总建筑面积30 000 m²。

青岛国际航海俱乐部景观设计

Qingdao International Sailing Club Landscape Design

项目地点：山东 青岛
用地面积：13 151.4 m^2

Location: Qingdao, Shandong
Site Area: 13,151.4 m^2

青岛国际航海俱乐部位于青岛海滨著名的八大关景区，面向海上明珠青岛第一海水浴场，拥有绝佳的自然及人文优势。俱乐部始建于1953年，是贺龙元帅提议兴建的项目。本项目荣获2011年全国人居经典建筑规划设计方案竞赛金奖、山东省优秀工程勘察设计评选三等奖、青岛优秀建设工程勘察设计评选二等奖。

青岛四方奥林匹克体育公园

Olympic Sports Park, Sifang District, Qingdao

项目地点：山东 青岛　Location: Qingdao, Shandong
用地面积：20000m²　Site Area: 20,000 m²
建筑面积：8000m²　Building Area: 8,000 m²

青岛四方奥林匹克体育公园建设在覆盖后的河道上，总体布局呈“丁”字形，总用地面积约20000m²，建筑面积8000m²。以“健康、运动、生活”为设计理念，集合了极限运动、休闲活动、体育健身、儿童娱乐、市民活动、文化娱乐等多项服务功能于一体，充分体现了“我运动我快乐”的主题精神。

青岛城建桃源山色住宅小区景观设计

Qingdao Chengjian Peach Garden Residential Community Landscape Design

项目地点：山东 青岛　　Location: Qingdao, Shandong

用地面积：69 237 m^2　　Site Area:69,237 m^2

青岛市城建桃源山色住宅小区总用地面积为69 237 m^2。建筑依山而建，气势宏伟。景观设计依就山势，打造了一座特色鲜明的西式台地景观小区。

青岛城建竹韵山色花园小区景观设计

Q.U.C.G. Bamboo Style Gargen Residental Community Landscape Design

项目地点：山东 青岛　Location: Qingdao, Shandong
用地面积：10 hm²　Site Area:10 ha

青岛市城建地产竹韵山色花园小区，总用地面积约10 hm²。建筑及景观风格体现青岛本土文化风貌。景观设计整体流畅，以欧式园林的造园基调并通过师法自然的软景搭配，营造了一座生活气息浓厚的花园景观示范社区。

扫描查看更多信息

山东大卫国际建筑设计有限公司

Shandong David International Architectural Design Co., Ltd.

山东大卫国际建筑设计有限公司，成立于1998年10月，是经国家建设部批准成立的甲级勘察设计股份制公司。公司主要从事民用建筑设计，尤其对房地产开发、工程前期咨询与策划、规划设计有较深入的研究。
公司主要创办人申作伟先生从事建筑、规划设计26余年，设计作品获全国性优秀设计奖15项，获山东省优秀设计奖50余项，申作伟先生历任临沂市规划设计研究院院长、山东省规划设计研究院院长、中国居住区规划专业委员会委员等职，是首批国家一级注册建筑师、国家注册规划师、山东省工程设计大师、山东省优秀建筑师、山东省十佳注册师。2003年申作伟先生被评为中国新本土居住设计名家；2007年，获“中国城市建设卓越人物”称号；2008年3月，被授予“山东省十大杰出青年勘察设计师”称号，2008年8月，被授予“中国建筑设计行业优秀管理人才成就奖”；2009年，被评为“人居十年·2009中国建筑规划十大影响力设计大师”；2009年，荣获“2009中国建筑业十大创新建筑设计师”称号，与此同时荣获2009年度“法国金牌建筑设计师”称号。2010年4月被山东省住房和城乡建设厅授予“山东省勘察设计行业管理优秀勘察设计院长”称号。申作伟先生现已成为建筑设计界的学术带头人。
公司从成立之日起即以“设计精品，服务社会”为口号，不断提高设计水平，坚持走品牌设计之路。公司现有设计师150余人，是一支具有高度职业荣誉感和良好职业道德的设计队伍。公司注重建筑设计创新，立足于中国本土文化，融汇国际先进的设计理念和技术，精益求精，完成了一批享誉省内外的精品设计项目，得到了社会各界的一致好评。公司每年均有设计作品在各类评比中获奖。
公司于2003年通过了ISO9000质量体系认证，设计项目已遍及山东、北京、海南、辽宁、重庆、四川、山西、宁夏、云南、河南、河北、广西、吉林等地。公司以创建百年设计公司为奋斗目标，以质量和服务求生存促发展，本着“团结、责任、真诚、创新”的设计理念，在激烈的市场竞争中，赢得了较高的声誉。

Shandong David International Architecture Design Co., Ltd. founded in October 1998, is a joint stock company and has the Grade A qualification of survey and design approved by the Ministry of Construction, P.R.C. The company is mainly engaged in civil construction design, has more depth research on real estate development, preliminary engineering consulting, planning, and planning and design.
Mr. Shen Zuowei is the main founder, he engaged in the construction, planning and design for more than 26 years, his design works won more than 15 national Excellent Design Award, more than 50 items of Shandong Province Excellent Design Award. Mr. Shen served as president of Linyi City Planning and Design Institute, president of Shandong Province, planning and design Research Institute, member of residential planning committee of China, The first batch of registered architect of China, registered planner of China, design master of Shandong province engineering, excellent architect of Shandong province, outstanding registered engineer. Mr. Shen was named Master of New Local Living Design in China, 2003, he received the title "China Urban Construction Excellence" in 2007, was awarded Ten Outstanding Young Survey Designer, Shandong Province, in March 2008, and in August, was awarded outstanding management talent achievement Award for China Architectural Design Industry, was named Habitat Decade • 2009 Top Ten Most Influential Chinese architectural design master in 2009, the same year, won the title of 2009 China top ten innovative building construction designer, at the same time, won the title of 2009 French Gold Architect. He was awarded the title of excellent president for management of survey and design in Shandong Province Survey and Design industry by Shandong Department of Housing and Urban-Rural Development in April 2010.Mr. Shen Zuowei has become the academic leaders of architectural design industry.
From the date of its establishment, with the spirit of design quality and service society, the company constantly improves the design standards, and adheres to the way of design brand. There are more than 150 designers, it is a design team with highly professional pride and ethic. The company focuses on innovative architectural design, based on the Chinese local culture, integrates advanced design concepts and technology, excellence, completed a number of renowned boutique design projects outside the province, and has won the praise from the society. The design works get awards in various competitions each year.
The company passed the ISO9000 quality system certification in 2003, the design projects are located in Shandong, Beijing, Hainan, Shenyang, Chongqing, Sichuan, Shanxi, Ningxia, Yunnan, Henan, Hebei, Guangxi, Jilin and other places. With the goal of creating a "century design company", survival and promote development of quality and service, the design concept of "solidarity, responsibility, sincerity, innovation", the company has gained a high reputation in the fierce market competition.

地址：济南市英雄山路166号
邮编：250002
电话：+86-531-82719857/82719858
传真：+86-531-82719879
邮箱：dawei_sheji@163.com
网址：www.daweiarch.com

Add: No. 166 Yingxiongshan Road, Ji'nan
P.C.: 250002
Tel: +86-531-82719857/82719858
Fax: +86-531-82719879
E-mail: dawei_sheji@163.com
Web: www.daweiarch.com

鸿顺观邸小区规划设计

Hongshun View Mansion Community Plan and Design

设 计 师：张冰、赵晓东、郭真、赵娟、朱宁宁、王泽东、徐以国、韩丽丽
项目地点：山东 济宁
用地面积：132 000 m^2
建筑面积：328 306 m^2

Designer: Bing Zhang, Xiaodong Zhao, Zhen Guo, Juan Zhao, Ningning Zhu, Zedong Wang, Yiguo Xu, Lili Han
Location: Ji'ning, Shandong
Site Area: 132,000 m^2
Building Area: 328,306 m^2

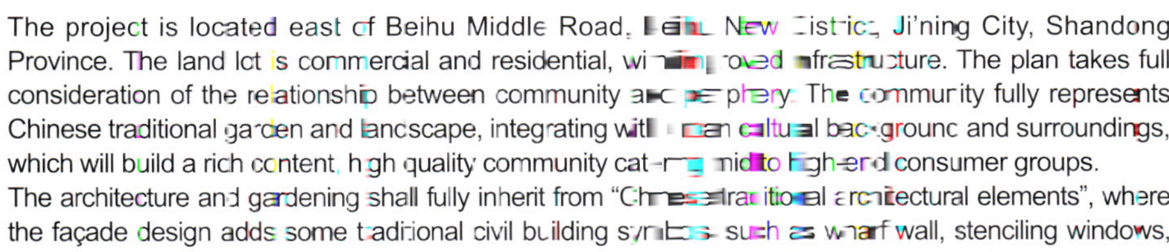

The project is located east of Beihu Middle Road, Beihu New District, Ji'ning City, Shandong Province. The land lot is commercial and residential, with improved infrastructure. The plan takes full consideration of the relationship between community and periphery. The community fully represents Chinese traditional garden and landscape, integrating with urban cultural background and surroundings, which will build a rich content, high quality community catering mid to high-end consumer groups.
The architecture and gardening shall fully inherit from "Chinese traditional architectural elements", where the façade design adds some traditional civil building symbols such as wharf wall, stenciling windows, light bricks and gray tiles, as well as modern building elements with full consideration of green, energy-saving and environment-friendly designs, so antique Chinese style is integrated with modern elements, forming a fresh Chinese-style residence harmonizing traditional civil residence with modern one.

项目位于山东省济宁市北湖新区北湖中路东，用地性质为商业和住宅，规划区内基础设施较为完善。规划充分考虑小区与周围环境的关系，充分体现中国传统园林景观的特点，结合城市的文化背景和项目的周边环境，努力打造一个富有文化内涵的面向中高端消费群体的高品质社区。
建筑形式和园林景观充分挖掘中国传统建筑元素，建筑立面设计中加入中国传统民居的建筑符号，例如马头墙、镂花窗、青砖灰瓦等，同时又加入现代建筑的元素，在建筑设计方面也作了绿色节能环保的考虑，使古朴的中式风格渗入现代元素，形成中国传统民居与现代风格完美结合的新中式住宅。

尚品 · 燕园修建性详细规划方案

Detailed Constructive Plan of Noble · Yanyuan

设 计 师：徐然
项目地点：山东 济南
用地面积：82 100 m^2
建筑密度：18%
绿 化 率：35%

Designer: Ran Xu
Location: Ji'nan, Shandong
Site Area: 82,100 m^2
Building Density: 18%
Greening Ratio: 35%

项目位于济南市历下区盛福片区，地块位于老城区边缘高新区西北向，距高新区不到3 km，到洪楼商圈仅4 km左右，且道路通畅，为住宅规划集中区。
规划中力求将都市特质与居住生活有机地结合，创造出回归自然、富有感召力的人居空间，大面积的中心庭院和大尺度的楼间距是本项目的最大特色。结合周边环境及用地现状，使整个规划与环境相协调；塑造优美宜人的绿色景观，努力营造家园感和归属感；以满足规划要求为前提，从实际状况出发，妥善解决交通、通风、防疫、环保等要求，合理组织人、车交通流线；从居住角度出发，充分利用现有条件，从景观、空间、户型等多方面提高居住质量，努力创造一个环境宜人的新世纪社区，体现以人为本的设计理念；重视街景及区域内部的城市新景观，塑造空间形态，体现高度发展的济南新城面貌。

The project is located in Shengfu block, Lixia District, Ji'nan City. The lot is in the margin of old downtown, less than 3 km northwest of High and New Tech District, and 4 km or so to Honglou Business Circle, with smooth traffic, and this area is a residential concentration area.
In the plan, we try to integrate metropolitan features with residential life, to create a charming habitat returning to the Nature. Large central courtyard and large floor space are the biggest features of this project. The surroundings and land occupation enable the harmony between the whole planning and the environment; it will create agreeable greening, home feeling and acculturation; preconditioning the planning requirements, it will properly meet traffic, ventilation, disease prevention, environmental protection and other requirements as practical, to organize pedestrian and vehicular traffic flows; from habitant perspective, it will make full use of existing conditions, to increase living quality by landscape, space, house type and other aspects, trying to create an agreeable, high level new century community representing human-based ideology; and it will pay great attentions to street sights and new landscapes within the area. It will shape spatial forms to represent new appearance of Ji'nan City in fast development.

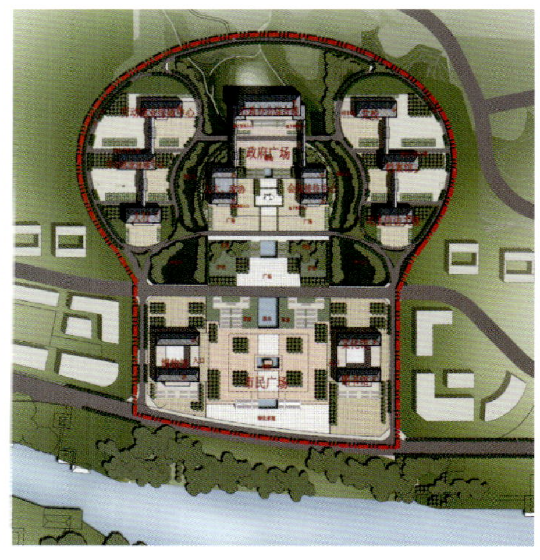

彭阳县行政中心规划

Pengyang County Administrative Center Plan

设 计 师：张冰、徐以国、赵娟、韩丽丽
项目地点：宁夏 彭阳
用地面积：309 800 m^2
建筑面积：65 119 m^2

Designer: Bing Zhang, Yiguo Xu, Juan Zhao, Lil Han
Location: Pengyang, Ningxia
Site Area: 309,800 m^2
Building Area: 65,119 m^2

彭阳县位于宁夏回族自治区南部边缘，六盘山东麓，地处黄土高原中部丘陵沟壑区，地貌类型复杂多样。彭阳历史悠久，文化源远流长，人文荟萃，文物蕴藏丰富，风景名胜众多。
项目位于彭阳县新老城区交接的位置，起到承接作用，使新城与老城均能方便享受行政中心的服务功能；规划区内地形复杂，位于悦龙大道北侧高地上，设计中充分考虑地形以及周边道路情况。简洁、开放、自由、和谐、以人为本的形象成为现代政务中心景观设计的特征，设计中遵循突出地域性、强调公共性的设计原则。
建筑风格结合彭阳县人文地理特征，融入现代设计手法，重点突出彭阳县行政中心庄重、大方、开放的特点。

Pengyang County is located in the south margin of Ningxia Autonomous Region, at east foot of Mt. Liuban, in the middle hilly trench area of Loess Plateau, with various landforms. Pengyang County is historic, cultural, with rich human and cultural resources, and many tourist destinations.
The project is in the very position of old town, inheriting from the old town, and makes the new town and old town both can enjoy the services from administrative center. The plan area with various landforms is on the highland north of Yuelong Avenue, and the design fully considers landforms and peripheral roads conditions. It is concise, open, free, human-based, harmonious, transparent and disciplined landscaping image will feature modern administrative center, whose design follows and highlights the principle of locality and publicity.
The architectural style combines with local human geography, integrates modern design approaches, and highlights the solemnity, generosity and openness of Pengyang County Administrative Center.

青铜峡市行政中心规划

Plan of Administrative Center, Qingtongxia

设 计 师：张冰、赵晓东、张生、徐以国、董恒祥、赵娟
项目地点：宁夏 青铜峡
用地面积：490 800 m^2
建筑面积：14 098 m^2

Designer: Bing Zhang, Xiaodong Zhao, Sheng Zhang, Yiguo Xu, Hengxiang Dong, Juan Zhao
Location: Qingtongxia, Ningxia
Site Area: 490,800 m^2
Building Area: 14,098 m^2

行政中心兼顾了城市景观中心的作用，为市民提供了大面积绿化环境和活动中心，以解决城市公共活动的问题，突出青铜峡市地域性的特点。每个城市都有各自的特色，任何区域环境对该区域空间的塑造和发展都会产生巨大影响，形成以地域环境为特征的空间场所。在设计中突出地域性有助于城市特色的表现，有助于城市中居民空间归属感的形成。建筑风格结合当地特色，融入新中式的概念，并控制建筑的整体高度，与周围环境充分统一。

The administrative center also serves as the city landscape center, providing large greening and activity center for citizens, which will solve public activity problems, and feature locality of Qingtongxia City. Every city has its own character, and every regional environment will have huge impact on the space shaping and development of the area, forming a space featured by local environment. The design highlights locality, which helps to express city features, and helps to form citizens' space homeliness. The architectural style combines with local features, integrates with neo-Chinese concept, and controls the overall height of buildings, in full unity with surroundings.

项目基地位于淄博市区西南部，处于城市老城与新城的连接地带，位于城市重要的主干道张周路南侧，沿路有淄博体育中心、区政府大楼等重要的公共建筑，区位优势明显。办公入口设置在西侧，货物主入口设置在北侧，使办公和仓库功能更加独立，互不干扰。卸货口集中设置在东侧，减少了对办公的影响，也使对南侧居民的影响减到了最小。

仓库区之间设置景观带，和办公前广场连成一片，发挥了更好的集中优势，使园区的环境更加优美，改善了工业园区呆板的形象，更加有利于融入周边的城市环境。外立面造型采用简洁的体块穿插，设计灵感来自于输变电变压器，使建筑的功能和形象有机结合。

The project is based in southwestern Zibo urban area, in the junction of old town and new town. It is south of arterial Zhangzhou Road, alongside which are Zibo Sports Center, District Administrative Building and other important public buildings with obviously advantageous geography. The office entrance is set in the west while goods entrance in the north, office and warehouse functions are more independent, without interference.

Between the warehouse zones are set landscaping zones, connecting with the plaza in front of office buildings, for better gathering, ameliorating industrial park's mechanical image and better integrating with the surroundings. Facade is formed by concise penetration, inspired by power transmission/distribution transformer, for the purpose of better integration of building functions and images.

淄博电力运维中心
Power OM Center, Zibo

设 计 师：李兴、张生
项目地点：山东 淄博
建筑面积：34 700 m²

Designer: Xing Li, Sheng Zhang
Location: Zibo, Shandong
Building Area: 34,700 m²

梁山客运中心
Passenger Terminal, Liangshan

设 计 师：张冰、贾志磊、王泽东、李冉
项目地点：山东 梁山
用地面积：110 000 m^2
建筑面积：22 000 m^2

Designer: Bing Zhang, Zhilei Jia, Zedong Wang, Ran Li
Location: Liangshan, Shandong
Site Area: 110,000 m^2
Building Area: 22,000 m^2

梁山客运中心建筑设计风格大气、浑厚，内外空间丰富、光影效果强烈，既现代又有厚重的历史感，颇具文化气质和品位。
将项目的设计与城市规划同时考虑，使项目合理融入社会，并成为梁山市新的核心与标志。充分展现梁山水浒文化的地域文化特色，成为梁山的一张城市名片。充分体现“无缝衔接、零距离换乘”等先进理念，实现客运交通一体化，成为购票易、换乘快、功能全的现代交通枢纽及新型的交通与商业综合体。

Liangshan Passenger Terminal is designed in generous and profound style, with strong light effect, rich in/outdoor space, modern and historic, featuring cultural character and taste.
The design concerts with municipal planning, integrating the project into the society, making it a new core and landmark of the city. It fully shows Liangshan Waterside culture and geography, making it a name card of this city. It fully represents the advance ideas of "seamless joint and one-stop transport", realizing passenger traffic integration. It will build a modern traffic hub with easy ticket service, rapid transfer and complete functions. It will create a new style traffic and commercial complex.

沂南县客运中心
Passenger Terminal, Yi'nan County

设 计 师：张冰、赵晓东、李兴、徐以国
项目地点：山东 沂南
用地面积：123 000 m^2
建筑面积：52 000 m^2

Designer: Bing Zhang, Xiaodong Zhao, Xing Li, Yiguo Xu
Location: Yi'nan, Shandong
Site Area: 123,000 m^2
Building Area: 52,000 m^2

项目位于沂南县城南部，朝阳路以东，澳柯玛大道以南。造型意喻城市文脉的继承，方案立面构思取自沂南的“崮”以及汉代的“门阙”。崮稳固、敦实，隐喻沂南人民质朴、坚强的精神；汉代的“门阙”即大门，隐喻沂南客运中心是人们进入沂南的大门，是沂南重要的展示窗口。立面造型处理借鉴了汉代的建筑形式，同时将汉石画像运用在立面上，是一种城市文脉的继承。
设计充分利用交通人流和商业之间的互动关系，达到人流带动商业、商业养护站房的目的。充分体现“无缝衔接、零距离换乘”的先进理念。在购票服务区和候车上客区之间设置通长的绿化带，营造良好的室内生态环境及高品质的交通建筑空间。

The project is located in the southern Yi'nan County, east of Chaoyang Road, south of Aucma Avenue. The shape intends to inherit from the city cultures: the facade takes origin in "cube" in Yi'nan area and "gate" in Han Dynasty. Cube is stable, solid, symbolizing Yi'nan people's straightforwardness and relentless spirits; while ancient "gate" is portal, metaphorizing Yi'nan Passengers Terminal is a portal to Yi'nan, a key window to Yi'nan. The facade adopts architectural forms in Han Dynasty, applying ancient portraits in the facade, to inherit from cultural deposit of the city.
Design makes full use of and combines the interactions between pedestrial flow and commerce, where pedestrian flow drives commerce, while commerce maintains station house, to fully represent the advanced ideas of "seamless linkage, and zero-distance transfer". Long greening zones are set between ticket service zone and waiting/boarding zone, to create good indoor ecology and high quality traffic building space.

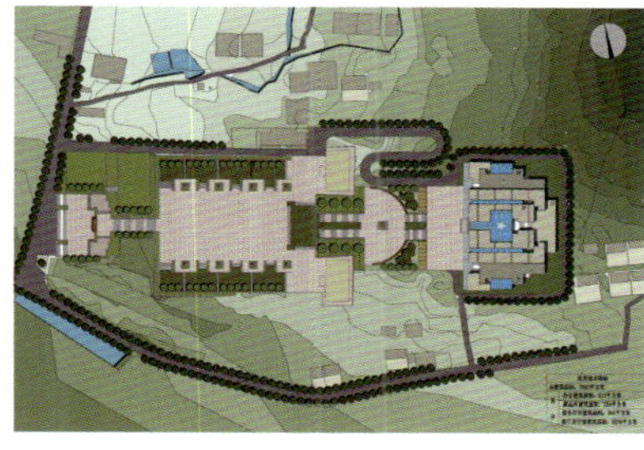
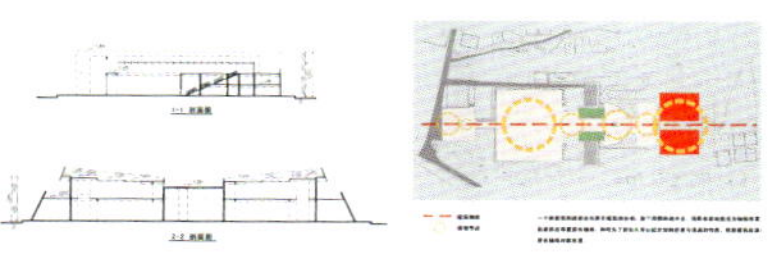

临沂费县大青山胜利突围纪念馆
Mt. Daqing Victorious Breakthrough Memorial Square, Fei County, Linyi

设 计 师：张冰、庞唐、王泽东、张科栋、徐以国、韩丽丽
项目地点：山东 临沂
建筑面积：8161m²

Designer: Bing Zhang, Tang Pang, Zedong Wang, Kedong Zhang, Yiguo Xu, Lili Han
Location: Linyi, Shandong
Building Area: 8,161 m²

1992年，费县县委、县政府在马头崖乡（现并入薛庄镇）李行沟建立了"大青山战斗突围旧址纪念碑"，并建立了多处永久性纪念设施。1997年11月，"大青山胜利突围纪念碑"、"大青山胜利突围纪念亭"等建起，2000年，共青团中央又将大青山革命纪念地扩建并命名为"全国青少年教育基地"。
此次项目设计是在原有纪念广场、纪念碑后方设计具有纪念意义的大青山胜利突围纪念馆。基地内原有建筑群体顺应山势，对称布局，新建筑尊重原有轴线布局模式，沿中心轴线作对称设计。由于建筑处在纪念广场的最后端，广场整体布局又顺应山势，呈平台空间高低错落，所以在设计新建筑高度与宽度时充分考虑参观者在参观过程中与新建筑物相互的视线关系。建筑整体造型舒展、平缓，呈现一种横向的构图，与垂直、挺拔的纪念碑相互融合。新建筑的设计使两个不同时间但相同功能的建筑物能够彼此对话。

In 1992, CCP Fei County Committee, Fei County People's Government established the "Mt. Daqing Battle Breakthrough Site" Monument and established many perpetual memorial facilities in Lixinggou Village of Matouya Town (now incorporated into Xuzhuang Town). In November 1997, here established "Mt. Daqing Victorious Breakthrough Monument" and "Mt. Daqing Victorious Breakthrough Memorial Pavillion" etc. In 2000, the Communist Youth League Central Committee renamed and established the Mt. Daqing Revolutionary Memorial Site to be "National Youth Education Base".
This design is a monumental Mt. Daqing Victorious Breakthrough Memorial behind the original memorial square and monument. Original buildings within the base follow the hilly landform, symmetric along the central axis. New buildings respect original axial layout, and are designed symmetrically along the central axis. Since the buildings are in the rearmost end of the memorial square whose layout follows the hilly landform, all terrace spaces are in different heights. The height and width of new buildings will fully consider visitor's sight of new buildings. The integral shape is extensive, gentile, showing a transverse structure, integrating with the vertical and tall monument. The design enables the two buildings from different times with the same functions to dialogue with each other.

济南市文化艺术中心
Cultural & Art Center, Ji'nan

设 计 师：张冰、张科栋、王泽东
项目地点：山东 济南
用地面积：18 500 m²
建筑面积：20 000 m²

Designer: Bing Zhang, Kedong Zhang, Zedong Wang
Location: Ji'nan, Shandong
Site Area: 18,500 m²
Building Area: 20,000 m²

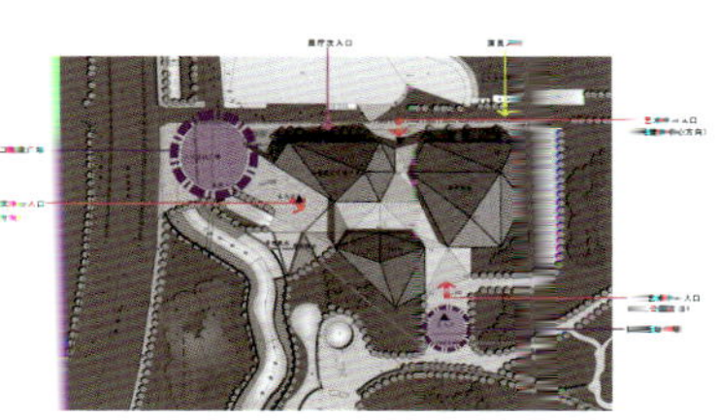

济南市文化艺术中心项目选址于济南市风景秀美、绿树葱葱的泉城公园西南角，紧邻山东省体育中心和济南市全民健身中心，项目周边交通便利，由十路、玉函路、舜耕路、马鞍山路构成项目周边城市路网，通畅的城市交通对市民文化艺术中心的可达性与疏导性起到关键作用。
建筑整体由三个简洁、现代的体块组成，中间为玻璃大厅。造型力求新颖别致，具有现代文化气息。三个建筑体块像三块石头，围合的玻璃大厅，象征文化艺术源泉；而玻璃幕向三个方向翻折，寓意艺术的泉水由北涌出，向外流淌，表现"清泉石上流"的意境。本项目恰当地处理文化艺术中心与周边的关系，形成整个区域良好的视线及景观环境，并通过文化中心的连接，在整体区域形成一个包含健身、文化、展览、游园等多种功能的市民文化艺术中心。

Ji'nan Cultural & Art Center is located in the southwest corner of Spring Park, a beautiful and green destination in Ji'nan City, near Shandong Provincial Sports Center and Ji'nan Municipal Fitness Center, with convenient traffic network, such as Jingshi Road, Yuhan Road, Shungeng Road, and Ma'anshan Road forming peripheral road network, and such smooth traffic conditions are crucial to accessibility and diversion of the Cultural & Art Center.
The project includes three concise and modern blocks, with a glass hall in the middle. The shape is novel with modern cultural senses. The three blocks are like three rocks, enclosing a glass hall, symbolizing the source of culture and art. While the glass curtain is folding in three directions, an artistic spring is flowing outside, corresponding to the poetic scenario of "spring flowing on rock". The project properly integrates with the periphery, and the whole area enjoys good sight and views, and the connection with Cultural & Art Center forms a civic cultural and art center integrating fitness, culture, exhibition, park and other functions.

聊城交大科技园国际交流中心

XJTU Science Park International Communication Center

设 计 师：申作伟、狄传广、王妍、李传运、王奎之、孙鸿昌
项目地点：山东 聊城
建筑面积：77 900 m²

Designer: Zuowei Shen, Chuanguang Di, Yan Wang, Chuanyun Li, Kuizhi Wang, Hongchang Sun
Location: Liaocheng, Shandong
Building Area: 77,900 m²

项目地处湖南路与南关街交叉口的东北角，是集高档星级酒店、公寓式酒店、商业于一体的大型商业地产。其功能主要为餐饮、住宿、会议、娱乐、康体、休闲、旅游、商业等。

为达到和摩天轮浑然一体，酒店设计强调造型体系的完整性。客房部分每个房间都配有可观湖的阳台或露台，这样不仅和摩天轮的动感呼应，同时也是对休闲酒店的特色表达。而裙楼部分则强调基础的敦实感，使其有足够的分量“托起”摩天轮和酒店客房。本项目将酒店与摩天轮完美结合，具有超强的震撼力。项目建成后将成为聊城市标志性的商业地产，为聊城市提供更美丽的城市景观。

The project is located in the northeast corner of the crossroad between Hunan Road and Nanguan Road, and it is large commercial land property integrating hi-grade star hotels, apartment hotels, and commercial buildings, with functions mainly including foods & beverage, accommodation, conference, entertainment, recreation, leisure, tourism, and commerce.

The design mainly starts from the nature, surroundings and architectural forms of city landscape and hi-grade hotels. In order to integrate with the Ferris wheel, the hotels design highlights system integrity. Every guestroom has a lake-viewing balcony or terrace, not only corresponding to the dynamic Ferris wheel, but also expressing the featured leisure hotels. Furthermore, the skirt buildings have a solid foundation, giving enough “support” to the Ferris wheel and hotel guesthouses. The project integrates hotels and the Ferris wheel perfectly, shocking people strongly. Upon completion, the project will become a landmark commercial property, serving also as a beautiful landscape of the city.

山东抗日民主政权创建纪念馆

Anti-Japanese Democratic Regime Memorial Museum, Shandong

设 计 师：朱宁宁、张生、赵娟
项目地点：山东 沂南
用地面积：44 000 m²
建筑面积：9 500 m²

Designer: Ningning Zhu, Sheng Zhang, Juan Zhao
Location: Yi'nan, Shandong
Site Area: 44,000 m²
Building Area: 9,500 m²

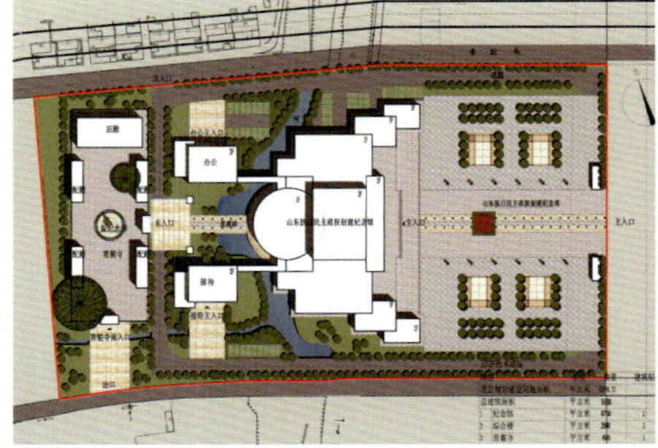

纪念馆的选址尊重历史，在原地复建青驼寺，其东侧设计建造山东抗日民主政权创建纪念馆及广场、山东抗日民主政权创建纪念塔、培训楼等其他附属建筑。规划设计将青驼寺内的原有纪念碑、新建纪念馆建筑及纪念塔串联在一条主轴线上，体现对历史的尊重，水流蜿蜒而行，人车分流，游览空间丰富宜人。

纪念馆的建筑单体设计体现本土化思想，采用环抱、欢迎的姿态，表达沂蒙人民性格热情的特点；建筑由层层的体块堆积，外墙整体采用毛石面，厚重、质朴的体块层次，象征着沂蒙的群山。

The project is located in the historic place, Qingtuo Temple (restored). On the east side we design to build Shandong Anti-Japanese Democratic Regime Memorial Museum and Square, Shandong Anti-Japanese Democratic Regime Founding Memorial Tower, Training Building and other auxiliary buildings. The design aligns original monuments, newly built memorial museum buildings and memorial towers within the temple into a main axis, showing our respect to history. The brook is winding, pedestrian flow and vehicle flow are diverted, and the sightseeing content is rich and agreeable.

The single building of the museum represents local ideas, in a embracing and welcoming gesture, to express a cordial character, which is Yimeng people's character; the building is piled up by blocks layer by layer, and the external wall is covered by rough stone, heavy, pure, representing Yimeng mountains.

临沂苍山万悦·名品商业综合体

Wanyue · Fame Commercial Complex, Cangshan County, Linyi

设 计 师：张冰、张科栋、王泽东
项目地点：山东 苍山
用地面积：15 400 m²
建筑面积：73 000 m²

Designer: Bing Zhang, Kedong Zhang, Zedong Wang
Location: Cangshan, Shandong
Site Area: 15,400 m²
Building Area: 73,000 m²

项目位于临沂市苍山县，处于沿海地区大开发与黄淮海平原大开发的交叉地带，位于苏南、胶东半岛两大经济带的中间地带，区位优势比较明显。
基地西部靠近银座超市方向设计为商场区域，商场与银座超市相互融合，资源互补，可以有效地带动本区域商业气氛的形成；基地东部靠近城市干道交叉口方向设计入口广场及商业街铺。商业休闲广场的设计不仅可以为城市居民提供日常休闲活动空间，还可以大量吸引购物人流，并通过广场上的喷泉、水景、绿化等手段，增强广场的休闲趣味性，增强商业人气。建筑整体立面以现代风格为主，虚实结合，沉稳大气。同时设计大量广告招贴，烘托商业环境。首层通透的玻璃橱窗可以最大限度增加商品展示空间。建筑外饰面材料以石材和玻璃为主，彰显高贵气息与时代气息。

The project is located in Cangshan Coounty of Linyi City, in the juncture of coastal development and Yellow River- Huai River – Hai River Basin development, in the middle of South Jiangsu and Shandong Peninsula, the two economic zones. It enjoys obvious regional advantages.
In the design, the western base near Ginza Supermarket will be mall area, integrating with the complementary supermarket, to effectively drive the formation of commercial climate in this area. The eastern base near crossroad of urban arterial roads is designed for entrance square and commercial stores. The commercial & leisure square will not only provide ordinary leisure space for citizens, but also attract shoppers flow; meanwhile the fountain, waterscape, greening and other means on the square will increase the interests of leisure, and increase the commercial atmosphere. The façade is mainly modern style, illusionary and practical, stable and generous. It also designs a lot of advertising posters, to boost commercial environment. The 1st floor glass show windows will maximize the goods display space. External finish materials are mainly stone and glass materials, exposing nobleness and modernism.

临沂汇宝湾商业综合体

Treasure Bay Commercial Complex, Linyi

设 计 师：张冰、张科栋、范昕
项目地点：山东 临沂
建筑面积：164 000 m²

Designer: Bing Zhang, Kedong Zhang, Xin Fan
Location: Linyi, Shandong
Building Area: 164,000 m²

本项目选址于临沂市核心商业区域，项目周边商业气氛浓厚。基地南邻涑河南路，北临城市河道景观带——涑河，东邻静雅酒店。项目周边交通便利，景观价值极高，建筑设计极富挑战性。
项目利用现有的沿涑河景观，建筑整体形态层层跌落，错落有致，大量的室外平台是绝佳的观河场所，增加了商业综合体的商业价值。同时室外平台上布置大量层次分明的绿化空间，使绿化退台与涑河景观交相辉映。商业布局采用相对开放式的布局方式，内街外街相互连通，开放的布局方式与景观互动，大大增强了购物的趣味性与观赏性。建筑形态以曲线为主，串联各个功能块，建筑整体活泼、动感强烈，立面丰富。开放式布局方式、叠错的景观搭配、动感时尚的建筑造型必将成为临沂商业综合体的新典范。

The project is located in the CBD of Linyi City, surrounded by mature business climate. The base is near Suhe South Road in the south, near Urban Riverview Belt – Su River in the north, and adjoining Jingya Hotel in the east. The traffic is convenient, the landscaping is valuable and the architectural design is very challenging.
Making use of existing Su River view, the building forms are retiring floor by floor, making a difference. Many outdoor terraces are good river-viewing places, adding commercial value to the complex. Many greening areas are set on outdoor terraces, in obvious layers, and green retiring corresponds with Su River landscaping. The commercial layout is open style, where inner streets interact with outer streets, and the open layout matches with landscape, largely increasing shopping interest and appreciation. The building forms are themed by curves, connecting with every functions, dynamic, energetic, with various façades. The open layout, overlaying landscape match and dynamic shapes will certainly make it the example of commercial complex in Linyi City.

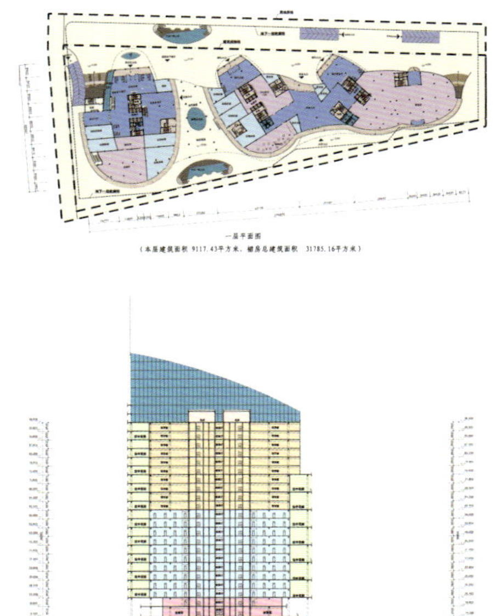

项目用地位于济南市小清河滨河新区内，为小清河综合治理建设的一部分。用地周边道路纵横交织，与济南城市交通网络联系紧密。项目用地东西狭长，共三个地块，其中东、西两端的地块一和地块二为公共绿地，中部的地块三为集商务办公、SOHO办公、公寓、餐饮及休闲于一体的商业用地。建筑设计的立意吸取了柳叶及古莲子的寓意。建筑裙房共4层，南部层层退台，曲线造型如清河水般流畅；主体共设计3个塔楼，成角度自由组合，平面形如三片柳叶漂浮水中；塔楼单体平面由柳叶变形后与古莲子形似。古莲子与水有关，与荷花有关，与泉城济南有关，更有沉睡百年而复苏的深刻意义，寓意小清河在历经近代工业污染的低潮期后，正焕发着昔日的光彩。三塔楼高峻挺拔，以西侧主楼为旗舰，如扬帆起航的舰船昂首向前。

The project is located in Little Clear Riverside New District of Ji'nan City, serving as a part of Little Clear River comprehensive governance program. The land is crossed by roads from every direction, connecting Ji'nan municipal traffic closely. The land is a strip long in east-west direction, including three lots, of which, east and west lots (lot 1 and lot 2) are public green lands, and middle lot (lot 3) is a commercial land integrating commercial, office, SOHO office, apartment, catering and leisure buildings. The intention of this design absorbs the implies of willow leaves and ancient lotus seeds: the skirt building has 4 floors, with south part retiring floor by floor, and its curved shape is as smooth as the clear water from the river. The main body includes three towers, forming flexible angles, flowing on the water surface resembling three willow leaves; a single tower transforms from willow leaf into ancient lotus seed in level view, which relates to water and lotus, and further to Ji'nan, the Spring City, with profound significance of a century dream and renaissance, implying the Little Clear River after the ebb of modern industry pollutions is emitting long-missing glories. The three towers are outstanding, with the west wing building like a flagship, sailing forth.

济南滨河商务中心

Ji'nan Riverbank Business Center

设 计 师：申作伟、张冰、肖艳萍、陈欣欣、张方建、邱金伟、孙华坤、王才华、郭飞、马尚勇、梁越
项目地点：山东 济南
用地面积：55 500 m^2
建筑面积：198 766.70 m^2
容 积 率：地上4.73，地下2.296
建筑密度：32%

Designer: Zuowei Shen, Bing Zhang, Yanping Xiao, Xinxin Chen, Fangjian Zhang, Jinwei Qiu, Huakun Sun, Caihua Wang, Fei Guo, Shangyong Ma, Yue Liang
Location: Ji'nan, Shandong
Site Area: 55,500 m^2
Building Area: 198,766.70 m^2
Plot Ratio: Aboveground 4.73, underground 2.296
Building Density: 32%

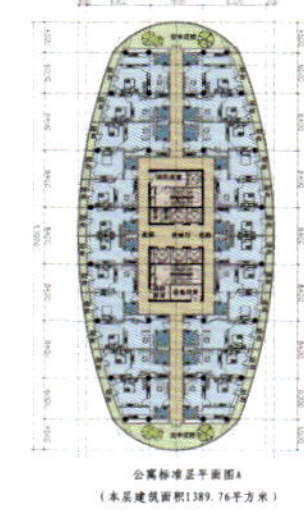

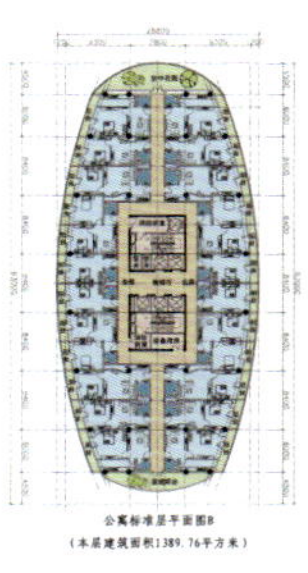

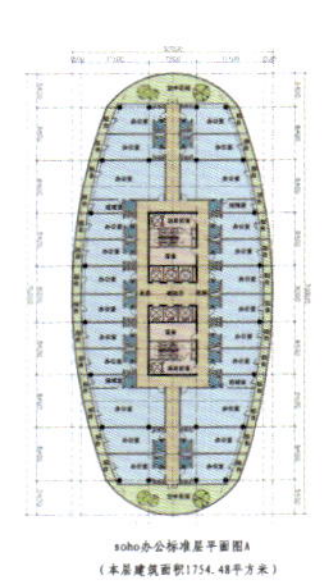

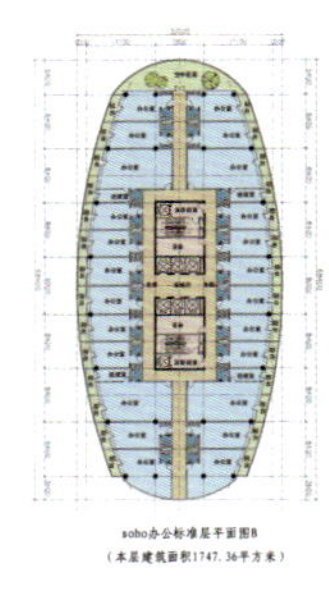

黄河天街民俗风貌商业综合体项目建筑方案设计

Yellow River Celestial Street Folks Commercial Complex Concept Design

设 计 师：申作伟、张冰、赵晓东、郭真、张生、张科栋、王泽东、朱宁宁、董恒祥、徐以国
项目地点：甘肃 兰州
建筑面积：270 200 m^2
五星级酒店面积：90 200 m^2
商业综合体面积：180 000 m^2

Designer: Shen Zuowei, Zhang Bing, Zhao Xiaodong, Zhang Zhen, Zhang Sheng, Zhang Kedong, Wang Zedong, Zhu Ningning, Dong Hengxiang, Xu Yiguo
Location: Lanzhou, Gansu
Building Area: 270,200 m^2
Including 5-star hotel Area: 90,200 m^2
Commercial complex Area: 180,000 m^2

项目地处兰州市区中部的徐家湾村，滨河中路以北。商业地块南侧即是母亲河——黄河，北侧为规划住宅区地块。整体共包括两个部分：靠近基地西南角方向拟规划设计一栋五星级酒店，高度暂定150 m，36层左右；靠近基地东侧方向拟规划设计黄河天街民俗风貌商业综合体，建筑主体高度为60 m，功能包括综合购物商场、主力店、民俗风貌小吃街、电影院、主题式购物店、商务风情酒店、花园公寓等。建筑整体采用层层退台的立面布局方式，造型活泼动感，立面丰富，轮廓线高低错落有致。考虑到兰州商业开发市场的业态模式以及商业市场开发力度，本设计规划以新型商业购物模式为出发点，以旅游、文化为主题，以花园景观绿化为亮点，打造兰州主城区新地标。

This project is located in Xujiawan Village, in middle of Lanzhou urban area, north of Riverside Middle Road. On the south side of the commercial lot is our mother river – Yellow River, and on the north side is a planned residential community with commercial buildings. The project includes two parts: near the southwest corner of the base is a planned 5-star hotel, 150m high, with 36 floors or so; near the east side of the base is a planned Yellow River Celestial Street Folks Commercial Complex, 60m high, with functions including comprehensive shopping center, flagship store, folks snack street, cinema, themed shop, business exotic hotel, and garden apartment. The buildings are dynamic with retiring façade, rich of forms, and elegant profiles. In consideration of industry model and power of commercial development, this design will start from new shopping model, featuring tourism and culture, and build landmark in downtown Lanzhou by means of gardening, landscaping and greening.

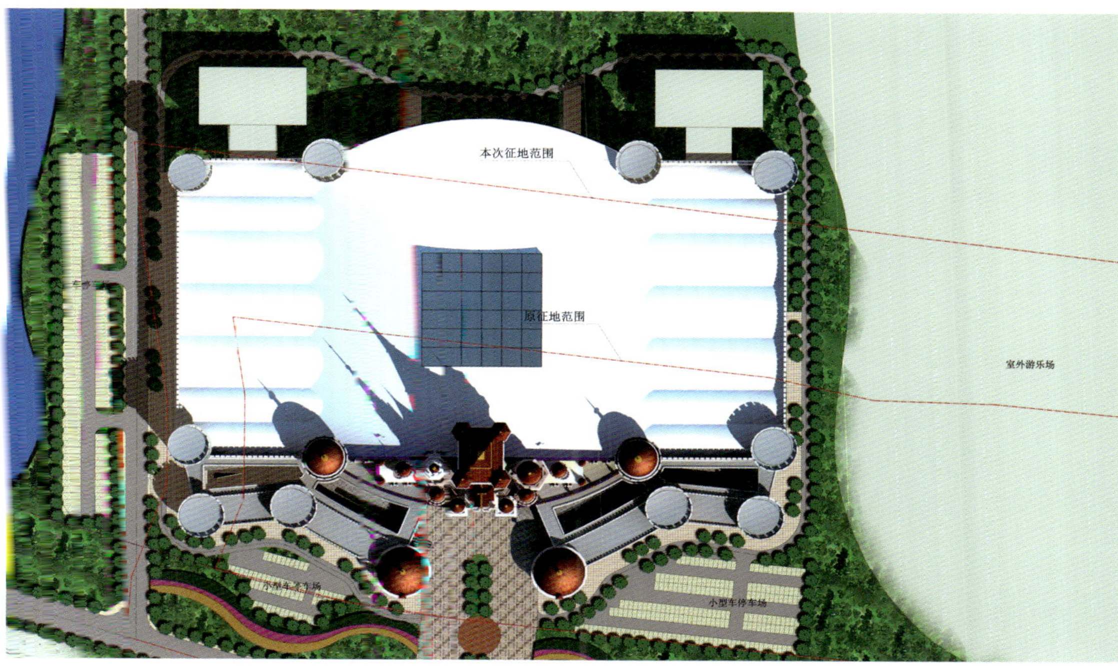

蓬莱海洋王国主题公园

Penglai Sea Themed Park

设 计 师：申作伟、邱英林、孙保朝、徐斌
项目地点：山东 蓬莱
用地面积：227 500 m²

Designer: Zuowei Shen, Yinglin Qiu, Baozhao Sun, Bin Xu
Location: Penglai, Shandong
Site Area: 227,500 m²

海洋王国主题乐园将建设在蓬莱美丽的滨海区域，它西邻千年古阁蓬莱阁，北接烟波浩淼的大海。宜人的风景、欢腾的浪花，赋予了海洋王国动感、欢乐的意境。所以，“欢乐的海洋王国”将成为本项目的主题。

海洋王国主题乐园项目采用独特的海洋游乐与陆地游乐相结合，游乐与休闲、度假及购物相结合的独特的新模式，打造出一个独具特色的、大型的、综合性的主题乐园。

海洋王国的规划目标是成为国内独一无二的、具有鲜明特色和超强震撼力的主题乐园，不但服务于蓬莱的游客，而且辐射蓬莱周边城市乃至整个北方地区，成为具有全国影响力的主题乐园之一。

Sea Kingdom Themed Park will be built in the beautiful coast of Penglai, adjacent to historic Penglai Pavilion in the west, and adjoining the great sea in the north. The agreeable scenes and joyous waves bestow a dynamic and happy sense of sea kingdom. So the "happy" "sea kingdom" becomes our theme positioning of this project.

The park combines unique sea recreation and land recreation, integrates tourism, leisure, resort, shopping and other functions, and creates a unique, large comprehensive themed park.

The park is planned to be a nationwide unique themed park with obvious features and strong shocks. The park not only serves for tourists from all over the country, but also radiates to peripheral cities or even the whole north China, to become a nationally influential themed park.

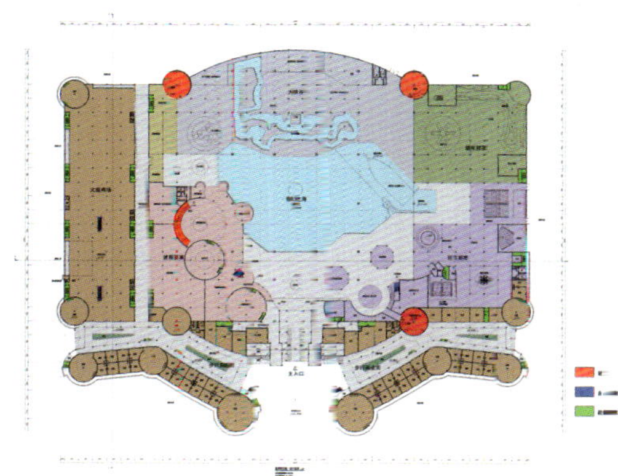

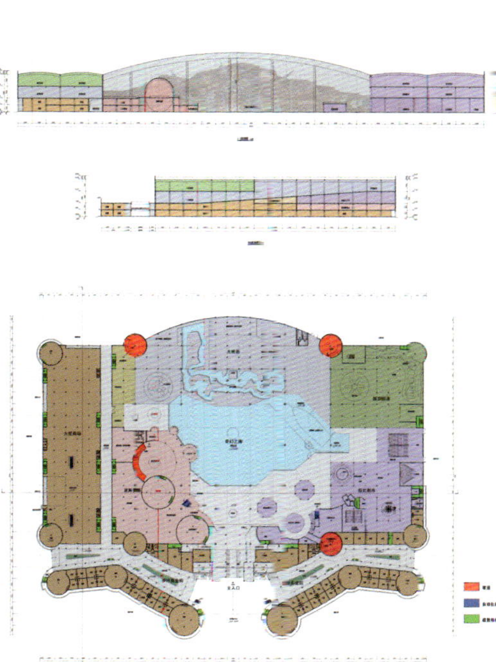

北京东方华脉工程设计有限公司青岛分公司
ChinaHumax Engineering Design Co.,Ltd. Qingdao Branch

北京东方华脉工程设计有限公司青岛分公司成立于2002年，具有建设部颁发的甲级建筑设计资质、乙级规划设计资质，现有国家一级注册建筑师、一级注册结构师、注册规划师、注册设备工程师等各类专业技术骨干人员80余名。公司位于青岛市经济技术开发区，主要经营范围有建筑、结构、给排水、电气、采暖通风、景观、市政设计。设计项目获得了社会和业内人士的高度好评，公司赢得了良好的口碑。公司与北京西林筑景设计公司及北京市建筑设计院、五合国际设计集团、中国电子工程设计院、清华大学等保持长期合作关系，能提供青岛、北京工作交流的机会，为新生代提供良好的提升平台，也为公司培养了优秀人才。

北京东方华脉工程设计有限公司青岛分公司坚持“敬业爱岗、创新设计、提高质量、真诚服务”的设计方针，以先进的技术、优异的质量和快捷的效率来保持公司旺盛的生命力。

地址：青岛市经济技术开发区长江中路519号建国大厦18层
电话：+86-532-68972799
传真：+86-532-68972799
邮箱：dfhmqd@163.com
网址：www.chinahumax.com

Add: Floor 18th,Jianguo Building,Changjiang Middle Road No.519, Economic and Technological Development Zone,Qingdao
Tel: +86-532-68972799
Fax: +86-532-68972799
E-mail: dfhmqd@163.com
Web: www.chinahumax.com

建国大厦
Jianguo Building

项目地点：山东 青岛
建筑面积：48 029 m²

Location: Qingdao, Shandong
Building Area: 48,029 m²

江山丽城
Jiangshan Licheng Community

项目地点：山东 青岛
用地面积：104 316 m²
建筑面积：354 939.11 m²

Location: Qingdao, Shandong
Site Area: 104,316 m²
Building Area: 354,939.11 m²

青岛软件园
Qingdao Software Park

项目地点 山东 青岛
用地面积 75 499.4 m²
建筑面积 123 608 m²
容 积 率 1.64

Location: Qingdao, Shandong
Site Area: 75,499.4 m²
Building Area: 123,608 m²
Plot Ratio 1.64

武夷山路社区改造项目
The Renovation Project of Wuyi Mountain Road Community

项目地点：山东 青岛
用地面积：131 193 m²
建筑面积：791 737 m²
容 积 率：6.03

Location: Qingdao, Shandong
Site Area: 131,193 m²
Building Area: 791,737 m²
Plot Ratio: 6.03

扫描查看更多信息

山东景城建筑规划设计有限公司
Shandong King City Architecture & Design

山东景城建筑规划设计有限公司成立于1996年，其前身为潍坊市奎文区建筑设计研究院。2002年率先成为潍坊市首批升级为国家乙级资质的设计院之一，并于2008年成功改制。经过16年的风风雨雨，16年的辛勤耕耘，在政府、各级主管部门以及各大房地产开发企业和兄弟单位的关心帮助下，如今已经发展成为拥有工程设计国家甲级资质的综合性科技服务企业。

公司拥有建筑、规划、园林景观、结构、给排水、建筑电气、智能化建筑设计、暖通空调等多种专业人才，专业配套齐全。设计类型包括大型居住区、酒店、大中学校、高级写字楼、大型厂房仓库等工程项目。实现了从规划、绿化到单体工程方案，再到施工图纸的一条龙设计，为广大客户提供了更为方便、更为综合的高品质服务。公司现有职工60余人，是由一批学识高、才思敏捷的中青年工程师和实践经验丰富的专家组成的建筑设计队伍。工程技术人员占90%以上，其中各专业注册工程师12人，高级职称15人，中级职称23人。

多年来，公司领导层不断深化内部改革，大力推进两个文明建设，实现了产值不断翻番的跨跃式发展。连续多年获得市区两级“文明单位”、“先进单位”等荣誉称号；获国家级优秀设计1项、省级优秀设计7项、市级优秀设计16项、受表彰个人更达到80余人次。

公司将把持续改进作为企业永恒的目标，不断提升员工素质，完善资源装备，改进质量管理，继续向顾客提供一流的产品和优质的服务。在建筑业飞速发展的今天，公司坚持“精心设计、创作精品、超越自我、创建一流”的奋斗目标，一如既往，不懈努力，创造美好的未来。

地址：山东潍坊樱前街虞河西岸瑞泰南郡写字楼13层
电话：+86-536-8799920
传真：+86-536-8799950
邮箱：SD-jingcheng@163.com

Add: 13F of Ruitai South Country Office Building, West Bank of Yu River, Yingqian Street, Weifang City, Shandong Province
Tel: +86-536-8799920
Fax: +86-536-8799950
E-mail: SD-jingcheng@163.com

国大·东方天韵
Guoda · Eastern Classic City

设 计 师：刘红波、孙俊杰、刘炳峰
项目地点：山东 潍坊
用地面积：233 591 m²

Designer: Hongbo Liu, Junjie Sun, Bingfeng Liu
Location: Weifang, Shandong
Site Area: 233,591 m²

国大·东方天韵小区位于潍坊市奎文区，北海路以西，玉清街以北，通亭街以东，卧龙街以南，地块近似正方形，规划总用地面积233 591 m²。

项目追求现代简约欧式的建筑风格，特色鲜明，简洁时尚，再现了简约欧式建筑的神韵，置身其中，可以体会到传统与现代结合所营造出的特殊韵味。住宅的立面材质与小区已建建筑相协调，主墙面材质采用浅咖啡色面砖，两层沿街立面采用暗红色的花岗石，细部线条采用浅咖啡色真石漆。每户的客厅、卧室、餐厅全部做了空调隐蔽设计；每户阳台的侧面采用太阳能管充当空调隐蔽的格栅，即解决了太阳能的安放问题，又做到了空调的隐蔽，一举两得。

东昊 · 托斯卡纳规划
Donghao · Toscana Plan

设 计 师：刘红波、徐宁、刘炳峰
项目地点：山东 潍坊
用地面积：14.84 hm^2
建筑面积：324 293 m^2

Designer: Hongbo Liu, Yujiao Xu, Bingfeng Liu
Location: Weifang, Shandong
Site Area: 14.84 hm^2
Building Area: 324,293 m^2

该项目位于奎文区，东临四平路，南临乐川街，西靠白浪河景观带，有着得天独厚的自然景观优势。该项目拟规划建设高档居住区，以高层、小高层为主，配有一部分多层住宅。规划上致力于营造社区内部优良的景观，以大型园林景观绿地、度假式的悠闲生活情调作为整个楼盘的支持点，创造出让人们远离都市喧嚣的感觉。采用自由的村落式布局，配合托斯卡纳风格的园林设计，创造亲切宜人的区内空间尺度。

白浪河

四平路

乐川街

金钻华庭
Gold Diamond Complex Building

设 计 师：刘红波、迟强、刘炳峰
项目地点：山东 潍坊
用地面积：10 484 m²
建筑面积：32 529 m²

Designer: Hongbo Liu, Qiang Chi, Bingfeng Liu
Location: Weifang, Shandong
Site Area: 10,484 m²
Building Area: 32,529 m²

金钻华庭项目位于诸城市开发区西部，纵二路以西，诸城市新华书店图书物流中心以北，吕兑社区以东，中国网通集团有限公司诸城分公司以南。
规划拟建一座集办公、商业、公寓为一体的现代化综合性楼群，是市民购物、娱乐、办公、居住的休闲商住中心。与北侧的民生医院，东侧的学校共同形成开发区的经济、文化中心圈。

清荷园12号楼
Building 12 of Qinghe Garden

设 计 师：刘红波、迟强、吕林
项目地点：山东 潍坊
建筑面积：51 580 m²（不含地下部分）

Designer: Hongbo Liu, Qiang Chi, Lin Lu
Location: Weifang, Shandong
Building Area: 51,580 m²(Excluding Underground)

项目位于东风街与金马路交叉口西南角。建筑面积为51 580 m²（不含地下部分），其中建筑高度达到35 m以下的建筑面积为18 134.65 m²，建筑高度达到35 m以上的建筑面积为33 445.35 m²，建筑层数为地下1层，地上28层，地上总高度为99.65 m，结构形式为框架剪力墙结构体系。

昌邑佳乐家中央商务区

Changyi Jialejia CBD

设 计 师：刘红波、许玉蛟、马世龙
项目地点：山东 潍坊
用地面积：17 779 m²
建筑面积：71 060 m²

Designer: Hongbo Liu, Yujiao Xu, Shilong Ma
Location: Changyi, Shandong
Site Area: 17,779 m²
Building Area: 71,060 m²

昌邑佳乐家中央商务区项目位于山东省昌邑市交通街与北海路交叉口东南角。昌邑佳乐家中央商务区整个项目的颜色以灰色和芝麻白为主调，突出尊容华贵的品质；在建筑细部的处理上，采用外窗和墙体交互凹凸的设计，增强建筑立面的层次感，完美地展现了新都市主义的简约风格，本现出大气、前卫，舒适、健康、自然的现代价值观。

昌邑佳乐家中央商务区项目是一个兼具大型商务、娱乐服务中心的高品质住宅。集高档商务写字楼、大型商业购物中心（其中包括餐饮、健身、电影院、购物中心、专卖店）等建筑功能为一体，建成后将成为昌邑市民娱乐休闲的新中心及新的地标性建筑也将为周边商业和房地产业发展提供契机。

青岛原创工程设计有限公司
Qingdao Yuanchuang Engineering Design Co., Ltd.

青岛原创工程设计有限公司是建设部批准的具有建筑工程甲级资质的综合性设计公司，并已通过IS09001：2000质量管理体系认证。青岛原创工程设计有限公司拥有建筑、规划、结构、给排水、暖通空调、电气、装饰等各类专业大批具有设计实力和创新能力的优秀设计师，并拥有先进的现代化专业设施和设备，具有较强的方案创作能力、施工图设计能力、装饰景观施工能力。在长期以来的工作中一向注重质量与诚信的结合，享有良好的社会声誉，是一支在市场经济中诞生并具有现代经营管理理念和超前意识的生力军。
近年来，公司秉承一贯的精品设计理念，创作设计了大量的居住、教育、医疗、公共建筑及花园式工业厂区等作品，尤其擅长总体设计协调，在力求规划合理、交通组织清晰、空间组织流畅、景观富有韵律和特色的前提下，竭尽所能为建设单位提升开发的社会价值和增加经济价值。并在工程设计中总结、积累了大量的设计经验。许多项目以其优良的设计品质、完善的后期服务，赢得了广大客户的信赖。
公司秉承创新的设计理念，在功能、经济、美观的基础上，奉献具有长久生命力和社会价值的设计作品。
公司经营理念是：诚信，互利，双赢
公司管理理念是：以人文本，科学管理
公司企业精神是：团结、创新、开拓、进取
青岛原创工程设计有限公司愿社会同各界人士广泛合作，共求发展，同创辉煌！

Qingdao Yuanchuang Design Co., Ltd. is a comprehensive design company with class-A construction qualification granted by the Ministry of Construction, and has adopted the IS09001:2000 Quality Management System. It has a group of excellent designers with design and creation capabilities in the fields of building, planning, structure, water supply and drainage, heating, ventilation and air-conditioning, electrical, decoration and others, with advanced modern specialized facilities and devices. It has strong conception creativity, construction drawing design capability, decorative and landscaping construction capabilities. For a long time, it is paying great attentions to the combination of quality and credit, with good social reputation, it is a new force of modern operation and management ideas and avantgarde senses, born in market economy.
Over recent years, we follow the exquisite design idea, to create and design a large number of residential, teaching, medical, public, building and industrial park works, especially proficient in general plan coordination; in addition to reasonable planning, smooth traffic and space organization, rich rhythm and landscape features, we try the best to increase and promote socioeconomic values for the developers; we are ready to conclude from project designs, and accumulate a great many design experiences. Many projects with good design quality and complete after services have won vast trust from customers.
We follow the creative design idea, to contribute long-living design works with social value on the basis of functionality, cost-effectiveness, beauty.
Our operating philosophy is “credit, mutual benefit, win-win”.
Our management philosophy is “human-based, scientific management”.
Our entrepreneurship spirit is “union, creation, expansion, progress”.
Qingdao Yuanchuang Design Co., Ltd. is willing to cooperate with all social communities, in seek for common development, and common victory!

地址：青岛市崂山区山-东头路58号盛和大厦1号楼1-3层
电话：+86-532-83950136/83950137/83950138/83950139
传真：+86-532-83950117
邮箱：qingdaoyc@vip.sina.com
网站：www.qdyc-arch.com

Add: Floor 1-3, Building 1, Shenghe Building, 58 Shandongtou Road, Laoshan District, Qingdao
Tel: +86-532-83950136/83950137/83950138/83950139
Fax: +86-532-83950117
E-mail: qingdaoyc@vip.sina.com
Web: www.qdyc-arch.com

1–5. 黄山头玉器城
6. 某商务大厦

青岛原创
QINGDAOYUANCHUANG

1

2

3

4

5

6

1-2. 宁阳某小区
3. 德州某商务区
4-6. 某军分区

1. 某乡镇规划方案
2. 某地块设计方案
3-5. 某现代创意传媒学院

1

2

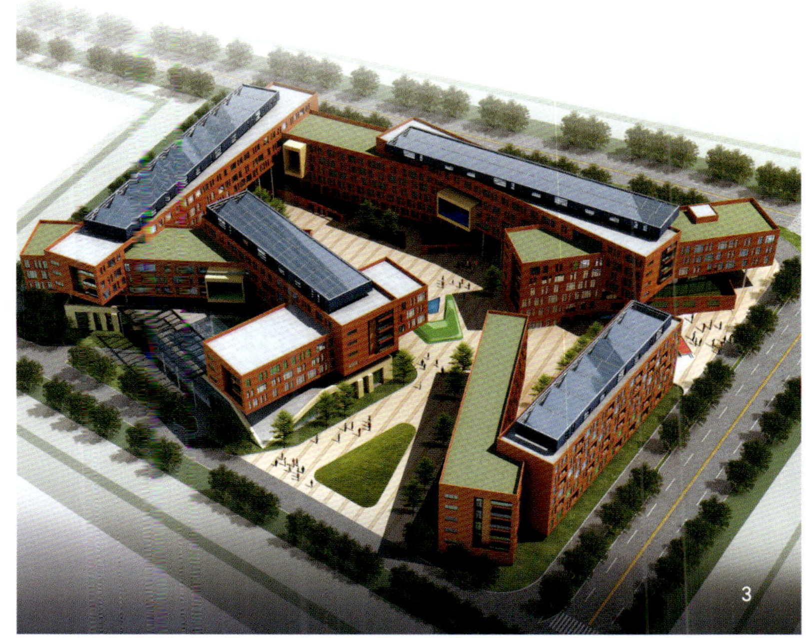

3

4

5

青岛原创
QINGDAOYUANCHUANG

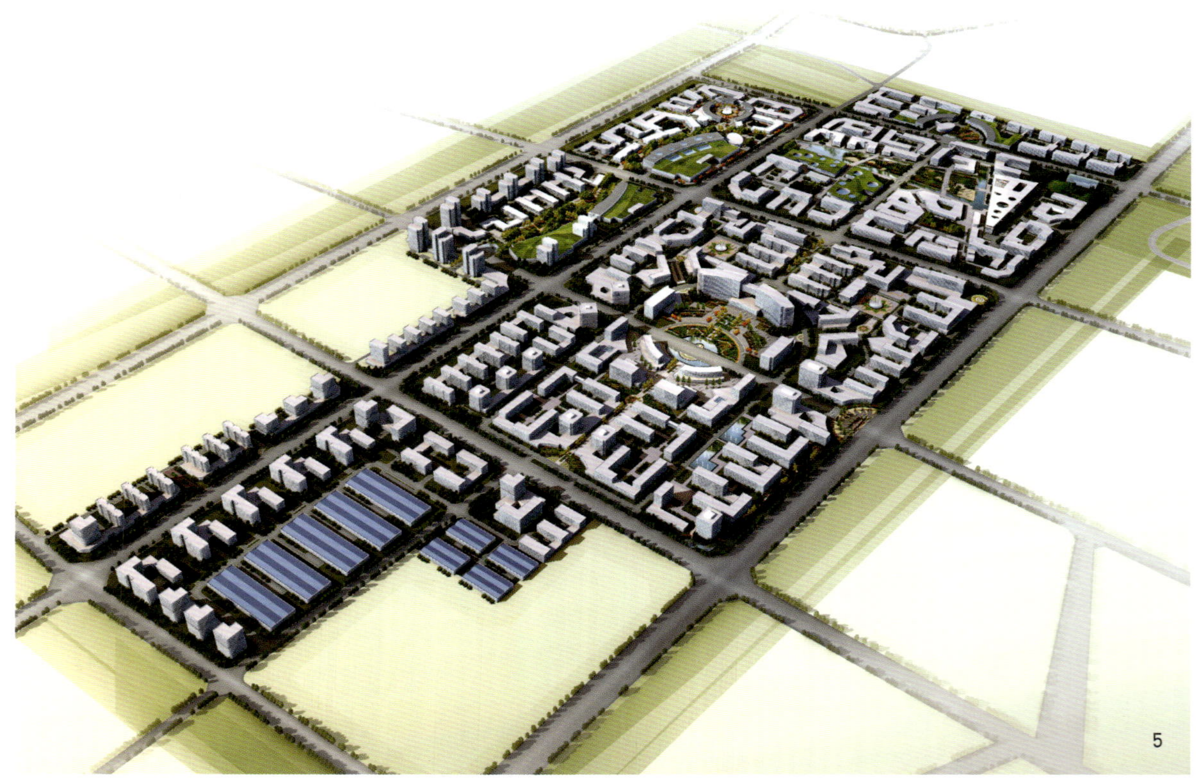

1–4. 某自行车商城
5. 硅谷鸟瞰

1-2. 太阳部落
3-6. 某玉石城

青岛原创
QINGDAOYUANCHUANG

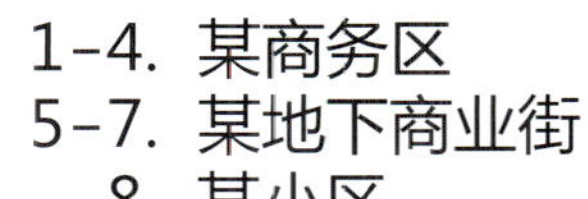

1-4. 某商务区
5-7. 某地下商业街
8. 某小区

TONTSEN 方大設計

TONTSEN 建筑设计事务所（美国）
上海方大建筑设计事务所

TONTSEN Architects Associate(USA)
Shanghai Fangda Architects Associate

TONTSEN方大设计集团前身为美国TONTSEN设计公司。于2000年进入中国，现已汇聚了200余位来自世界各地的业内设计精英。集团总部坐落于中国上海，专业从事城市规划、建筑设计、景观设计等业务。核心成员包括上海方大建筑设计事务所、上海方大建筑设计有限公司、美国TONTSEN建筑设计事务所（中国）、TONTSEN伦敦设计中心、TONTSEN香港景观部。

公司的核心创作团队，由TONTSEN方大的中国创始人齐方博士及多位国内外设计专家组成。TONTSEN方大秉承名校治学态度，合璧中西创作理念，作品遍布全国各地，业绩斐然，其中包括：上海耀江国际广场、上海紫园、上海佘山月湖山庄、上海鹏欣假日酒店、北京京贸国际公馆、天津爱家星城、重庆家纺城、温州香缇半岛、温州中梁首府、南昌恒茂国际华城、无锡奥林匹克花园等众多优秀作品，多个项目获得建设部及上海市各类建筑设计奖项。

2011年TONTSEN方大迎来了建筑创作新里程，南通观音山商业综合体、上海皇冠假日酒店、盐城中南购物中心、常州中润商业广场、安阳师范学院等多个项目的成功，印证了TONTSEN方大在城市综合体、商业办公、文教建筑领域的超强设计实力。

TONTSEN方大设计集团一直遵循着自己特有的理论——空间价值论，这是TONTSEN方大的设计之本，其核心内涵是指优秀的设计源于对市场需求表象之后的建筑本质的深刻理解。

正是基于空间价值论，TONTSEN方大不但在住宅设计领域硕果累累，而且在公共建筑领域也创作出一个又一个的杰作。由齐方博士提出的TONTSEN方大空间价值论起源于对建筑空间的深度剖析和对市场的深刻理解，在众多实际案例上得到了充分的验证。随着TONTSEN方大在商业、办公、酒店、教育各个建筑领域的不断成功，使空间价值论更加完善，为TONTSEN方大下一个10年创作规划提供了扎实的理论基础。

在过去的10年里，TONTSEN方大已具备巨大的资源整合能力和超强的核心竞争力。TONTSEN方大作为国际设计机构，既代表一种高水准的设计品质和设计能力，又是多种资源的集合体，在TONTSEN方大，可以找到建筑领域各种相关事务的全方位解决方案。TONTSEN方大的未来将要整合最优秀的设计资源，融汇古今中外，从科学、哲学和艺术高度创造出更多具有TONTSEN方大设计特色的优秀作品。

TONTSEN方大致力于成为中国建筑创新的领导者，力图通过发挥自身设计能力和技术优势，提供增值和高品质服务，为促使客户的成功发挥直接推动作用。

TONTSEN, formerly known as America TONTSEN Design Company has already absorbed more than two hundred of global industry design elites since its first presence in China in 2000; The headquarter places in Shanghai, China and is engaged in the professional fileds of the urban planning, architectural design and landscape design. The croup consists of Shanghai Fangda Architecture Design Office, Shanghai Fangda Architecture Design Co., Ltd, TONTSEN Architecture Design Office (US), TONTSEN London Design Center and TONTSEN (Hongkong) Landscape Design Company.

Our core design team consists of Doctor Qi Fang who is founder of TONTSEN China, as well as multiple international and domestic design experts. TONTSEN has inherited the attitude towards scholarly research of famous universities, and sino-western design philosophy; our works are throughout China, gaining brilliant achievements, among which Yaojiang International Square Shanghai, Ziyuan Shanghai, Yuehu Villa Sheshan Shanghai,Jingmao International Mansion Beijing, Aijia International Xingcheng Tianjin, International Textile City Chongqing, Xiangti Peninsula Wenzhou, Zhongliang Capital Wenzhou, Hengmao Shopping Center Nanchang,and Wuxi Olympic Garden etc. have been honored with architecture design awards from Ministry of Construction and Shanghai Municipality. TONTSEN embraces its new milestone for architecture design projects in 2011, such as Guanyin Mountain Complex Nantong, Crowne Plaza Shanghai, Zhongnan Shopping Mall Yancheng, Zhongrun Plaza Changzhou, Normal College Anyang and etc. It reflects TONTSEN's ultra strong design abilities in the fields of complex, commercial and education buildings.

TONTSEN follows its unique theory i.e. "value of space" from its founding. It is the ultimate source of TONTSEN's design and the core meaning is the comprehensive understanding of the architecture nature on the basis that good design is inspired by the market demands.

Found on the theory "value of space", TONTSEN succeeds not only in the field of residential design, but also in public building design. The theory "value of space" proposed by Doctor Fangqi originated from the deep analysis of architectural space and comprehensive understanding of market. It is proved in many projects. The theory "the value of space" is improving along with the success TONTSEN gained in the design of commercial, office, hotel and education projects, which has formed the theoretical basis for the next ten years of TONTSEN.

In the past ten years, TONTSEN has possessed resource integration capability and core competitiveness. As an international design group, TONTSEN is the representative of high level design quality and ability as well as the integration of multiple resources. All-round solutions for architecture related issues can be found in TONTSEN. The future aim TONTSEN in pursuit of is to integrate the best design resources and digest Chinese and Western, ancient and modern. It aims to create more TONTSEN featured great designs in the level of science, philosophy and art.

TONTSEN commits itself to be the innovative leader in Chinese architecture field and makes the most of the advantages of design abilities and technology to provide value-added and high-end service.

地址：上海市浦东新区东方路971号钱江大厦27层
电话：+86-21-50582111
传真：+86-21-50583009
邮箱：tontsen@tontsen.com
网址：www.tontsen.com

Add: 27th Floor,Qianjiang Plaza,Dongfang Road No.971,
Pudong New District,Shanghai
Tel: +86-21-50582111
Fax: +86-21-50583009
Mail: tontsen@tontsen.com
Web: www.tontsen.com

温州南湖国宾 1 号

Southlake No.1 Ambassador, Wenzhou

项目地点：浙江 温州
建筑面积：164 000 m^2

Location: Wenzhou,Zhejiang
Building Area: 164,000 m^2

温州南湖国宾 1 号项目位于温州市瓯海区，是 TONTSEN 方大为温州中梁集团设计的又一创新产品。

根据市场需求、规划要求等多项设计指标，在确定"以景观和文化为出发点"的总体布局原则后，设计师重点在户型和建筑立面造型上进行了多轮方案比较，最后选出了既符合市场需求，又完整地体现了豪宅设计意图的方案。

温州具有一种精神，即海纳百川、兼收并蓄，放眼世界、为我所用。所以设计师借鉴了法国古典建筑风格，来体现南湖项目特有的尊贵品质和价值观。

现代人的需求越来越多元化，单一型的空间越发不能满足人们的生活要求，现代人需要能够体现价值的居住空间。TONTSEN 方大的超强研发能力，不但延续了新型花园洋房、"空中别墅"系列产品的设计理念，还进一步细化和升级相关豪宅的设计手法和建筑细节，使居住者能拥有更为灵活的复合型居所。

South Lake No.1Ambassador located in Wenzhou Ouhai District is another innovative design from TONTSEN for Wenzhou Zhongliang Group .

Landscape and culture are set to be the initial point of the master plan on consideration of the market requests, planning requirements and other design specifications.The designers have compared several options for the house type and façade and then selected the best option, which meets the market needs and reflects the design intent ideas for luxury apartments.

Wenzhou is featured in the spirits"All Inclusive and everything applicable". Thus, the designers have introduced the French classic style to reflect the distinguished quality and value of South Lake.

Due to the different lifestyles of people now,always one single typology of space can never satisfy the life requirements anymore. Each living space needs to be able to reflect the value of living space.TONTSEN not only inherits the new type of garden villa, but also creates"hanging villa".And it commits itself to refine and upgrade the design techniques for luxury apartments and architecture details,so that we can have a more flexible housing.

上海皇冠假日酒店

Crowne Holiday Hotel, Shanghai

项目地点：上海　　Location: Shanghai

建筑面积：143 000 m²　　Building Area: 143,000 m²

上海皇冠假日酒店位于青浦区北侧一综合商务区，整个区域分为两个片区：西区为皇冠假日酒店，东区是4栋公寓式办公楼。设计师采用建筑与景观的一体化理念：在中央景观区域靠近酒店处设计了一个充满活力的下沉广场，在酒店一层全日餐厅处设计了一个大露台，通透的全景玻璃使室外的景观成为建筑空间的一部分，优美的建筑线条与静谧的花园融为一体，互相映衬。

酒店大楼建筑单体采用大面积玻璃幕墙设计，用金属分割条进行竖向分割，整齐而有序，尽显竖向挺拔之感。建筑顶端的弧线构架装饰以及顶板依大楼主体样式圆角向内，使建筑棱角不那么坚硬，建筑物本身也变得柔美起来。建筑外观层层退台的横向线条处理，将整体形象烘托得更为现代、时尚。

Crowne Holiday Hotel is located in Shanghai Qingpu District, north to a comprehensive commercial district. The entire region is divided into two areas. Crowne Holiday Hotel is in the west area and east area is hosting four SOHO towers. The designers have demonstrated the integration concept of architecture and landscape. A vital sunken plaza is designed in the central landscape area next to the hotel. There will be a large terrace in the all day dinning on the first floor of the hotel. Transparent viewing glass makes the outdoor landscape as part of the indoor space and beautiful architectural lines blend with the garden, setting each other off beautifully.

Hotel tower is designed to have large areas of facade with vertical metal division bar, neat and orderly and enjoying the sense of the vertical upright. At the top of the tower, The arc frames for decoration and the circular bead of the top slab turns inward like the main building style, to make the building profile gentle. The horizontal lines of the multilayer set-back models of the tower profile has set off the hotel image as modern, fashion .

安阳师范学院文博楼

Wenbo Buiding of Normal College, Anyang

项目地点：河南 安阳
建筑面积：26 000 m²

Location: Anyang,He'nan
Building Area: 26,000 m²

文博楼位于安阳师范学院黄河大道校区内，包括院属博物馆、汉推基地、交流中心、研究基地、历史学院及文学院等功能，是安阳师范学院的标志性建筑物。

文博楼位于安阳师院主轴线林荫大道西侧，与规划行政楼及中轴线上的图书馆形成三足鼎立之势，文博楼两个建筑体块之间通过次轴线将人行出入口也联接起来，整个大楼协调统一。主入口前设置大面积的水面景观，建筑物如同漂浮在水面上，夜景效果梦幻而绚丽，仿佛穿越了远古时空，将历史与现代文明联结起来。文博楼建筑外观造型根植于历史与文化的深厚底蕴，现代的建筑科技和设计手法阐释出了木化石是"历史的雕琢印迹"这一概念，象征着安阳师范传承历史文化的同时又引领教育时代潮流的魄力和气概。

This project is located in the Huanghe Avenue Campus of Normal College Anyang. The Wenbo Building serves the museum under the institute, Han Culture Promotion base, communication center, research base, History Faculty and School of Art, which enjoys the iconic status in Normal College Anyang.

Wenbo Building, on the west of the Boulevard – the main axis of Normal College Anyang, stands like the legs of tripod together with Administration Building and library. The two components of Wenbo Building are connected by secondary axis, which gathers the pedestrian. The design of the main entrance employs mass water landscape to make the architecture "floating" on the water. The night sight is dreamy and flowery, making people like walking through the past and connecting history and modern civilization. The facade of Wenbo Building introduces the rich history and culture background. Meanwhile, the modern architectural technologies and design methods complement the historical elements. It implies that Normal College Anyang not only inherits the culture and history, but also guides the trends of the education.

南昌西格玛广场

Sigma Plaza Nanchang

项目地点：江西 南昌

建筑面积：31 000 m²

Location: Nanchagn,Jiangxi

Building Area: 31,000 m²

南昌西格玛广场位于南昌市区洪都大道，主要功能为办公和商业

根据功能需要，建筑裙房为功能基座，建筑单体设置购物中心和办公区域大堂，合理的功能分区，能够有效组织不同的人流和车流，引入相应的功能空间。建筑主体的一层、二层为写字楼大堂及对外商业娱乐空间，地下一层为车库和设备用房。整体造型犹如镶嵌在托座上的钻石，在白天和夜晚分别呈现出不同的迷人光彩。

整个建筑用其单纯的体量、简洁的形态与周围城市空间和秀美的青山湖融为一体，创造出“全球化、多元化”的建筑，力求达到视觉、流线、功能上的多重互动，最终提升项目所在区域的景观和市场价值。

Sigma Plaza, located in Hongdu Avenue, Nanchang, is consisted of office and commercial parts.

According to the functional needs, the plan proposed to create a base functional podium consisted of a shopping center and office lobby in each building. Designers planned the rational function zoning, which can effectively organize the different pedestrian and vehicle flows and introduce the corresponding auxiliary function space. The first and second floors are office lobbies and external entertainment spaces, while the basement level one is mainly for parking and equipment houses. The architecture is shaped like a diamond on pedestal, charming in daytime and night.

The architecture's size, simple form is merged together with the surrounding urban space and beautiful Qingshan Lake, creating a "global, diversified" high-tech building and striving to achieve the vision, flow lines, functional interaction. It ultimately enhanced the landscape and market value of the vicinity.

南通观音山城市综合体

Guanyin Mountain City Complex, Nantong

项目地点：江苏 南通
建筑面积：347 000 m^2

Location: Nantong ,Jiangsu
Building Area: 347,000 m^2

南通观音山选址于南通市新城区发展主轴线上。南通东部新区的快速发展，以及观音山镇的纺织工业中心的发展，为本项目打造以纺织行业为商业契机的高品质商业办公综合体创造了机遇。南通观音山项目涵盖中国家纺名品城、中国家纺研发基地、特色商业区、超高层精英商务楼、品牌企业总部基地、企业总部办公、公寓式办公等几大功能，它们组成了以商务和商业为主要功能的城市综合体。

总体规划设计首先将整个项目依据核心品牌办公区、高层商务区、企业形象展示区进行划分，在每个功能区细分地块功能，做到功能配套完善、便捷。其次是高起点规划景观环境，设计师整合现有生态资源，形成网状生态格局，实现生态化办公。第三是采用空间集约化原则，根据基地各个区域的不同价值，因地制宜地设置不同功能分区。

Guanyin Mountain City Complex, located on the urban development axis of the new district of Nantong. The rapid development of the eastern district in Nantong and the developing direction for the textile industry in Guanyin Mountain town has created a favorable space and opportunities for this project. Nantong Guanyin Mountain project includes the Famous Textile Brands Center, Textile Research and Development Base, Featured Goods Center, Super High-rise Office, Famous Enterprise Headquarters Base, Headquarter Office, SOHO and etc, which has formed into the business-oriented mixed-use project.

Firstly, for the master plan, the entire project will be designed based on the core brand offices, high-rise commercial district and corporate image promotion. In each domain, subdivision will be made according to different land use and layout in order to ensure the facilities to be comprehensive. Secondly, plan the landscape with higher standards and integrate the existing ecological resources to form the netlike ecological layout and achieve ecological office. Thirdly, intensify the spaces and set up functions according to different values of different regions.

上海耀江国际广场

Yaojiang International Plaza, Shanghai

项目地点：上海
建筑面积：105 000 m²

Location:Shanghai
Building Area: 105,000 m²

耀江国际广场的设计运用了新城市主义的手法，力求建筑组群与城市的和谐统一。项目沿袭了上海外滩区域整体建筑风格，使之成为区域景观与城市滨江建筑的有机组成部分，达到与城市交融的目的，成为外滩延伸段的群体建筑之一。项目裙房高度为7～8层，与外滩建筑群高度与轮廓保持呼应。外立面采用灰色玻璃幕墙，虚化建筑体量，虚实结合。新建筑以谦虚的姿态与百年外滩融为一体。

In the design of Yaojiang International Plaza, “New Urbanism" has been adopted to integrate the architecture groups and the city on the base of the entire architectural style on the Bund so the landscape can be part of the waterfront architecture group and the extension of the bund.

The skirt building of project is height for 7～8 floors, echoing height and outline of the Bund complex. The façade adopts the gray glass curtain wall to blur the building mass, combining with actual situation. The new building by modest attitude and hundred –years Bund is as a whole.

PDG INTERNATIONAL
泛太平洋设计集团有限公司

扫描查看更多信息

泛太平洋设计集团有限公司（加拿大）于1994年在加拿大多伦多成立，并于1995年在上海成立设计分支机构。公司主要以城市规划及建筑设计为主，包括酒店、办公、商业文化设施、住宅、学校、体育场所设计等，并重视发展城市景观设计、建筑设计与室内设计的衔接与渗透。

公司依托国际化背景，以"国际化＋本土化"的专业团队，形成自己的创作和服务特色。在创作过程中，公司全面认知每个项目的土地价值、开发定位、文脉延续、艺术感知及心理体验，以开放、理性和积极的态度，去寻求创作中的每一个机会和最佳答案。公司还重视实施过程的技术控制和提升，逐步发展了一套适应于当代市场多样变化的服务体系。

公司成立至今完成了九寨天堂国际会议度假中心、上海烟草集团科教中心、成都世纪城洲际酒店、海南万宁石梅湾威斯汀酒店、成都新国际会展中心等众多国内重大工程的设计，并赢得了"建设部全国人居经典"综合金奖、"建设部建筑形态"金质奖、"建设部创新风暴中国建筑设计示范住宅"单项奖、"上海市优秀住宅小区工程设计项目"一等奖、"上海市优秀勘察设计"二等奖、"上海市住宅小区优秀规划奖"等近30个设计奖项。公司先后获得2004、2005年度中国建筑二十大品牌影响力规划建筑设计事务所（公司），2006年度亚洲建筑规划设计机构100强，2007引领中国建筑规划设计十大品牌设计机构，2011年中国商业地产设计领跑企业，2011年中国旅游地产设计领跑企业以及"2012年度中国酒店业金星奖"最佳酒店设计机构等荣誉称号。

泛太平洋设计集团有限公司（加拿大）于2001年1月成立上海泛太建筑设计有限公司，并于2005年9月成立上海泛巢建筑设计事务所。

地址：上海市浦东新区张江高科园区
伽利略路11号伽利略商务公馆6单元
电话：+86-21-31265800
传真：+86-21-50814700
邮箱：ppddg@ppddg.com
网址：www.ppddg.com

Add: Unit 6, Calileo Mansion, 11 Jialilue Road,
Zhangjiang Hi Tech Park, Pudong District, Shanghai
Tel: +86-21-31265800
Fax: +86-21-50814700
E-mail: ppddg@ppddg.com
Web: www.ppddg.com

万宁石梅湾威斯汀五星酒店
Westin Shimei Bay Resort, Wanning

项目地点：海南 万宁
用地面积：117 680 m²
建筑面积：111 350 m²

Location: Wanning, Hainan
Site Area: 117,680 m²
Building Area: 111,350 m²

遵循以人为本的理念，探寻资源的可持续利用，建设环境友好型度假村，除了满足旅游者豪华度假需求的同时，亦要促进当地的社会经济发展和环境保护。因此，设计师确定了海南万宁石梅湾威斯汀五星级酒店项目的基本设计目标：以自然风光为主线，以生态技术为主要手段，以东南亚式风格与现代建筑相结合为基础，力求达到人与自然的和谐，旅游者与当地社区的和谐，酒店建筑风格与当地原有建筑风格的和谐，以促进酒店及其所在地区的可持续性发展，并使之成为2012年度中国酒店业金星奖最佳休闲度假酒店。

This is an environment-friendly resort following the "human-based" idea and exploring the sustainable use of resources, which not only meets the demands of luxury holidays for visitors, but also promotes the local socioeconomic development and environmental protection. Therefore, we confirm the basic design principles of the Westin Shimei Bay Resort project in Waning City of Hainan Province as follows: we seek to achieve the harmony between human and nature, visitors and local community, the hotel style and the original architecture style, by using the natural scenery as main line, the eco-technology as the primary method and combining the Southeast Asian style and the modern architecture, to promote the sustainable development of the hotel and the region.

芜湖冶芳园

Yefang Garden, Wuhu

项目地点：安徽 芜湖
用地面积：110 537 m²
建筑面积：29 977 m²

Location: Wuhu, Anhui
Site Area: 110,537 m²
Building Area: 29,977 m²

本项目为五星级宾馆，位于芜湖精品景区——神山公园之中，建筑群形体依山势展开，带有徽派民居特色的马头墙进退有序，构成生动活跃的画面，与周围的自然风景紧密结合。
庭院空间充分运用"掩""映""转""借"等手法，由厅到庭，由庭到院，由院到园，一层层渐次展开；各种功能观庭围院以厅、廊相接，步移景异。
建筑细部采用现代的建筑材料表现，并结合从传统建筑语言中提炼而出的建筑符号，在古意中透出时代感。

This project, a five-star hotel, is located in the excellent scenic spot of Wuhu City – Shenshan Mountain Park. The building cluster follows the mountainous landform, with firewalls featuring Anhui residential style in a vivid and vibrant picture, integrating closely with the surrounding natural scenery.
The courtyard space, making full use of contrasting and leaning method, extends layer by layer from hall to court, court to yard, and yard to garden; all functions are present in the courtyard, connected with each other by halls and corridors, and the scenery varies at every step.
In details, modern building materials express the architectural symbol extracted from traditional architectural language, which implies a modern sense from the ancient impression.

婺源旅游度假小镇总体规划

Tourism Town Master Plan, Wuyuan

项目地点：江西 婺源
用地面积：1 140 816 m²
建筑面积：461 171 m²

Location: Wuyuan, Jiangxi
Site Area: 1,140,816 m²
Building Area: 461,171 m²

本项目是作为整个婺源旅游区的配套设施建设的，主要功能是为来婺源的旅游者提供一个可供停留、休憩、休闲、消费的新型旅游集散地，集合了居住、投资、旅游、生态和文化等多种功能元素。

This project is built as the supporting facilities for the whole tourism of Wuyuan City, the main function of which is to provide the tourists with a new-type tourist terminal for stay, rest, leisure and consumption, integrating habitat investment, tourism, ecology, culture and many other functions.

绿地郑州新都会

The Greenland Urban Complex, Zhengzhou

项目地点：河南 郑州
用地面积：93 077 m²
建筑面积：727 740 m²

Location: Zhengzhou, He'nan
Site Area: 93,077 m²
Building Area: 727,740 m²

绿地新都会地块项目定位为政务区核心地段大型商务商业综合体，业态涵盖甲级办公、五星酒店、购物中心商业、商务配套型商业等，建成后将成为政务区乃至郑东新区的新地标。绿地新都会项目的整体建筑群沿金水东路展开，空间轴线指向东风南路与金水东路交汇处，空间层次上由东向西逐渐增高，并以西南角甲级办公楼为最高点，引领整个建筑群，气势宏大，具有强烈的标志性，形成强烈的序列感，充满张力。

The project of Greenland International Plaza is positioned as a large business and commercial complex in the heart of central governmental district covering first-grade offices, five-star hotels, shopping centers and supporting business, which upon completion will become a new landmark of the central governmental district and even Zhengdong New District. The integral building cluster of this project extends along Jinshui East Road to the junction of Dongfeng South Road and Jinshui East Road. The space level is becoming higher from east to west, and the first-grade office building at the southwest corner, as the commanding height, will lead the whole spectacular building cluster with remarkable identification, strong sense of sequence, and full of tension.

武汉百瑞景中央生活区商业街

The Bridge Living Capital, Wuhan

项目地点：湖北 武汉
用地面积：42 401 m²
建筑面积：25 765 m²

Location: Wuhan, Hubei
Site Area: 42,401 m²
Building Area: 25,765 m²

考虑基地位于1954年老厂区，且处于武昌区中心，具有独特的武汉近代建筑的特征，同时考虑商业街本身的商业特性，使用现代风格中穿插武汉近现代历史老建筑的形式，让人们用现代的眼光审视历史建筑。通过设计使得商业街更加亲切，突出以人为本的设计理念。设计中注意建筑的细节，使得建筑更加精致，展示出商业街的“表情”。设计灵活且具有韵律感的平面和天际线，用广场节点打破结构，打造浓厚的商业氛围，使商业价值最大化。

Considering that the base is located in the old plant area since 1954, the center of Wuchang District, with very special architectural texture of Wuhan City, and meanwhile given the commercial characteristics of streets, the project uses the modern style of buildings interlaying with the historical buildings of Wuhan City. This multi-style integration offers people with new tastes in old buildings, from a modern view. The design makes commercial streets more friendly and highlights the people-oriented idea. It puts much emphasis on the architectural details, which can make the architecture more delicate and show the “expression” of commercial streets. Moreover, it adopts flexible, rhythmic planarity and skyline and breaks the structure by introducing the plaza, to create dense commercial conditions and maximize the commercial value.

大连新星艾维尼小镇
Xinxing Avignon Town, Dalian

项目地点：辽宁 大连
用地面积：122 700 m^2
建筑面积：147 800 m^2

Location: Dalian, Liaoning
Site Area: 122,700 m^2
Building Area: 147,800 m^2

在地块中营造自然生态的健康住宅，以住宅的围合形成景观上的大庭院。将大型的景观绿化作为本地块的核心，使人们拥有开阔的景观视野，形成良好的住区小气候，塑造出完整统一的立面，突出了集团的品牌形象。
本项目获“2011 年中国人居范例最佳方案设计”金奖。

In the lot build the healthy housing of nature and ecology. The landscape surrounded by the housing will form a great courtyard, in which the core task is the greening of large-scale landscape, to make people have a broad vision, form a good microclimate in residential community, shape an integrated façade and highlight the brand identity of group.
This design won the gold award of “The Best Design of China Habitat Models in 2011”.

郑州联创国际中心
United Design Group Center Zhengzhou

项目地点：河南 郑州
用地面积：32 062 m^2
建筑面积：101 073 m^2

Location: Zhengzhou, Henan
Site Area: 32,062 m^2
Building Area: 10[illegible] m^2

郑州联创国际中心项目位于郑东新区西北侧，项目用地面积为 32 062 m^2。项目包含了 5～7 栋高端办公建筑及 1 栋独立商业建筑。
设计提出一个“四水归堂”的概念，从一种最小的单元开始，在演变的过程中形成的不同组合。“四水归堂”，最大特点是“田”，是一个凝聚天地人气的风水大格局，会使企业公司更有向心力和凝聚力。

This design is started from the conception of “waters converging in the hall”, which has formed different combinations in the evolution process from the smallest unit. The distinguished characteristic of ‘waters converging in the hall”, just as its name implies, is “converging”. It develops a large pattern of geomancy gathering the Spirit of Land, which can enhance the prestige and the leadership of the leaders.

成都青羊工业集中发展区

Qingyang Industrial Development Zone, Chengdu

项目地点：四川 成都
用地面积：155 939 m²
建筑面积：277 138 m²

Location: Chengdu, Sichuan
Site Area: 155,939 m²
Building Area: 277,138 m²

原有建成地块提出了“森林峡谷”的理念，什么是峡谷？峡谷是大自然，是一种气质，是一种精神。设计师将延续这样的精神，并从大自然中提取出新的精神元素。将非规则、非人造、扭转而延绵的形态元素运用到设计中，把随意的曲线形态不刻意地撒布在地块中，营造自然的、生态的、浪漫的新型总部办公园区。

On the established lot put forward an idea of “Forested Gorges”. What is the gorge? It is the nature, a temperament and a spirit. Desigers will continue this spirit and extract the new spiritual elements through the understanding of natural. The design absorbs the irregular, inartificial, long and winding form elements and intersperse the free curve patterns naturally among the land so as to build a natural, ecological and romantic head office park of new type.

扫描查看更多信息

上海经纬建筑规划设计研究院有限公司

Shanghai LongiLat Architectural Design & Research Institute Co., Ltd.

上海经纬建筑规划设计研究院有限公司是一家以城乡规划、建筑设计为主的现代科技型企业，同时拥有城乡规划甲级、建筑工程设计甲级资质。业务范围涵盖城乡总体规划、城市发展研究、居住区规划、建筑设计、室内装饰设计、景观设计和房地产开发技术咨询等。上海经纬建筑规划设计研究院有限公司是上海市建筑学会理事单位，获上海市文明单位称号

地址：上海市杨浦区长阳路1568号
（宁国路503号）复地四季广场10-12号楼
电话：+86-21-65039009
传真：+86-21-65638325
邮箱：jwjz@china.com
网址：www.sladi.com.cn

Add: No.10-12 Building, Fudi Four Season Square(Ningguo Road No.503), Changyang Road No.1568, Yangpu District, Shanghai
Tel: +86-21-65039009
Fax: +86-21-65638325
E-mail: jwjz@china.com
Web: www.sladi.com.cn

保时捷总部大楼

Porsche HQ

设计师：陈铁峰、杨毅、徐伟华
项目地点：上海
用地面积：19 900 m²
建筑面积：90 640 m²
建筑密度：39%
绿化率：20%
建筑高度：99.25 m

Designer: Tiefeng Chen, Yi Yang, Weihua Xu
Location: Shanghai
Site Area: 19,900 m²
Building Area: 90,640 m²
Building Density: 39%
Green Ratio: 20%
Building Height: 99.25 m

项目位于浦东软件园陆家嘴分园区内，是整个园区内的最高建筑之一。建筑通过简洁的方形体块围合而成，玻璃与铝板虚实的搭配，使整个建筑形体既具有有力的线条，又营造出错落有致、生动有趣的空间感。建筑外墙的设计概念取自于“工业经典建筑”。墙面以银灰色铝板为主，配合竖向的金属框架形成强烈的垂直感。细部则采用金属构件与玻璃窗的组合，强调了研发建筑的工业性。大楼外立面的设计以规整、黄金比例分割的凹窗为特色，整个立面虚实分明，营造出丰富的阴影效果，并采用银灰色作为主楼的主要颜色，使建筑于现代中蕴涵传统特色。结构采用框架核心筒形式，框架采用劲性混凝土柱，各层均采用钢梁加钢质楼层板组合形式，起到有效减小框架柱截面的作用，同时，能更有效地提高结构的抗震性能。

The project is located in the Lujiazui Branch Park of Pudong Software Park, and is one of the highest buildings in the whole park. Enclosure of simple rectangular blocks and collocation of void glass and solid aluminum plate give a strong linear appearance and a sense of space, orderly and interesting. The concept for the outer wall of the building comes from "Industrial Classic Architecture". The wall panel is mainly set with silver gray aluminum plate, forming a strong sense of verticality with the vertical metal frames. The details are the combinations of metal components and glass windows, emphasizing the industrial property of the R&D buildings. The design for the outer facade of the building is featured with regular concave windows in golden section proportion. The whole facade has clear void-solid distinction, creating a rich shadow effect and utilizing silver gray as the main color of the main building, implicating the building with traditional features in modern style. The structure is in a frame-core wall mode, and the frame adopts stiff concrete column. Each floor consists of girder combining with steel floor deck, thereby promoting seismic resistance of the structure as well as reducing the effect from section of the small frame column efficiently.

江西省九江职业技术学院濂溪新校区规划建筑设计

Lianxi New Campus of Jiujiang Vocational and Technical College Architectural Design

设 计 师：陈铁峰、丁文涛、陈雅龄
项目地点：江西 九江
用地面积：546 900 m²
建筑面积：253 200 m²

Designer: Tiefeng Chen, Wentao Ding, Yaling Chen
Location: Jiujiang, Jiangxi
Site Area: 546,900 m²
Building Area: 253,200 m²

九江职业技术学校前身为九江船舶工业学校，创建于1960年，是全国首批开办的高等职业教育的10所学校之一。本项目规划以核心景观为中心，充分分析庐山核心景观区及周边景观要素，并将其引入校园，形成山景与校园之间的视线通廊。九江技术学校以船舶见长，校门的设计理念来源于船舶的形态，并寓意扬帆起航；入口轴线上的景观雕塑象征从校园走出的学生将拥有大鹏展翅的未来。考虑到教育建筑的属性，在设计上注重场所的多元化，实用、典雅，并具有文化与学术气氛。

Jiujiang Vocational and Technical College (preceded by Jiujiang Shipping Industry College) was founded in 1960, one of the first 10 schools conducting higher vocational education. The project is centered with core landscape, adequately analyzes the Lushan Mountain core landscape area and of its peripheral landscape elements, leading to the campus entry, forming the sight passage between the mountain scene and the campus. This college is specialized in shipping, so the design idea of the school gate originates from the form ships, meaning to make sail; the landscaping sculpture on the entry axis symbolizes the vigorous future of the students graduating from the school. Considering it is an educational architecture, the design focuses on the diversity of the places, practicability, elegance, and the atmosphere of culture and academy.

芜湖第二人民医院

No. 2 People's Hospital, Wuhu

设 计 师：魏轶侃、陈铁峰、张述诚
项目地点：安徽 芜湖
用地面积：35 000 m²
建筑面积：110 000 m²
容 积 率：2.71

Designer: Yikan Wei, Tiefeng Chen, Shucheng Zhang
Location: Wuhu, Anhui
Site Area: 35,000 m²
Building Area: 110,000 m²
Plot Ratio: 2.71

项目地处九华中路与黄山中路交汇的西南方，地块总面积约6.8 hm²；在设计中将标志式的主楼视作一颗强劲跳动的“心脏”，象征城市主要功能的健全和有序运作。竖向变化发展的构图体系，象征蒸蒸日上的城市发展，通过医院的救死扶伤，给人们带来健康美满生活。五段式的组合方式改变了以往长板式的造型，解决了因为建筑体量大而产生的建筑面宽过长问题，减少了主楼对城市的压迫感，在满足功能要求的同时，为城市提供了一个标志性的景观建筑。项目曾获得“2011年上海市工程设计”优秀奖；“2011年度上海市优秀工程设计电气专业”二等奖。

The project is located southwest of the intersection of Jiuhua Middle Road and Huangshan Middle Road, and the total area of the land lot is about 68,000m²; in the design, the landmark main building is seen as four flying wings, connecting the heart-shaped central hub. A strong and vibrating heart symbolizes the sound and orderly operation of the main functions of the city. The vertically changing and developing structural system represents the prosperous urban developing rhythm, constantly saving lives and bringing a healthy and happy life. The 5-section combination is a change from the normal long plate shape, solves the problem of overlong building width due to huge building volume, reduces the oppression to the city from the main building, and provides a landmark landscape architecture while meeting the functional requirements. Meanwhile, this architecture adopts the new technology of chilled water storage in the energy saving & environmental protection. Awards of the project: Excellent Engineering Design Award in Shanghai in 2011; 2nd Prize of Excellent Engineering Design (Electric Work) in Shanghai in 2011.

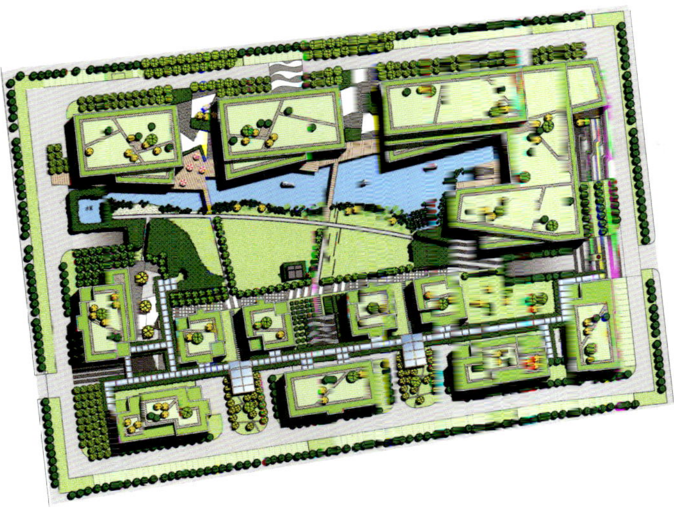

上海张江高科技开发园区

ZJ InnoPark IBM HQ, Shanghai

设 计 师：白蔚、叶松青
项目地点：上海
用地面积：50 000 m²
建筑面积：90 000 m²
容 积 率：1.6
建筑密度：20.32%
绿 化 率：36%

Designer: Wei Bai, Songqing Ye
Location: Shanghai
Site Area: 50,000 m²
Building Area: 90,000 m²
Plot Ratio: 1.6
Building Density: 20.32%
Green Ratio: 36%

张江高科技园区产业区——创新园工程位于上海市浦东新区张江高科技园区产业区A2-12地块，主要用途是引进国际大型企业的研发机构及高科技孵化项目。设计上力求超前性、先导性、科技性，贯彻以人为本，走可持续发展的道路，以体现“新科技、新技术、新工艺、新设备”四新概念，使之真正达到“高起点、高水平、高质量、高标准”。

区内分为高层建筑和多层建筑两大区域，高层沿基地北面布置，层沿基地南侧布置，形成南低北高，西低东高的天际线。整个基地通过建筑的围合形成一个中央共享景观空间，使每栋房屋都有着良好的景观视野。

2008年上海张江高科技园区产业区——创新园，曾荣获“第二届上海市建筑学会建筑创作奖”佳作奖。

The project is located at A2-12 plot in ZJ InnoPark in Pudong New District, Shanghai, and it is a research & Development and Incubation Building project. The main purpose is to introduce the research & development organization of international major industry and hi-tech incubation project; the design tries to be advancing, leading, technical, human-based, and takes on sustainable development, to reflect the four new concepts of "new science and technology, new skills, new techniques, new equipments" and realize the idea of "high starting point, high level, high quality, high standard".

The buildings in the area divided into two sections of high rise buildings and mid rise buildings, and the high rise buildings are arranged along the north side of the base, forming a skyline of low in the south & west and high in the north & east. The buildings on the base enclose a central sharing landscape space, keeping a good landscape vision for every building.

The project won Excellent Work Prize of 2nd Architecture Creation Award by Shanghai Academy of Architecture in 2008.

天津日月湾国际总部商务园
Sun-Moon Bay International HQ Business Park, Tianjin

设 计 师：魏轶侃、张榜、张述诚、张琴
项目地点：天津

Designer: Yikan Wei, Bang Zhang, Shucheng Zhang, Qin Zhang
Location: Tianjin

项目位于天津市东丽区华明示范镇东区D-E街坊，规划地块距离天津站15 km、天津滨海国际机场6 km、天津港35 km，南临津汉公路以及规划中的地铁2号线，是新老城区相互联系的核心区域。
天津日月湾国际总部商务园集商业中心、总部商务办公、五星酒店为一体，为高集聚、强辐射的多功能商业复合体，以高档次综合总部商务模式为东丽区的发展提供支撑。结合园区景观走廊，组织整个园区的空间形式，达到景观绿化的延续与融合。

The project is located in Neighborhood D-E of East Area of Dongming Demonstration Town in East Tongmou District, Tianjin. The planned land lot is 15km from Tianjin station, 6km from Tianjin Binhai International Airport, 35km from Tianjin Harbor, and beside Jin-Han Highway and the planned Metro Line 2 in the south. It is the core area of the interconnecting new and old city areas.
The Business Park of Tianjin Sun-Moon Bay International Headquarters integrates a business center, HQ office buildings, 5-star hotel into a multifunctional commercial complex of high concentration and strong radiation, supporting the development of Dongli District in the business mode of high-grade comprehensive headquarter. The landscape passage in the park is combined, the space forms of the whole park are organized, so as to achieve the continuity and integration of the landscape greening.

长沙大河西先导区枫林路两厢城市设计

Urban Design of Fenglin Road Surrounding Area, Dahexi ,Changsha

设 计 师：张榜、胡永武、唐剑晖、陈术
项目地点：湖南 长沙
用地面积：1214 hm^2

Designer: Bang Zhang, Yongwu Hu, Jianhui Tang, Shu Chen
Location: Changshan, Hu'nan
Site Area: 1,214 ha

长沙大河西先导区枫林路两厢城市设计位于长沙市大河西先导区启动区中部，东接五一大道，西至三环线，北邻河西商务区、麓谷园区及长沙高新技术开发区，南邻岳麓山风景区及梅溪湖国际功能区，是大河西先导区东西向重要发展轴。本项目的规划面积约1214 hm^2。枫林路两厢的建设对于长沙多中心发展及向西拓展的城市整体格局的形成具有至关重要的意义，它将是联系城市主中心、河西副中心及星马组团的重要通道，缓解主城区入口、用地、资源压力并向河西疏解的主要通道，是长沙今后重要的交通枢纽地段。枫林路未来将承担轨道交通、城市快速交通和城际客运、市区公共交通等任务。

Urban design on both sides of Fenglin Road's in Changsha forerunner area of Dahexi is located in the promoter region of central Dahexi pilot area east of Wuyi Avenue, west of the Third Ring, north of the business district of Lugu Park in Hexi and Changsha High-tech Development Zone, south of the Mount Yuelu scenic area and Meixihu international zones,which is the major east-west development axis in forerunner area of Dahexi. The scope of planning is formed by the both sides of Fenglin Road, east to Juzizhou West bridgehead, west to West Third Ring Road. It is 300 to 400 meters range from the north to the south, which is an area of about 1214 ha. It is very important for Changsha to build a polycentric development and westward expansion of the city as a whole pattern formation, which has a crucial significance. It will be an important corridor which contacts the primary center, Hexi deputy center and star groups. The project, relieving the pressure of urban population, land, resource and being the main channel to Hexi,which comes into an next important transport hubs in Changsha. As time goes, Fenglin Road will undertake urban and cities rapid transit and intercity rail transit, urban public transport, and more.

泰州市城西大门城市设计

Urban Design for West Gate in Taizhou

设 计 师：付予光、陈晨杰、杨星福、陶修军
项目地点：江苏 泰州
用地面积：13 500 m^2

Designer: Yuguang Fu, Chenjie Chen, Xingfu Yang, Xiujun Tao
Location: Taizhou, Jiangsu
Site Area: 13,500 m^2

随着城市的发展，以工业产业为特色的西大门片区在继续发挥产业重要作用的同时，完成以重点培育先进制造业、精细化工、现代物流和现代服务业等"综合功能发展"为重要特征的区域产业转型；通过产业升级、功能配套、兼容服务与新兴产业引进等途径，构建现代服务业的高地，有效增强地区经济的活力，提高街区的吸引力。这些新的产业将重新汇聚人气，使片区焕发出全新的生机。
规划设计强调"活力"，提出了"绿色活力西门"的概念。
使门户商业活力与景观魅力和谐统一。
2011年11月"泰州市城市西大门城市设计"荣获"第四届上海市建筑学会建筑创作奖"佳作奖。

Along with the urban development, the area with industrial feature continues to play its important role and further transforms to regional industry featured with "multifunctional development", such as key leading manufacturing industry, fine chemistry industry and modern logistics, modern services industry. By means of industry upgrade, function support, compatible service and new industry introduction, the project is built as upland of modern service industry, effectively enhancing the vitality of the regional economy and increasing the attraction of the blocks. These new industries regain the popularity and revitalize the area.
The design focuses on "vitality", and proposes the idea of "green and vital West Gate".
It emphasizes the harmonious integration of commercial vitality and landscape charms.
The urban design for West Gate in Taizhou City won Excellent Work Prize of 4th Architecture Creation Award by Shanghai Academy of Architecture in November 2011.

唯士国际设计与发展有限公司

W.I.T.H. International Design and Development Co.,Ltd.

WITH·唯士国际设计与发展有限公司（简称唯士国际）是由上海都易建筑设计有限公司（都易国际）和上海齐越建筑设计有限公司（齐越国际）共同组成的。唯士国际参股并管理的上海宝厦建筑设计有限公司拥有建设部颁发的建筑工程甲级资质，使得唯士国际能完成更多的优秀设计作品，为客户提供更多优质服务。

唯士国际在高端住宅社区、城市商业综合体、办公楼、酒店等设计领域形成了自己的专业优势，并在新古典高端住区设计领域处于全国领先地位，并与万科、绿地、华润、建业、协信等国内一线房企，建立了长期合作关系。

唯士国际拥有专业设计人士150余名，是一个富有活力、充满激情和创造力的设计团队。竭诚为客户提供基于市场需求、营销、成本数据分析和专业理想的设计服务，并通过严格的项目管理体系来实现产品的精细化设计、成本控制和高效率实施。

贵州·驰扬·商业综合体

池州·东池·东壹号商业广场

广州 · 万科 · 华府

广州 · 万科 · 红郡

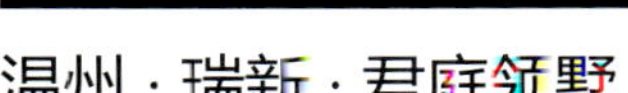

温州·瑞新·君庭领墅

丽江·驰扬·国际商贸城酒店

杭州 · 万科 · 西溪蝶园 · II

成为2010年万科全国四大样板楼盘
荣获“2011年中国人居典范建筑设计方案竞赛”金奖

杭州 · 万科 · 金色家园

重庆·协信·协信公馆

荣获

“2011年中国人居典范建筑设计” 金奖

“詹天佑设计奖重庆优秀建筑设计” 金奖

“詹天佑设计奖重庆优秀环境设计” 金奖

“詹天佑设计奖重庆优秀规划设计” 金奖

义乌·名品城

扫描查看更多信息

上海加合建筑设计有限公司
Shanghai Team+ Design Architectural Design Co.,Ltd.

加合建筑是一个充满创造力和进取心的设计团队，以探索中国当代建筑设计之路及发展本土精品事务所为目标。倡导团队精神、敬业精神及探索精神，在设计上以理性而有创造力的工作方式对每个项目进行深入的研究，从项目条件、城市文脉、开发经营模式、技术材料运用及生态环保节能等方面寻找适合的切入点。力求在城市、使用者、开发者之间找到最佳平衡点，以共生共赢实现整体价值最大化。

加合建筑的核心设计观为"联结、链接"，力图通过设计将人与自然、人与城市、人与历史、人与人有机联系在一起。在不断实践的过程中，针对项目的不同特点寻找相应最佳的链接方式。通过持续的探索，逐步建立起对于环境、城市、建筑与人的系统性认知，并积极尝试各种能将它们有机联结的方式。城市的可持续发展需要有一个健康的生态链，包含功能链、环境链、经济链、技术链及社会链。通过对每个项目条件的深入解析，从项目的前期定位便开始全方位考虑其内在生态链系统，力求设计能使项目在各方面达到平衡共生和可持续发展。

好的设计构思往往来自于朴实的动机和简单的原则，对基本问题全面、深刻的认知和探讨是设计工作的重点，从而形成清晰的设计观念与理性的设计方法。

加合建筑对每一个实施项目均保证全方位的投入：从方案设计的反复推敲、全程跟进的深化设计和掌控到深度介入项目实施的各个环节，有效确保设计的完善度及实施效果，努力使其成为精品项目。在发展本土精品事务所的同时，注重探索以事务所培养优秀设计师的有效方式，力求形成促进设计团队成长的良好环境。

在多年的设计实践过程中，加合形成了既汲取西方设计体系的精华，又与中国本土现状相联结的思想体系、设计方法和工作流程。在与国内外一线开发集团、有特色的中小型开发机构及高新企业的长期合作中不断提升设计品质及综合协调管理能力，做到专业而有序的全程服务。这也保证了在中国经济高速发展的时代背景下，加合建筑在作品中倾注的理性思考与激情创作都能得以最大程度地实现。

地址：上海市莫干山路50号21号楼2层
邮编：200060
电话：+86-21-62773208
邮箱：teamplus@teamplus.cn
网站：www.teamplus.cn

Add: Floor 2, Building 21, 50 Moganshan Road, Shanghai
P.C.: 200060
Tel.: +86-21-62773208
E-mail: teamplus@teamplus.cn
Web: www.teamplus.cn

Team+ Design Studio is a creative and endeavoring design team, committed to exploring the roadmap of Chinese modern architectural design and developing local elite design firms. It advocates teamwork spirits, professionalism and adventurism. In design, it makes in-depth research on every project with its rational and creative work style, seeking for a suitable breakthrough point among project condition, urban context, development & operational mode, technical material application and ecologic environmental protection & energy-saving etc. It tries to seek the best balance between the city, users and developers, to maximize the integral value by coexistence and win-win relationship.

Its core outlook on design is "link", trying to link human and nature, human and city, human and history, human and human organically through design. In continuous practice, Team+ seeks the best linking way for every individual project. Through continuous exploration, it gradually builds up a systematic cognition on environment, city, architecture and human beings, and actively attempts various ways of linking them organically. Urban sustainability needs a healthy eco-chain, including functional chain, environmental chain, economic chain, technical chain and social chain. Through in-depth analytics on every project condition, Team+ considers the inherent eco-chain system ever since the preparatory positioning of the project, seeking the best design of balance, symbiosis and sustainable development in all respects.

A good design concept always comes from pure motif and simple principle, and Team+ emphasizes the overall, profound cognition and exploration on fundamental problems, to form clear design concept and rational design method. Team+ ensures full range of inputs in every project in execution: from elaboration of scheme design, seamless following-up in-depth design and control, to in-depth involvement in every process of the implementation, in order to effectively ensure the perfection and benefit of design, and realize a series of exquisite projects. Meanwhile, in developing local exquisite design firm, it also creates a good environment for team growth, by effective means of nurturing excellent designers.

In design practice of many years, Team+ has formed a special ideological system, design method and work process, integrating the essence of Western design system into Chinese locality. Through cooperation with domestic and international development groups, characteristic small to medium-sized development organizations and hi-new tech enterprises, it has been increasing its design quality and comprehensive coordination & management capabilities, offering seamless professional services in order. The rational thinking and passionate creation poured into the designs will be realized to the largest extent in fast development of Chinese economy.

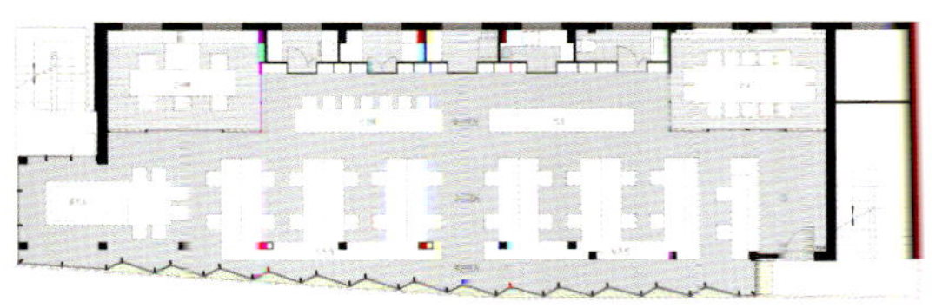

加合建筑办公室改造设计

Office Renovation of Team Plus Design Studio

设计师：程洪波、李霞、权伍贤、林良纳　　Designer: Hongbo Cheng, Xia Li, Wuxian Quan, Liangna Lin

本办公室的改造是加合建筑理念的一次小规模实践。对原建筑做了小心的改动，将沿河界面做得更为通透，折线形的玻璃幕墙与苏州河的灵动呼应，室内因此也顿感开阔。建筑与绿化的结合带来了自然的气息，加合建筑的设计师们会一起参与绿植的维护。希望营造开放、有序、温暖的内部空间氛围，与“阳光、健康”的团队文化相契合。简洁的细节处理对选材、施工提出了很高要求，幕墙、家具、灯具等许多特殊设计承蒙施工方的合作，基本实现了设计的想法。这个小小的项目虽然经历了不少曲折，但最终基本未改初衷，为设计师们创造了一个理想的工作环境。

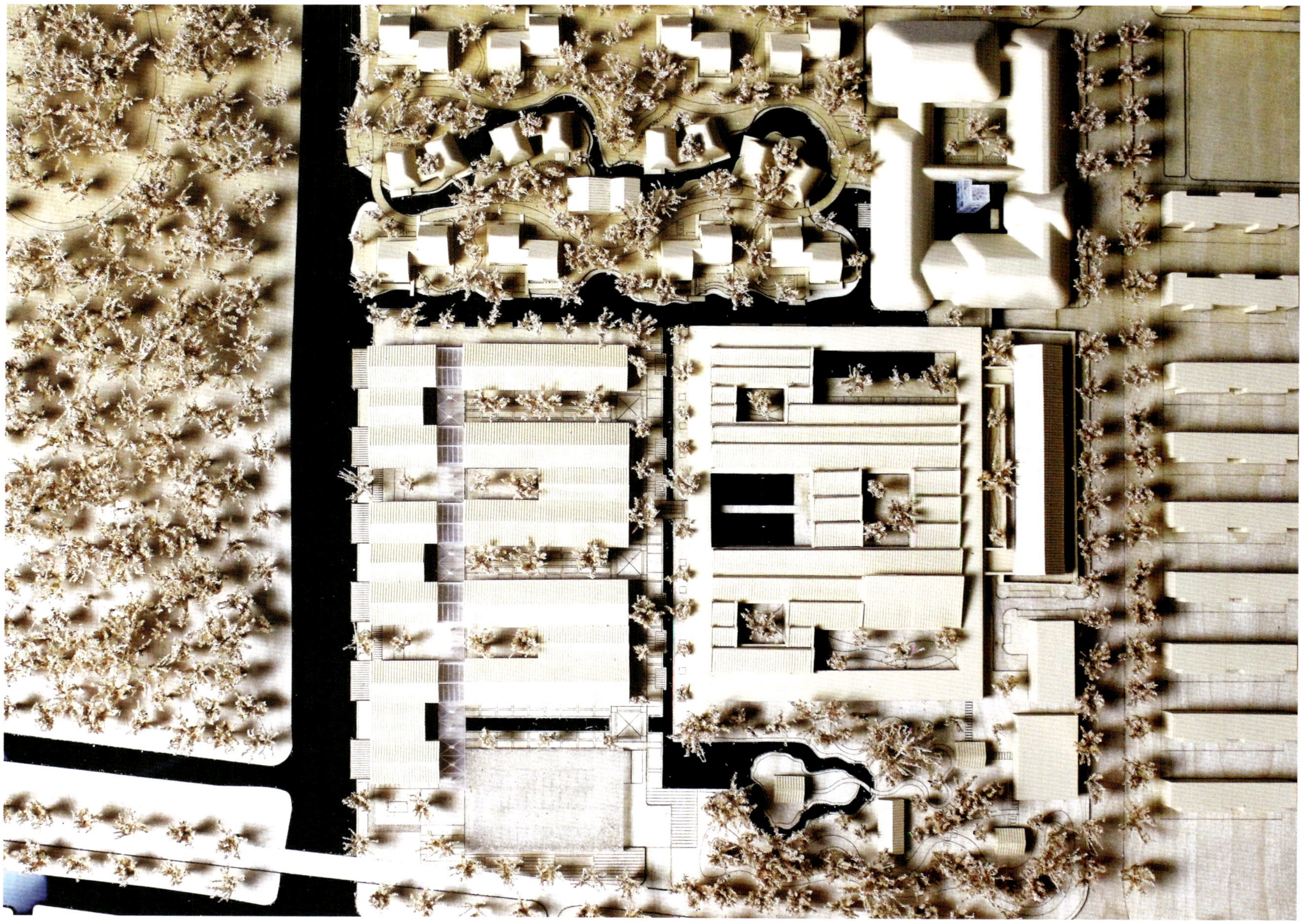

崇明养生精品酒店改扩建

Renovation & Extension of Boutique Hotel, Chongming

设 计 师：李霞、程洪波、柳光叶、杨文若、台君君、马力涓、权伍贤、林良纳、陈福全
建筑面积：35 000 m²

Designer: Xia Li, Hongbo Cheng, Guangye Liu, Wenruo Yang, Junjun Tai, Lijun Ma, Wuxuan Quan, Liangna Lin, Fuquan Chen
Building Area: 35,000 m²

该项目位于上海崇明岛中部，隔河为东平国家森林公园。地块内部树木葱郁，环境优越；现状建筑为20世纪中后期典型工业建筑，具有浓郁的时代特征，主体建筑空间形式独特，结构完好，可加以改造利用。产业转型导致工厂失去往日辉煌，设计尝试用生态的方式使得该区域重新焕发活力。设计理念为"引入自然资源、创造生态环境、低碳建造和节能环保、可持续发展"。具体方式为通过利用森林公园的景观辐射，引入水资源，结合基地内部保留的树木营造内部森林；采用低碳环保的改扩建方式，合理利用原设施，结合绿色技术，达到节能环保；寻找适合的项目定位，制定合理的规划方案，选择合理的经营模式，达到项目的可持续发展。

南京金地名京

Gemdale Mingjing, Nanjing

设 计 师：李霞、程洪波、邵明良、徐青青、杨文若、朱江海、高静、柳光叶
建筑面积：200 000 m²
合作单位：南京民用建筑设计院（施工图设计）
HMA GROUP（景观设计）

Designer: Xia Li, Hongbo Cheng, Mingliang Shao, Qingqing Xu, Wenruo Yang, Jianghai Zhu, Jing Gao, Guangye Liu
Building Area: 200,000 m²
Partners: Nanjing Civil Architectural Design Institute(Drawings Design);
HMA GROUP (Landscape Design)

金地名京项目位于南京市河西新区，占地9ha。设计旨在创造一个都市中的花园社区。为了营造开阔的中心庭园，住宅采取三排折线型格局，自然形成数个现代园林式的围合空间，收放有致，序列清晰。北侧中心花园结合下沉庭院布置会所，将地上、地下的景观和活动有机结合，提升地下空间的环境品质。高层建筑为了表达内部庭园空间而呈现如山脉般绵延的态势，营造"横看成岭侧成峰，远近高低各不同"的空间感受。建筑设计源于对空间环境的关注，以舒展的形体处理将建筑融于外部环境中，使得空间更具张力，为人们创造具有感染力和归属感的空间体验。

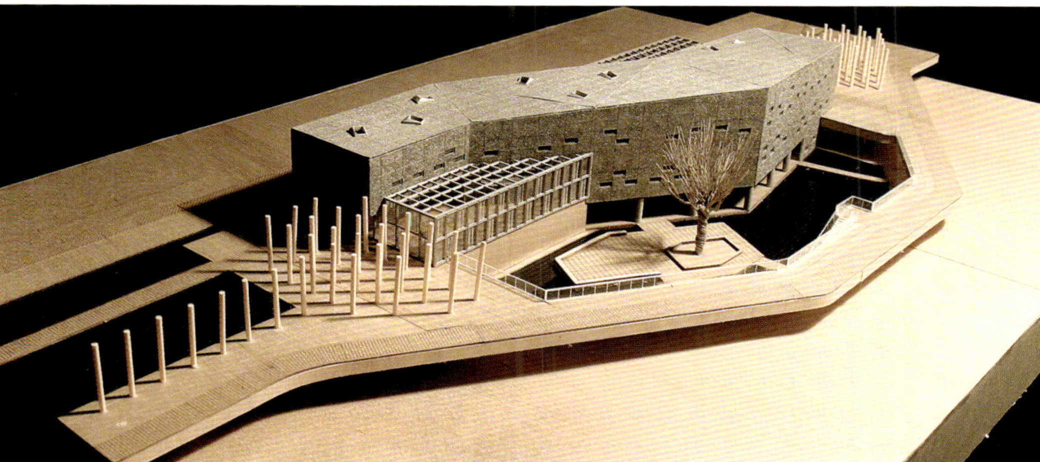

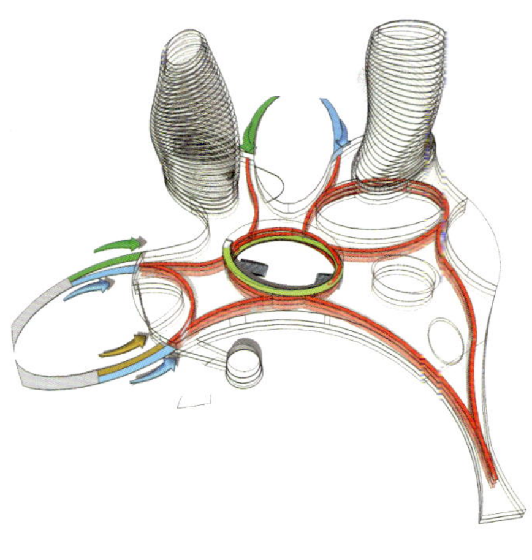

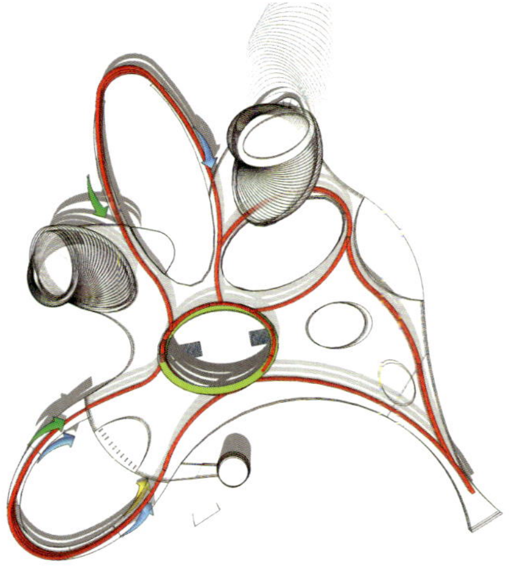

立体交通体系

上海唐镇综合体

Tangzhen Complex, Shanghai

设 计 师：李霞、程洪波、台君君、权伍贤、林良纳
建筑面积：300 000 m²

Designer: Xia Li, Hongbo Cheng, Junjun Tai, Wuxian Quan, Liangna Lin
Building Area: 300,000 m²

唐镇综合体由地铁枢纽、购物中心、办公、酒店、公寓等组成。项目位于地铁站的上方，区域辐射力强。设计旨在将城市公共交通、公共绿地及相邻地块有机联结，形成区域人流汇集枢纽、城市公共空间枢纽和绿色生态枢纽。各功能人流的汇集、各公共空间的紧密联结、多层次的立体花园形成了开放、多元、生态的城市节点，为城市创造未来的活力中心。

南京所街商业

SuoJie Commercial Street, Nanjing

设 计 师：李霞、程洪波、柳光叶、李为、马力涓
建筑面积：40 000 m^2

Designer: Xia Li, Hongbo Cheng, Guangye Liu, Wei Li, Lijuan Ma
Building Area: 40,000 m^2

该项目用地面积2 hm^2，地上建筑面积40 000 m^2，地下建筑面积25 000 m^2。商业业态包含休闲娱乐、特色餐饮、办公、精品商业及精品酒店。项目主题定义为慢生活，强调未来的消费方式将由简单满足物质需求演变为生活方式的消费、体验式的消费，进而寻找慢生活与现代商业结合的最佳方式。设计结合沿河展开的狭长地形组织一系列适宜停留、体验的城市生活空间，与不同功能及河道景观互动组成立体多元的开放区域。城市舞台、商业步行街、艺术广场、河畔餐饮步道、退台花园等为人们提供亲近自然、开放交流的场所。自然环境与休闲商业紧密结合，创造全新的休闲生活体验。

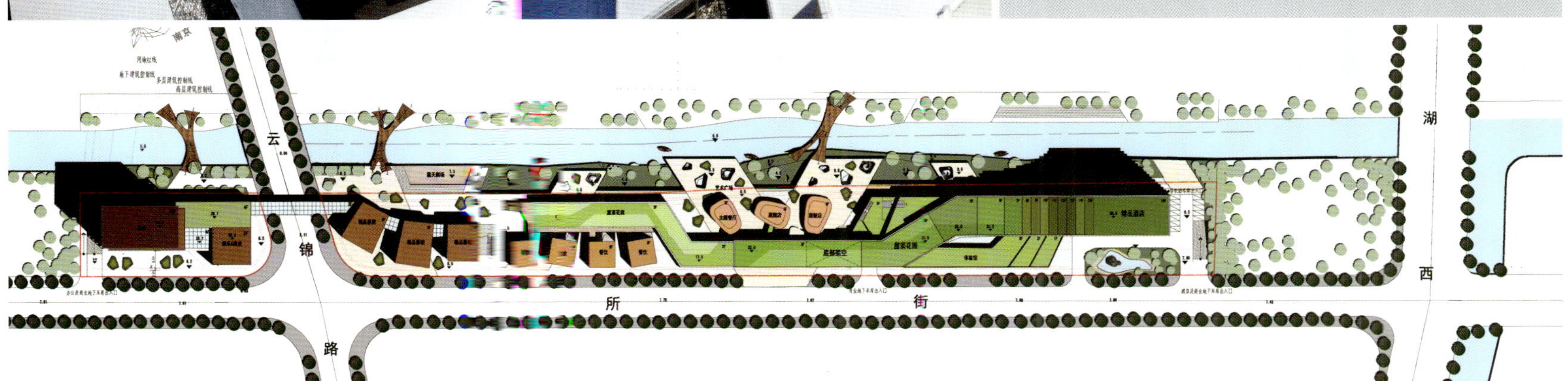

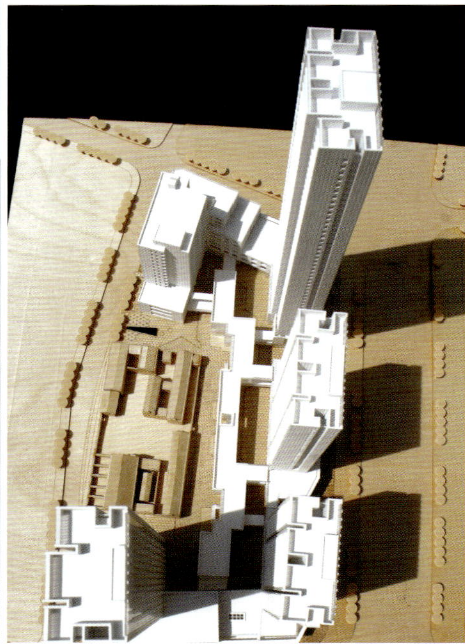

常州莱蒙时代

Landmark Time Square, Changzhou

设 计 师：李霞、程洪波、邵明良、杨文若、徐青青、李为、柳光叶、马力涓
建筑面积：130 000 m²
合作单位：东南大学建筑设计研究院（施工图设计）；深圳易道（景观设计）

Designer: Xia Li, Hongbo Cheng, Mingliang Shao, Wenruo Yang, Qingqing Xu, Wei Li, Guangye Liu, Lijuan Ma
Building Area: 130,000 m²
Partners: Architectural Design and Research Institute of Southeast University(Drawings Design); Shenzhen Yidao(Landscape Design)

该项目位于江苏常州市核心区，为莱蒙都会大型商业综合体的一部分。地块内有一处文物保护建筑"传胪第"，砖雕精美、院落空间宜人。中国城市目前的更新大多以"覆盖"的方式进行，摧毁了许多城市记忆的载体。该设计关注城市更新、商业开发与城市记忆之间的关系，提出的解决方式是将老建筑的院落空间进行延伸，创造出具有舒适尺度的商业"院街"和提升生活品质的"院宅"。以开放的空间面向城市和保护建筑，新老建筑之间既有对话又有对比，使这里成为具有独特空间体验和文化内涵的城市魅力街区。

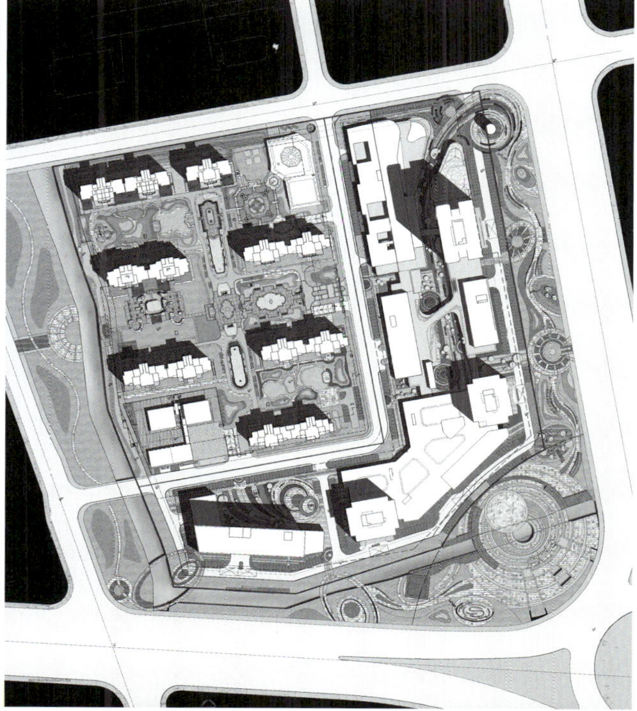

南通星光耀广场

ORStar City, Nantong

设 计 师：李霞、程洪波、邵明良、台君君、杨文若、李[illegible]、徐青青、姚勇、柏松、权伍贤
建筑面积：440 000 m²
合作单位：中建上海院（施工图设计）；HWA（景观设计）

Designer: Xia Li, Hongbo Cheng, Mingliang Shao, Junjun Tai, Wenruo Yang, Wei Li, Qingqing Xu, Yong Yao, Song Bai, Wuxuan Quan
Building Area: 440,000 m²
Partners: China Architecture Shanghai Institute (Drawings Design); HWA(Landscape Design)

该项目地处南通市崇川新城中心，占地约12 hm²，为具有商业、文化产业、居住等复合功能的城市综合体。未来城市的新兴区域需要更多开放空间，更趋多元复合的城市功能、更加丰富的生活选择。为此设计重点赋予这个综合体三个特点：运动和生态为特色的健康休闲环境；艺术、影视、教育、创意、展示等与商业及居住空间交融互动带来时尚活力；多重主题、尺度宜人的开放立体街区。这些特色的营造将城市与人们的生活紧密联结，产生富有活力的城市聚集点。

珠海巨人集团南方总部

Giant Group South China HQ, Zhuhai

设 计 师：李霞、程洪波、台君君、李为、杨文若、权伍贤、
林良纳、马力涓、李志东、李强
建筑面积：220 000 m^2

Designer: Xia Li, Hongbo Cheng, Junjun Tai, Wei Li, Wenruo Yang,
Wuxuan Quan, Liangna Lin, Lijuan Ma, Zhidong Li, Qiang Li
Building Area: 4220,000 m^2

巨人集团南方总部位于珠海唐家湾高新区，背倚青葱翠绿的凤凰山麓，面向浩瀚的中国南海。项目由巨人南方总部、独立总部研发楼、高层研发楼、精品酒店、企业会所和休闲商业等功能组成。规划格局以弧线形与区域规划形态及海岸线相呼应，创造山与海的景观视线通廊，以开放式格局营造与山、海、岛、花园景观环境相融合的滨海总部商务花园。设计中通过内外结合的空间设计来提升办公环境的舒适性，并采用多种生态、节能技术手段来创造环保、可持续发展的绿色园区。多元复合的配套功能、开放共享的空间格局、绿色生态的建筑和环境、独特的滨海休闲气息创造出充分人性化的新一代高新产业园。

中船第九设计研究院工程有限公司

China Shipbuilding NDRI Engineering Co., Ltd.

中船第九设计研究院工程有限公司由原中船第九设计研究院改制而成，隶属于中国船舶工业集团公司。公司是一家多专业、综合性的大型工程公司，主要从事工程咨询、工程设计、工程项目总承包等业务。

公司将继承发扬中船第九设计研究院“经营是前提，质量是生命，技术是基础，人才是关键”的治院方针，不断丰富“创新、拓展、诚信、敬业”的企业精神内涵，不断提高设计咨询能力，拓展工程管理、工程总承包能力。公司将奋发努力，竭诚为广大客户提供全方位服务，为国防工业和社会主义现代化建设做出新的贡献。

地址：中国上海市武宁路303号
电话：+86-21-62549700
传真：+86-21-62573715
邮箱：sxy@ndri.sh.cn
网址：www.ndri.sh.cn

Add: 303 Wuning Road, Shanghai, China
Tel: +86-21-62549700
Fax: +86-21-62573715
E-mail: sxy@ndri.sh.cn
Web: www.ndri.sh.cn

上海交通大学海洋深水试验池

Shanghai Jiaotong University Deep Seawater Test Pool

中交南方总部基地（A区）总部大厦

CCCC South HQ (Zone A) Building

项目地点：广东 广州
建筑面积：520 000 m²

Location: Guangzhou, Guangdong
Building Area: 520,000 m²

项目位于广州海珠区珠江新城南端中轴线起点的珠江后航道之滨，为广州市重点工程。方案以“城市启航、中交启航”为设计理念，营造出积极开放的城市空间和锐意进取的企业文化精神。设计以稳健的群体建筑形态，理性的建筑布局，升腾向上的天际轮廓线，形成千帆竞发、百舸争流的建筑形象，从而塑造出城市南端的新地标，打造出飘逸的城市滨水建筑风景线。
总部大厦设计为二星绿色建筑，同时采用了结构叠合柱、水源热泵、变制冷剂流量多联式空调等多项先进技术。

The project is located in the shore of Pearl River rear canal serving as the starting point of the central axis at the southern end of Pearl River New Town, Haizhu District, Guangzhou City. The project is a key project of Guangzhou City. It follows the design idea of “sailing from the city, and sailing from the CCCC”, to create an active open city space and endeavoring corporate spirits. The design has steady cluster building forms, rational layout, and rising skyline, to form a sailing and competing image, thus create a new landmark in the southern end of the city, and create a flexible urban waterside building scenery.
The HQ building design is a 2-star green building, adopting the structure of laminated columns, water source hot pump variable refrigerant flow multi-split air-conditioning and other advanced technologies.

上海新发展亚太万豪酒店

New Development Asia-Pacific Marriot Hotel, Shanghai

项目地点：上海
建筑面积：70 860.65 m²

Location: Shanghai
Building Area: 70,860.65 m²

上海新发展亚太万豪酒店是一座现代化、高智能、高舒适度、配套完善的高档五星级酒店建筑，设计遵循万豪国际酒店设计标准，为客人提供温馨而舒适的入住感受。
客房层采用“工”字形平面，充分利用了长风公园、苏州河的良好景观；客房平面与底部方形裙房结合，利用“工”字形凹口布置大空间，满足酒店公共空间的功能要求，妥善解决了局促场地上的建筑布局问题。内外空间连贯布局形成东西轴向景观空间系统，从东侧入口水景、大挑檐雨棚、大开间大堂、大堂吧的系列空间变化，延伸至西侧的园区中心绿地，景观序列与主楼形态互相呼应、相辅相成。玻璃幕墙结合整齐的黑色窗框和杏黄色石材格框，强调塔楼的形态，挺拔且有节奏感。

The hotel is a modern, well-equipped, intelligent and comfortable hi-end 5-star hotel building. The design follows the standards of Marriot International Hotel, to provide guests with warm and pleasant experience of accommodation.
The guesthouse floors are H-shaped, to make full use of the good landscapes of Changfeng Park and Suzhou River. The guesthouse floors combine with bottom square skirt buildings, and arrange large spaces at the H-shaped grooves, to meet the functional requirements of hotel public space, and properly mitigate the negative effects of narrowed site. The internal and external spaces penetrate the layout to form a east-west axial landscaping space system, from the space sequence of east side entrance waterscape, projecting-eaves canopy, large-bay hall, large lobby, to the west side park center green belt, corresponding to and complementing with the main building forms. The glass curtain walls combine with tidy black window frames and yellowish stone grids, highlighting the tower form, rational and rhythmic

1. 河北省质量技术监督局-新建实验室
   New Lab of Hebei Provincial Administration of Quality and Technical Supervision

2. 中船龙穴造修船基地
   CSSC Longxue Shipbuilding & Repairs Base

3. 江南长兴造船基地
   CSSC Changxing Shipbuilding Base

4. 启东新城区高级中学艺术馆
Art Museum of Qidong New District Senior Middle School

5. 中石油青岛海工建基地
CNPN Qingdao Marine Engineering & Construction Base

扫描查看更多信息

HSarchitects
上海汉思建筑设计事务所

上海汉思建筑设计事务所是一家具有建筑设计甲级资质的合伙制设计公司（简称 HSa）。HSa长期专注于大型城市商业综合体项目，已完成和正在建设中的大型购物中心、超高层写字楼、豪华酒店、大型娱乐设施等商业综合项目总面积超过800万 m²，业务涵盖从规划方案至施工图至现场配合的全过程，积累了十分丰富的经验。尤其对大型商业建筑的招商、营运、消防设计、与公共交通设施相关的设计等方面具有十分深刻的理解。HSa致力于为客户提供规划、建筑、结构、机电、室内、景观等全方面的服务。公司内各专业技术人员崇尚创新与协作精神，在项目初期就开始三维协同工作，为项目赢得品质和时间。

HSarchitects Shanghai (HSa) is a partnership architecture firm with Class-A Architecture Design Certificate, providing master planning, architecture, structure, electronics, interior and landscape design service. Focusing on commercial complex design over a long-term, HSa has a deep understanding of commercial investment, operation, fire protection and transport organization. The whole process covers planning, construction and coordination in the field. Innovation, creative and teamwork are the spirits of each firm members, and 3D design make sure each HSa building has its own personality.

地址：中国上海长宁路1018号龙之梦大厦901室
邮编：200042
电话：+86-21-33727290
传真：+86-21-33727291
网址：www.HSarchitects.cn

Add: Room 901, No.1018 Changning Road, Cloud Nine Plaza, Shanghai, China
P.C.: 200042
Tel: +86-21-33727290
Fax: +86-21-33727291
Web: www.HSarchitects.cn

上海特福隆水世界购物中心

TFL Water World Shopping Center, Shanghai

设 计 师：胡志文、詹晨
项目地点：上海
用地面积：85 327 m²
建筑规模：600 000 m²

Designer: Owen Hu, Chen Zhan
Location: Shanghai
Site Area: 85,327 m²
Building Area: 600,000 m²

本方案以水元素为主题，提取弧形符号，精心设计和营造出“雨林主题”、“温泉主题”、“溪谷主题”、“河畔主题”等特色主题购物步道，为顾客带来探险般的购物体验。入口的大型瀑布与喷泉表明“水”的主题、聚集人气，表演广场可提供各种节日庆典与商业活动。贯穿商场始末的弧形铺装带来犹如水流般的流线感，具有良好的导向性。以圆形为基本元素的铺装与构筑物就像从建筑上洒落的水滴紧密地与主题建筑相呼应。

The project is themed by water element. It extracts arc symbol, to design and create characteristic themed shopping walking streets, such as “rainforest theme” “hot spring theme” “valley theme” and “riverbank theme” providing venturous shopping experience to customers. Large waterfall and fountain at the entrance express the theme of "Water", bringing visitors together, and the performing square can provide venue for various celebrations and commercial events. The arc decoration brings water flow of streamlines throughout the market, serving as a good guide. As the round for basic and structures look like falling water drops closely corresponding to the themed buildings.

上海莘庄凯德龙之梦

Capita Dragon Dream Shopping Center, Xinzhuang Town, Shanghai

设 计 师：胡志文、杨若伟
项目地点：上海
用地面积：11 827 m²
建筑面积：197 211 m²
建筑密度：38.8%
绿 化 率：20%

Designer:Owen Hu, Yang Phil
Location: Shanghai
Site Area: 11,827 m^2
Building Area: 197,211 m^2
Building Density: 38.8%
Green Ratio: 20%

该项目由一幢4层大型购物中心与一幢32层酒店综合楼组成，主楼建筑高度170.5 m，建筑面积为197 211m²（其中地上部分为91 392 m²，地下部分为105 819 m²），建筑属于一类建筑，主楼结构形式为框筒结构，裙房为框架结构，为部拥有大面积开放绿地和广场，环境优美，建成后将形成莘庄镇规模最大、功能最全的复合式商业中心及莘庄地标式建筑。

The project includes a 4-floor large shopping center and a 32-floor hotel complex. The main building is 170.5m high with total floor area 197,211 m^2, including 91,392 m^2 aboveground and 105,819 m^2 underground. This A-class building is a frame tube structure, and the skirt building is a frame structure, with large area open green land and square inside, very beautiful. Upon completion, it will become the largest and most functional complex commercial center and landmark of Xinzhuang Town, Minhang District, Shanghai.

成都东客站龙之梦城
Dragon Dream Town, Chengdu High Speed Railway East Station

设 计 师：胡志文、詹晨、林晨
项目地点：四川 成都
用地面积：164 872.26 m²
建筑面积：1 930 682.79 m²
建筑密度：60%

Designer: Owen Hu, Chen Zhan, Chen Lin
Location: Sichuan, Chengdu
Site Area: 164,872.26 m²
Building Area: 1,930,682.79 m²
Building Density: 60%

项目位于成都市东部即将建成的新客站西面，地处成—德—绵—眉山—乐山城市经济带及成渝经济带的交汇处。依托成都高铁东客站，项目将承担娱乐购物、旅游休闲等商业服务功能。成都龙之梦城与东客站结合成紧密的组团，一方面，铁路网络辐射全省，乃至全国；另一方面基地周边地铁辐射全市。利用东客站的交通集聚作用，盘活区内优势资源，使其与站前公共区整体发展。在公共交通系统的支撑下，该站前地区将成为一种新型的社会、文化和城市的活力中心。站场周围的商业商务区和便捷的交通集散场将和谐共存，并相互促进。

The project is located in the new railway station in the eastern part of Chengdu City, which is pending completion. It is in the junction of Chengdu-Deyang-Mianyang-Meishan-Leshan urban economic zone and Chengdu-Chongqing economic zone. Based on this station, Chenghua District will undertake entertainment, shopping, tourism, leisure and other commercial service functions. Dragon Dream Town will combines closely with the station in a group. The railway network will radiate whole Sichuan Province or even whole China; on the other hand, peripheral metro lines will radiate whole Chengdu City. It mobilizes favorable resources by advantage of the traffic hub, and integrates them with public area in front of the station. Supported by public traffic system, such front public area will become a new dynamic center of community, culture and city. Surrounding commercial business area and convenient traffic terminal will coexist in harmony and develop together.

沈阳龙之梦亚太城
Dragon Dream Asia Pacific Town, Shenyang

设 计 师：胡志文、詹晨、任侠	Designer: Owen Hu, Chen Zhan, Xia Ren
项目地点：辽宁 沈阳	Location: Shenyang, Liaoning
用地面积：557 000 m²	Site Area: 557,000 m²
建筑面积：4 370 000 m²	Building Area: 4,370,000 m²

项目位于沈阳大东区原中捷矿山机械厂地块，地理位置优越，该项目具有较高的容积率，包括一栋430 m高标志性塔楼、四栋250 m高甲级办公楼、超高层住宅小区以及大规模高层商业裙房。并且充分利用地下空间，设置了大型地下商业空间、大型地下车库，货运系统也集中在地下。
在整个龙之梦亚太城中央区域，开辟出了6 hm²的大型城市中央水景公园，并对外开放。能在城市中心高容积率地块中开辟出这样大面积的开放型公共休闲场地，靠的正是周边高密度的商业建筑布局，而大面积的城市景观设施也提升了整个地块的商业品质。在城市综合体中整合和优化城市公共交通资源，集聚丰富的商业类型，是对城市综合体新型商业模式的探索和创新，并且形成清晰多样的城市综合体建筑风貌，达到复合型商业建筑形式和功能统一、营造城市综合体内浓郁商业氛围、串联各大商业设施的城市商业景观设计效果，实现了大型城市综合体对城市的景观贡献的目标。

The project is located in the land lot of former Zhongjie Mining Machinery Plant, Dadong District, Shenyang City. The geographic conditions are good. This project has high plot ratio, and it includes a 430 m high symbolic tower, four 250 m high class-A office buildings, super high-rise residential community and large scale high-rise commercial skirt building. It makes good use of underground space, to set up large area underground commercial space, large underground garage, and cargos transporting system is also mainly underground.
In the middle, there is a 60,000 m² city center water view park open to public. Such large public leisure place in such a high plot-ratio land is a miracle which depends heavily on surrounding commercial building layout that is high density. Such public facility also promotes the whole commercial quality of the land lot. It integrates and optimizes city public traffic system in city complex, and gathers a rich, leading metropolitan life. It also forms a clear variety of city complex images by exploring and innovating new business model in city complex, and thus large city complexes make contributions to city landscape by the integration of complex commercial building form and functions, strong commercial climate within city complex, and a string of city commercial landscaping designs of various large commercial facilities, achieving the target, which is the contribution of the city complexes to the city Langscape.

上海工程勘察设计有限公司

Shanghai Engineering Survey & Design Co.,Ltd.

上海工程勘察设计有限公司，是上海国泰创业（集团）有限公司旗下的一家由建设部批准的具有建筑工程设计甲级资质的综合性设计公司，拥有IAF和CNAS所颁发的全面质量管理ISO9001：2008双重认证，也是“全国优秀勘察设计企业”之一。公司自1992年成立以来，专注于建筑工程领域的设计和研究，现已发展成为集丰富的设计经验、强大的技术力量和完善的管理制度为一体的现代化综合设计公司。时至今日，公司所承担的项目总量超过5000个，受到社会各界和相关部门的肯定并屡获嘉奖，为城市建设和社会的经济发展做出了卓越的贡献。

地址：上海市武宁南路318号
电话：+86-21-62310080
传真：+86-21-62317080
邮箱：info@saec.cc
网址：www.saec.cc

Add: No 318 South Wuning Rd,Shanghai, China
Tel: +86-21-62310080
Fax: +86-21-62317080
E-mail: info@saec.cc
Web: www.saec.cc

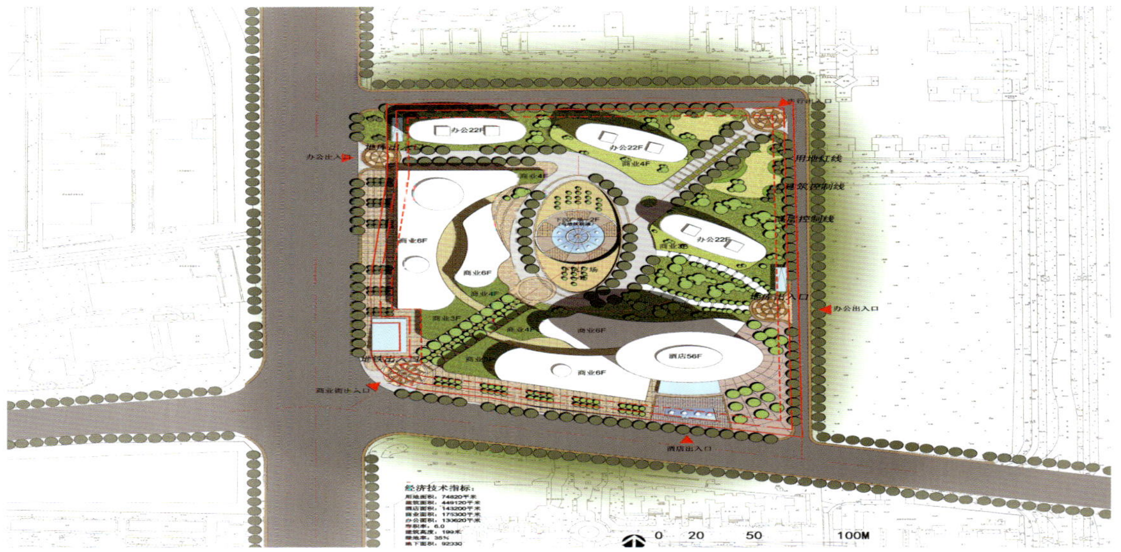

佛山南海城市广场

上音附中

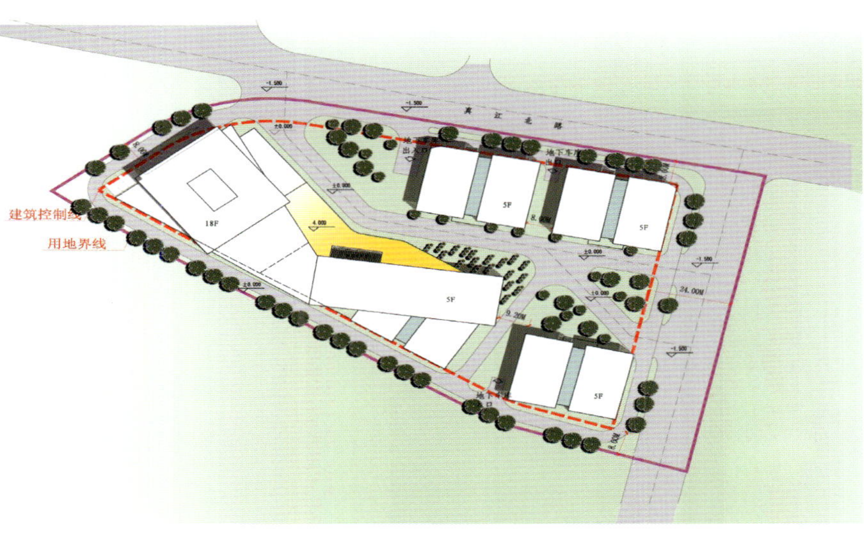

未来岛

扫描查看更多信息

MOD
德中建筑·摩德

摩德工程咨询（上海）有限公司

MOD Engineering Consulting (Shanghai) Co., Ltd.

MOD·摩德工程咨询（上海）有限公司是一家外商独资（德国）。设计咨询专业公司，由旅德建筑师、德中建筑协会主席陈冰先生于2004年创立，公司经营的范围包括城市规划、建筑设计、园林景观等。设计团队来自中德两国。

公司的优势首先在于其国际化的理念、视野与中国建筑实际的紧密结合，力争为业主提供既具有国际化水准，同时又符合中国实际情况的设计作品，追求具有时代感的中国本土文化气息的设计文化。

公司的一大特色是规划、建筑、景观三位一体的设计。MOD的理念是通过建筑设计来整合城市空间，同时又在城市空间环境相互联系的基础上设计单体建筑。设计师充分运用建筑在环境中生长的理念，将生态景观的打造放到至高的地位，以人为本，创造人性化的居住、生活与工作环境。

MOD奉行“以先进理念吸引业主、以优秀作品打造品牌”的宗旨，遵循服务至上的理念，为业主提供具有国际水准的创意性设计作品，打造21世纪生态人居环境。

绿地汇创国际广场生态·节能 · 星级办公楼

Green Land Huichuang International Plaza, Ecological · Energy Saving · Star Rating Office Building

设 计 师：陈冰、石远清、朱宇哲
项目地点：上海
用地面积：19 917.6 m²
建筑面积：72 550 m²

Designer: Bing Chen, Yuanqing Shi, Yuzhe Zhu
Location: Shanghai
Site Area: 19,917.6 m²
Building Area: 72,550 m²

项目是建设部首批公布的获得“绿色建筑评价标志”的6个项目其中之一，获得二星标志认证，与众多国家级项目荣列一席。该项目标志着绿地汇创国际广场三大业态完全呈现，更标志着上海五角场高端商务圈即将开启生态办公的新时代。

本地块位于上海市东北区副中心五角场核心区的边缘，紧邻主干道翔殷路和国定东路。基地北侧设一19层准甲级办公楼，南侧设一组14层LOFT办公楼，中间为2～3层主题商业酒吧区。LOFT办公楼中庭有14层的跨度，高耸而现代，其中若干楼层设计天桥横跨整个中庭，犹如悬于空中的雕塑，丰富了整个中庭空间。天桥不但能形成较为活跃的人流动线，对行走在桥上的人们也是一种全新的体验。

The project, as one of the six projects with “Green Building Label” awarded by the Ministry of Construction for the first time, wins the 2-star label, in parallel with many state-level projects. This project marks the complete display of three major building in Green Land Huichuang International Plaza, and moreover, marks a new time of eco-office emerging in Shanghai Pentagon Plaza hi-end business circle.

This plot is located in the margin of Pentagon core area of sub-center in northeastern Shanghai. It is close to Xiangyin Road and Guoding East Road. On the north side of the site sets a19-floor class-A office building, on the south side set a group of 14-floor LOFT office buildings, and in the middle sets a themed commercial bar area of 2-3 floors. The atrium of the LOFT office buildings spans 14 floors, high and modern, of which some floors have crossover spanning the whole atrium, resembling a sculpture hanging in the air, which enriches the whole atrium space. The crossover does not form an active pedestrian, but also offer a fresh experience to the people on the crossover.

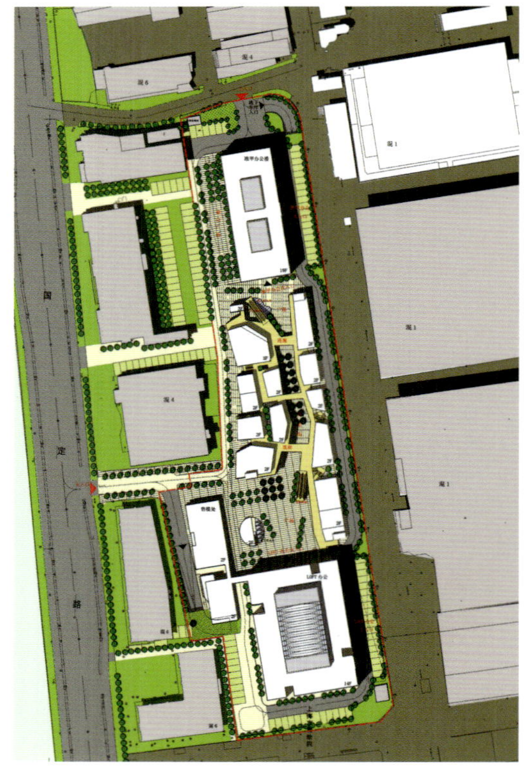

总平面图

商业内街

LOFT办公楼

LOFT内庭

中星昆山“外滩印象”

Zhongxing Kunshan "Bund Impression"

设 计 师：陈冰、石远清、朱宇哲
项目地点：江苏 昆山
用地面积：54 814 m²
建筑面积：137 035 m²

Designer: Bing Chen, Yuanqing Shi, Yuzhe Zhu
Location: Kunshan, Jiangsu
Site Area: 54,814 m²
Building Area: 137,035 m²

项目位于昆山市，北临庆丰路，西临青阳港，东临已建成的景枫嘉苑，南近沪宁铁路线。
立面与剖面设计：建筑单体立面设计中，采用了ART-DECO建筑风格，强调古典与现代风格的统一，力求多样化及功能与形式的统一。建筑的外部设计充分利用材料的肌理、光影效果及建筑的形体组合来塑造建筑物的个性。北侧的沿街商铺立意于ART-DECO建筑风范的情节，用古典外装饰线条表现其尊贵的气质。色彩素雅，以温暖色调为主，以独特的气质与周围建筑的形式及色彩形成对比，具有既柔美又富阳刚的时代气息。

The project is located in Kunshan City, adjacent to Qingfeng Road in the north, Qingyang Port in the west, the completed Jingfeng Jiayuan in the east and Shanghai-Nanjing Railway in the south.
Façade and profile design: in the façade design of single buildings, Art Deco is adopted as the architectural style. The façade design also emphasizes the unity of classical and modern style and seeks diversification as well as the unity of function and form. The external design of buildings makes the best of the texture of materials, the light effect and the of building forms to shape the character of the buildings. Based on the Art Deco style, the stores along the street in the north side show a quality of dignity with classical external moldings. Simple but elegant and mostly in warm color, this project contrasts with the building types and color forms nearby, with a unique quality and culture, showing the flavor of the era, which is feminine and also masculine.

总平面图

小区主入口

沿庆丰路街景

入口商业

入口大喷泉

水岸景观

中央景观湖

金屋·秦皇半岛——城市中央生态城区

Golden House · Qinhuang Peninsula – Central Ecologic Urban Area

设 计 师：陈冰、石远清
项目地点：河北 秦皇岛
建筑面积：4 000 000 m²
容 积 率：2.8
绿 化 率：40%

Designer: Bing Chen, Yuanqing Shi
Location: Qinhuangdao, Hebei
Building Area: 4,000,000 m²
Plot Ratio: 2.8
Green Ratio: 40%

项目充分利用600万 m²原生态森林公园的高负氧离子含量给身体带来健康特色，同时新建266 666.7 m²中心湖景公园配有优美的景观和便捷的高品质休闲空间，打造临渤海湾的国际领先近海生活理念。设计采用国际先进的低碳环保的理念，在建筑中充分使用低碳环保建材及环保节能技术，为城市营造绿色核心。

在中央生态区的基础上，融商业体验、旅游观光、自然人文、高档居住为一体，引领秦皇岛城市住宅发展新趋势，为消费者创造出充分体验自然生活和人文文化的生态复合人居生活环境，打造秦皇岛最高品质的居住社区。

The project makes full use of a primitive forest park (> 6,000,000 m²) with rich content of negative oxygen ions that are healthy to human bodies, and meanwhile builds a central lakeshore park (266,666.7 m²) with beautiful landscape and convenient high quality leisure space, and creates the world-leading seaside lifestyle near Bohai Bay. It adopts world-class low-carbon environment-friendly ideas, to apply low-carbon environment-friendly building materials and environment-friendly energy-saving technologies in construction, to create a green core for the city at huge cost.

Based on central ecologic area, the project integrates commercial experience, tourism sightseeing, nature & culture, and hi-end residence. It will lead the trend of urban residential development in Qinhuangdao City, create an ecologic composite habitable life that fully represents natural life experience and cultural values for consumers, and build the highest quality residential community in Qinghuangdao City.

秦皇岛商业银行总部大楼

欧式风情商业街

商业中心

超高层酒店式公寓

总平面图

虹桥临空10-3、11-3地块方案设计

Hongqiao Linkong Plot 10-3 & Plot 11-3

设 计 师：陈冰、Sabine Wutzlhofer、石远清
项目地点：上海

Designer: Bing Chen, Sabine Wutzlhofer, Yuanqing Shi
Location: Shanghai

本项目中的11-3地块与10-3地块分别位于北临空福泉路东西两侧，属于外环线以内临空园区的核心公共中心地区。园区毗邻虹桥国际机场。
本设计方案肌理从周边多样的城市环境中脱颖而出，整体化结构中布置的内庭空间提供各单元的交通连接以及各功能区的自然采光。5个不同主题的内庭形成的空间序列引人入胜，将步行者引入安静的内庭空间中去。
整个建筑群体采用统一的立面系统，同时又能提供或通透、或秩序、或生动的多种变化。统一的立面体系使多个建筑体块形成一个鲜明的、个性化的城市景观；同时多种变化的可能性又满足了不同功能的需求。

Plot 11-3 and Plot 10-3 of this project are located in the core public center on the east and west sides of Fuquan Road in northern Linkong Park within the Outer Ring Road, adjacent to Hongqiao International Airport, Minhang District, Shanghai.
The design is outstanding among the diverse urban conditions in periphery. The atrium spaces in the integral structure provide traffic connections between all units, and natural sunlight for every functions. The atrium space sequence in five different themes attract pedestrians into a quiet atrium space.
The whole building cluster is in a facade system, while providing transparent, orderly and lively diversity of changes. The uniform facade system integrates a number of building volumes into a distinct personalized urban landscape, while satisfying functional requirements with a diversity of possibilities.

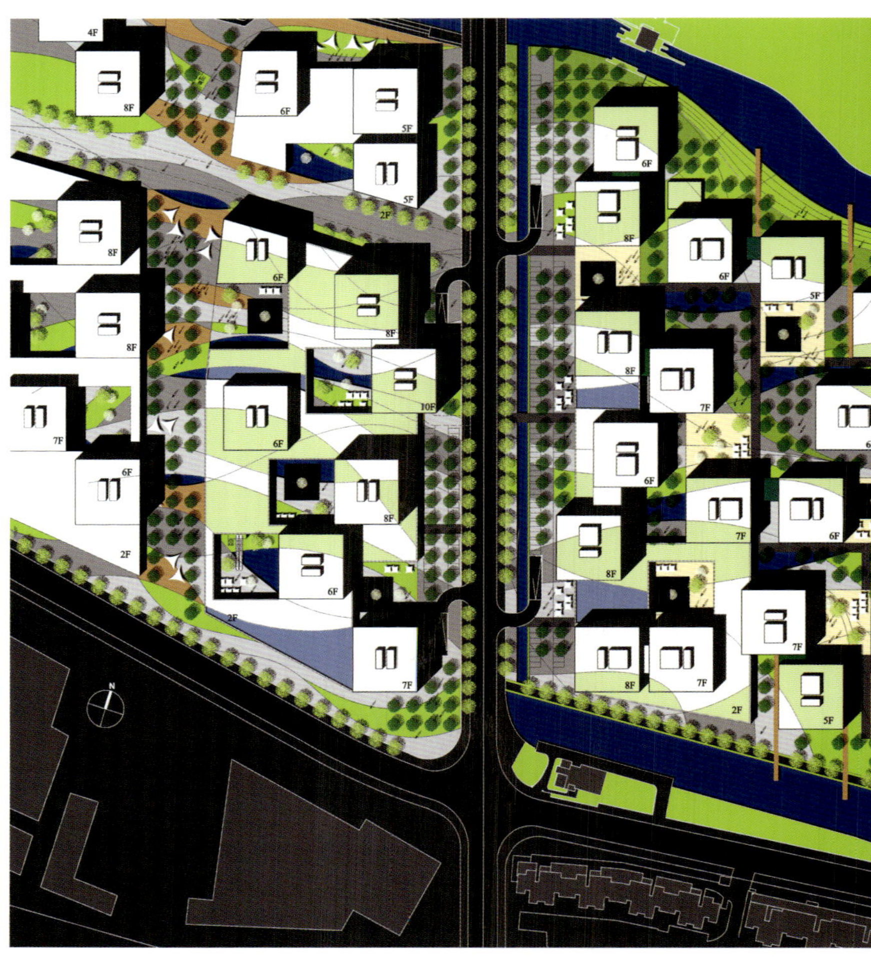

5F
5F
1F
5F
5F

酒店建筑 HOTEL DESIGN

華墨國際 HM International

黄山高尔夫酒店三期
Huangshan Golf Hotel and Resort(Phase 3)

华墨国际是由美国HM 国际有限公司及上海华墨建筑设计事务所有限公司共同创立的一家全方位、多元化的设计公司。在美国得州和中国上海分别注册。公司的理念是注重建筑实践的全过程，包括全面的建筑工程、设备以及相关服务如规划设计、建筑设计、城市设计、室内设计、景观设计。华墨国际为许多地产集团、政府部门完成众多成功的作品，并和国内大型设计集团及国际著名设计公司有着广泛的业务合作。公司以城市设计、城市综合体、星级酒店、高级办公楼、博物馆、医疗建筑、教育建筑、大规模居住区规划、商业地产、旅游地产、高科技厂房等方面见长。

扫描查看更多信息

扬州皇冠假日酒店
Growne Plaza Hotel in Yangzhou

福建漳州芗江酒店
Xiangjiang Hotel in Zhangzhou

拉萨圣·瑞吉酒店二期
Holy Raj Resort in Lhasa

嘉兴市穆湖度假酒店
Muhu Holiday Resort in Jiaxing

城市设计 URBAN DESIGN

新昆山市中心中央活力区城市设计
The Urban Design of Central Activity Zone in Kunshan

昆山高新区综合商务园城市设计
Hi-tech Business Park in Kunshan

昆山可再生能源产业主题公园城市设计
Kunshan Low-carbon Industry Theme Park

安徽宣城市“中国·茶城”城市设计
"TEA City"--Xuancheng Urban Design

華墨國際 HM International

城市综合体 HOPSCA

济南西客站站前综合体
Jinan West Railway Station

义乌中心区 10、11 号社区
No.10 and No.11 Neighborhood in Yiwu Central Area

上海杨浦"渔人码头"二期
The Fisherman's Wharf in Shanghai

成都锦尚国际
Jinshang International in Chengdu

住宅建筑 RESIDENTIAL BUILDING

六安市天盈上城
Tianying Shangcheng in Liuan

江苏靖江人民南路地块
Renmin South Rode Block in Jingjiang

黄山御宾国际
Yubin International in Huangshan

上海外高桥宝雅 50 公馆
Waigaoqiao Baoya 50 Mansion in Shanghai

上海交通大学微电子学院楼
College of Micro Bectronic Building at Shanghai Jiaotong University

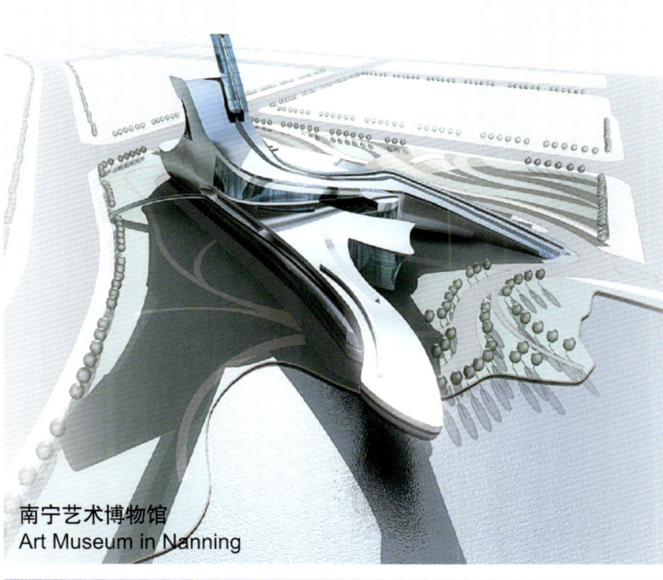
南宁艺术博物馆
Art Museum in Nanning

昆山低碳科技馆
Low-Carbon Technology Pavilion Architectural Design Of Kunshan Low-Carbon industry Theme Park

上海长征医院医教综合楼
New Changzheng Medical & Ednication Comprehensive Building

安徽省皖投置业园
Wantou Investment Office Park in Anhui

昆山新能源技术检测与发展中心
Kunshan Detection of New Energy Technologies and the Development Center of the Design

上海怡亚通物流加工基地
Yitatong Software Development and Distribution Center — Shanghai

六安市公安局指挥中心
Public Safety Bureau Command Center in Liuan

上海都冀建筑设计有限公司

Shanghai DUGEE Design Co.,Ltd.

都冀建筑成立于2007年，设计范围包括城市设计、建筑设计以及景观设计，包含办公、商业、住宅、酒店和学校设计。成立5年来，公司凭借专业建筑设计知识，丰富的设计经验以及全程掌控的能力，一直致力于为客户提供优质的创新建筑设计服务。

都冀建筑重视城市景观、建筑设计、室内设计的相互间的关联与渗透，并秉承"热情忠诚"的服务信条与"融会创新，追求卓越"的设计理念，去寻求能使项目土地价值、市场定位、文脉、艺术感以及人的体验感等相互交融的空间形态。

地址：上海普陀区陕西北路 1388 号上海银座企业中心
电话：+86-21-51688771
传真：+86-21-51685271
邮件：biz@dugee.cn
网址：www.dugee.cn

Add: 1F North Shaanxi Road No. 1388 Ginza business center
Tel: +86-21-51688771
Fax: +86-21-51685271
E-mail: biz@dugee.cn
Web:www.dugee.cn

美好愿景 · 金银潭

Wonderful Vision · Jinyintan

项目地点：湖北 武汉
用地面积：12.45 hm^2
建筑面积：35.6万 m^2
容 积 率：2.62

Location: Wuhan, Hubei
Site Area: 124,500 m^2
Building Area: 356,000 m^2
Plot Ratio: 2.62

合理的空间整体布局：小区建筑体量外高内低，内部空间疏朗、视野开阔、景色优美；布局采用人车完全分流系统，健康步道围绕景观花园而设；总体风格为极简主义的设计风格。

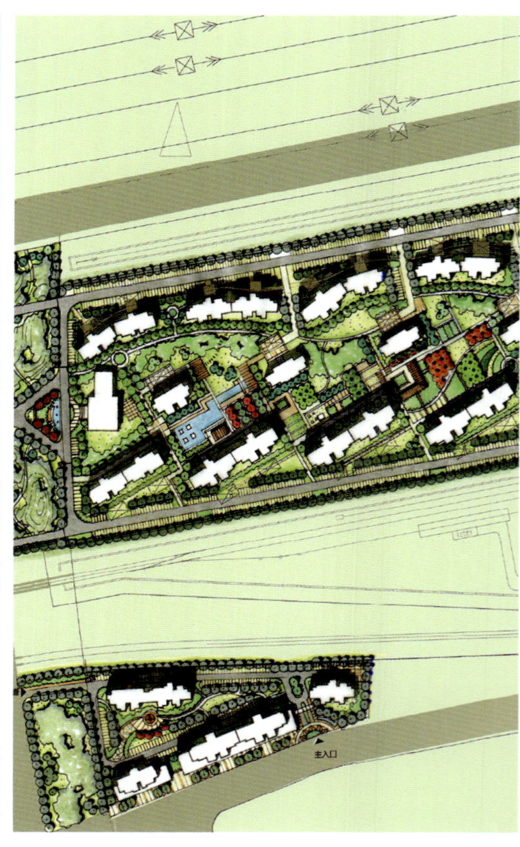

壹 · 未来城一期商业街

One · Future City Commercial Street Phase I

项目地点：湖北 咸宁	Location: Xianning, Hubei
用地面积：4.4 hm²	Site Area: 4.4 ha
建筑面积：62 000 m²	Building Area: 62,000 m²
容 积 率：1.41	Plot Ratio: 1.41
绿 化 率：15%	Green Rate:15%

规划的布局不仅考虑了与周边道路的自然衔接，同时考虑了内部的有机组合。在每个城市道路路口附近设置广场入口，丰富了未来的城市空间形象，同时，购物的人通过城市道路路口可以用最短的距离进入步行街内部，可以方便购物的人流，并有效减少横穿马路的行为。入口广场、中央广场和轴线性步行街空间连接整体街区。斜向的布局用最用效的方式将周边道路衔接起来，并吸引穿过的行人，增加步行街内部人流量。

直线型商业街布置形式是传统商业空间方式之一。设计中通过连接各体量的空中走廊把直线型的空间划分成了不同大小的围合空间，也有效地削弱了空间的视觉穿透性。

关注城市的形象并兼顾造价的经济性一直是设计需要考虑的问题。步行街立面窗户和阳台的错列布置，使立面形式简洁、具有时代性；造价符合业主要求，顶层的收分处理以及檐口的出挑，使建筑形象别致、与众不同。

上海大椽建筑设计事务所

Shanghai Dachuan Architectural Design Office

上海大椽建筑设计事务所的多位核心成员拥有15年专业背景，曾任职于各大设计机构，积累了丰富的工程经验、非常具有团队默契，确立了共同的价值观和判断力。公司重视对设计的宏观控制力并专注于微理的研究，希望创造更为契合城市的设计作品。目前该公司为客户提供商业综合体及高端住宅项目规划设计、建筑方案、建筑初步设计，以及施工图建筑设计的配合控制服务。同时通过协调完善各专业，进一步提升项目的整体品质。公司成立以来，已为上海地产集团、上海爱家集团、上海绿地集团、吉盛伟邦集团、融创集团、吉林大众置业、四川恒邦等多家国内大型地产集团提供设计服务。我们期待与更多拥有共同理想的业内人士携手共进。

SHANGHAI DACHUAN ARCHITECTURAL DESIGN OFFICE has a number of key members with 15 year professional experience. They used to work at different large design institute, thus accumulated rich project experience and established good team work relationship with common idea on value and judgment. The office focus on general design control and special research, creating design works adapt to the city. We design commercial complex project and high standard residential building, including planning, architectural schematic design, preliminary design and construction drawing coordination and control. We strive to promote integral quality through good coordination of individual disciplines. Since establishment of the office, we have worked for a number of large real estate companies, including Shanghai Land Group, Shanghai Aijia Group, Shanghai Green Land Group, JSWB Group, Rongchuang Group, Jilin Dazhong, Sichuan Hengbang etc. We expect to march forward with more people of common ideal.

地址：上海市复兴中路369号13层
电话：+86-21-63737771
传真：+86-21-63731118
邮箱：1554973486@qq.com
网址：www.dca.sh.cn

Add: Floor 13, No. 369 Fuxing Middle Road, shanghai
Tel: +86-21-63737771
Fax: +86-21-63731118
E-mail: 1554973486@qq.com
Web: www.dca.sh.cn

武汉金融城A04
International Financial City A04, Wuhan

项目地点：湖北 武汉
用地面积：81 000 m^2
容 积 率：4.7

Location: Wuhan, Hube
Site Area: 81,000 m^2
Plot Ratio: 4.7

项目位于武汉金融城核心区，紧邻606 m超高层地标建筑，与多栋超5A级写字楼和超星级酒店为邻，旨在打造大武汉地区的金融及办公中心，汇集湖北省最高端的办公、酒店和商业购物中心。开发目标希望在符合城市空间的整体发展要求的基础上，为武汉提供具有国际视野及满足高端人士需要的、能让人充分感受武汉文化底蕴和现代时尚气息的高端居住社区。

在空间上尝试通过规划现有用地，形成对内私密含蓄的空间形态，同时对外营造带些许经典文化风情的综合型城市新社区形象。售楼处风格采用经典比例划分，赋以现代元素；商业街区通过居饕界、艺饕界、食饕界等不同主题空间构建多种表皮风格相融合和渗透的氛围，以促进交流、规避不利因素。

居住板块由3个组团组成，建筑高度布局从和平大道至滨江地块为80 m、90 m、140 m，不同高度的组合能够完善社区形象，丰富城市天际线，更充分利用景观资源。设置多中心开放式集中绿地景观，结合各区域形象设计入口大堂前区景观，创造灵活宜人的灰空间，给社区带来活力，成为具有标志性特征的社区景观空间。

建筑把新古典的元素揉合进现代建筑的表情之间，结合武汉的文化底蕴，适当运用局部坡屋面、本土建筑石材、简洁的线脚，塑造一个蕴涵古典韵味同时充满现代气息的人居空间。

The project is located at the center of Wuhan financial city, adjacent to the 606 m landmark building and a number of 5A office buildings and hotels. This district will be developed into a financial and office centre in large Wuhan. Office buildings, hotels and retails of the highest standard in Hubei province concentrate here. The development goal is to comply with the general developing requirements of urban space as a base, to establish international standard modern residential atmosphere with inner cultural content of Wuhan which meet requirements of high end users.

The land plot is planned into private space inside and new community complex outside with some classical cultural feeling. The sales building adopts classical ratio plus modern elements. The commercial street consists of different themes such as residential top, cultural top, eating top etc. Various skin styles gather together, promoting communication and solving disadvantages.

The residential zone is made up of 3 groups. Building height vary from 80 m, 90 m to 140 m between the peace avenue and the riverside. Different height mix enriches community image and urban skyline, taking full advantages of landscape. The flexible connection of multiple greening centers and the landscape areas before individual lobby entrance bring vitality to the community and make symbolic community landscape space.

New classical elements are incorporated into the expression of modern building, with Wuhan culture, slope roof in part local stone and simple line, a humane and modern space with some classical charm is created.

乐山翡翠国际社区四期
Jade International Community Phase IV, Leshan

项目地点：四川 乐山
用地面积：185 000 m²
容 积 率：5.0

Location: Leshan, Sichuan
Site Area: 185,000 m²
Plot Ratio: 5.0

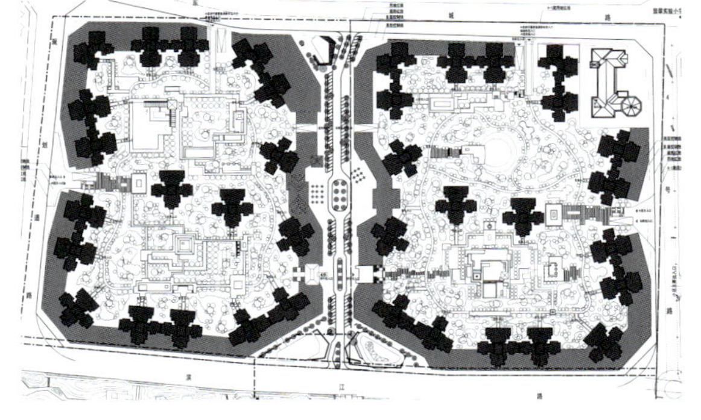

恒邦翡翠国际社区四期位于为乐山岷江东岸，其沿江岸线长达500 m。项目立志通过打造具有打动力的情景商业街区（小镇式商业街），将配套商业的角色转换成为主角，让休闲的主题商业成为崭新的生活方式、生活品质的启动者，最终形成"滨江活力休闲街区社区——乐山生态宜居社区"标准的大盘品质。

在地块中间部位规划一条情景商业街，贯穿滨江路与城东路，商业街平面空间开合有序，建筑高低错落，规划形态与尺度形成真正的风情商业街。高层住宅因商业街而分成左右两个200 m×150 m的组团，并使组团空间形成流动感。滨江采用点板结合且对称的形态，形成既有通透感又体现端庄的的滨江形象。高层住宅造型设计采用简约ART-DECO，商业建筑采用简约古典风格，既保证了品质感，也使功能与形式完美契合。

Hengbang Jadeite International Community Phase 4 is located at east river side of Leshan Minjiang. The plot covers a riverside distance of 500 m. It aims to build a vivid commercial district (town street), turning secondary commercial facilities into major role and leisure commercial activities into high quality living style, achieving high standard project of "vivid riverside leisure community, Leshan ecological residence"

A vivid commercial street is planned to be placed in the middle of the plot. The street will go through Binjiang road and Chengdong Road. Space keeps changing, closed or open, high or low. The form and size is just for a leisure business street. High residential building is divided by the street into 2 groups (200 m×150 m), with dot and group in flow. The riverside adopts a mix of dot and plate form, symmetric, not solid and elegant. The residential building is designed in Art Deco style while the commercial facility is designed in simple classic style. It assures the quality as well as the perfect realization of function and form.

长春绿地中央墅

Greenland Central Villa, Changchun

项目地点：吉林 长春
用地面积：72 000 m²
容 积 率：1.03

Location: Changchun, Jilin
Site Area: 72,000 m²
Plot Ratio: 1.03

项目基地位于南部新城核心区，距离市政府1.5 km，北望森林公园，坐拥南部新城绝佳居住环境。建筑整体布局沿中心轴线对称展开，以楼王形成空间轴线的对景和收束，体现法式风格的严谨和大气。景观设计以庄重凝练的几何对称形式为主，局部辅以灵活变化。开阔、大气、修剪整齐的草坪，几何图案的灌木及花草，营造出气势不凡的法式园林效果。

立面细部以法式公寓作为风格参考，借鉴古典三段式手法，划分为基层、中部墙体及上部屋顶三部分；外立面采用壁柱、宝瓶、山花、以及各种形式的窗套、壁龛等细节增加层次性，强化整体风格；在建筑色彩上，一层电梯洋房以米黄色的天然石材为基调，二层以上采用同色系真石漆，辅以深灰色石板瓦屋面，整体色彩细腻典雅。联排别墅和双拼别墅以天然石材为基调，局部采用同色系真石漆，凸显法式建筑的精致与典雅。

The project is located at the New City in the south, 1.5 km from the municipality. It enjoys excellent living environment with a forest park to the north. The general layout design is symmetric. The king building on the axis leads the mirror and restrain, displaying French style strictness and magnificence. Landscape design shows serious geometric symmetry in whole and flexibility in part. Open and wide grassland is well cut with bush and flower in geometric pattern, thus building a grand French Garden.

The apartment façade design refers to French style. The façade is classically divided into 3 parts, including base, middle and top. The exterior wall is designed with all kinds of column, vase, flower, window frame and recess so as to enrich details and enforce integral style. The color of the 1st level of the lift villa is light yellow of natural stone. The above level is stone paint in same shadow with stone roof of dark grey. The general color is delicate and elegant. The color of town house and 2 unit villa is generally natural stone color with some stone paint of same shadow.

上海筑誉设计及工程集团
Shanghai Zhuyu Architectural Design & Engineering Group

上海筑誉设计及工程集团 2007 年在上海创建。集团由 6 家独立的法人机构合并组建，开展在中国大陆的设计及施工工程业务。
集团依托国际化背景，以国际 + 本土的专业化团队，形成自己的创作和服务特色。在创作观上，强调设计是一种行为，一种过程，一种服务，而宗旨在于发现及寻找实现特定项目价值最大化的机遇及解决方案。业务范围包括城市设计、酒店、办公、商业、住宅、学校、体育场所等，并重视发展建筑设计服务的延伸。
近年来集团涉足包括室内精装修设计、景观设计、幕墙设计等专业设计及施工工程在内的优势门类，大力推进绿色生态建筑的发展。同时，推进设计与工程相结合的战略，逐步实现设计及施工工程一体化发展的目标。

地址：上海市静安区安远路 555 号 806 室
电话：+86-21-62981500
传真：+86-21-62986711
邮箱：mail@zhuyudesigngroup.com
网址：www.zhuyudesigngroup.com.cn

Add: Room 806, An Yuan Road No.555, Jing An District, Shanghai
Tel: +86-21-62981500
Fax: +86-21-62986711
E-mail: mail@zhuyudesigngroup.com
Web: www.zhuyudesigngroup.com.cn

上海东航金叶苑
China Eastern Airlines Golden Leaves Garden, Shanghai

项目地点：上海
用地面积：247 954 m^2
建筑面积：700 000 m^2
建筑密度：23.6%
合作单位：加拿大泛太平洋设计集团

Location: Shanghai
Site Area: 247,954 m^2
Building Area: 700,000 m^2
Building Density: 23.6%
Partners: Pan-Pacific Design Group(Canada)

项目位于上海市徐家汇 WS5 地块滨江商务区内，与世博区隔江相望，距龙华历史文化风貌区仅 2 km 左右、到上海重要的对外交通枢纽上海南站 4.5 km 左右。
本项目用地东起云锦路（规划），西至天钥桥南路，南起龙耀路，北至龙华机场办公区龙恒路。集居住、办公、商业、教育、医疗、社区服务等为一体，作为滨江 CBD 城市功能向西的延伸与完善，项目创造了大型高品质景观楼盘，城市生活尊贵府邸。建筑以 ART-DECO 和古典洋房风格为主，形成简约、高雅、尊贵的建筑形象，同时也体现了时代感及上海城市文化与历史底蕴。

The project is located within the riverbank CBD of Lot WS5 in Xujiahui District, Shanghai. It is opposite to World Expo Zone on the other bank of the river. It is only 2 km or so from Longhua Historic and Cultural Zone, and about 4.5 km from Shanghai South Railway Station, an important traffic hub from external world to Shanghai.
The land lot extends from Yunjin Road (planning) in the east, to Tianyue Bridge South Road in the west, from Longyao Road in the south, to Longheng Road in Longhua Airport Office Zone in the north. It is an integration of residential buildings, office buildings, commercial buildings, education, medical services, community services and other services. As an eastward extension and improvement of riverbank CBD urban functions, it creates a large high quality landscaping building cluster, and also a city life venerable mansion. The architectural image is keynoted by Art Deco and classic house style, concise, elegant, respectable, epochal while representing Shanghai urban culture and historic deposit.

万宁石梅湾威斯汀五星级度假酒店

The Westin Shimei Bay Resort (5-star hotel), Wanning

项目地点：海南 万宁
用地面积：110 321 m²
建筑面积：82 565 m²
合作单位：加拿大泛太平洋设计集团

Location: Wanning, Hainan
Site Area: 110,321 m²
Building Area: 82,565 m²
Partner: Pan-Pacific Design Group(Canada)

前湖迎宾馆（方案）

Qian Lake Guest House(Concept)

项目地点：江西 南昌
建筑用地：1 144 500 m²
建筑面积：132 069 m²
建筑密度：3.63%

Location: Nanchang, Jiangxi
Site Area: 1,144,500 m²
Building Area: 132,069 m²
Building Density: 3.63%

上海浦东发展银行南昌办公楼

Shanghai Pudong Development Bank Nanchang Office Building

项目地点：江西 南昌
建筑用地：14 570 m²
建筑面积：48 500 m²
建筑密度：35%

Location: Nanchang, Jiangxi
Site Area: 14,570 m²
Building Area: 48,500 m²
Building Density: 35%

南昌红谷滩金融大厦

Honggu Beachland Fianacial Tower, Nanchang

项目地点：江西 南昌
建筑用地：10 820 m²
建筑面积：67 655 m²
建筑密度：38.6%

Location: Nanchang, Jiangxi
Site Area: 10,820 m²
Building Area: 67,655 m²
Building Density: 38.6%

台州世界贸易中心

Taizhou World Trade Centre

项目地点：浙江 台州
建筑用地：11 727 m²
建筑面积：109 000 m²
建筑密度：60%

Location: Taizhou, Zhejiang
Site Area: 11,727 m²
Building Area: 109,000 m²
Building Density: 60%

上海明捷万丽酒店

Mingjie Renaissance Hotel, Shanghai

项目地点：上海

Location: Shanghai

酒店位于上海真如副中心，是一座集高档住宅、办公楼、酒店、商业于一体的大型多业态的都市综合体，与 7 号地铁线岚皋路站直接相连。酒店拥有时尚雅致的客房及套房，拥有国际风味的创意美食令客人的每一餐都充满惊喜，另有 24 h 健身中心、室内游泳池、蔓达梦水疗。1029 m² 的创意组合会议空间可满足客人不同的会务需求，专业的宴会统筹团队由始至终为客人带来完美的会议体验。

The hotel is located in Shanghai Zhenru Sub-center, directly adjoining Metro 7 Lan'gao Road Station. It is a large urban complex integrating hi-end residential buildings, office buildings, hotel and commercial buildings. The hotel has modern and elegant guestrooms and suite rooms, its creative food with international flavors makes each meal full of surprise, and in addition, it also has 24-hour gym center, indoor swimming pool and Mandara SPA. Its creative conference space of 1,029 m² can satisfy a variety of conference demands, and its professional dinner arrangement team can bring guests with perfect conference experiences all the way.

上海浦东机场 2 号航站楼、磁悬浮列车站台幕墙工程

Magnetic Levitation Trains Station Curtain Wall of Pudong Airport Terminal 2, Shanghai

项目地点：上海
建筑面积：485 500 m²

Location: Shanghai
Building Area: 485,500 m²

二号航站楼位于一号航站楼东侧 400 m，由主楼、连接廊、候机长廊 3 部分组成，建筑面积 48.55 万 m²，几乎是一号航站楼面积的 2 倍。已经运营的磁悬浮车站和在建的轨道交通 2 号线车站坐落在两楼之间。

The terminal 2 is located 400 m east of terminal 1. It is composed of three parts, namely, main building, connecting corridor and waiting corridor. It covers a total building area of 485,500 m², approximately 2 times larger than terminal 1 floor area. The operating magnetic levitation trains station and the metro 2 station in progress are locating between the two terminal buildings.

北仑小浃江片区城市设计

Xiaojiajiang Area of Beicang District Urban Design

项目地点：浙江 宁波
用地面积：1 670 000 m²

Location: Ningbo, Zhejing
Site Area: 1,670,000 m²

政务新区名城地产项目

Mingcheng Real Estate Project, New Municipal and Cultural District

项目地点：安徽 合肥	Location: Hefei, Anhui
用地面积：447 408 m²	Site Area: 447,408 m²
建筑面积：2 005 814 m²	Building Area: 2,005,814 m²
建筑密度：24.53%	Building Density: 24.53%

上海景悦喜来和酒店项目

Jingyue Xilaihe Hotel, Shanghai

项目地点：上海	Location: Shanghai
用地面积：49 696 m²	Site Area: 49,696 m²
建筑面积：25 020 m²	Building Area: 25,020 m²

上海综星置业工业综合办公楼项目

Zongxing Property Industry General Office Building, Shanghai

项目地点：上海	Location: Shanghai
用地面积：73 508 m^2	Site Area: 73,508 m^2
建筑面积：67 699 m^2	Building Area: 67,699 m^2
容 积 率：0.86	Plot Ratio: 0.86
建筑密度：37.59%	Building Density: 37.59%

该项目位于上海市奉贤区环城东路以西，肖湾路以北，作为上海综星置业有限公司投资筹建的办公楼一定要突显企业形象和实力，体现公司的文化与品牌，树立办公楼建筑设计的新形象。完善的功能布局，合理的交通流线，简洁的外观形象，使该建筑从整体到细节都蕴涵着理性的设计感。ART-DECO建筑风格的外立面，简洁大方、稳重挺拔。作为双子形态的两栋高层建筑，一左一右，一张一合，相互呼应。该建筑的裙房设计延续主楼外立面的风格，与高层建筑组合形成丰富多变的天际线，使建筑群体更好地融入城市的整体风貌之中。

This project is located west of Ring East Road and north of Xiaowan Road, Fengxian District, Shanghai. As an office building invested and built by Shanghai Zongxing Property Co., Ltd. ("Zongxing Property"), it will represent the corporate culture and brand, to build a new image of office building design. This architecture is full of rational designs from integrity to details, such as complete functions, reasonable traffic flow, and concise appearance. The façade in Art-Deco style is designed to be concise, generous, steady and straight. The twin towers are hi-rise buildings, complementing each other. The skirt building design continues the design of main building façade, to form a variable skyline with hi-rise buildings, so that the building cluster is better integrated into the cityscape.

沃得句容宝华山酒店别墅综合项目

Ward Baohuashan Hotel & Villa, Complex, Jurong

项目地点：江苏 镇江	Location: Zhenjiang, Jiangsu
用地面积：80 004 m^2	Site Area: 80,004 m^2
建筑面积：40 970 m^2	Building Area: 40,970 m^2
容 积 率：0.6	Plot Ratio: 0.6
建筑密度：19.4%	Building Density: 19.4%

崇明县港西镇 133/2 宗地酒店

Zongdi Hotel on Plot 133/2, Gangxi Town, Chongming County

项目地点：上海
用地面积：84 126.13 m²
建筑面积：36 478.9 m²
容 积 率：0.3
建筑密度：14.8%
合作单位：加拿大泛太平洋设计集团

Location: Shanghai
Site Area: 84,126.13 m²
Building Area: 36,478.9 m²
Plot Ratio: 0.3
Building Density: 14.8%
Partners: Pan-Pacific Design Group(Canada)

项目坐落于崇明岛崇中分区，规划中的崇明新城以北。崇明岛内地势低平，河汊纵横，具有"水清、土洁、气净"的优越生态环境和丰富的滩涂、湿地、森林资源，是上海市域"最后的净土"。国家沿海大通道、上海深水港等一系列重大工程设施的建设将在未来的 5 ~ 10 年内极大地凸现崇明的战略区位优势。在上海建设国际性生态化都市和上海城市发展趋向临海化的新背景下，崇明未来的发展将日益紧密地融入上海都市区的建设中，实现成为世界级城市上海生态岛和"海上花园"的重大战略转变。

The project is located in Chongzhong Sub-district of Chongming Island, besides the Chongming New Town in planning to the north. With low and flat terrains and the weaving rivers, Chongming Island has superior ecological environment with clear water, clean soil and fresh air, as well as rich tidal land, wetland and forest resources, which is known as the "last pure land" in Shanghai. The construction of national coastal highway, Shanghai deepwater port and a series of important project facilities will greatly highlight the strategic area advantage of Chongming Island. As Shanghai is being built to be an international ecological city and tends to sea-bordering development, the future development of Chongming Island will increasingly integrate into the urban construction of Shanghai, and realize its important strategic change to be the ecological island and oceanic garden of Shanghai the world-class city.

邢台辰光国际酒店

Chenguang International Hotel, Xingtai

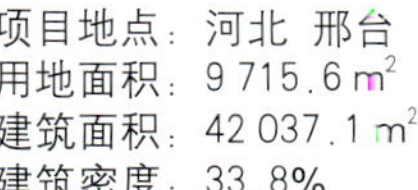

项目地点：河北 邢台
用地面积：9 715.6 m²
建筑面积：42 037.1 m²
建筑密度：33.8%

Location: Xingtai, Hebei
Site Area: 9,715.6 m²
Building Area: 42,037.1 m²
Building Density: 33.8%

项目用地位于邢台市中心城区西部，南临中兴西大街，北临李演庄南街。基地现状为现有辰光大酒店的餐饮中心。现有辰光大酒店位于用地的西南角，出入口面向东侧，将与新建酒店共用入口广场。建筑外墙采用实墙构造结合外包铝塑板幕墙和独立窗体系，这样不仅从风格上保证了典雅高档，同时相对于纯玻璃幕墙而言更能降低建筑造价；在建筑节能、保温方面也有极大的好处，能够适应北方的气候环境。在建筑顶部采用有限面积的玻璃幕墙装饰面，强调挺拔的建筑体量，加强建筑高耸向上的风格态势。

The project is located in the western downtown Xingtai, besides Zhongxing West Street in the south and Liyanzhuang South Street in the north. The temporary building of restaurant center of Chenguang International Hotel is now located on the site. Chenguang International Hotel is currently located in the southwestern corner of the site, its entrance and exit faces the east, and it will share the entrance square with its new hotel. The exterior wall adopts solid wall structure and combines out-covering aluminum and plastic glass curtain wall and independent window system. Therefore, the façade not only ensures elegant and hi-end character, but also has lower building cost than that of pure glass curtain wall. It is also good at energy-saving and heat preservation, and can fit northern climate. The roofing adopts glass curtain wall decorative face of limited area to highlight its tall and straight building volume and enhance its towering image.

扫描查看更多信息

上海广万东建筑设计咨询有限公司
HMA Architects & Designers

HMA是一家全日资公司，创立于2002年末，于2004年完成了上海8号桥的设计。该项目的一期工程在2005年竣工，旋即就以其独特的建筑立面风格和丰富的空间感，得到了广泛认可和高度评价。8号桥项目成为了国内旧厂房改建的标杆、上海城市的新地标、创意产业园的典范，并由此奠定HMA在国内建筑改造设计领域的领先地位。

随后几年内HMA陆续完成了多个建筑改造项目，如X2创意空间、2010上海世博会浦西最佳城市实践区B4展馆改建、大宁中心广场（第一机床厂改建）、复地四季广场（上海第四制药厂改建）等，这些项目都以丰富多变的空间感和雅致内敛的立面风格获得了很高的评价。其中世博会B4地块的设计获得了2010年公开国际竞赛一等奖。

HMA并未止步于建筑改造设计。2007年HMA有幸加入日本福冈博多运河城的原班策划、设计团队，共同进行了南京水游城的设计。联合团队凭借在商业建筑方面已有的设计经验及对优秀商业设施的执著追求，最终出色完成了南京水游城的设计工作。功夫不负有心人，目前此项目商业运营情况良好，更成为南京当地的标志性商业设施。南京水游城不仅通过其开放式的空间、丰富多彩的互动活动，给予消费者舒适的空间体验、难忘的文化社交感受。特别是全面开放的地下商业空间、贯通全设施的运河水系及音乐喷泉系统，颠覆了以往地下商业空间的压抑感和单调感。南京水游城已经成为国内开放式MALL的开创性典范。

南京水游城开启了全新的商业建筑设计和商业运营模式，HMA此后相继主持了盘锦水游城、天津水游城、武汉水游城、湘潭水游城、锦州水游城等的建筑设计。正是此类商业模式设计经验的积累，为HMA赢得了良好的商业设计口碑。HMA在上海、北京、成都、苏州、徐州、武汉等地都留下了或将留下自己的设计作品。

地址：上海黄浦区建国中路3号建国大厦3层
电话：+86-21-54669966
传真：+86-21-64152060
邮箱：hma@hmadesign.com
网址：www.hmadesign.com

Add: 3F, Jianguo Building, Middle Jianguo Road No. 3, Huangpu District, Shanghai
Tel: +86-21-54669966
Fax: +86-21-64152060
E-mail: hma@hmadesign.com
Web: www.hmadesign.com

城市最佳实践区
中部系列 B4 展馆

City Best Practice Zone Middle Series B4 Show Space

项目地点：上海
用地面积：15 840 m²
建筑面积：9 520 m²

Location: Shanghai
Site Area: 15,840 m²
Building Area: 9,520 m²

B4 展馆区由 3 栋厂房组成，设计师在 1 号、2 号楼之间加了风雨顶棚，把他们连接在一起。以几何状的钢架为结构体系，覆盖编织状的膜结构作为顶棚，可辅助遮风挡雨，形成半开放的内庭同时可兼作室外展示区。考虑到展览建筑对采光的特殊要求，把原厂房的窗用砖封堵。重新设计了屋架，增加了屋顶的侧向采光天窗，而使屋顶在形式上极其富有老上海石窟门里弄的韵味。3 栋厂房中靠黄浦江最近的是 3 号楼，它独特的结构形状被保留下来。设计师把它和人流动线接近部分的外墙全部拆除，完整地保留原来的梁柱体系，并在人的高度以上部分装上固定玻璃百叶，可以有效地防止风雨的侵袭。在原来的结构体内部做了一个雕塑感极强的酒瓶状内胆作为展示区，展示区室内的自然光源来自 3 个天窗。

B4 Show Space includes 3 factory buildings, and we add canopy between building 1 and building 2, to connect them. The geometric steel frame is the structure, the knitted cover is the canopy, to shade us from wind and rain, and form a semi-open atrium, also serving sa an outdoor display zone. The windows are blocked with bricks to meet special sunlight requirements for exhibition. The roof trusses are redesigned to add lateral skylights, and on the rooftop are flavors of ancient Shanghai cluster civil residences called "stone-enclosed gate style".

The building 3 of the plant is nearest to Huangpu River, whose special structure is well maintained. We have removed all external walls between the building and pedestrian flow, remained its former beams system, and fixed glass shutters above human height, to effectively protect the building from rains and winds. We have made a bottle-like inner space for display, which is strongly sculptural with natural lights from three skylights.

大宁中心广场

Daning Central Plaza

项目地点：上海
用地面积：78 737 m²
建筑面积：100 000 m²
摄 影 师：Masato、曾江河

Location: Shanghai
Site Area: 78,737 m^2
Building Area: 100,000 m^2
Photographers: Masato, Jianghe Zeng

本项目是对始建于 1994 年的上海第一机床厂的改建，将旧厂房改造成为"创业产业园"。项目主要分为商业区和办公区，商业业态基本集中分布于灵石路、万荣路的街角（基地东北角），部分商业沿灵石路分布在基地周边。休闲娱乐的配套设施位置在基地的最南侧。基地中部为创意办公区，该区域北部和南部分别为别墅型办公、创意办公，面积由小到大、由北向南递增，满足企业成长发展所需的空间拓展需求。基地广场主要设置在万荣路周边，位于南北向的中部，是商业和办公的交接处，既是标志性的空间，也是北侧办公展示区和北侧商业区人流引导的场所。

The project is to rebuild Shanghai No. 1 Machine Tool Factory which was established in 1944. In the project, old factory buildings will be rebuilt into "Creative Industry Park". The project includes commercial zone and office zone. The commercial zone is concentrated on the corner of Lingshi Road and Wanrong Road (northeastern corner of the base), some commercial buildings are distributed alongside Lingshi Road around the base. Recreational facilities are located in the south most side of the base. The central part is creative office zone which comprises of villa office, creative office from north to south, where the number of small office spaces is increasing, to meet the growth requirements for emerging enterprises. The plaza is set near Wanrong Road, in the middle of this south-north section, the juncture of office and commerce, serving as not only a landmark, but also a diversion of pedestrian flow fro/to north office or commercial zone.

盘锦水游城
Panjin Aqua City

项目地点：辽宁 盘锦
用地面积：80 896.06 m²
建筑面积：230 000 m²（商业、酒店部分）

Location: Panjin, Liaoning
Site Area: 80,896.06 m²
Building Area: 230,000 m² (stores, hotels)

整个工程由水游城、商业步行街、酒店、销售住宅、回迁住宅和办公建筑组成，是集休闲、购物、餐饮、娱乐、居住、酒店、办公等多功能于一体的城市商业综合体。水游城由数栋建筑物组成，为充分利用土地的形状，采用南低北高的布局，以获得最优的日照条件。

商业以水为主题，结合建筑的功能与空间，展现水的整个生命历程中的各个阶段的特点。从冰山开始经过瀑布、溪流、江河一路蜿蜒流转到达大海，红色的海滩代表盘锦的知名景观。建筑内部营造的空间因水的元素而灵动，充满阳光明媚、空气清新、流水潺潺的自然韵味，提升商业空间的驻足时间与回游性，并创造全新的购物体验。

The whole project is composed of aqua city, commericial walking street, hotels, rediences for sale, residences for resettlers and office buildings, serving as a city complex integrating leisure, shopping, catering, entertainment, residence, hotel, office and other functions. The project includes several buildings, in the layout low in south and high in north, following the landform, to gain optimal sunlight.

The commercial buildings are themed by water whose life cycle with features is demonstrated with building functions and spaces. From iceberg to waterfall, stream, river, and winding to the sea, where red beach is metaphor of famous landscape of Panjin. The interior space is inspired by water elements, full of bright sunlight, fresh air and flowing water, promoting visits and returns of customers, while creating a brand new shopping experience.

天津水游城

Tianjin Aqua City

项目地点：天津
用地面积：39 800 m²
建筑面积：181 079 m²
摄 影 师：Nacasa&Partners

Location: Tianjin
Site Area: 39,800 m²
Building Area: 181,079 m²
Photographers: Nacasa&Partners

整个工程由酒店、商业、办公三部分组成，是集休闲、购物、餐饮、娱乐等功能于一体的城市商业综合体。
依据设计指导思想，设计中力求突出重点、强调秩序与节奏，地块内建筑西高东低以减小对城市街道的压力。由于地块呈南北向狭长形态，因此总体布局以点线结合，突出重点。在两条城市主干道大丰路与芥园道的交角布置一栋21层点式高层酒店，并结合地铁上加建的裙房，使之成为水游城地标性建筑。在临大丰路的商业街的中部，结合内部空间设球状造型，打断过长的线性沿街立面，突出重点。基地西侧设4幢板式高层写字楼，围绕商业中心形成2幢东西朝向、2幢南北朝向的布局，丰富沿街景观。北侧的5层高独立商业成为地块的收尾建筑。

The whole project is composed of hotel, commercial and office buildings, it is a city complex integrating recreation, shopping, catering, entertainment and other functions. According to guidelines, the design seeks for order and pace with highlights, and the buildings are high in west and low in east to reduce the pressure on streets. The land lot is a strip south to north, so the general layout is linear with highlighted spots. The two main roads, Dafeng Road and Jieyuan Avenue cross at a 21-floor pointed high-rise hotel wh ch marks the land with the subway underlying the group houses. In the middle of the 3-floor commercial street buildings near Dafeng Road, sphere is shaped inside the space, breaking up the façade on street which is too long, so as to mark the highlight. On the west side of the base are four plate high-rise office buildings, around the CBD, two of which are east to west, and the other two are south to north, to enrich streetscape. North 5-floor high-rise independent commercial building is end of the lot.

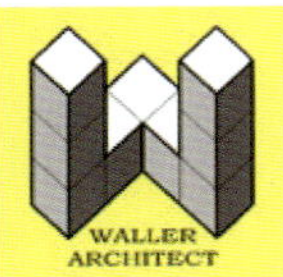

上海沃尔建筑设计有限公司
Waller Architecture Design Co.,Ltd. Shanghai

上海沃尔建筑设计有限公司创办于2003年，公司凭借专业的团队、创新的设计、完善的服务已成功地为众多的政府机关、跨国企业集团、上市公司和个人提供整体的建筑和室内创意设计，拥有丰富的设计和服务经验。
上海沃尔工作领域涉及建筑设计、室内设计、项目管理和古迹维护，上海沃尔一直站在世界建筑设计业和建筑工程业的最前沿，自成立以来已经完成50多项设计，包括办公大楼、银行和金融机构、政府建筑、会展中心、大型剧院、高级酒店、购物中心、地铁车站、私人豪宅等。

地址：上海市静安区新闸路831号丽都新贵18-C
电话：+86-21-52340632
传真：+86-21-52340630
邮箱：caozhenfu@vip.sina.com
网址：www.taliesinwest.com.cn

Add: 18-C, Cosmopolitan Priority Building, 831 Xinzha Road, Jing'an District, Shanghai
Tel: +86-21-52340632
Fax: +86-21-52340630
E-mail: caozhenfu@vip.sina.com
Web: www.taliesinwest.com.cn

包头国际会展中心
International Convention and Exhibition Center, Baotou

项目地点：内蒙古 包头
用地面积：120 085 m^2
建筑面积：90 000 m^2
建筑高度：45 m

Location: Baotou, Inner Mongolia
Site Area: 120,085 m^2
Building Area: 90,000 m^2
Building Height: 45 m

项目位于包头市建华南路西侧、建设路南侧、成吉思汗生态园东侧。会展中心建筑群的功能是综合和多元的，具有展览交易、会议、文化交流、科学教育、商务办公及酒店等综合功能。项目是包头对外开放的一个新的窗口和平台，向世界和国人展示包头改革开放的成果。

整个会展中心的建筑群体犹如一只振翅高飞的草原雄鹰，寓意包头这座著名的草原钢城在新时期将获得更大、更迅猛的发展，像草原上的雄鹰，一飞冲天。建筑形体舒展，立面造型强劲有力，是名副其实的包头城市“新名片”。

The project is located on the west side of South Jianhua Road, on the south side of Construction Road, and on the east side of Genghis Khan Eco-Park. The building cluster of the center has diverse and comprehensive functions, including exhibition and trading, convention, cultural exchange, scientific education, commercial office, and hotel. As a new window and platform of Baotou opening to the outside, it shows the results of Baotou's reform and opening-up to the world.

The building cluster is like an eagle flying high above the grassland, which implies that Baotou the famous steel city on grassland will make greater and faster development in new times. As an eagle in grassland flutters and roars to the sky, the architecture with extending forms and strong façade shapes can serve as a “new name card” of Baotou City.

天目湖国际饭店

Tianmu Lake International Hotel

项目地点：江苏 溧阳
建筑面积：53 200 m²
建筑高度：10 m

Location: Liyang, Jiangsu
Building Area: 53,200 m²
Building Height: 10 m

项目位于江苏省溧阳市天目湖风景区内，是一座五星级度假酒店，拥有300个多种类型的客房、22栋别墅及各种设施用户。建筑设计着眼于创造雅致、自然的休闲空间，重视建筑同自然环境的呼应，将建筑融入湖光山色之中，设计具有原创性，让山水为建筑增色，建筑为山水添秀。

The project is located in Tianmu Lake scenic zone of Liyang City, Jiangsu Province. It is a five-star resort that has 300 guestrooms in various types, 22 villas and various facilities. With an eye to create elegant and natural recreation spaces, the design pays great attentions to the correspondence between architecture and natural environment. The design integrates the architecture with the landscape. With the originality of this design, the landscape and architecture complement each other.

西塔里埃森建筑师之家
Taliesin West Architects House

项目地点：江苏 南京
建筑面积：256 m²
建筑高度：4 m

Location: Nanjing, Jiangsu
Building Area: 256 m²
Building Height: 4 m

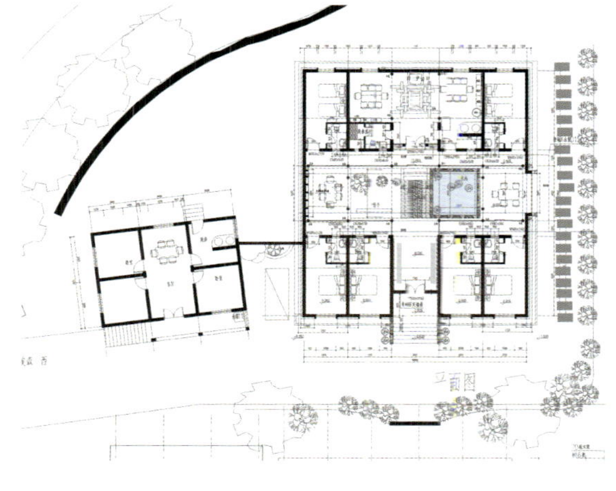

该项目是在南京郊区臼湖山林区设计建造的供建筑师交流和休息的场所，由3位设计师共同出资，拥有6个标准客房、活动室、茶室和餐厅各1个，用于建筑师作品展览和交流。

建筑采用地方材料和工艺，结构、地板和天棚均采用杉木材料，外墙采用清水砖墙，加1 mm厚的黄泥保温，中心庭院为日式旱沙铺地，大部分设施均为建筑师自己设计和施工，在地面和围墙大量使用旧砖、旧瓦和木材，使得资源可以再利用，室外地面采用石子铺地，为透水性地面，雨水得到收集并用于浇地，极大地保护了生态环境。

The project is built in the Jiu Lake forest in the suburban Nanjing, and used for architects to communicate and rest. It is jointly invested by three designers, and has six standard guest rooms, three activity rooms, teahouse and restaurants for architects' artworks exhibition and exchange.

The building adopts local materials and process, uses fir log structure, fir floor and fir ceiling. The external wall is brick wall with thick yellow mud for heat preservation. The central courtyard is paved with Japanese dry sand. Most facilities are designed and constructed by architects themselves. A lot of old bricks, old tiles and wood materials are used in floors and enclosing walls for resources recycling. Outdoor ground is paved with cobblestones to make the ground permeable, and rainwater can be gathered for irrigation, which protects local ecology and environment greatly.

包头奥林匹克公园三大体育馆

Three Gymnasias in Olympic Park, Baotou

项目地点：内蒙古 包头
建筑面积：83 200 m²
建筑高度：33 m

Location: Baotou, Inner Mongolia
Building Area: 83,200 m²
Building Height: 33 m

设计灵感来源于大自然的草地和落叶，通过提炼自然曲线并结合左侧的高尔夫球场，3座体育馆仿佛落在奥林匹克公园的榉树树叶上，在满足使用功能的前提下，体现了生态的和谐和对自然的敬畏。

The design is inspired by grassland and fallen leaves in the nature. By refining natural curve and combining the golf course on the left, the three gymnasia are just like the Zelkova leaves. Apart from its practical use, the design also shows its reverence for the nature and its harmony with the ecology.

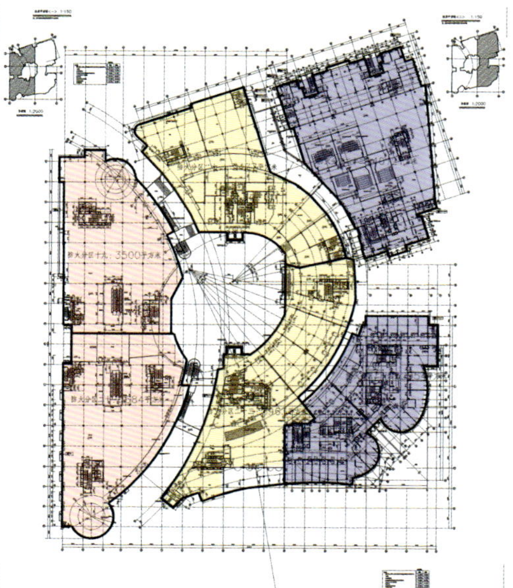

平陵广场设计

Pingling Plaza Design

项目地点：江苏 溧阳
用地面积：42 900 m^2
建筑面积：270 000 m^2
建筑高度：99 m

Location: Liyang, Jiangsu
Site Area: 42,900 m²
Building Area: 270,000 m²
Building Height: 99 m

工程位于溧阳市平陵中路和西大街交叉处，规划设计商业购物广场、商务办公、餐饮和住宅。项目设计中，为提高基地的商业价值、充分利用基地，通过弧形商业街和空中连廊将一系列的建筑组群合成一体，形成了以周围城市道路、两条主轴线、空中连廊为主的多条商业街，完成了水平围合与立体交叉相结合的商业环境。

The project is located in the junction of Middle Pingling Road and West Street of Liyang City. It is planned to design commercial shopping plaza, commercial offices, restaurants and residences. In the design, to develop the business value of the land and make full use of it, aerial corridors which integrate a series of building clusters, to form a number of commercial streets featured by surrounding urban roads, two main axes and aerial corridors, and create the commercial environment combining horizontal closure and vertical crossing.

扫描查看更多信息

BDI 上海柏创建筑设计有限公司
Boarch Design International

BDI（Boarch Design International）柏创国际注册于美国洛杉矶，2001年在上海设立事务所开拓中国市场，在10年多的发展历程中，BDI完成了大量国内知名项目，并赢得了项目业主的尊重和赞誉。BDI长期致力于提供城市规划、大型居住区、城市综合体、交通建筑、办公、酒店、商业、学校等相关专业的设计服务，在建筑咨询、城市规划、建筑设计、室内设计、景观设计、平面图形设计及数字辅助设计等领域建立了广泛的合作客户群体，积极为业主提供市场分析及产品策划等辅助服务，并在城市综合体、集群商业、高档写字楼、居住区规划、高档住宅社区等方面形成了专业优势。BDI力图营造轻松、愉悦、平等、人性化的工作环境，团队的成员有来自国内各著名设计院经验丰富的建筑师、工程师、规划师、园林建筑师、室内设计师、建筑施工管理专家和各类专业顾问，以及来自美国、芬兰、马来西亚、菲律宾等国家的优秀设计师，他们热爱生活、激情洋溢，充满理想，在快乐的工作中实现着个人的人生价值。BDI一贯追求品质并始终以"设计改变生活"为核心理念。

BDI (Boarch Design International) was registered in Los Angeles, USA. In 2001, it set up an office in Shanghai, to operate in Chinese market. Since its entry into Chinese market, BDI has completed a large number of domestic famous projects, winning respects and reputations from the owners. BDI is always committed to providing professional design services for urban planning, large residential communities, urban complexes, traffic buildings, office, hotels, commercial buildings and schools, establishing extensive customer base in architectural consulting, urban planning, architectural design, interior design, landscaping design, planar graphic design, CAD and other fields. It actively provides market analysis, product planning and other auxiliary services, and forms its own professional advantages in urban complex, commercial cluster, hi-grade office buildings, residential community planning, hi-grade residential communities and other fields. BDI tries to create easy, happy, equal and human-based work conditions, and its team members include experienced architects, engineers, planners, gardening architects, interior designers, construction managers and various professional consultants from domestic famous design institutes, as well as excellent designers from the USA, Finland, Malaysia, Philippines and other countries. They are loving life, energetic, idealistic and realizing their life value by happy work. BDI is always in pursuit of high quality, with the core philosophy of "design changing life".

地址：上海市浦东新区浦东南路528号证券大厦北塔17楼
电话：+86-21-68824686
传真：+86-21-68824687-8001
邮箱：bdi@vip.163.com
网址：www.boarch.com

Add: 17th Floor,the Securities Building,Pudong South Road No.528, Pudong New Area, Shanghai
Tel: +86-21-68824686
Fax: +86-21-68824687-8001
E-mail: bdi@vip.163.com
Web: www.boarch.com

新华重机
Xinhua Heavy Machinery Community

设 计 师：郭三锁、刘次圣、李超、祝坚、曹宝华、刘俊、金波
项目地点：河南 郑州
用地面积：66 003.80 m²
建筑面积：198 466.66 m²

Designer: Sansuo Guo, Cisheng Liu, Chao Li, Jian Zhu, Baohua Cao, Jun Liu, Bo Jin
Location: Zhengzhou, Henan
Site Area: 66,003.80 m²
Building Area: 198,466.66 m²

项目位于郑州市东侧20 km的中牟县，以高档高楼、花园楼房、精品商业打造国道——商都大道周边高品质小区。总体规划以一条主轴贯穿其中，使中央区域的景观得到很好的渗透，两侧为高层和洋房，空间高低错落，结合下沉庭院、景观植被、水体营造，给人以良好的视觉效果。
建筑高层采用ART-DECO的设计风格，立面整体强调竖向线条，色彩深浅交错，使建筑尊贵而不失现代感。花园洋房采用英式风格，大量运用褐色砖石，整体色彩以深色为主，辅以少量暖灰涂料中和，以园林绿化为衬，使得整体和谐，彰显英伦格调。

The project is located in Zhongmu County, 20 km east of Zhengzhou City. It is a high quality community close to National Highway & Shangdu Avenue, comprising of high-end high-rise buildings, garden houses, and fine commercial buildings. The general plan is penetrated by a main axis, so the central landscape can be shared by the whole community. The combination of the high buildings and houses alongside, high and low contrast in space, the sinking courtyard, landscaping vegetation, and water body gives a good visual effect.
The high floors are designed in Art Deco style. The façade emphasizes vertical lines, dark and light in color, to make the building respectable yet modern. The houses are designed in English style, with massive use of brown masonries, main in dark color, neutralized by a small amount of warm grey paints. Gardening and greening complement the architecture, to create an integral harmony, highlighting English yuppie style.

郑州电子商贸城
Electronic Commercial City, Zhengzhou

设 计 师：郭三锁、刘次圣、祝坚、李超、曹宝华、郁翠凤、傅庆好
项目地点：河南 郑州
用地面积：15 322 m²
建筑面积：151 941.37 m²

Designer: Sansuo Guo, Cisheng Liu, Jian Zhu, Chao Li, Baohua Cao, Cuifeng Yu, Qinghao Fu
Location: Zhengzhou, He'nan
Site Area: 15[illegible] m²
Building Area: 1[illegible]1.37 m²

项目位于郑州市中心火车站附近，地理位置十分优越，交通[illegible]
地块南北两侧各设置一栋高层办公楼，前后错落布置，间[illegible]大以减少对周边现有建筑的影响。在这两栋高层办公楼中，北侧一栋最高达45层，南侧为37层，[illegible]呼应形成一组壮观的建筑群落，使本项目在周边地区成为区域性标志。
结合周边城市道路及人流方向，裙房大型商业布置成内街[illegible]在中心部位设计下沉式广场。下沉式广场将各种人流汇集到一起，形成商业的聚集效应，同时也增[illegible]及城市的可识别性。

The project is located near Zhengzhou Railway Station in downtown area, with advantageous geography and convenient traffic.
The south and north sides of the site are each designed with a high-rise office building, and the two buildings are staggered fore and aft so that the space between the buildings is maximized to reduce the impacts on the surrounding architectures. Of the two high-rise office buildings, the north one has 45 floors and the south one has 37 floors, forming a spectacular building cluster, and making this project a regional landmark in the periphery.
Considering peripheral urban roads and pedestrian flows, [illegible] commercial skirt buildings are enclosing an inner street with a square sinking from the central part. The sinking square gathers pedestrian flows, to form a business cluster, and meanwhile increase the identification of the architecture and the city.

廊坊花园5号
No.5 Garden, Langfang

设 计 师：郭三锁、祝坚、李超、曹宝华、金波、傅庆好、郁翠凤
项目地点：河北 廊坊
用地面积：76 686 m²
建筑面积：203 481 m²

Designer: Sansuo Guo, Jian Zhu, Chao Li, Baohua Cao, Bo Jin, Qinghao Fu, Cuifeng Yu
Location: Langfang, Hebei
Site Area: 76,686 m²
Building Area: 203,481 m²

项目地处廊坊城市总体规划的城市标志地带核心位置。建筑将新古典主义风格的典雅与现代建筑设计手法完美结合，强调建筑由下至上的层层收分，立面的对称性及立面开窗、阳台所形成的洞口比例。造型上力求沉稳、大气、经典、挺拔。以沿艺术大道高层为代表，整体强调竖向线条，破除沿街大体量的敦实厚重感，使建筑挺拔且具有强烈秩序感。建筑立面通过窗的不同比例、外墙的凹凸变化和楼高的落差来增加建筑立面的变化。

The project is located in the core of urban landmark zone of Langfang urban overall planning. The architecture integrates the elegance of neoclassic style and the approaches of modern architectural design, emphasizing bottom-to-up retiring architecture, symmetric façade and the window-balcony proportion. The shape tries to be sedate, generous, classic and upright. The design is represented by the high-rise building along Art Avenue, integrally emphasizes the vertical line, removes the heavy impression of large buildings along the streets, and makes the architecture upright and gives it strong sense of order. The façade of the architecture has different proportions of windows, uneven outer walls and different building heights, highlighting the changing rich contents of the architecture outer walls.

穆斯林商贸城

Muslim Commercial & Trade Town

设 计 师：刘次圣、李超、祝坚、郁翠凤、曹宝华、傅庆好
项目地点：宁夏 银川

Designer: Cisheng Liu, Chao Li, Jian Zhu, Cuifeng Yu, Baohua Cao, Qinghao Fu
Location: Yinchuan, Ningxia

中国穆斯林国际商贸城规划投资160亿元，占地2 066 666.67 m²，一期完成投资36亿元，已建成5A级商贸城42.6万 m²，仓储区10万 m²，副食百货商贸城20万 m²。商贸城雄踞银川商贸物流核心地带，毗邻京藏和银青高速公路出入口，紧靠银川长途客运站及城市公交总站。依据便捷的交通优势，辐射300 km消费地域范围，构建涵盖宁夏全境及内蒙古、甘肃、陕西等省内50余个市县、总量2230万人口的大商圈。

The project investment amounts to RMB16 billion, covering a site area of 2,066,666.67 m². Phase I project has been invested with RMB3.6 billion. Phase I has completed a class-5A commercial town covering an area of 426,000 m², a storage area covering 100,000 m², and a food product & department store covering 200 000 m². The project is located in the core of Yinchuan commerce logistics zone, adjacent to the exit of Beijing-Tibet Expressway and Yinchuan-Qingdao Expressway, and close to Yinchuan Coach Station and City Bus Terminal. With the convenient traffic, and a consumption area of 300 km radiating outward, the project is made a large business circle for a population of 22,300,000 all over Ningxia, and over 50 cities and counties in Inner Mongolia, Gansu, Shaanxi and other regions.

廊坊华夏幸福城朗园

Lang Garden, China Fortune Town, Langfang

设 计 师：祝坚、邹峻青、曹宝华、王雪梅
项目地点：河北 廊坊
用地面积：79 200 m^2
建筑面积：230 665 m^2

Designer: Jian Zhu, Junqing Zou, Baohua Cao, Xuemei Wang
Location: Langfang, Hebei
Site Area: 79 200 m^2
Building Area: 230 665 m^2

项目结合地块道路特点，以十字路口办公楼为重心，以新古典主义设计风格，向两翼辐射，坐拥地块优势，彰显尊贵气息。规划设计上不仅创造开放性的公共城市空间，同时在空间上围合出一块大型居住休闲中心景观，坚持以人为本，提供城市花园的休憩空间。
项目单体在立面上由下而上分成3个层次，上部以新古典主义建筑形象，进退有序，庄严不失活泼；中部立面柱子及窗套处采用较为精致的处理手法点缀出建筑的精致感；底部商业色彩上以深色为主，外侧采用柱廊形式形成趣味性的灰空间，材质为真石漆和石材结合，大大提升了商业氛围及商铺的价值。

Combining the roads on site, centering with the office building at the crossroad and radiating sideways, the neoclassic project embraces its geographic advantage and appears gracious. The planning and design not only create an open public urban space, but also enclose a large space of residential & leisure central landscape, insisting on human-based idea and providing urban garden recreation space.
The façade of a single building is divided into 3 layers up to down; the upper layer refines the architectural appearance with neoclassicism, orderly, venerable and lively; the middle layer embellishes the delicacy of the architecture with façade pillars and window casings by delicate means; the bottom layer is for retail shops decorated in dark color, the outer side is an interesting gray space formed by porch and pillars, and in the meantime the materials are the combination of real stone lacquer and stone materials, greatly enhancing the commercial atmosphere and upgrading the value of the retail shops.

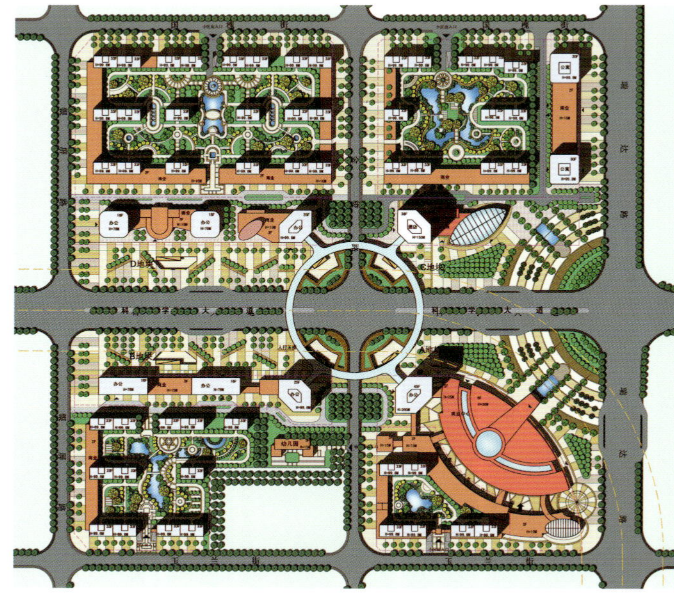

郑州高新数码港
High-tech Cyberport, Zhengzhou

设 计 师：祝坚、邹峻青、曹宝华
项目地点：河南 郑州
用地面积：240 400 m²
建筑面积：1 237 166 m²

Designer: Jian Zhu, Junqing Zou, Baohua Cao
Location: Zhengzhou, He'nan
Site Area: 240,400 m²
Building Area: 1,237,166 m²

本项目位于郑州西开发区科技城核心规划区域，东部有连接中心城区的科学大道穿过地块，它承担着成为桥头堡门户的重要作用。
结合地块的交通优势和教育、科研等综合职能，项目方案以科学大道为主轴，合理组织大型购物中心与特色步行街、商务办公与餐饮休闲、酒店与公寓、地下与地上等诸多关系，将商业功能、居住功能通过规划有机的结合起来，建筑布局点、线、面相结合。
以两栋超高层酒店办公建筑为总体标杆，下沉广场为局部核心，绿色生态为焦点方向，在建筑单体上，充分将建筑品质感和商业气氛感有机融合，创造连续的城市界面和丰富的天际线，统筹设计广告位置，维护沿街形象完整。
在规划上，坚持组团发展、产业互动，通过对人群的生活习惯、流动线路、综合需求的分析，合理规划住宅、办公、酒店、商业、购物等总体布局。

The project is located in the core planning area of Tech Town in Western Zhengzhou Development Zone. The land lot is penetrated by Science Avenue connecting eastwards with downtown, signifying as a bridgehead portal.
Considering traffic, planning and other space advantages, as well as education, research and other functions, the project tries to organize large shopping centers, featured pedestrian streets, commercial/office & catering/leisure, hotel & apartment, underground/aboveground among many other relations along the primary axis of Science Avenue, to integrate commercial functions, residential life by planning, and integrate point, line and plane in the layout.
The two overhigh-rise hotel office buildings serve as the general benchmark, the sinking square is the partial core, and the green eco is the focus. For a single building, the building quality integrates with the commercial atmosphere, to create continuous city interface and rich skylines. The project designs the advertising positions from a strategic height, to maintain roadside image integrity.
The planning insists on cluster development and industry-city interaction, to make a reasonable general layout of residential, office, hotel, commercial, shopping and other buildings, according to the life custom, flow lines, and comprehensive requirements of the population.

扫描查看更多信息

山西省建筑设计研究院

Shanxi Architectural Design and Research Institute (SADI)

山西省建筑设计研究院创建于1953年，是山西省第一家国家综合甲级设计研究院，并获得国际ISO9001：2000、ISO14001：2004和GB/T28001—2001三个体系认证。全院现有职工641人，有专业技术人员[illegible]人，其中教授级高工13人，享有政府特殊津贴专家10人，高级建筑师、高级工程师124人，建筑师、工程师225人，有各类国家注册人员162人，山西省青年设计专家6人，具有一批学术造诣较高、省内外知名的专业技术带头人。设综合设计所（10个）、建筑方案创作所、工程勘察所、惠州办事处、工程咨询中心、新型结构研究中心和岩土工程研究所等设计及科研机构。

服务项目有城市规划、工程咨询、工程设计、工程勘察等。半个多世纪的精心设计、锐意创新，为山西的城乡建设做出了突出的贡献。在医疗建筑、体育建筑、商业建筑、文化教育建筑、大型公共建筑、居住建筑和邮电、电力、餐饮、宾馆、工厂等工业与民用建筑工程设计及建筑节能设计中积累了丰富的经验。多年来累计完成工程设计8000多项，建筑面积达7000余万 m²。项目遍布三晋大地，南疆北国，远及亚、非、欧诸洲。获国家、部、省级奖励150余项，研究、推广、应用新技术、新工艺280余项，并注重理论研究与学术交流，享有良好的社会声誉、广受业界好评，获得国建筑设计行业诚信单位荣誉称号。

精致建筑、精彩生活。山西省建筑设计研究院将更加关注用户需求，为用户提供更好的服务，为社会做出更多的贡献。

地址：山西省太原市府东街5号
邮编：030013
电话：+86-351-3285917/3285361
传真：+86-351-3073613
邮箱：sxy-khfw@126.com
网址：www.sxjzsj.com.cn

Add: 5 Fudong Street, Taiyuan City, Shanxi Province
P.C.: 030013
Tel: +86-351-3285917/3285361
Fax: +86-351-3073613
Email: sxy-khfw@126.com
Web: www.sxjzsj.com.cn

灵石裕园世家别墅
Yuyuan Clan Villa, Lingshi

项目地点：山西 晋中
用地面积：900 m²
建筑面积：540 m²

Location: Jinzhong, Shanxi
Site Area: 900 m²
Building Area: 540 m²

枣园新区E-05地块项目
PLot E-05 Project in Zaoyuan New Area

项目地点：山西 太原
用地面积：156 300 m²
建筑面积：220 000 m²

Location: Taiyuan, Shanxi
Site Area: 156,300 m²
Building Area: 220,000 m²

山西建筑职业技术学院新校区

New Campus of Shanxi Architectural Technical College

项目地点：山西 晋中
用地面积：336 800 m^2
建筑面积：285 100 m^2

Location: Jinzhong, Shanxi
Site Area: 336,800 m^2
Building Area: 285,100 m^2

山西职工医学院教学行政综合楼
Teaching & Administration Complex Building of Shanxi Workers Medical School

项目地点：山西 晋中
用地面积：1785 m²
建筑面积：24 083 m²

Location: Jinzhong, Shanxi
Site Area: 1,785 m²
Building Area: 24,083 m²

山西职工医学院图书馆
Library of Shanxi Workers Medical School

项目地点：山西 晋中
用地面积：6273 m²
建筑面积：13 979 m²

Location: Jinzhong, Shanxi
Site Area: 6,273 m²
Building Area: 13,979 m²

清华炉研发及中试基地项目
R&D and Pilot Base of Tsinghua Furnace

项目地点：山西 太原
用地面积：45 914 m^2
建筑面积：190 000 m^2

Location: Taiyuan, Shanxi
Site Area: 45,914 m^2
Building Area: 190,000 m^2

迎泽四季城
Four Seasons Town, Yingze

项目地点：山西 太原
用地面积：56 600 m^2
建筑面积：325 000 m^2

Location: Taiyuan, Shanxi
Site Area: 56,600 m^2
Building Area: 325,000 m^2

洪洞大槐树文化中心
Hongtong Dahuaishu Cultural Center

项目地点：山西 临汾
用地面积：98 600 m²
建筑面积：49 400 m²

Location: Linfen, Shanxi
Site Area: 98,600 m²
Building Area: 49,400 m²

晋中市博物馆
Jinzhong Museum

项目地点：山西 晋中
用地面积：28 400 m²
建筑面积：33 500 m²

Location: Jinzhong, Shanxi
Site Area: 28,400 m²
Building Area: 33,500 m²

山西省综合展览馆
Shanxi Comprehensive Exhibition Center

项目地点：山西 太原
用地面积：13 333.33 m²
建筑面积：18 100 m²

Location: Taiyuan, Shanxi
Site Area: 13,333.33 m²
Building Area: 18,100 m²

太钢博物馆
TISCO Museum

项目地点：山西 太原
用地面积：15 000 m²
建筑面积：9950 m²

Location: Taiyuan, Shanxi
Site Area: 15,000 m²
Building Area: 9,950 m²

太原铁路局职工文化活动中心

Cultural Activities Center for Workers of Taiyuan Railway Administration

项目地点：山西 太原
用地面积：20 000 m^2
建筑面积：77 906 m^2

Location: Taiyuan, Shanxi
Site Area: 20,000 m^2
Building Area: 77,906 m^2

沁水县星级酒店

Star-rated Hotel, Qinshui

项目地点：山西 晋城
用地面积：9000 m^2
建筑面积：22 663 m^2

Location: Jincheng, Shanxi
Site Area: 9,000 m^2
Building Area: 22,663 m^2

长风西山高尔夫会所

Changfeng Xishan Golf Club

项目地点：山西 太原
用地面积：9000 m^2
建筑面积：5430 m^2

Location: Taiyuan, Shanxi
Site Area: 9,000 m^2
Building Area: 5,430 m^2

沁源县体育中心
Qinyuan Sports Center

项目地点：山西 长治
用地面积：16 427 m²
建筑面积：7440 m²

Location: Changzhi, Shanxi
Site Area: 16,427 m²
Building Area: 7,440 m²

山西体育中心运动员公寓及管理办公楼
Athletes' Apartment and Administration Office of Shanxi Sports Center

项目地点：山西 太原
用地面积：39 817 m²
建筑面积：5250 m²

Location: Taiyuan, Shanxi
Site Area: 39,817 m²
Building Area: 5,250 m²

北营村城市综合体
Urban Complex in Beiying Village

项目地点：山西 太原
建筑面积：430 000 m²

Location: Taiyuan, Shanxi
Building Area: 430,000 m²

扫描查看更多信息

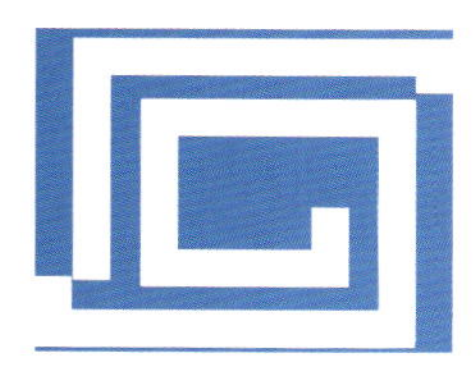

西安利群建筑工程设计事务所
Xi'an Liqun Associates,Inc.

西安利群建筑工程设计事务所于2003年4月成立，持有建设部颁发的建筑工程设计甲级资质，是西北地区唯一一家专业从事医疗建筑设计的企业。在医院新建及改扩建设计方面具有较强的专业优势，业务涵盖国内数十个省、市、自治区。

西安利群遵循质量第一、服务至上的经营理念，强调质量是设计的灵魂，服务是品质的保证，通过创新的手法和有效的管理，努力实现每个项目社会效益和经济效益的最大化。

事务所主要技术骨干多年从事医疗建筑设计，参与了多所医院的新建及改扩建工程，有着丰富的设计经验。事务所的作品山西省运城市中心医院获得“陕西省第十六届优秀设计”一等奖，同时获得“[illegible]年度全国优秀工程勘察设计行业奖建筑工程”三等奖；由利群团队担任主要设计人的第三军医大学西南医院门诊大楼获得“2008年度中国医院建筑优秀设计”一等奖，西安交通大学第一附属医院医疗综合大楼获得“陕西省第十五届优秀工程设计”一等奖。

事务所在长期与境外著名医疗建筑设计大师的合作过程中，学习、吸收了境外先进的医疗管理 、医疗服务及医疗建筑设计理念，并将之用于本土医疗建筑设计实践中。结合中国国情扬长避短、整合力量，设计业绩个性鲜明，特色突出，受到业主广泛好评。

地址：西安市高新一路5号正信大厦B座2603室
电话：+86-29-83151599
传真：+86-29-83151599-816
邮箱：Xian_liqun@163.com
网址：www.xaliqun.com

Add: Room 2603, Block B, zhengxin Building, Gaoxinyilu No.5, Xi'an
Tel: +86-29-83151599
Fax: +86-29-83151599-816
E-mail: Xian_liqun@163.com
Web: www.xaliqun.com

1

2

3

4

5

6

1	山西省运城市中心医院	125 000 m^2	方案、初设、施工图
2	陕西省宝鸡高新医院	122 000 m^2	方案、初设、施工图、可研报告
3	山西省侯马市人民医院	75 000 m^2	方案、初设、施工图
4	陕西省西安141医院	72 000 m^2	方案、初设、施工图
5	山西省河津市人民医院	60 000 m^2	方案、初设、施工图、可研报告
6	宁夏回族自治区儿童医院	48 000 m^2	方案、初设、施工图、可研报告

7. 陕西省西安临潼秦皇医院 130 000 m^2 方案、初设、施工图
8. 甘肃省康复中心医院 100 000 m^2 方案、初设、施工图
9. 陕西省咸阳市第一人民医院 120 500 m^2 方案、初设、施工图、可研报告
10. 中国人民解放军第159医院外科楼 26 000 m^2 方案、初设、施工图
11. 山西省长治县大医院 55 000 m^2 方案、初设、施工图
12. 湖南省益阳市中心医院总体规划、外科住院楼 50 000 m^2 方案、初设、施工图
13. 甘肃省兰州大学第二医院医疗综合楼 112 000 m^2 方案、初设
14. 吉林省吉林市中心医院总体规划及一期工程 72 000 m^2 方案、初设、施工图
15. 甘肃省兰州军区乌鲁木齐总医院医技外科楼 405 000 m^2 方案、初设、施工图

西安樊广智建筑设计事务所
Xi'an Fan Guang Zhi Architects Inc.

在“著名建筑师事务所”铭牌下诞生 2005 年起执业，历时 6 年，业绩丰厚。
西安樊广智建筑设计事务所是由著名建筑师樊广智自行创建的甲级资质建筑设计企业。于 2004 年获得国家建设部批准，为国家首批“著名建筑师设计事务所”。注册地在西安。2005 年启动建筑设计业务活动。2009 年经中国勘察设计协会全国勘察设计企业资质审查换证，设计资质证号：A161006654。
西安樊广智建筑设计事务所拥有各类个人执业注册资质和工程设计资质工程技术人员、技术管理人员 16 人，其中：执业一级注册建筑师 5 人，二级注册建筑师 1 人，英国谢菲尔德大学景观学硕士毕业景观设计师 1 人，建筑设计 5 人，高级结构工程师 1 人（聘用）；档案管理、行政后勤管理人员 3 人。
业务：各类建筑工程设计，小区级规划设计、高品质环境景观设计，装饰设计及编制建筑工程项目建议书，可行性研究等多种业务。

地址：西安市高新三路财富中心二期 B 座 2512 号
电话：+86-29-65676928
传真：+86-29-65676928-8822
邮箱：Fgzhsws@163.com

Add: No.2512, Tower B, Fortune Plaza Phrase II, Gao Xin San Road, Xi'an
Tel: +86-29-65676928
Fax: +86-29-65676928-8822
E-mail: Fgzhsws@163.com

南山庭院
South Mountain Courtyard

项目地点：陕西 西安
用地面积：98 289 m²
建筑面积：76 032 m²
容 积 率：0.77
建筑密度：23%
绿 化 率：42%

Location: Xi'an, Shan'xi
Site Area: 98,289 m²
Building Area: 76,032 m²
Plot Ratio: 0.77
Building Density: 23%
Green Ratio: 42%

项目位于西安城南长安区内，秦岭脚下环山路以北，承袭关中三合头院落式民居的地方性特色；建筑采用现代美学元素及色彩，成功地表达出“传统文化的神韵，时代的精神风貌”。
南山庭院的建筑设计在形式、功能、空间、文化方面成功地有机结合，在西安第一个塑造出“秦人文化”的现代版。在当地业界已经产生了较大示范性效应。
建筑的安保性特别好。每户附有车位（库），户门三防，门窗采用节能环保的塑钢双玻门窗；屋面、外墙为挤塑板节能构造设计。
项目建成后设计先后获：“陕西省第十五次优秀工程设计奖”二等奖。“2009 年度全国优秀工程勘察设计行业住宅与住宅小区”二等奖

The project is located north of Ring Road at the foot of Qin Range within Chang'an District, southern Xi'an City. It follows the local expression of Western Shaanxi Triplet Courtyard Residence: The architecture is treated with modern aesthetic elements, and improved with modern elegant colors, to successfully express the special effects of “traditional cultural soul, and epochal spiritual style”.
The architectural design is an integration of form, function, space and culture, to successfully reproduce a modern version of “Qin culture” in Xi'an City. It has produced large demonstrative effects in local industrial community.
Good security. Every residential unit has a parking lot (garage), a frost/theft/fireproof gate, energy-saving and environment-friendly double layers of PVC glass doors and windows; plastic extrusion energy-saving structures are designed on the roof and external walls.
The design of project upon completion has won the following awards: the provincial second prize of the 15th Excellent Engineering Design of Shaanxi Province, the second prize of National Excellent Engineering Survey & Design Industry – Residential Building & Residential Community in 2009.

中国建筑西南设计研究院有限公司

CHINA SOUTHWEST ARCHITECTURAL DESIGN AND RESEARCH INSTITUTE CORP.LTD.

扫描查看更多信息

中国建筑西南设计研究院有限公司始建于1950年，是中国同行业中成立时间最早的大型甲级建筑设计院之一，隶属世界500强企业中国建筑工程总公司。建院60多年来，本院设计完成了近万项工程设计任务，项目遍及中国各省、市、自治区及全球10多个国家和地区，是中国拥有独立涉外经营权并参与众多国外设计任务经营的大型建筑设计院之一。2009-2012年该院连续4年被亚洲建筑师协会评为“中国十大建筑设计公司”，并获得“全国工程勘察设计百强”企业称号。

地址：四川省成都市天府大道北段866号　Add: No.866, North Section of Tianfu Avenue, Chengdu City, Sichuan Province
电话：+86-28-62550866　Tel: +86-28-62550866
传真：+86-28-62550900　Fax: +86-28-62550900
邮箱：xnyyb@vip.163.com　E-mail: xnyyb@vip.163.com
网址：www.xnjz.com　Web: www.xnjz.com

成都七中高新校区

Chengdu No. 7 Middle School Hi-Tech Zone

设 计 师：刘艺
项目地点：四川 成都
建筑面积：108 927.00 m^2

Designer: Yi Liu
Location: Chengdu, Sichuan
Building Area: 108,927.00 m^2

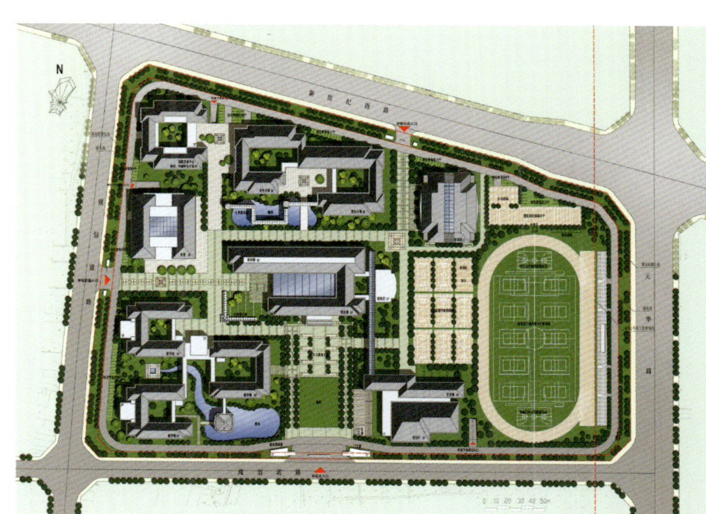

整体规划中教学区、运动区、生活区构成“品”字形布局，彼此之间联系便捷。校园交通设计遵循“人车分流”的原则，使内部功能组团形成完整的步行区域。
校园建筑采用传统结合现代的手法：四坡屋顶，红砖墙和石材勒脚的设计来自于对老校区历史建筑特征的提炼与再现，金属构架与玻璃砖等材料的运用又体现了现代的气息。新校区参照绿色校园建设标准，合理的规划布局保证了良好的朝向与自然通风。整个项目采用了雨水收集系统，回收雨水作为绿化灌溉与景观用水的水源，体现了生态环保的理念。

The teaching zone, sports zone and living zone are laid out in a triangle, very convenient for intercommunication. The campus is separating vehicle flow from pedestrian flow, and internal functions provide a complete walking space.
The buildings are complex of traditional and modern manner: hipped rooftop, red brick wall, and stone plinth are extraction and reproduction of old campus elements; the metal structure and glass bricks among other materials are modernistic. New campus is a green campus, whose reasonable layout ensures good direction and natural ventilation. The campus is equipped with rainwater collection system which provides a water source of greening irrigation and waterscape use, keynoting eco-environmental protection

四川华电办公大楼

Office Building of China Huadian Corporation, Sichuan

设 计 师：刘艺
项目地点：四川 成都
建筑面积：56 289 m^2

Designer: Yi Liu
Location: Chengdu, Sichuan
Building Area: 56,289 m^2

四川华电办公大楼是汇集办公、调度、档案等多功能为一体的企业总部大楼。设计在用地南北侧密植树林，树林的中央设置一片宁静的水面。轻盈的玻璃塔楼从水中升起，类似一座岛屿，水面与玻璃相互映射，塑造光景变幻的迷人景象。

办公楼分为南北两座板楼，两楼之间围合出近 50 m 高的室内中庭，创造出全体员工共同分享与体验的空间。空中会议厅从中庭横跨而过，如同发电机的巨大线圈飘浮在半空中。这个被称为“能量圈”的会议厅设计充满了未来感，塑造激动人心的建筑场景。

This office building is a corporate HQ building with integration of office, dispatch, archiving and other functions. In the design, a dense forest is planted on south and north sides of the land lot. In the center of the forest is a tranquil pond, a glass tower rising from where, resembling an isle, whose reflections on the water surface can produce a charming illusion of lights and views.

The office building is composed of two towers, enclosing a 50 m high indoor atrium in between, as a sharing space and experience for entire office staff. The conference hall spans across the atrium in the air, like a huge coil of electric generator floating in the air, dubbed “energy circle”. This circle excites people of this futuristic marvel.

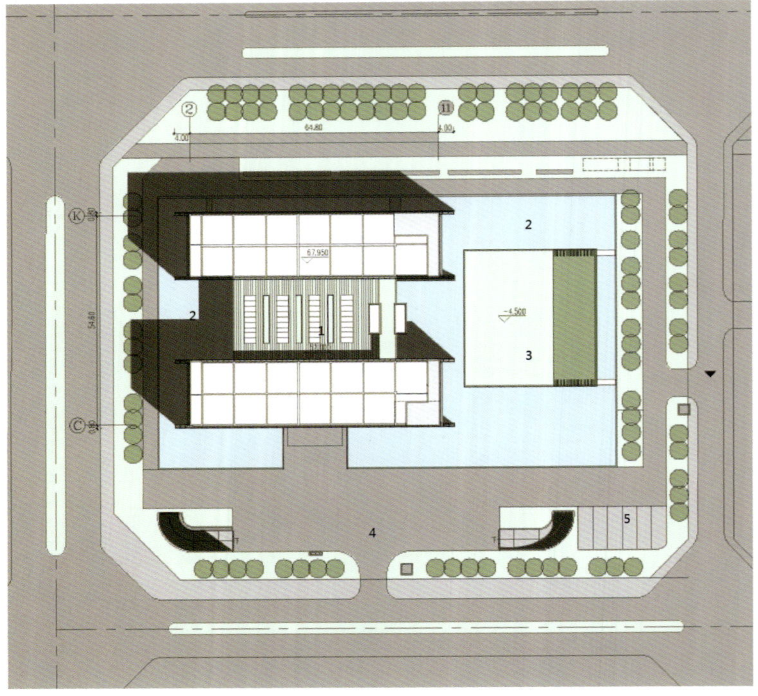

总平面图

玻璃电梯井
空中花园
景观水池
空中会议厅
下沉式庭院

大楼的设计取消了通常意义上的裙房，主体建筑形象单纯完整。将通常设于裙房的大跨度会议厅、餐厅和配套设施分别提升到空中或设置在地下。大会议厅“悬浮”在空中，餐厅和健身中心位于负一层，环绕在种植有高大树木的下沉庭院的周边区域。大楼的另一个庭院位于空中，利用12层中庭的屋顶创造了一片私密性的花园， 供13至17层的公司高管办公区专享。中庭内的6部玻璃景观电梯将垂直方向的各个中庭和庭院串联在一起，上下之间“梯移景异”。

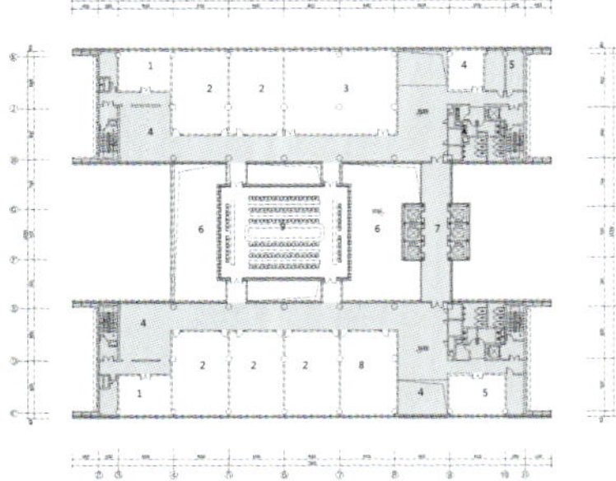

五层（会议中心）平面图

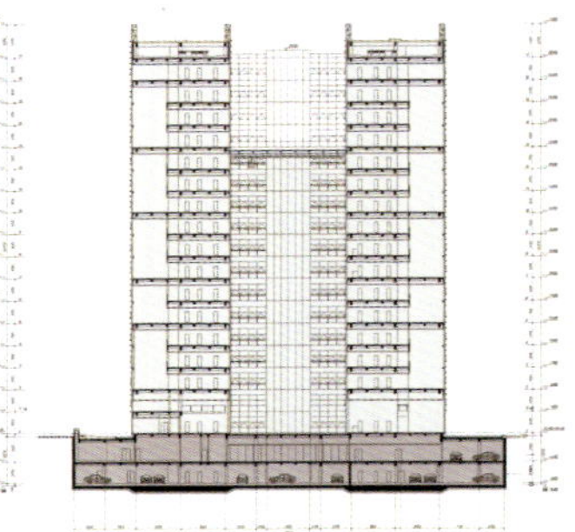

剖面图

中国建筑西南设计研究院有限公司第二办公区办公楼

Office Building in the Second Office Area of China Southwest Architectural Design and Research Institute Co.,Ltd.

设 计 师：邱小勇
项目地点：四川 成都
建筑面积：86 545 m^2

Designer: Xiaoyong Qiu
Location: Chengdu, Sichuan
Building Area: 86,545 m^2

项目充分考虑了所处位置的特殊性，建筑布局遵循城市设计的原则，融入城市肌理；同时，基地毗邻城市交通干道，作为"快速阅读"的对象，具有恰当的形体抽象美感及视觉冲击力。
形体构成着重于在"服从"与"叛逆"的矛盾中，寻找平衡的契合点，运用形体合理组合，打破单一格局，从空间结构和视觉效果上更好地平衡城市环境，并突出自身标志性。

The project takes full consideration of specific location, so the layout follows urban design to integrate with urban texture, meanwhile the base is adjacent to arterial road, and at a quick glance, its abstract figure has proper visual impact.
The figure is in a paradox between "service" and "landmark", and it is important to balance the two. Organic formation will break up the monotonous layout, and better balance the urban conditions by structure and visions, with obvious marks.

神仙树社区服务中心
Shenxianshu Community Service Center

设 计 师：钱方
项目地点：四川 成都
建筑面积：13 834.38 m²

Designer: Fang Qian
Location: Chengdu, Sichuan
Building Area: 13,834.38 m²

神仙树社区服务中心是政府为民办实事、创建和谐社区的示范工程。
设计针对狭长的基地特点结合多样化的功能需求，继承传统建筑的院落空间特点，以院落组织串联并分隔不同的功能块。设计打破传统建筑与城市公共空间分离的状态，在建筑中贯穿了一条给市民休闲的空间立体散步道，即从本质上提供了使建筑成为公共性空间的可能。这条"道"串联了建筑的多种功能，充分体现了公共空间的开放性。设计"全生命周期控制"的理念促使该建筑有良好的完成度，并产生很好的经济效益和社会效益。

This center is a public service project for the people, for the harmony of the society. The land lot is a strip, so the design follows traditional courtyard characteristic, to connect different yards and separate different functions. The design breaks the isolation between traditional architecture and city public space, offering a walking street among the buildings, which essentially provides the possibility of publicizing all the buildings to citizens. This street connects many functions, fully representing the openness of public area. The "total life cycle control" idea of design procures better completion, and produces better economic and social benefits.

成都金沙艺术剧院

Jinsha Art Theatre, Chengdu

设 计 师：刘艺
项目地点：四川 成都
建筑面积：39 895.8 m²

Designer: Yi Liu
Location: Chengdu, Sichuan
Building Area: 39,895.8 m²

“金色面纱”的概念源于著名的金沙黄金面具与太阳神鸟的启发。
剧院的公共休息大厅的外表皮被设计成“金色面纱”的核心，独特的流畅轮廓富于视觉标志性。剧院的门厅与休息厅被设计成弯曲的平面空间，拥有 180° 宽银幕式的观景视野，体现出“透明面纱”的设计构想。
起伏的曲面造型与观众舞台空间高度吻合，半弯顶的杂技剧场屋面倾斜而下，与景观水池相连。斜屋面处设计为露天剧场，成为金沙剧院室内观演厅堂之外的第三个舞台，为市民提供交往空间与非正式的表演场所。

The concept of “golden veil” is originated from unearthed gold mask and phoenix paintings in Jinsha.
The surface of public lounge is designed as core of the “golden veil”, whose unique lines are symbolic. The entrance hall and the lounge are designed into bent flat space, with 180° sightseeing which represents the conception of “transparent veil”.
The curved surface perfectly meets the height of audience stage. The semi-curved rooftop of acrobatics theatre is inclined downwards to the waterscape pond. The inclined roof is designed as an open air theatre, serving as the third performing platform beyond indoor theatres, which is an informal venue of performances and communications for citizens.

瓦努阿图国际会议中心

Vanuatu International Conference Center

设 计 师：郑勇
项目地点：瓦努阿图 维拉港
建筑面积：6573.72 m²

Designer: Yong Zheng
Location: Port Vila, Vanuatu
Building Area: 6,573.72 m²

方案构思由该国岛屿的存在形状演化而来，Y 字形墙体将建筑划分为三个功能区。主门厅将三部分联系起来，并向大海开敞，形态舒展，表现出瓦努阿图人民对外国友人的欢迎姿态。考虑气候湿热的特点，将外廊、院落、敞厅等结合使用，在调节建筑小气候的同时形成丰富的内外空间效果。大礼堂体量取自当地著名的“火山”造型。而螺旋上升的虚实划分又有财富图腾“猪牙”的意象，寓意国际会议会使岛国经济迅猛发展。屋面细节划分则由纳卡丽叶抽象而来。外廊兼有本地杆栏草屋和中国亭榭的意味，隐喻这是中瓦友谊的象征。建筑主色调和红色坡屋面与近邻议会大厦协调。

The conception is derived from the formation of this island nation. Y-shaped wall divides the building into three function zones. The entrance hall connects the three zones and is open to the sea, in agreeable exposure, representing welcoming posture of Vanuatuan citizens. The climate is moist and hot so the outer corridor, courtyard and open air hall are combined in use, to condition atmosphere while enriching in/outdoor space effects. The auditorium takes shape of local famous “volcano”. The spiral illusion of the virtual and real world also figures the totem of a “boar's tusk” implying economic boom of the island boosted by international conference. The roof details are abstracted from namele leaves. The outer corridor is in Chinese flavors but it also tells the form of local thatched house with balusters, implying Vanuatu-China friendship. The red color and red sloping roof concert with the Parliament building.

瓦努阿图国际会议中心
Vanuatu International

瓦努阿图 维拉港
Vila Vanuatu

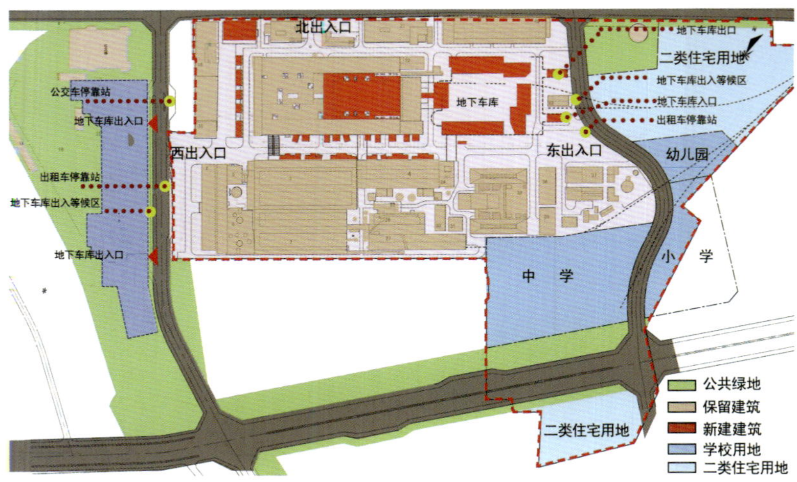

成都东区音乐公园

Chengdu East Music Park

设 计 师：张远平
项目地点：四川 成都
建筑面积：184 348.11 m²

Designer: Yuanping Zhang
Location: Chengdu, Sichuan
Building Area: 184,348.11 m²

设计对（原红光电子管厂）不同年代的老建筑群进行保护、加固、改造和创新。传承工业文明、保留现代遗址、以音乐文化为导入产业，力求营造出一个历史与现代，工业与音乐交汇的活力创意园区。

根据现存建筑的不同现状分别采用相应的改造方式。结构改造设计采用预应力，阻尼支撑，型钢构造柱及圈梁等方法以提高其综合抗震能力，并按遗址保护原则，保留加固痕迹，展现独特美感。

The project is located on the historical site of Red Light Electronic Tube Plant. The design maintains, reinforces, reconstructs and renovates the old building clusters dating back to different ages, it inherits industrial civilization, remains modern relics, and introduces industries and business through music culture, striving for a dynamic creative park that integrates history and the present, industries and music.

It reconstructs in different manners according to different conditions of the buildings. The structural design will consider prestress, damper support, formed steel structural column and ring beam etc. to increase its aseismic performance and present unique beauty in a reinforced conservation of the relics.

遵义市科技馆

Science and Technology Museum, Zunyi

设 计 师：李雄伟
项目地点：贵州 遵义
建筑面积：16 297 m²

Designer: Xiongwei Li
Location: Zunyi, Guizhou
Building Area: 16,297 m²

科学技术源于生活，服务生活；而承载科技的建筑则来源于自然，最终回归于自然。方案以回归为主题，力图打造一个别具特色的科技馆形象。设计以湿地公园为出发原点，采用覆土建筑的形式，将建筑融入环境中，保留一系列环境要素，体现地域特色又具有创新的建筑风格，强调功能合理的同时更好地体现了建筑与自然“和谐共生”的思想。

Science and technology come from life and return to life; while this architecture comes from nature and returns to this natural park, bearing the social responsibility of science popularization. The museum hides in the hills and rocks. The theme is "return", so the museum tries to build a unique scientific image, starting from the wetland park, where earth sheltered architecture integrates with the environment, while remaining a series of environmental elements. This design represents regional features while innovating architectural styles, and organizes multiple functions while "harmonizing" with the nature.

成都东客站

Chengdu East Railway Station

设 计 师：金旭炜、郑勇
项目地点：四川 成都
建筑面积：108 000 m^2

Designer: Xuwei Jin, Yong Zheng
Location: Chengdu, Sichuan
Building Area: 108,000 m^2

成都东客站是我国西部最重要的铁路客运中心之一，设计提炼三星堆文明及金沙文化元素，通过一定抽象处理来增加建筑的易读性：舒展的屋面来源于金沙太阳神鸟张扬的火焰形态，结构造型融入独特的青铜面具艺术元素，有力的结构支撑与屋面构成厚重、大气的建筑形象。候车大厅的顶棚宛如神鸟的羽毛又似片片竹叶，使建筑更具天府之国的神韵。

成都东客站率先采用地铁、车站、铁路轨道桥、高架道路桥一体化的整体结构体系，承轨层为桥建合一形式。同时针对站房平面尺寸较大的特点，对结构的多点输入地震响应和防震缝布置对屋盖结构动力特性的影响作了专项研究。

Chengdu East Railway Station is one of the most important railway passenger terminals in southwest China. The design adopts some cultural elements extracted from ancient and tribal localities in Sanxingdui and Jinsha relics, to increase architectural readiness through certain degree of abstraction: the extensive roof takes its origin from the flaming phoenix in the paintings unearthed at Jinsha Village; the structure takes its shape from unique bronze mask elements, with strong supports forming a thick, generous image with the roof; the canopy of station hall resembles phoenix feathers or bamboo leaves, implying the flavors of Sichuan Province, the legendary State of Heavens.

It is an integration of subway, station, rail bridge and overhead highway bridge, with rail-bearings are built in the bridge. Meanwhile, due to large planar size of the station house, the design has made special researches on the multi-input seismic response of the structure, and the dynamic impact of aseismic joints deployment on rooftop structure.

成都仁恒置地广场
Yanlord Land Plaza, Chengdu

设 计 师：郑勇、郑国英
项目地点：四川 成都
建筑面积：212 482 m²

Designer: Yong Zheng, Guoying Zheng
Location: Chengdu, Sichuan
Building Area: 212,482 m²

项目定位为成都市 CBD 中心标准最高、设施一流、管理服务国际化、具有城市标志性的城市综合体建筑，力求成为成都写字楼、公寓、高端商业的“制高点”。
项目以成都得天独厚的城市人文氛围以及丰富的周边自然环境因素为灵感，采撷“天府之国”山水地域文化元素，与现代建筑的简洁、明朗相结合。主体建筑以大尺度整体玻璃幕墙一气呵成，表达“高山流水”的广阔意境，充分展现了高新科技表现力与深厚地域人文主义的浪漫融合。

The project is positioned as a landmark complex in Chengdu CBD, with highest standard, first-class facilities, and international property management, taking the “height” of office, apartment and hi-end commercial buildings in Chengdu.
Inspired by unique local cultures and rich natural elements, as well as local landforms in the “state of heavens”, the architecture is concise and modernistic, whose main building is made of a large-size whole piece of glass wall, representing the vastness of “high mountain and flowing water”, making the hi-tech expressionism and romantic local cultures.

金牛区茶店子街区城市设计

Municipal Design of Chadianzi Block, Jinniu District

设 计 师：刘刚
项目地点：四川 成都
用地面积：3 000 000 m²

Designer: Gang Liu
Location: Chengdu, Sichuan
Site Area: 3,000,000 m²

金牛区茶店子街区位于老成灌路与三环路交接处，片区未来将是城西人文、商业、绿化、人气与活力集聚的门户和中心，具有重要的城市地位，城市设计是旧城更新的重要课题。茶店子街区的构想，寄托了设计者对城市更新的思考：将城市的中心区域塑造成精彩的生活舞台，使城市中的各个场所成为交流共享的开放平台，将空间还给城市，将生活还给市民。茶店子街区城市设计在规划理念与设计实践上寻求突破创新，致力于实现“公园中的城市”“超级街区”及“生活与漫游的舞台”三大目标理念，以一系列多层次、多向度的空间场所呈现出独具魅力的立体街区形态，以新颖的群像塑造出崭新的城市形象和区域地标，为城市更新树立全新目标。

This block is located in junction of Laochengguan Road and the Third Ring Road. It is a future downtown of western Chengdu, a new hub of cultures, commerce, greening, population and energies. For this important urban position, the design will involve renovation of old city. The conception carries our fresh ideas for the city to build the core of city into a colorful life platform, where every corner can be a shared platform open to citizens who will embrace the city space as a real residence. The design in mindset and practice is seeking breakthroughs, trying to realize “city in park”, “super-block”, “living and roaming platform”, the three pictures, as well as 3D figures featuring its multi-level and multidirectional space, and this novel architectural cluster will mark a brand new city image and regional landmark, to create a new objective of the city.

综合体街区
中央绿带
商业西街
居住社区

制高点≥120米
一般性塔楼≤100米
商业西街临街建筑≤30米
周边塔楼≤80米

扫描查看更多信息

四川山鼎建筑工程设计股份有限公司
Sichuan Cendes Architecture Engineering Design Co., Ltd.

Cendes

四川山鼎建筑工程设计股份有限公司（Cendes）具有国际背景的公司决策层和国际化项目管理运营体系，国际水准的专业技术服务团队为项目提供整体解决方案。
Cendes 以国际智慧融合当地经验，秉承“筑就城市理想”的理念，其各地运营公司的500多名国内外专业人士，在建筑、规划、环境、能源等领域为政府及私人机构提供专业咨询和综合工程设计服务。
Cendes 拥有各地的丰富经验和本土知识，通过不断的技术创新和项目实践，持之以恒地改善并维护各地的建筑品质、城市环境和社会环境的可持续发展。
Cendes 以全面和超越的视野、整合的解决方案，应对各个领域日益复杂的挑战，为公司品牌发展创造持久影响力。
山鼎设计（Cendes）各地运营公司：
· 四川山鼎建筑工程设计股份有限公司（住建部甲级资质）
· 山鼎设计（北京）
· 山鼎设计（上海）
· 山鼎设计（西安）

Sichuan Cendes Architecture Engineering Design Co., Ltd. (Cendes) provides overall project solutions with international decision-making layer, project management and operation system and professional technical service team.
Combining the international wisdom and local experience, upholding the concept of “Creating City Ideal”, Cendes and its more than 500 experts from home and abroad in its operating companies provide professional consulting and complex engineering design service for government and private organization in such fields as architecture, planning, environment and energy etc.
Cendes owns and possesses abundant experience and local knowledge, through the continuous technical innovation and project practice, it keeps to improve and maintain the local architecture quality, city environment and the sustainable development of social environment.
Cendes faces more and more complicated challenges in all fields with thorough and superior view and integrated solution to create lasting influence for the company brand.
Cendes Operating Companies:
Sichuan Cendes Architecture Engineering Design Co., Ltd. (Housing Ministry A-Class Qualification)
Cendes (Beijing)
Cendes (Shanghai)
Cendes (Xi’an)

锦绣森邻
Splendid Forest Neighborhood

项目地点：四川 成都
用地面积：123 333.33 m²
建筑面积：220 000 m²
建筑密度：27.4%

Location: Chengdu, Sichuan
Site Area: 123,333.33 m²
Building Area: 220,000 m²
Building Density: 27.4%

项目地处成都市居住卫星城温江县规划新区内，周边城市交通状况良好，140 m宽高标准的光华大道从温江经过本项目地块南侧并直通成都市区，同时项目地块西临“花博会”主场馆，北望江安河，整个区域环境宜人，适合居住。
设计以“森”——自然生态的社区景观，“邻”——和谐融洽的邻里生活为主题。
规划以4个“森林”岛组成建筑群体，围而不合的退台式花园洋房构成各具特色的邻里单元，精心设计的不同尺度、风格各异的林带将邻里单元包裹。小高层以点式景观建筑的姿态有机地融入，形成丰富的空间形体。户型以极具创新性和人情味的设计在业内赢得广泛的赞誉。项目获得住建和城乡建设部人居规划设计优秀奖。

This project is located in the planned new area of Wenjiang County, a satellite town of Chengdu City. The traffic condition of the surrounding cities is good. The 140m broad Guanghua Road of high standards runs directly from Wenjiang to Chengdu downtown through the south part of this project. Meanwhile the project is besides the main venue of “Flower Expo” in the west, and facing Jiang’an River in the north. The whole area is agreeable and habitable.
Forest – A community landscape with natural ecosystem
Neighborhood – A harmonious and friendly neighborhood
Four “forest islands” make up a building cluster according to the planning. The enclosing yet not closed retiring garden houses form different neighborhood units of unique features which are surrounded by well-designed forest belts of different scale and style. Small High-rise houses integrate into the environment well as landscape architecture, creating a rich space form. The innovative and humanistic design of house type has been spoken highly of in the architecture industry. This project won the MOHURD Habitat Planning & Design Award.

龙湖长桥郡

Longfor Long Bridge Shire

项目地点：四川 成都
建筑面积：270 000 m^2
容 积 率：0.52
建筑密度：22.82%

Location: Chengdu, Sichuan
Building Area: 270,000 m^2
Plot Ratio: 0.52
Building Density: 22.82%

别墅紧临风景优美的牧马山，拥有独一无二的生态环境。在设计中根据项目的地理环境特点，始终强调别墅区的生态社区感，强调人与环境的关系，使得别墅成为环境的一部分。将会所布置在基地临道路的中部，增加配套的商业以弥补周边商业的缺乏。

每幢别墅既有前院，又有内庭和私家后花园。在总体规划和建筑功能的设计上，参照北美式别墅建筑的风格。形成组团式别墅布置模式，每个组团内部建筑摆放及朝向布置均经过精心考量，根据不同区域和景观条件，布置不同的别墅类型，达到建筑与环境的高度和谐。

Close to the beautiful Muma Mountain, the villa has unique ecological environment. Based on the geographical environment features of this project, the design stresses that the villa is an ecological community, emphasizes the relationship between human and nature, and makes the villa integrated into the environment. The club is placed in the central part of the base which faces the road, to increase the commercial level of the supporting facilities and make up for the business shortage.

Every villa has a dooryard, an inside courtyard and a private back garden. The project adopts the modern style of North American villas in layout planning and architectural functions. The group villa arranging pattern is formed according to the planning structure. The interior architectural settings and orientations of each group have both been carefully considered. Different types of villas are arranged differently based on different region and landscape, to make the buildings highly integrated into the environment

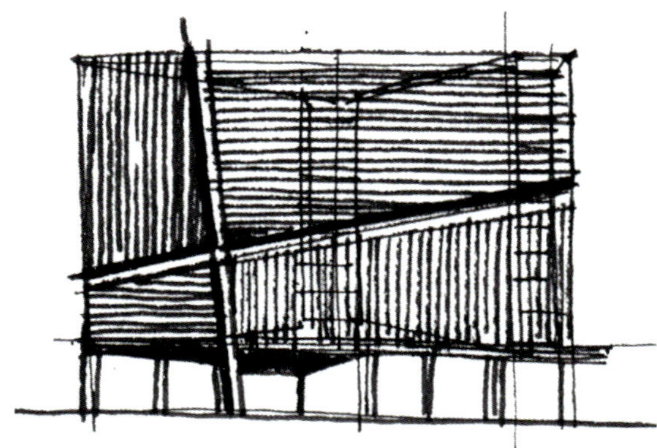

尚都
Fashion City

建筑面积：104 000 m²
容 积 率：10.08
建筑密度：74.29%

Building Area: 104,000 m²
Plot Ratio: 10.08
Building Density: 74.29%

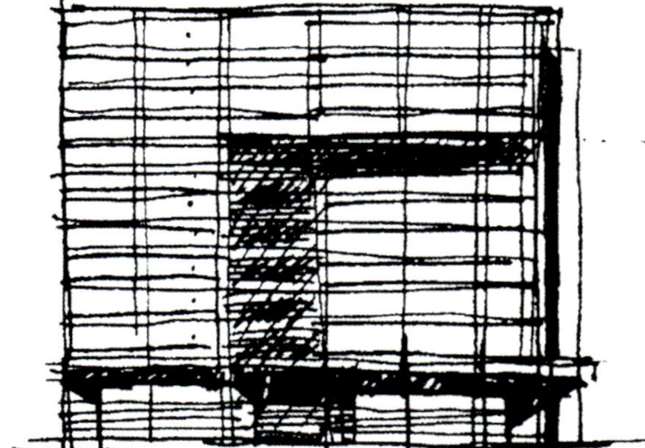

项目着重于商业氛围的营造，广场空间、街道线性空间与建筑本身的结合，有效地引导人流进入商场且起到了调节大量人流的功能；建筑内部精心设计了一条商业流线，室内商业中庭不受室外环境影响而又可以享受阳光和景观，创造出更为舒适的以人流动线为设计重点的购物休闲环境。
由于本案位于繁华的春熙路商圈，周边商城、写字楼林立，故造型和色彩上的设计手法反其道而行之，以传统的方式在细节处理上使本案明显有别于周边建筑，同时注重外立面整体效果，行人在繁花似锦的楼群中能迅速被其低调、大气的特质所吸引。

This project tries to create a commercial atmosphere. The combination of square space, street linear space and the building itself effectively guides the pedestrian flow into the mall and diverts the crowds. Inside the building is a well-designed commercial streamline. The commercial atrium inside the mall can enjoy the sunshine and landscape without being affected by the outside environment, creating a more comfortable shopping and relaxing environment. Due to the fact that this project is located in the prosperous Chunxi Road business circle, where a forest of shopping malls and office buildings can be seen, the style and color design of this project is quite different from the surrounding buildings by using traditional method in details treatment. Meanwhile the exterior façade emphasizes the integral effect, thus people can quickly tell apart this mall from great building clusters and get attracted by its low-key yet grand quality.

第一城
Top City

项目地点：四川 成都

Location: Chengdu, Sichuan

本案位于成都传统的春熙路商业圈的中心地带，特殊的地理位置决定了该建筑的性质是以高档购物中心为主，集餐饮、娱乐、写字楼功能为一身的标志性商业综合体。设计上充分合理地利用土地资源，以步行街引导人流，以广场和柱廊骑楼吸引人流，盘活每一寸商业面积。步行街、购物中心、写字楼之间人流相互穿插，通过大道带动整个基地的商机。商业流线简洁、清晰，通过多个商业广场和斜跨地块的步行街设置，极大限度地争取了商铺的临街面，大大提升了商业价值且有效规避了该地块本身道路窄小、拥挤的弊端。

This project is located in the heartland of tradit onal Chunxi Road business circle in Chengdu city. The special lccation determines that this building is a remarkable business complex, mainly functioning as a high grade shopping mall and as restaurants, entertainment places and office buildings too. The design makes full use of the land resources. The pedestrian flow is attracted by the square and arcade and guided by the walking street, thus every inch of business area is effectively used. The pedestrian flows are intersecting between the walking street, shopping mall and office building, and the avenue makes the whole place prosperous. The commercial streamline is simple and clear. By arranging the walking streets between a couple of commercial squares and diagonal blocks, this project expar ds the roadside of the shops as much as possible, increases commercial interests and effectively avoids the demerit of this land lot that the roads are narrow and crowded.

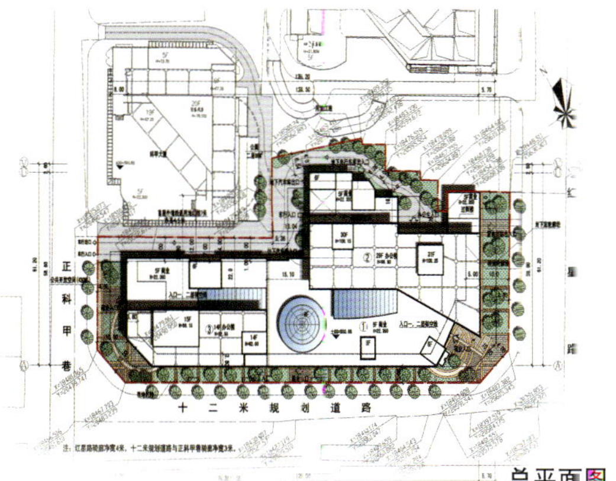

总平面图

远大荷兰水街
Yuanda Holland Water Street

建筑面积：37 000 m^2 Building Area: 37,000 m^2
容 积 率：1.55 Plot Ratio: 1.55
建筑密度：0.5% Building Density: 0.5%

本方案为集餐饮、娱乐、休闲和商业等功能为一体的综合性建筑，其所处地理位置及建筑本身的规划和设计决定该建筑将成为本区域的标志性建筑。建筑风格以现代欧式为主，注重建筑的细部刻画和人性化的尺度控制，讲究时尚的品味、独特的格调、丰富的表现力，通过建筑与环境的有机结合渲染出一片浓郁的商业文化氛围。

This project is a catering, entertainment, leisure and business complex. Its geographical location and layout planning determines that it will be a landmark of this region. The architectural style represents the modern European style, stressing on the details of the building and humanistic scale controlling. The style also has fashionable tastes, unique patterns as well as rich expressiveness, and creates a strong commercial culture by integrating the building into the environment.

扫描查看更多信息

成都基准方中建筑设计事务所

Chengdu Jizhunfangzhong Architectural Design Associates

创新、技术、服务是基准方中的设计服务理念。
创新——作为开拓市场的先行者，在过程中谋求设计与市场的最佳结合点，将最新的设计理念恰当地运用于产品之中，以完成业主、市场以及社会均能接受的设计作品。
技术——三者中最为厚重的一环，它不仅是设计事务所赖以生存的重要保障，也是设计出好的产品的基础，扎实过硬的技术功底是一个成功的设计团队不可缺少的资本，是团队核心竞争力的重要体现。
服务——民营设计企业相比传统国营设计企业的一大优势，是设计以人为本，尊重业主、尊重社会的客观需要，良好的服务执行与完善的服务体系一方面可以稳固市场的开拓成果，展示事务所的管理水平，同时也是事务所创造具有恒久价值的建筑产品（作品）的必要保障。

成都事务所（总部）
地址：成都市西玉龙街 6 号新世纪广场 10F-16F，20F，29F
电话：+86-28-86582121
邮箱：KF@jzfz.com.cn

Chengdu Architects Inc.(Headquarter)
Add: 10F-16F, 20F, 29F, New Century Square, Xi Yu Long Street No.6, Chengdu
Tel: +86-28-86582121
E-mail: KF@jzfz.com.cn

中国现代五项赛事中心

China Modern Five Sports Events Center

为了确保赛事顺利举行，对时间、成本和施工难度的有效控制成为了设计的核心。同时，设计亦充分体现国际上先进的体育建筑设计理念，注重形式与功能的完美统一，在严格限定的建设规模内灵活安排不同的功能置换空间和应急变更计划，充分考虑赛时和赛后的不同要求与用途，使场馆在功能上可适应更多其他比赛和活动的要求，提高利用率。此外，在节能技术，中央空调技术和大跨度网壳屋面结构技术等方面，谨慎地选择使用了国内较为成熟的先进技术产品，提高项目的科技水平。
设计将建筑置于周边环境之中，与河水交相辉映。建筑立面造型简洁、优雅，与周边自然景观协调统一。

For ensuring smooth running of sports events, effective control of time, cost and construction difficulty become the core of design. At the same time, the design also fully represents international leading sports architecture design concept, focuses on the perfect unity of forms and functions, flexibly arranges different function replacement space and emergency plans in strictly limited building scope, and fully considers different requirements and usages during and after competition to make the center functionally applicable in many other competitions and activities and increase use ratio. In addition, as for energy-saving technology, central air-conditioning technology, large-span reticulated roof structure technology and other aspects, the project prudently uses domestic mature and advanced technology product to develop the scientific & technologic level of the project.
The design places buildings in neighboring environment and match them with rivers. The facades of buildings are in simple and elegant shapes, which harmoniously unite with surrounding natural landscapes.

武海中华青城
Wuhai China Green City

项目地点：四川 都江堰
建筑面积：260 000 m²

Location: Dujiangyan, Sichuan
Building Area: 260,000 m²

项目以“邻里社区，人性空间”作为设计的指导思想，以多种类型的组团合院式布局为规划的基本单元，通过组团间的相互错动和衔接，构建多种层次的立体空间；区域内高低错落，疏密有致，避免了传统住区平铺直叙式的天际线；公共区域—组团入口—组团庭院，形成层次鲜明、丰富有趣的空间和景观体系。
项目积极探索合院型住宅的空间和设计特点，立面上采用自然地中海风格。建筑中融入了阳光和活力，采取更为质朴温暖的色彩，建筑随光影而变化，质朴而又充满人性化。通过户与户之间空间的围合，以及二层平台处的连接，构建一个具有合院特质的生活空间，从而构建一个真正意义上的“和谐住区”。

To make a neighboring community and human space is the guiding concept of the project design. With multiple types of group courtyard layout as the planned basic unit, the project builds multilevel 3D space through mutual displacement and connection between groups; the area with high and low buildings in even density avoids the straightforward skyline of traditional residential communities; from public area – entrance of group – group courtyard house, it forms distinct levels and abundant and interesting space and landscape system.
The project actively explores space and design characteristics of courtyard houses, and the facades adopt natural Mediterranean style. The buildings integrate with sunshine and vitality and adopt unvarnished and warm colors, which change with shadows, showing its simplicity and humanization. Through space closure between residential units and connection of platforms on the second floor, it builds a space with the courtyard features, thereby building a “harmonious residential community” which is what really matters.

华润成都万象城

CRG The Mix City, Chengdu

项目地点：四川 成都
用地面积：46 671.33 m²
建筑面积：317 639.21 m²

Location: Chengdu, Sichuan
Site Area: 46,671.33 m²
Building Area: 317,639.21 m²

项目位于万年场双庆路与二环路交叉处的东南角，西临二环路，北临双庆路。
项目功能上集高端商业、餐饮娱乐、办公于一体，设计上利用原有的商业和城市资源，在已日趋成熟的东二环路商圈基础上，通过地产品牌效应和丰富的商业组合类型扩大和延伸了成华区的商业影响力，为未来形成更大规模的城市商业副中心积蓄了充分必要的前提条件。
本项目的空间设计充分利用成都当地地理气候和项目场地周边的有利条件，结合当地社会人文形态和生活节奏，引进国际高端办公休闲购物娱乐综合开发的成功经验，力图构建一片与全球时尚联动、贴近地方民生、高效利用空间及低碳节能的城中乐土。

The project is located in the southeastern corner of the junction of the Second Ring Road and Shuangqing Road in Wannianchang Sub-district, besides the Second Ring Road in the west and Shuangqing Road in the north.
In terms of functions, it is an urban comprehensive project that integrates high-end business, restaurants, entertainment and offices into one. In terms of design, on the basis of the maturing East Second Ring Road Business Circle, the project uses the original business and urban resources to expand and extend the commercial influence of Chenghua District through the brand effect of real estate and various business combination types, which accumulates necessary and abundant preconditions for forming larger second urban commercial center in the future.
In terms of space design, the project fully exploits the local geography, climate, project location and surrounding advantages, combines local social and cultural forms and life pace, and introduces international successful experiences on comprehensive development of high-end offices, leisure, shopping and entertainment, aiming at building a paradise which keeps pace with global fashion, gets close to local lives, efficiently uses space and nurture lives in environmental-friendly ways.

朗诗成都绿色街区

Landsea Green Residential Area, Chengdu

项目地点：四川 成都
建筑面积：85 000 m²

Location: Chengdu, Sichuan
Building Area: 85,000 m²

项目是成都第一个绿色住宅项目。项目基地位于成都市成华区，所处的迎晖路南北两侧均为高档居住区和商务中心。项目根据用地形状和所处环境布置了高层住宅及低层商业两种产品类型。住宅户型设计中以动静分区及干湿分区为设计原则。建筑体型简洁、明快，高层外立面底部采用浅咖啡色石材，顶部采用淡黄色面砖；底层外立面以淡黄色石材为主，配以少量青灰色槽钢点缀。阳台均采用玻璃栏杆，形成简约的现代建筑风格。在材质色彩上形成和谐统一的视觉效果。以淡黄色和暖灰为主的立面材质给居住者带来一种色彩温馨、亲切宜人的居家感受，同时其精致的建筑细部，增加了建筑的价值感。

The project is located in Chenghua District of Chengdu City. The surroundings of Yinghui Road where it lies are all top grade residential areas and business center. According to the land shape and environment, the project designs two product types, namely high-rise residential buildings and low-rise commercial buildings. The residential units are designed in the principles of activity zoning and dry zoning. The building shape is featured by simple and straightforward style. The bottom of the high-rise facades is made of light coffee stone materials, the top is made of light yellow face bricks and the low-rise facades are made of light yellow stone materials embellished with small amount of steel-grey channel steel. All balconies use glass railings to form modern architectural style of simplicity. Therefore the project forms a harmoniously unified visual effect in materials and colors. The façade materials mainly in light yellow and warm grey give people warm and pleasant living feelings, and at the same time the delicate building details which are dealt in extremely meticulous ways add buildings feelings of value.

首创国际城商业综合体

First City Commercial Complex

项目地点：陕西 西安　Location: Xi'an, Shaanxi
建筑面积：621 000 m²　Building Area: 621,000 m²

首创国际城商业综合体是一个集超高层办公、四星级酒店、公寓、街区式商业为一体的综合开发项目。
项目用地位于西安市北客站南侧，离市政府仅1.2km，是进入西安市的第一门户，项目规划结合甲方意图、市场环境、城市展示等多方面要求进行综合设计，力争将项目打造成本区域的地标性建筑群，为城市、为社会做贡献。
目前，本项目部分楼已封顶，各界反应均甚好。

First City Commercial Complex is a comprehensive development project which integrates super high-rise offices, 4-star hotel, apartments, and street-type business into one.
The project is located on the south side of North Railway Station of Xi'an City, only 1.2km away from local government seat. It is the front door into Xi'an City. The project is planned to make comprehensive design combining with party A's intentions, market environment, urban exhibition and many aspects, strive to be a landmark building cluster in the area and make contributions to the city and the society.
Now some buildings of the project have finished topping and the feedbacks from all communities are all good.

渝兴 黄茅坪产业基地

Huangmaoping Industrial Base, Yuxing

项目地点：重庆
建筑面积：550 000 m^2

Location: Chongqing
Building Area: 550,000 m^2

项目功能定位是打造集办公、商业、酒店、汽车产业配套于一体的大型中高端项目。
基地位于重庆渝北区南部，属于黄茅坪工业区，周边涵盖了工业、住宅、商业和园博园等多种业态。
设计目标是通过为汽车产业服务的汽车研发中心、销售中心、产业会议中心、综合办公楼、酒店、汽车产业主题商业等复合业态的有机的联系和过渡将本项目打造成区域的商业中心。这个多功能的中心将会是信息交流最密集的地方，它不停的聚集城市的社会能量，并辐射到周边更大的区域。通过人气的不断聚集和自我发展的不断更新定位，它将成为真正意义的核心区域并能带动区域经济提升。

The project is positioned to be a large and middle high-end project which integrates offices, business, hotel, and auto supporting industry into one.
The base is located in the south of Yubei District, Chongqing. It is in Huangmaoping Industry Park and the periphery covers industry, residences, business, expo gardens and many other business types.
The design aims to make the project the commercial center of the area through organic connection and transition of auto research center, sales center, industry meeting center, complex building, hotel, auto industry theme business and other composite business types which serves for auto industry. The multi-functional center will be the place with the densest information exchange. It is gathering up social energy in the city and radiating larger peripheral areas. By getting increasingly noticed and continuous updating and positioning of self-development, it will become a significant core area that drives regional economy.

扫描查看更多信息

VTR(美国)国际设计研究有限公司（中国机构）

VTR (USA) International Design Institute Ltd. (Chinese Organization)

四川中维工程设计有限公司

Sichuan Zhongwei Design Engineering Co.,Ltd.

“建筑设计，反映着每个时代科技、经济与文化的内涵。设计的意义也随着时间而演变，成为一种新的国际语言。”

VTR（美国）国际设计研究有限公司（中国机构）（简称：VTR），是一家拥有建筑工程甲级设计证书、甲级工程咨询证书的专业设计单位，并在2008年通过了国家质量管理GB/T19001–2000 idt ISO9001：2000标准认证。

VTR（美国）国际设计研究有限公司（中国机构）自2002年进入中国以来，一直致力于融合东、西方文化，强调“人”、“从”、“众”的规划设计理念，同时遵从“建筑以人为本”的人性化共识。近年来VTR又完成了一大批城市标志性建筑，如：成都九龙国际城、成都精亿名城、都英·新世纪广场、简阳东城国际、江西九江柴商国际商业广场、交大科技公园、金堂职业技师学院、莫斯科春天、斑竹园竹韵水街、重庆壹号公馆等项目。

继往而拓进，传承而创新是VTR一直坚持的建筑设计理念。创作是创造的基础，创新是创作的灵魂。VTR团队一直专注把握这一稍纵即逝的设计灵感，并结合个体项目特征，为业主们创作出一个又一个达标作品。

地址：四川省成都市高新区天府大道北段1480号9号楼F座5层

电话：+86-28-86786662

传真：+86-28-86788919

邮箱：vtr@vtr-sc.com

网址：www.vtr-sc.com

Add: 5F, Block F, Building No.9, Tianfu Avenue North 1480.Gaoxin District, Chengdu

Tel: +86-28-86786662

Fax: +86-28-86788919

Email: vtr@vtr-sc.com

Web: www.vtr-sc.com

都英 · 新世纪广场

金堂职业技师学院
Jintang Professional Technician Institute

项目地点：四川 成都　Location: Chengdu, Sichuan
用地面积：14 908.03 m^2　Site Area: 14,908.03 m^2
建筑面积：56 325.43 m^2　Building Area: 56,325.43 m^2

规划设计在原有教学大楼和实训楼的边界上进行延续与变化，达到与原有教学楼和实训楼规划的和谐统一。方案立面设计采用具有浓郁校园色彩的建筑风格，新建建筑在形体和色彩的处理上更加大气、现代，特别是对建筑主要立面的处理，加入了大量的体块穿插、咬合和虚实结合的手法，能很好地体现学校建筑的时代感。对原有建筑也作了一定的改造，为了不改变原有建筑功能和结构，对原有建筑的改造主要集中在外装材料上，使用了跟新建建筑相呼应的暗红色面砖和乳白色、深灰色涂料，使得整个学校的现代感和校园氛围更加强烈。

The design extends the boundary of the original teaching building and training building and makes some changes, to achieve a dialectic unity between the original teaching building and training building in planning. The façade is designed in strong campus architectural style. The form and color of the new building are more magnificent and modernistic, especially on the treatment of the main building façade. By means of a large number of inserts, interlocks and void-solid combinations, the design fully represents the sense of times of campus architecture. The original building has also been transformed to some extent. In order not to change the architectural functions and structure of the original building, the changes focus on exterior decorating materials. Dark red brick and milk white, deep grey paints are used to agree with the new building, making the whole campus more modernistic and making the campus atmosphere much stronger.

简阳东城国际·港湾丽都
East Town International · Harbor Lido

项目地点：四川 简阳　Location: Jianyang, Sichuan
用地面积：19 196 m^2　Site Area: 19,196 m^2
建筑面积：112 000 m^2　Building Area: 112,000 m^2

本项目致力于打造出具有浓厚韵味的中式建筑群，并赋予建筑星火相传的文化氛围，让规划既能体现当地历史文化，又具有现代化居住环境的优越性。

This project tries to build Chinese style building clusters with dense lingering charm, and endows the architecture with continuing cultural atmosphere. The project is planned to present the local historical culture, and have advantageous modern living environment.

重庆壹号公馆
No. 1 Residence, Chongqing

项目地点：重庆　Location: Chongqing
用地面积：14 356.38 m^2　Site Area: 14,356.38 m^2
建筑面积：135 073.68 m^2　Building Area: 135,073.68 m^2

项目位于重庆市大足区双桥经济技术开发区，交通便捷。
建筑立面设计采用英伦风格，通地建筑退台、坡屋面等建筑手法使立面层次丰富、进退有序。采用老虎窗、欧式铁花等元素丰富建筑立面的细节。同时通过石材、面砖的合理设计搭配，建筑整体色彩以暗红、金色为主，坡屋面则使用灰色，以体现雍容华贵的建筑形态。

This project is located in Shuangqiao Economy & Technology Development Zone, Dazu district, Chongqing, and the traffic here is smooth and convenient. The façade is designed in English style, by means of retiring, sloping roof etc. to make the façade variable and orderly. The façade is enriched by dormant window, European iron openwork and other details. Meanwhile the building is mainly dark red and golden through the reasonable design and arrangement of stone materials and bricks, while the sloping roof is grey, to represent the decent and stylish architecture.

成都九龙国际城

Chengdu Kowloon International Town

项目地点：四川 成都
用地面积：46 549.88 m²
建筑面积：291 835.17 m²

Location: Chengdu, Sichuan
Site Area: 46,549.88 m²
Building Area: 291 835.17 m²

该项目建筑采用半围合的布局方式，分为A、B两区，它们之间由一条U形商业步行街串联，形成围合互通式街区规划。考虑到双华路为主要人流进入的主要道路，故将U形步行街的入口以及两个大型商业中心布置在双华路北侧，起到吸引人气的作用。步行街两侧减少建筑形体本身的遮挡，最大可能地保证可视性以及人流可达性。两个大型商业中心之间由天桥相连，完善购物流线。

The project adopts the half-closed layout, divides into A&B sections that are connected by a U-shaped commercial pedestrian street in the form of enclosing yet interconnecting block planning. Considering Shunghua Road as the main road for major pedestrian flow to enter, the architects intentionally put the entrance to the U-shaped pedestrian street and two large commercial centers on the north side of Shuanghua Road so as to raise the popularity. At either side of the pedestrian street, the obstruction from the building shape is minimized to ensure the visibility of the buildings and the accessibility of pedestrian flows. The two commercial centers are connected by an overpass, completing the shopping streamline.

斑竹园 · 竹韵水街

Mottled Bamboo Garden · Bamboo Rhyme Water Street

项目地点：四川 成都
用地面积：61 518 m²
建筑面积：16万 m²
容 积 率：2.6

Location: Chengdu, Sichuan
Site Area: 61,518 m²
Building Area: 160,000 m²
Plot Ratio: 2.6

本项目作为重要的城市标本，承载着巨大的历史使命。以独特的码头水文化、竹文化为引领，以水林生态为基底，以川西水街建筑为特色，以高端餐饮娱乐业态为核心"，实现文化、生态、形态、业态的四态完美融合。

营造现代中式风格的情景街区。灰砖与玻璃和钢的对比塑造了硬朗、光洁、通透、利落而不失趣味的建筑个性和文化品味。引入现代商业和文艺业态，打造心灵沉淀之地，创意灵感之地，品味生活之地，引领文化消费和时尚消费。

This project, as an important urban representative, carries a huge historical mission. "Leading with the unique harbor water culture, bamboo culture", "based on the water & forest ecology", "featuring West Sichuan water street architecture", "centering on the hi-end F&B entertainment industry", the project realizes the perfect integration of the culture, ecology, morphology and industry.

The project builds modern Chinese-style scenario blocks. The contrast between gray bricks and glass & steel creates such a strong, bright, penetrating, agile yet interesting architectural character and cultural taste. The modern commercial industry and the cultural industry are introduced to make it a place for soul purification, inspiration source, life taste, and lead cultural consumption and fashion consumption.

九江柴桑国际商业广场

Chaisang International Commercial Square, Jiujiang

项目地点：江西 九江
用地面积：3967.62 m²
建筑面积：53 924.86 m²

Location: Jiujiang, Jiangxi
Site Area: 3,967.62 m²
Building Area: 53,924.86 m²

设计过程中用了大量"方体块"，通过这些体块穿插、交错、凹凸形成建筑丰富的体量组合关系，使得该建筑体量变化丰富、空间层次清晰。通过对塔楼大面积的玻璃幕墙的设计形成与建筑裙楼实墙的虚实对比，从而使建筑轻盈、通透、灵动，给人以与时俱进、舞动青春的建筑形象；在材质及色彩上与地方建筑相呼应，以使得该建筑充分融入周边环境中。

The design utilizes massive "square masses", and these masses interlay, stagger, and bump, forming a variable combination of building volumes, to enrich the building volume variations and clarify the space layers. The contrast between the tower designed with large glass curtain wall being void and the annex designed with solid walls being solid gives the building a light, penetrating, agile, up-to-date and young image, and the materials and colors of the buildings coordinate with the local architectures so the project integrates well into the peripheral environment.

成都精城国际
Chengdu Elite International

项目地点：四川 成都
用地面积：8040.11 m^2
建筑面积：38 369.70 m^2

Location: Chengdu, Sichuan
Site Area: 8,040.11 m^2
Building Area: 38,369.7 m^2

项目位于成都市金牛区三环路羊犀立交外侧。
在设计中，充分利用基地特征与基地产生对话，顺应基地和城市肌理，力争使设计与城市空间相匹配，形成内外空间相联系的城市景观。地上地下通盘考虑，合理组织人流、车流，使内外交通简捷、畅通、互不干扰。
在建筑立面设计上，强调竖向线条，使建筑呈现出强劲的挺拔感与力量感，建筑整体色调呈暖灰色。为使建筑达到新颖、活泼、雅致的效果，立面利用玻璃幕墙和铝塑板幕墙的虚实对比，体现现代建筑的艺术性格与科技感，使建筑更具现在艺术魅力。

The project is located outside of the Yangxi Flyover on the Third Ring Road, Chengdu City.
The design adequately utilizes the base features, follows the relation between the base and the city, and tries to match up with the urban space, so the urban landscape is linked to the inner and outer spaces. In specific design, all aboveground and underground spaces are taken into consideration to appropriately arrange the pedestrian flow and traffic flow so that the inner and outer transportations are respectively simple and direct and smooth.
The design for building façade emphasizes on vertical lines, which gives the building a strong sense of uprightness and power, and the overall tone of the building is warm gray. In order to have the fresh, active and elegant effect, the façade adopts the contrast between glass curtain wall being void and aluminum-plastic panel curtain wall being solid, presenting the artistic style and technological sense of modern architecture, and standing it out with better modern artistic charms.

西南交通大学科技公园
The Science and Technology Park of Southwest Jiaotong University

项目地点：四川 成都
用地面积：145 497.93 m^2
建筑面积：1 544 271 m^2

Location: Chengdu, Sichuan
Site Area: 145,497.93 m^2
Building Area: 1,544,271 m^2

项目位于九里堤片区西南交通大学以西，府河以东，处于成都“北改”片区“西”与“北”的交接地带。本方案从城市空间构成角度入手，深入研究城市肌理和周边现状，结合商业办公酒店等业态，打造集休闲、娱乐、购物为一体的大型城市综合体。
在外立面设计上采用高技派建筑风格，钢结构与混凝土技术相互结合使建筑表皮现代、新颖。酒店全新玻璃幕墙设计，将水流艺术形态和现代建筑技术完美结合，将科技时尚主题体现得淋漓尽致。

This project is located in Jiulidi block, west of Southwest Jiaotong University and east of Fu River. It is in the transition zone of the west part and north part of Chongqing's "north-reform" block. This project starts from the structure of city spaces and researches into the city texture and peripheral conditions. The project aims at creating a large-scale metropolitan complex of relaxing, entertaining and shopping.
The façade is designed in the High-Tech architectural style. The surface of the building is modern and novel, and steel structure is combined with the concrete technology. The modern, high-tech and fashionable design style represents the character of times. The brand-new design of glass curtain wall in hotels perfectly combines the artistic form of water flow and modern building technology, which fully represents the theme of technology and fashion.

莫斯科春天
Spring in Moscow

项目地点：四川 成都
用地面积：24 677 m^2
建筑面积：533 006 m^2

Location: Chengdu, Sichuan
Site Area: 24 677 m^2
Building Area: 533,006 m^2

本项目位于成都一环路和解放北路交汇处，隶属于成都"北改"项目金牛区华西集团片区地块。项目定位为融主题购物中心、高端住宅、俄罗斯风情商业街为一体的大型城市综合体项目。
本案外立面以俄罗斯风情为主题，通过对俄罗斯建筑风格的深入研究及对其建筑形式的高度提炼，将古典建筑元素和现代建筑技术完美融合并在办公楼、酒店及住宅中充分展现。商业街采用较为复古的外立面处理手法，打造具有文化氛围的商业街区。整体风格现代中体现俄罗斯古典神韵，成为成都"北改"项目中最具特色的商业文化新地标。

This project is located in the intersection of Chengdu First Ring Road and Liberation North Road, and the land belongs to Huaxi Group's land of Chengdu's "North-reform" project in Jinniu District. The project is designed as a large-scale metropolitan complex of theme shopping mall, high grade residence and Russian style business street.
The façade of this project is themed by Russian style. After a deep research of Russian architectural style and a high refinement of Russian architectural form, the classic architectural elements and modern architectural technology are perfectly integrated. The project simplifies and refines the typical Russian architectural elements, which are fully displayed in office buildings, hotels and residences. The business street adopts a relatively ancient treatment of façades to create a business block full of culture. The integrated style represents modernity as well as classic Russian taste, making the project the most special business culture landmark of Chengdu's "North-reform" project.

都英·新世纪广场
Duying · New Century Square

项目地点：四川 内江
用地面积：65 563.36 m^2
建筑面积：272 640.54 m^2

Location: Neijiang, Sichuang
Site Area: 65,563.36 m^2
Building Area: 272,640.54 m^2

新世纪广场项目位于成渝经济区中心内江市隆昌县黄土坡工业园，地块交通便捷，直接与隆昌县主城区相连。
在住宅功能区的设计上，采用围合式大中庭布局，将主要人行入口设置在重庆路一侧，形成超大中庭景观空间，有效地提高了小区品质。建筑体量处理充分结合城市肌理，对建筑的层数进行适当的处理，形成错落有致、构成丰富的城市天际轮廓线。在设计中采用与片区规划建筑风貌相协调的新古典主义建筑风格。通过挺拔有力的竖向线条、凹凸有秩的立面以及纯正而不失现代感的古典元素，有效地提升了项目自身品质与居住环境。

The project is located in Huangtupo Industry Park in Longchuang County, Neijing City which is the center of Chengdu-Chongqing Economic Area. The land is directly connected to the downtown of Longchuang County and the traffic here is smooth and convenient.
Residential functions are designed in an enclosing courtyard layout. The main pedestrian entrance is set on one side of Chongqing Road, to form a huge courtyard landscape space that effectively improves the quality of the residential community. The design fully combines the building volume with the city structure and properly controls the number of building floors, to form a diversity of city skylines in picturesque disorder. The Neoclassic architectural style is adopted in the design, which agrees with the planned architectural style of the region. Strong vertical lines, façades with orderly concave and convex, and pure modern classic elements effectively enhance the quality and living environment of this project.

扫描查看更多信息

四川国鼎建筑设计有限公司

SICHUAN GUODING ARCHITECTURE DESIGN CO.,LTD.

四川国鼎建筑设计有限公司是一家具有国家住房和城乡建设部颁发的建筑行业（建筑工程）甲级资质、市政行业（给水、排水、道路、环境卫生工程）专业乙级资质、城市规划编制乙级资质且通过ISO9001质量认证的企业，从事资质证书许可范围内相应的建筑工程总承包业务以及项目管理和相关的技术与管理服务。

公司定位为从事技术型现代服务行业的企业。以"为用户提供有价值的服务"为企业宗旨，以"获得用户最高满意度"为服务目标，以"建筑历史经典"为公司的质量使命。四川国鼎用专业的眼光、创新的思维面对每一项任务，四川国鼎也始终换位于用户的立场，以用户的利益引领我们的服务。四川国鼎急用户所思，行用户所愿。

在公司完善的专业配备下，主要开展城市规划、民用建筑、公共建筑、工业建筑、园林景观、建筑装饰、市政（给水、排水、道路、环境卫生工程）等设计类服务，以及项目管理、项目代建等技术咨询类服务。

地址：四川省成都市高新区科园南路88号
天府生命科技园A1办公楼905室
电话：+86-28-85315296/7/8
传真：+86-28-85315296-818
邮件：zggdjz@126.com
网址：www.zggdjz.com.cn

Add: Room 905, Life Science Park Office Building A1, High-tech Zone South Road No.88, Chengdu, Sichuan
Tel: +86-28-85315296/7/8
Fax: +86-28-85315296-818
E-mail: zggdjz@126.com
Web: www.zggdjz.com.cn

科创岳池酒店

Kechuang Yuechi Hotel

设 计 师：罗永春
项目地点：四川 成都
用地面积：61 535.1 m²
建筑面积：294 979.32 m²

Designer: Yongchun Luo
Location: Chengdu, Sichuan
Site Area: 61,535.1 m²
Building Area: 294,979.32 m²

建筑设计理念及特色是以中式院落作为主体，将各栋建筑功能通过院落进行有机组合。

Architectural design philosophy and feature: The main body of this design is a Chinese-style courtyard, and all buildings are integrated by the courtyard.

新鹿医药产业园

Xinlu Medicine Industry Park

设 计 师：罗永春
项目地点：四川 都江堰
用地面积：128 735.25 m²
建筑面积：214 591.34 m²

Designer: Yongchun Luo
Location: Dujiangyan, Sichuan
Site Area: 128,735.25 m²
Building Area: 214,591.34 m²

项目规划采用基本对称的大中轴线贯穿的布局形式，以一条中轴线和一条次要轴线将整个区域划分为三大板块，办公部分临近道路，其余为药业生产与动物养殖。园区内各栋建筑，通过过街楼式的廊道相连接形成组团，各组团区域独立存在、互不干扰，方便管理，从而满足了每个区域独特的需求条件。
整个厂区采用看似规整却灵活自然的布局形式，利用该地块独特的水资源打造中心湖泊式的超大景观带，以水为中心布置，围和成不同层次的广场和院落。
设计中充分利用地形，运用借景、对景、分景、隔景等多种手法来组织空间，形成园区中曲折多变、小中见大、虚实相间的景观艺术效果。

In the project planning, the major central axis penetrates the layout in symmetry. The central axis and a secondary axis divide the city into three blocks. The office block is near roads, and the rest blocks are for medicine production and animal breeding. In the park, all buildings are clustered with arcade-style corridors. The clusters are independent and easy to manage, more catering to specific requirements of every cluster.
The whole plant area is in a regular yet flexible layout. Its unique water resource is built into a central lake as an overlarge landscaping belt, and this water-based deployment encloses different levels of plazas and courtyards.
The design makes full use of landforms, and a variety of manners like sight borrowing, contrasting, division and separation, to create a landscaping art of winding changes, small to large analog, and virtual-real interval in the park.

都江堰逸岭滨江
Yiling Riverside, Dujiangyan

设 计 师：罗永春
项目地点：四川 成都
用地面积：26 500 m²
建筑面积：174 000 m²

Designer: Yongchun Luo
Location: Chengdu, Sichuan
Site Area: 26,500 m²
Building Area: 174,000 m²

项目位于成都市都江堰新城区，其产品包括五星级酒店、高端住宅等业态。
本方案采用多种艺术装饰风格，配合现代建筑的体量及新型材料，使得酒店简约而华贵，住宅大气且雍容。该项目整体布局灵活、通透，最大化地保证住宅视线丰富且开阔，酒店的地标性强烈，为城市形象起到很好的提升作用。
该项目虽为地产开发项目，但方案却最大化地保持了原始地貌及自然环境，最大化地利用了河道及河道周边环境。

The project is located in Dujiangyan New Town of Chengdu City. Its products include 5-star hotel, hi-end residential building and other properties.
The design adopts Art-Deco manners, together with modern building volume and new materials, to make the hotel concise and luxury, the residential buildings are spacious and graceful. The integral layout is flexible and transparent, to ensure the best sights and broad vision of the residential area, and the hotel as an obvious landmark will better improve the city image.
Although this project is a development project, the design reserves the original landform and natural conditions, and utilizes the river and surrounding environment to the largest extent.

酒店正立面、侧立面图

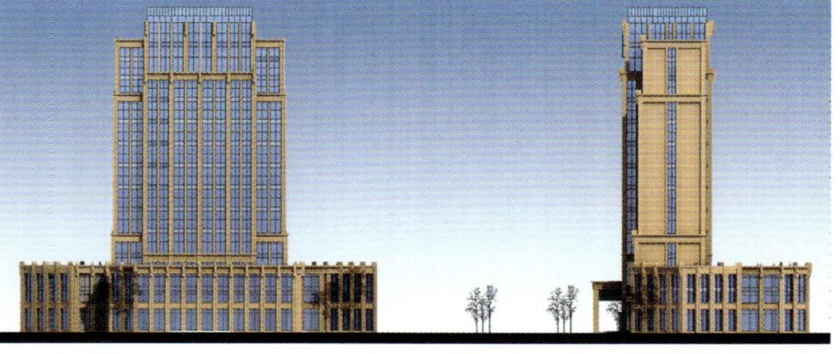

酒店背立面、侧立面图

住宅立面图

成都龙潭寺
Longtan Temple, Chengdu

设 计 师：刘洪
项目地点：四川 成都
建筑面积：200 000 m²

Designer: Hong Liu
Location: Chengdu, Sichuan
Building Area: 200,000 m²

成都龙潭寺是含五星级酒店、综合商业、办公、商务公寓等于一体的城市综合体，作为超高层建筑，建筑最高高度约为120 m。

LongTan Temple is a city complex integrating 5-star hotel, commercial, office, apartment hotel and others. As an over-high-rise building, it is about 120 m high.

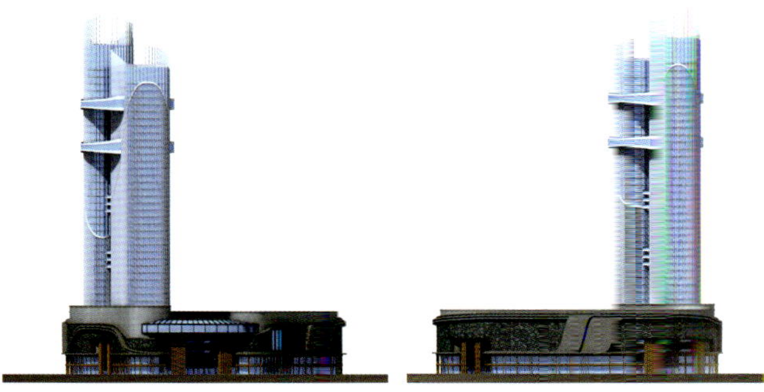

武侯科技创业城

Tech Entrepreneurship Town, Wuhou

设 计 师：刘洪
项目地点：四川 成都
用地面积：41 052.58 m²
建筑面积：32 409.61 m²

Designer: Hong Liu
Location: Chengdu, Sichuan
Site Area: 41,052.58 m²
Building Area: 32,409.61 m²

规划设计中根据国家有关政策、法规及建设方的要求，结合用地条件，致力于创造一个适用、经济、美观、现代化的居住环境。本方案首先保证城市规划的要求，并满足建设单位的需要，根据单体不同的功能，进行合理的分区；结合场地内现有条件进行规划，使建筑组群呈现良好空间效果。

各单体设计均沿袭现代建筑设计手法，以功能为主导，建筑空间为设计主体，平面紧凑、高效、合理；立面简洁，使用现代建筑材料，体现了建设单位所从事行业的性质和内涵。

The planning design, according to state policies, statutes and builder's requirements, considering land conditions, seeks for a suitable, economical, beautiful and modern habitat. This project will satisfy urban planning requirements, and satisfy builder's requirements, to make reasonable zoning at different functions; the project will make conditional planning with onsite conditions, to show good space effects of building cluster.

Every unit design will follow modern architectural design manner, dominated by functions. The building space is the main body of design. The layout is compact, efficient, and reasonable; the façade is concise, with modern building materials, to represent builder's professional nature and significance.

同聚和润国际

Tongju Herun International

设 计 师：刘洪
用地面积：72 500.33 m²
建筑面积：51 282.07 m²

Designer: Hong Liu
Site Area: 72,500.33 m²
Building Area: 51,282.07 m²

项目周边的环境包括自然环境、人文环境和市场环境等诸多方面。设计既要满足现实的市场需要，也要满足更高层次的设计要求，既使功能与周边环境相契合，又能起到应有的先导性。

对于“特色”的解释：一是通过合理规划，使小区的整体环境具有鲜明的形象和品位特征；二是通过精心设计使各小区与周边的住宅区相比，在建筑形式、户型设计上具有特色；三是在小区中引入符合信息时代特征的新技术、新设备，赋予整个小区设计以时代特征。通过对建筑单体、空间形式的创意和新技术的运用，在整体格调统一的前提下，使形体创造更具魅力，强调住宅环境与建筑、单体与群体、空间与实体的整合性。

The environment includes natural environment, cultural environment and market environment among many other aspects. The project will satisfy actual market demands, and satisfy higher level design requirements as well, to integrate the functions with surrounding environment, with due leading effects.

Interpretation of “feature”: 1) Through reasonable planning, the integral environment of the community has distinct image and characteristic; 2) Through elaborate design, the project is different from surrounding residential communities, both in architecture and unit design; 3) By introducing new technology and new equipment featured by IT epoch into the community, the community design is bestowed with epochal features. With single building creativity, space creativity and the application of new technology, new process, new material and new equipment, in unified style, the shape creation is more charming, with great emphasis on the integration between residential environment and architecture, between single building and building cluster, and between space and objects.

成都圣采置业2号地块

Plot 2 of Shengcai Property, Chengdu

设 计 师：罗永春
项目地点：四川 成都
用地面积：61 535.1 m^2

Designer: Yongchun Luo
Location: Chengdu, Sichuan
Site Area: 61,535.1 m^2

本次设计既注重对新型城市居住景观社区的打造，力求做到住宅舒适性与城市景观的双向升华，又要注重对小区内部的合理设计，使整个小区的品质得到全面的提升。总体采用围而不合的布局，经过规划设计，将本项目打造成为集自然、艺术、和谐、舒适为一体的高品质居住小区。

小区内宜人的绿化空间、舒适的人行路径、别致的景观小品贯穿其中，商业采用大商业形态与临街商业相结合的形式，营造浓厚的商业氛围，为城市更添一抹时尚的色彩。

This design tries to build new city residential landscaping community, to improve both residential comfort and city landscaping on one hand, and make reasonable design inside the community, to improve the overall quality. The general layout is enclosing but not closed, and the project plans to build a residential community integrating the nature, arts, harmony and comfort.

Greening space, comfortable pedestrian path and special landscaping works penetrate the community. Large commercial patterns combine with roadside commercial buildings to create dense commercial climate, and add some fashionable colors to the city.

设 计 改 变 世 界

扫描查看更多信息

天津大学建筑设计规划研究总院
Tianjin University Research Institute of Architectural Design & Urban Planning

天津大学建筑设计规划研究总院创建于1952年，是国家甲级建筑设计、甲级城市规划设计、甲级文物保护工程勘察设计、甲级工程咨询及甲级旅游规划单位，天大设计总院依托国家重点大学人才优势、学科优势和技术优势，凭借先进的技术装备和严谨的工作作风，近年来在全国范围内完成的各种类型、各种规模的建筑设计、城市规划设计中屡获各类奖项，受到项目委托单位及社会各界的好评。

天津大学设计总院现有一线设计人员470余人，其中45人为国家一级注册建筑师，35人为国家一级注册结构师，17人为国家注册城市规划师，拥有硕士学位的员工占全院总人数的40%以上。天津大学设计总院专业配套齐全，人员素质较高，在教育和科研建筑、旅游和疗养建筑、办公和商贸建筑以及大型医院建筑、体育建筑设计和城市规划、古建筑保护规划等方面都有不俗的成就。

地址：天津市南开区鞍山西道192号
邮编：300073
电话：+86-22-27404753
传真：+86-22-27401845
邮箱：td_design@126.com
网址：www.aatu.com.cn

Add: No.192, West Anshan Road, Nankai District, Tianjin
P.C.: 300073
Tel: +86-22-27404753
Fax: +86-22-27401845
E-mail: td_design@126.com
Web: www.aatu.com.cn

1895天大建筑创意大厦
1895 Architecture Innovation Building of Tianjin University

项目地点：天津
建筑面积：37 000 m²

Location: Tianjin
Building Area: 37,000 m²

项目位于鞍山西道科贸街东部，南临鞍山西道，东临卫津路，西临天大天财公司，北临天津大学北五村住宅区。项目重点在建筑行业再生能源利用、生态建筑技术转化、建筑科技咨询、建筑新技术方面进行研究及推广应用；引进如建筑规划软件开发公司、高端效果图制作公司等创意产业；引进与建筑规划创意产业相关的正版化设计软件，构建软件服务平台，充分实现网络资源高度共享；实现包括建筑项目前期策划、立项咨询、科研编制、方案设计、项目代建、工程招标代理、房地产评估、环境评价、节能设计等环节在内的建筑创意产业"一条龙"服务，形成高技术产业集群，同时，加强国际交流合作，提高天津建筑水平。此外，大厦中将设有国际开放工作室、教授工作室和艺术沙龙等，使之成为创意产业的孵化器，成为天津大学产学研的重要基地。

The project is located in the east of Science and Trade Street, Anshan West Road, to the west of Anshan West Road, to the east of Weijin Street, to the west of Genius Co., Ltd., and to the north of Tianda North Fifth Village residential community. The planning emphasizes the research, promotion and application of the renewable energy, eco-construction techniques, architectural technology consulting and new architectural technology in construction industry. It's aimed at introducing creative industries such as IT companies in architectural planning software development, companies in high-end architectural renderings drawing. We also need to introduce legal design software related to creative industries in architectural planning, establish a software service platform and fully realize the sharing of network resources. We're planning to provide one-stop service in creative architectural industries covering preliminary planning, project consulting, research team building, planning & design, project contracted, project bidding agency, real estate evaluation, environmental evaluation and energy-saving design etc. and form a high-tech industrial cluster. Meanwhile, we will strengthen international communication and cooperation to improve Tianjin's architectural design skills. Moreover, inside the building, there will be international open studio, professor's studio and art salon etc. to make this project become the incubator for creative industries and an important base for production, education and research.

天津工业大学新校区行政中心

Administrative Center in New Campus of Tianjin Polytechnic University

项目地点：天津　　Location: Tianjin
建筑面积：18 000 m^2　　Building Area: 18,000 m^2

顶部采用横向钢结构横杆遮阳，与走道的栏板形成丰富的光影效果。
外立面墙面上竖线条的窗，使建筑显得厚重挺拔。
凹进的窗洞与主体清水混凝土砌块墙面形成虚实对比，使建筑具有较强的雕塑感。
项目获2012年天津市“海河杯”优秀设计一等奖。

The project uses a horizontal steel structure beam on the top for sunshade, together with the tailgate in the aisle to form a rich light & shadow effect.
The mullion windows on the exterior façade wall make the buildings look tall and straight.
The recessed window openings form a strong virtual-real contrast with the bare concrete main walls, to create a strong sense of sculpture.
The project won the First Prize in 2012 Tianjin “Haihe Cup” Excellent Design

天津广东会馆修缮设计

Renovation of Guangdong Clan Association, Tianjin

项目地点：天津　Location: Tianjin

设计始终坚持“保护为主、抢救第一、合理利用、加强管理”的前提，尊重文物本体的“真实性”，整修设计方案遵循“最小干预”、“可逆性”的原则，最大限度延续和呈现文物本体的历史信息。
项目获2012年天津市“海河杯”优秀设计一等奖。

The design has always been sticking to the premise of “protection priority, rescue first and strengthening management” and respecting the “authenticity” of cultural relics. In accordance with the principles of “minimal intervention and reversibility”, the renovation planning tries to deliver and present the historical information of cultural relics as much as possible.
The project won the First Prize in 2012 Tianjin “Haihe Cup” Excellent Design

无锡灵山禅修中心

Ling Mountain's Meditation Center, Wuxi

项目地点：江苏 无锡　　Location: Wuxi, Jiangsu

建筑面积：8867.73 m^2　　Building Area: 8,867.73 m^2

项目获教育部2011年度优秀设计评选一等奖。

Awards: The First Prize in "2011 Excellence Design Competition by Ministry of Education"

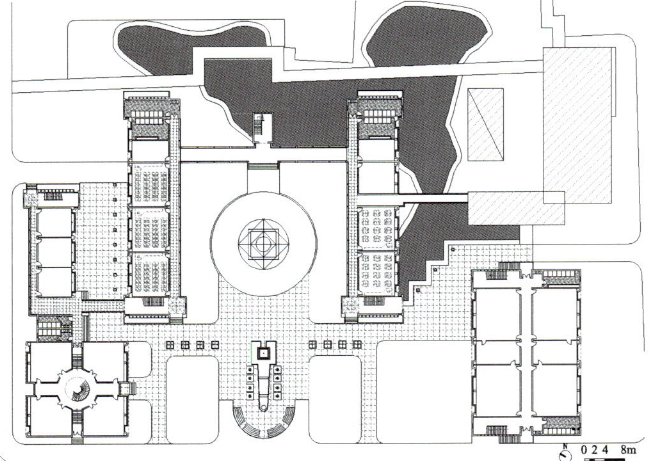

教学区首层平面图

舟山市沈家门小学
Shenjiamen Primary School, Zhoushan

项目地点：浙江 舟山
建筑面积：8818 m²

Location: Zhoushan, Zhejiang
Building Area: 8,818 m²

项目获全国第十一届工程设计项目评选银奖、中国第三届建筑学会建筑创作评选优质奖、教育部优秀勘察设计评选二等奖、建设部优秀设计评选二等奖、浙江省建设工程钱江杯评选优质工程奖
浙江省舟山市沈家门小学是全国五百强示范小学之一，由彭一刚院士设计。

Awards: Silver Medal for the 11th National Engineering Design Competition, Excellence Prize in the 3rd Architecture Creation Award of Architectural Society of China, Second Prize Excellent Exploration Design by Ministry of Education, Second Prize Excellent Design Competition by Ministry of Construction, Excellent Project Prize in Construction Project Qianjiang Cup of Zhejiang Province
Shenjiamen Primary School, one of National Top 500 Primary Schools, is designed by Academician Peng Yigang.

整个建筑群体的设计新颖独特，建筑风格现代、简洁、典雅；充分体现高新技术产业开发区的特点；建筑实现了与周边环境的互动，营造出了具有区域标志性的建筑群体。
郑板桥诗云："新竹高于旧竹枝，全凭老竿为扶持；明年再有新生者，十丈龙孙绕凤池。"鉴于本工程是青岛高新区作为起步区的建筑，其建筑形象对后续建设的项目起到引导作用。立面造型以竹为母题，取竹子"雨后春笋"、"节节高升"、"蓬勃向上"的美好寓意。
项目获2012年天津市"海河杯"优秀设计三等奖。

The building cluster has a creative and unique design with a modern, simple and elegant architectural style, which fully reflects the features of high-tech industrial development zone. We also realize the interaction between architecture and surroundings to create a cluster of landmark buildings.
There is a Zheng Banqiao's poem: With new bamboos outgrowing old ones, whereas old branches support new ones, as newcomers may shoot up again, then bamboo groves surround the pond". Since this is the first batch of buildings in Qingdao high-tech zone, its architectural image will play a leading role in the following projects. Bamboo is the theme of the façade design, delivering a good wish of "abundant (bamboo shoots after spring rain), ambitious (nodes on the bamboo are higher and higher), unyielding (bamboos stand tall and straight).

The project won the Third Prize in 2012 Tianjin "Haihe Cup" Excellent Design

青岛创业中心
Qingdao Incubation Center

项目地点：山东 青岛
建筑面积：107 500 m^2

Location: Qingdao, Shandong
Building Area: 107,500 m^2

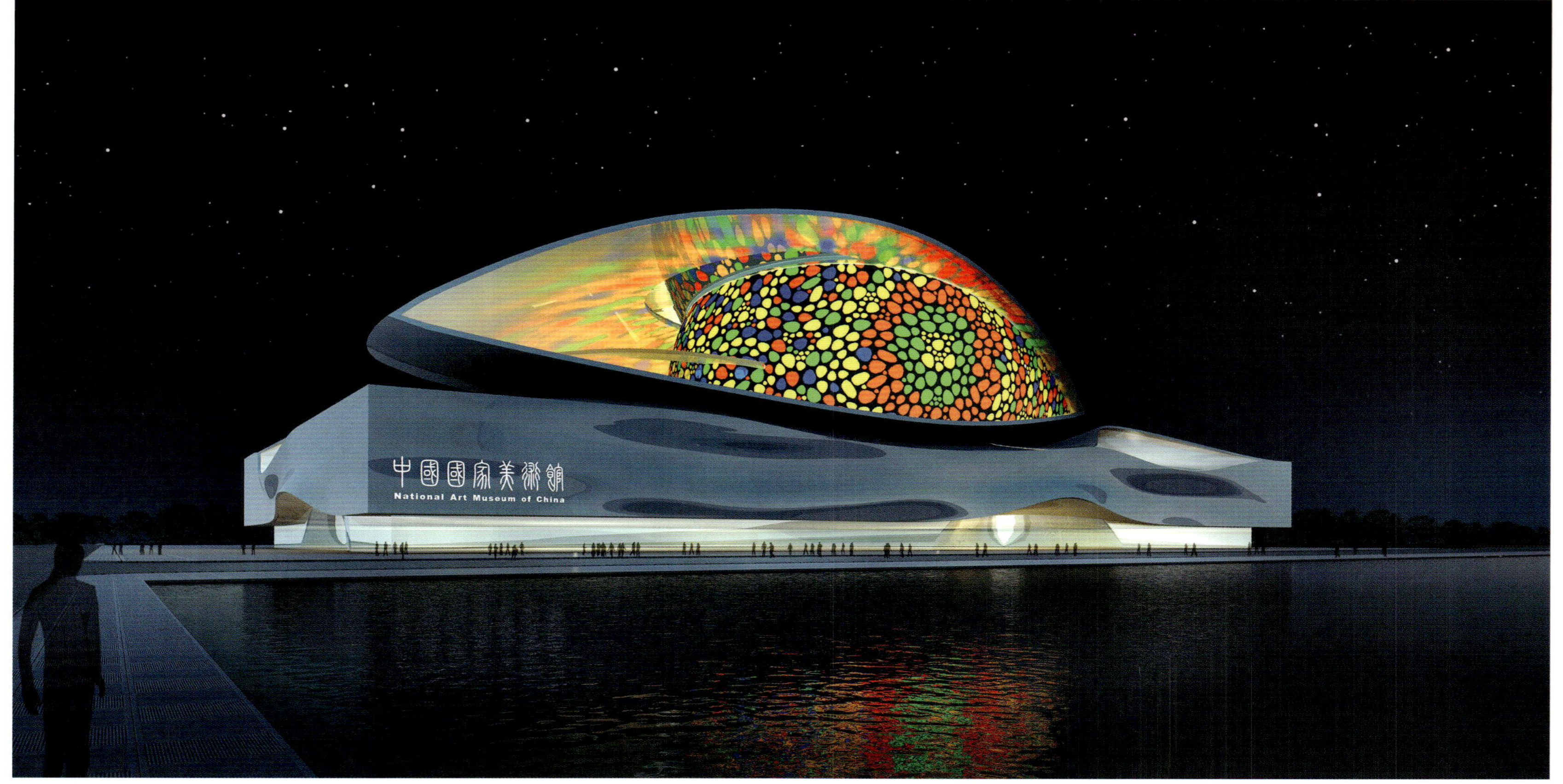

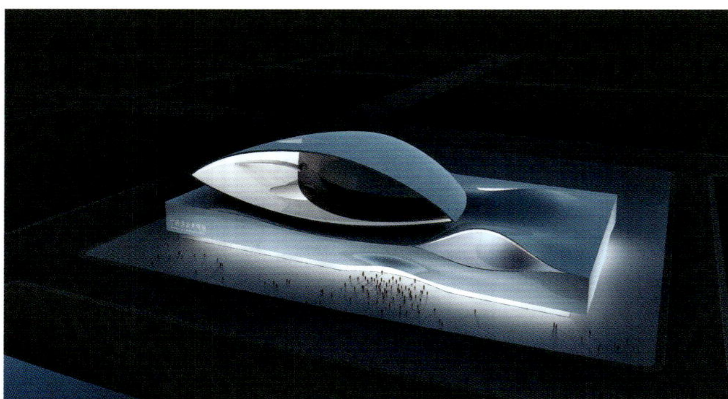

中国国家美术馆

National Art Museum of China

项目地点：北京
建筑面积：128 600 m²

Location: Beijing
Building Area: 128,600 m²

中国国家美术馆是美术展示的最高殿堂，设计师希望建筑本身能够体现人类欣赏美术作品时从视觉冲击到心灵沟通的感知过程，体现人类永不停息地发现美，寻找美的心灵冲动。
眼睛是心灵的窗口，许多卓越的艺术家对此有着精彩的诠释。"黑眼睛"，具象一个简单而丰富的"看"的概念，形成了中国国家美术馆设计鲜明的形象。

National Art Museum of China is the best palace for art exhibition. We hope that it can reflect the perception process from visual impact to spiritual communication when human beings are appreciating those fine arts. It can reflect human desire to seek for and find beauty unremittingly.
Eyes are the window to the soul. As regards to this metaphor, many distinguished artists have an excellent interpretation.simple yet rich content concept of "look" into a distinctive image designed by National Art Museum of China.

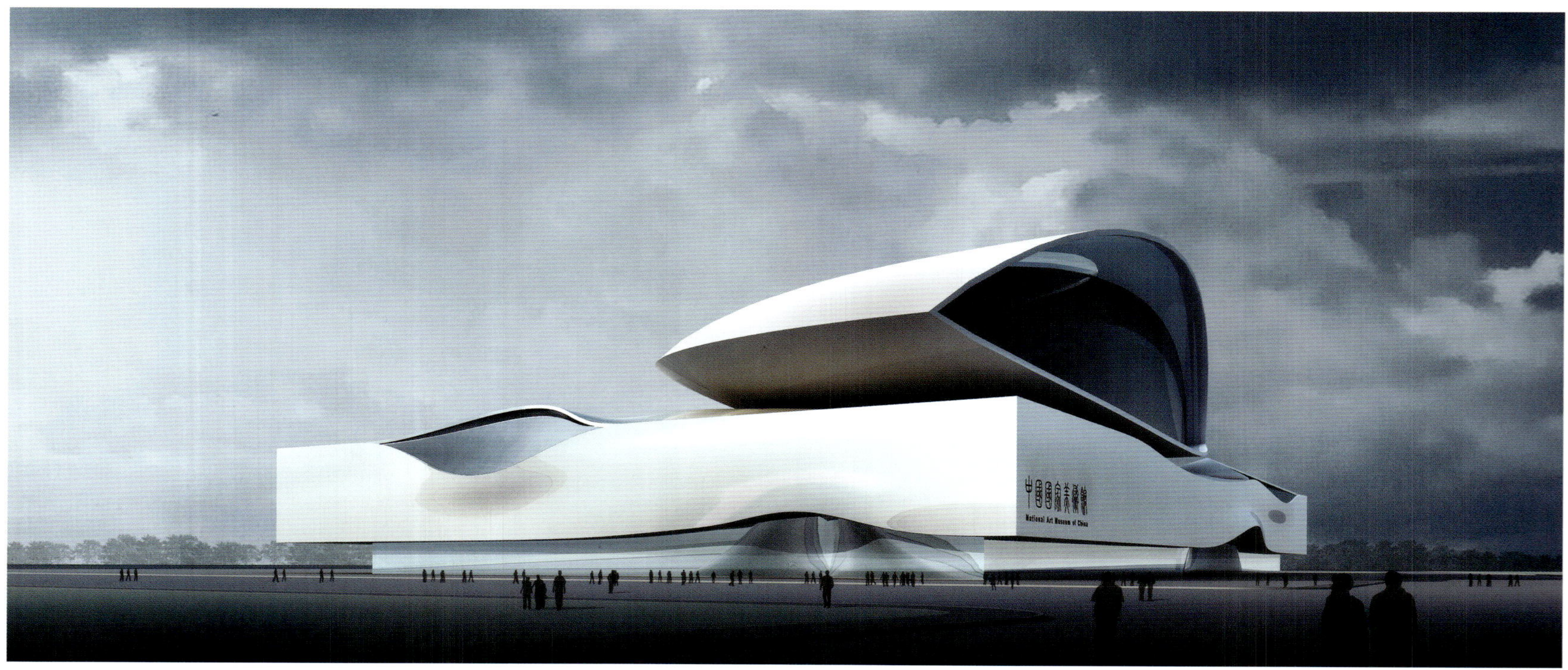

天津市环湖医院迁址新建项目

Relocation for Tianjin Huanhu Hospital

项目地点：天津
建筑面积：109 538 m²

Location: Tianjin
Building Area: 109,538 m²

新建医院拟在用地内建设一座集医、教、研、康复为一体的医疗建筑。为突出环湖医院的特点，体型流畅、简约、大气，设计中以横线条为主，建成后将成为为患者及职工提供最佳环境的现代化专科医院。

The project is planning to build a new hospital within the designated area for medical treatment, teaching, research and rehabilitation. In order to highlight the characteristics of the lakeside hospital, the design embodies simplicity and elegance mainly with horizontal lines. In the future, it will become a modern specialized hospital to provide the best conditions for patients and workers.

沧州市京沪高铁站前综合楼

Comprehensive Building in Front of Cangzhou Station of Beijing-Shanghai High-speed Railway

项目地点：河北 沧州
建筑面积：33 000 m²

Location: Cangzhou, Hebei
Building Area: 33,000 m²

沧州像海鸥一样依海而生，又拥有一双可以自由翱翔的翅膀。设计以此为概念，形体简洁、大方而不咄咄逼人。立面强调水平线条与韵律感，力求和谐地与站前广场融为一体，为旅客提供亲切的交流空间。厚重石材像极了海鸥的双翅，象征着这座追求自由的城市的信念与力量，与玻璃幕墙形成了强烈的虚实对比，塑造了与时俱进的时代感，水平线条的无限延伸感让人联想到当代铁路高速高效的特点的同时也暗示着沧州市无限的发展潜力。
项目获河北省“十佳”公共建筑奖。

Like a seagull, Cangzhou City embraces the sea with a pair of wings free to fly, which is our planning concept. The architectural form is very simple and generous but not rough. In façade design, this project puts emphasis on horizontal lines and rhythm. We strive to integrate façade design into the square in front of the station so as to provide space for travelers' communication. The heavy stone materials bear a strong likeness to the wings of a seagull, which symbolize this city's faith and power in pursuit of freedom. In contrast with glass curtain wall, it shapes a strong virtue-real comparison and a sense of contemporary times. The horizontal lines in a seemingly unlimited extension remind us of the modern railways' high-speed and high-efficiency and also suggest that Cangzhou City has an infinite development potential.
The project won the Hebei "Top Ten" Public Works Prize.

土地利用规划图

空间结构规划图

滨州北海经济开发区2011—2030城市总体规划

Overall Urban Plan of Beihai Economic Development Zone, Binzhou

项目地点：山东 滨州　Location: Binzhou, Shandong
用地面积：504 km^2　Site Area: 504 km^2

本规划以港、城、区联动为思路，围绕港口打造滨州北部新的经济增长点。以港口为龙头，产业为核心，生活配套设施为补充，塑造滨州高效、生态示范产业区。依托套尔河、新河、潮河、郝家沟等4条主要水系，打造新区特色景观廊道，形成水城相依的城市特色。

This planning regards the integration of harbor, city and zone as the main concept to develop a new economic driver surrounding the harbor in Northern Binzhou City. With the harbor first, the project focuses on developing the industries with assisted living facilities to establish an ecological and high-efficient industrial zone in Binzhou City. Relying on the main water systems of Taoer River, Xin River, Chao River and Haojiagou, we will build up a distinctive landscape corridor in the new zone to form a specialized waterside city.

天津滨海新区大沽船坞文化创意园城市设计

Urbanization Design of Creative Parkin Taku Dockyard, Tianjin Binhai New Area

项目地点：天津　Location: Tianjin
用地面积：3 000 000 m^2　Site Area: 3,000,000 m^2

该项目西邻响锣湾，北望于家堡，独踞滨海新区的核心区位；依托大沽船坞遗产，挖掘历史文化资源，打造环境特征明显、特色产业集聚的城市亮点地区；区内应积极发展文化创意、创作、交易以及媒体等相关产业，打造区域文化创意产业发展的首善之地。

This project is located in the center of Binhai New Area with Conch Bay in the west and Yujiabao in the north. Relying on the Taku Dockyard heritage, our planning is to explore the historical and cultural resources and make it become a specialized area with environmental advantages gathering characteristic industries. Within the area, we should actively develop the competent industries involving cultural creativity, creation, trading and media to build up the best place for regional creative industry development.

颐和园文物保护规划

Planning on Preservation of Cultural Relics in Summer Palace

项目地点：北京　　Location: Beijing
用地面积：3 350 000 m²　　Site Area: 3,350,000 m²

颐和园是世界文化遗产，是对中国风景园林造园艺术的一种杰出的表现，将人造景观与大自然和谐地融为一体。

The Summer Palace, one of the world cultural heritage sites, is an outstanding presentation of Chinese gardening art, integrating artificial landscape and the nature harmoniously.

北京市圆明园文物保护规划

Planning on Preservation of Cultural Relics in Old Summer Palace, Beijing

项目地点：北京　　Location: Beijing
用地面积：352 hm²　　Site Area:: 352 ha

圆明园是中国古典园林的杰出代表，在中国和世界园林史上占有重要地位。

The Old Summer Palace is an outstanding representative of classical Chinese gardens, which plays an important role in Chinese or even the world gardening history.

扫描查看更多信息

元正（天津）建筑设计有限公司
Yuanzheng (Tianjin) Architectural Design Co., Ltd.

元正（天津）建筑设计有限公司在城市规划设计、建筑设计、旅游地产及酒店设计、景观园林设计、工程咨询等专业拥有众多优秀设计师，在国内及海外均有重要工程业绩。由该公司设计的天津奥林匹克花园二期工程分别荣获“中国经典示范楼盘”综合奖以及“双节双优杯住宅方案竞赛”金奖。

该公司作为世界酒店联盟常务理事单位主持的世界酒店联盟设计中心，针对国内外酒店及旅游度假项目进行了多项工程的规划设计，同时，作为对外经济合作委员会理事单位，在旅游、绿色社区、生态智能城市等方面与国内外诸多项目展开积极合作。

该公司同时与欧盟及国内外科研机构在生态宜居城市社会服务设施配置标准研究、绿色社区标准与评价体系研究、生态智能城市深化设计及指标体系研究等方面进行广泛合作，并在多个城市展开相关标准体系制定及实施工作。

元正一贯秉承从策略导入的设计观念和高品质的设计服务，提供包括环境规划、城市设计、建筑设计、室内设计在内的全方位设计服务。元正致力于建筑空间的创造与体验，营造以人为本的城市环境，将艺术理想与实际建设加以融合，同时适应不同环境的认知感受而灵活变化。元正设计师在参与项目的整个设计流程中，始终立足市场并从使用者的角度进行“适宜设计”，强调环境、人文、艺术与市场的整合，创造建筑空间环境与人的和谐统一。

主要规划及建筑设计项目有韩国济州岛黎湖乐园度假酒店工程、斐济GPH大太平洋酒店、汤加王国Dateline国际日界线酒店、京西原始森林旅游度假区暨京台新城、西柏坡宾馆改扩建工程、悦来湖度假酒店、三亚小洲岛七星级国际度假酒店、天津奥林匹克花园、天津市河西区青少年宫、天津北辰道综合开发项目（300 hm^2）、天津北辰商业中心、天津1号地超高层建筑群综合体、天津瑞景3号地江南春色项目、中新（天津）生态城动漫园美星影视传媒乐活广场、天津和远领居、天津华尔国际大厦、天津滨海新区森林公园居住区规划、天津蓝岸森林商业综合体、廊坊新朝阳百万平方米商业综合体、廊坊大中广场、廊坊龙河中心区、廊坊建业大厦、廊坊·晓廊坊规划建筑设计、月色廊坊商业广场、华北油田万庄小区、廊坊市林业局办公楼、廊坊大学城青年港·自由郡、廊坊一中改扩建工程、廊坊第十四、第十五小学设计、廊坊隆福市场规划、廊坊卫生学校、廊坊望京华府规划设计、秦皇岛开滦路步行商业街区与道南片区设计、北戴河东山宾馆、北戴河东经路宾馆、秦皇岛北城一号项目、秦皇岛天洋新城四期、秦皇岛水榭云间、秦皇岛青馨家园TownHouse社区、山海关人民医院、秦皇岛市海港区房管局办公楼、秦皇岛市海港区第三街区南部片区修建性详细规划设计、滦南鹏程实业办公楼、唐山瑞德花园、邯郸迎宾路A地块、邯郸中道都市广场综合项目、邯郸中央公园、邯郸鸿基花苑、邯郸颐景蓝湾、邯郸大悟国粹嘉苑、邯郸滏阳河沿岸景观规划设计、承德茅荆坝裕景国际森林温泉城规划设计方案等百余项。

地址：天津市河西区友谊北路广银大厦1601
电话：+86-22-23558166/7/8
传真：+86-22-58780084
网址：www.dreambuilt.com.cn/www.imaap.com.cn

Add: No.1601, Guangyin Building, Youyi North Road, Hexi District, Tianjin
Tel: +86-22-23558166/7/8
Fax: +86-22-58780084
Web: www.dreambuilt.com.cn/www.imaap.com.cn

天津和远领居
Heyuan Leading Habitat, Tianjin

项目地点：天津
用地面积：103 054.84 m^2
建筑面积：111 771 m^2
容 积 率：1.35
绿 化 率：41.66%

Location: Tianjin
Site Area: 103,054.84 m^2
Building Area: 111,771 m^2
Plot Ratio: 1.35
Green Ratio: 41.66%

项目位于宝坻区东北部，规划设计以“打造亲情社区，构建和谐社会”理念为指导，使“亲情”有机地融入规划设计之中。在绿化方面采用多种手法，组合各种场所，包括老人、儿童活动场地，体育运动场等，创造了邻里亲情的氛围，以满足居民对功能使用和精神享受的要求。建筑室内风格采用新古典主义设计手法，体现了高贵、典雅的建筑气质。

北城一号
North Town No. 1

项目地点：河北 秦皇岛
用地面积：106 376.5 m^2
建筑面积：213 418 m^2
容 积 率：2.05
绿 化 率：38.73%

Location: Qinhuangdao, Hebei
Site Area: 106,376.5 m^2
Building Area: 213,418 m^2
Plot Ratio: 2.05
Green Ratio: 38.73%

项目基地地势北高南低，规划在满足一般居住区规划中强调的功能的合理、交通方便、配套齐全等要求的同时，着眼于高品质空间和使用功能的创造，力求处理好人、建筑及环境三者的关系，努力创造出有地方特色的现代化的居住小区。建筑形态上，采用新古典主义建筑风格，顶部为坡屋顶。依地势和建筑层数的不同，建筑高低错落，丰富小区轮廓线。本案主要设计为6～7层的花园洋房及18～28层的高层住宅。方案整体设计采用中轴对称式的设计手法，通过合理的规划设计来体现中央景观道的丰富种植特色与建筑整体风格的有机统一，同时充分有力地表达和谐、共融的居住区概念，轻松、休闲的居住功能以及宁静、素雅的居住观感的主题理念。

新朝阳广场
New Chaoyang Square

建设地点：河北 廊坊
用地面积：174 900 m²
建筑面积：1 024 821 m²

Location: Langfang, Hebei
Site Area: 174,900 m²
Building Area: 1,024,821 m²

项目地处廊坊市广阳区中心位置，用地174 900 m²。项目分为两期，一期为地块东北角已建成的朝阳购物商场，其余为二期用地。该项目处于城市重要商业节点——新朝阳商圈，除本项目外，还有天汇丰、君泰等商城。项目除一期的新朝阳购物商场外，拟在二期地块建设大型购物中心、商业精品店、五星级酒店、酒店式公寓、商务快捷酒店及办公写字楼等，共同构成的规模庞大的城市商业综合体，为廊坊及周边的人群提供一站式的商业服务。

本项目西北地块为集中式大型购物中心。南面两个地块1～4层设置独立跃层式的店铺。高层建筑均匀布置在两个地块上面。分别为两栋24层的酒店式公寓，3栋30层的酒店式公寓，3栋29层的酒店式公寓，两栋28层的酒店式公寓，1栋26层的快捷酒店，1栋26层的五星级酒店以及一栋26层的办公楼。建筑立面采用简欧风格，项目方案建筑高度均按限高规定，限制在100 m以内。

秦皇岛开滦路步行商业街区与道南片区规划

Plan of Pedestrian Business Block and Daonan Block, Kailuan Road, Qinhuangdao

项目地点：河北 秦皇岛
用地面积：24.25 hm^2

Location: Qinhuangdao, Hebei
Site Area: 24.25 ha

本项目包括开滦路商业步行街及开滦路两侧，道南解放里、缸砖里和吉星里小区危旧房改造工程。用地面积约24.25 hm^2。地块东西长约650 m，南北长约980 m。开滦路道南片区商业区北起三角花园，南至海滨路，其中的开滦路长约700 m。规划力求在保护开滦路“洋街”历史风貌的基础上，尽可能保留原有历史建筑，并通过整修、复原、改造、新建、迁移等方式，以开放式步行街的手法，营造一条集历史文化展示、旅游观光、商业、居住、休闲等综合功能于一体的21世纪“新洋街”。

精武商业广场
Jingwu Commercial Plaza

项目地点：天津
用地面积：16 642 m²
建筑面积：53 219 m²
绿 化 率：10%

Location: Tianjin
Site Area: 16,642 m²
Building Area: 53,219 m²
Green Ratio: 10%

项目位于天津西青区精武镇，荣华道与富兴路交汇处。项目的主要功能为商业、办公及酒店。受规划条件及场地内原有建筑、通讯塔等的限制，总平面呈矩形。建筑东、西退建筑红线8 m，北退建筑红线6 m，南面退建筑红线15 m。建筑主要入口位于建筑东面，北面设有疏散及货物出入的次入口。
本建筑设地下1层，地上5层。其中，地下1层为停车库及设备用房，地上1–4层是商业用房，5层为办公用房及酒店。
建筑立面采用简欧风格，使得本项目成为当地备受瞩目的地标性建筑。

滏阳河景观及酒吧街规划设计

Fuyang River Landscape & Bar Street Plan and Design

用地面积：116 000 m² Site Area: 116,000 m²
南北岸线：900 m South-North Shoreline: 900 m
水面面积：29 000 m² Water Area: 29,000 m²

滏阳河南至南湖，北至梦湖，全长约16 km。从滏阳河对于邯郸市的重要意义出发，充分发掘滏阳河的城市景观价值。通过打造具有独特风格与特色的滨水空间，以酒吧街的形式为特色，将原有的自发的自然式游玩变成一种自觉而有序的旅游；通过对河岸绿化，滨水道路的规划，码头的配置和软硬地面的组织，充分完善和提高滨水空间的整体环境及配套设施，形成优良的城市景观与景观休憩公共绿地。

中新生态城天津美星影视大厦

Tianjin Meixing Film & Media Plaza, Sino-Singapore Ecology City

项目地点：天津
用地面积：5007.8 m²
建筑面积：13 580 m²
绿 化 率：35%

Location: Tianjin
Site Area: 5,007.8 m²
Building Area: 13,580 m²
Green Ratio: 35%

项目紧邻动漫园中环路与内环路，与其他4块地块共同围合出中央的大型水面、绿地。在形体设计上严格按照规划前低后高的原则，积极维护中央景观，并且沿内环设置一连串的柱廊空间，创造出一个建筑与环境的过渡性灰色空间。立面设计采用现代、简洁的设计手法，并结合动漫园的产业特点，活泼、生动而又不失理性与秩序，力求把握主题动漫园孵化器的内在气质，追求整体秩序下的适度变化，合理利用陶土板、金属、玻璃石材等建筑材料，塑造出简洁、舒适、现代、典雅的集商业与办公为一体的建筑。

在外部交通组织上，契合两侧道路要求，在用地内部留出一条步行系统，两侧适量设置与动漫产品相关的店面，增加区域的活力。

本项目是按照《中新天津生态城绿色建筑评价标准》和《中新天津生态城绿色建筑评价技术细则》设计并经有关部门和专家评估和认定的绿色建筑。

扫描查看更多信息

天津市城市规划设计研究院
Tianjin Urban Planning & Design Institute

天津市城市规划设计研究院，是具有城市规划和建筑设计甲级资质的专业性设计团队，现有员工70人，其中：博士1人，硕士14人；国家一级、二级注册建筑师、国家一级注册结构工程师、国家注册公用设备工程师、国家注册电气设备工程师共25人，高级职称人员33人。
建筑分院做为天津市规划建筑设计院的重要组成部分，提倡整体设计、多专业优化结合的一体化全程专业服务模式，实现了各专业无缝对接。建筑分院在整体综合的设计理念指导下，使方案创意全程实现，保证设计达到最高品质。

地址：天津市河西区黄埔南路81号万顺大厦B座13层
电话：+86-22-28012350
传真：+86-22-28012350
邮箱：Buildingbranch_tj@126.com
网址：www.tjcityplan.com

Add: B-13 Wanshun garen, No.81 HuangPuNan Road, He Xi District, Tianjin
Tel: +86-22-28012350
Fax:+86-22-28012350
E-mail: Buildingbranch_tj@126.com
Web: www.tjcityplan.com

天津市文化中心修建性详细规划
Detailed Constructive Plan of Tianjin Culture Center

项目地点：天津
用地面积：900 000 m^2
建筑面积：1 000 000 m^2

Location: Tianjin
Site Area: 900,000 m^2
Building Area: 1,000,000 m^2

天津文化中心图书馆

Tianjin Cultural Center Library

项目地点：天津
用地面积：40 000 m^2
建筑面积：50 000 m^2
合作单位：（株）山本理显设计工场

Location: Tianjin
Site Area: 40,000 m²
Building Area: 50,000 m²
Partners: Riken Yamamoto & Field Shop (Japan)

宁强县天津中学

Tianjin Middle School, Ningqiang County

项目地点：陕西 宁强
用地面积：130 000 m^2
建筑面积：60 000 m^2

Location: Ningqiang, Shaanxi
Site Area: 130,000 m^2
Building Area: 60,000 m^2

泰安道二号院

Tai'an Avenue No. 2 Yard

项目地点：天津
用地面积：10 000 m²
建筑面积：20 000 m²

Location: Tianjin
Site Area: 10,000 m²
Building Area: 20,000 m²

泰安道四号院

Tai'an Avenue No. 4 Yard

项目地点：天津
用地面积：20 000 m²
建筑面积：100 000 m²

Location: Tianjin
Site Area: 20,000 m²
Building Area: 100,000 m²

天津电子信息职业技术学院

Electronic Information Vocational Technology College, Tianjin

项目地点：天津
用地面积：400 000 m^2
建筑面积：210 000 m^2

Location: Tianjin
Site Area: 400,000 m²
Building Area: 210,000 m²

城投大厦

Chengtou Building

项目地点：天津
用地面积：10 000 m^2
建筑面积：30 000 m^2

Location: Tianjin
Site Area: 10,000 m^2
Building Area: 30,000 m^2

天津人口和家庭公共服务中心

Tianjin Planning Birth & Reproductive Health Service Center

项目地点：天津
用地面积：20 000 m²
建筑面积：70 000 m²

Location: Tianjin
Site Area: 20,000 m²
Building Area: 70,000 m²

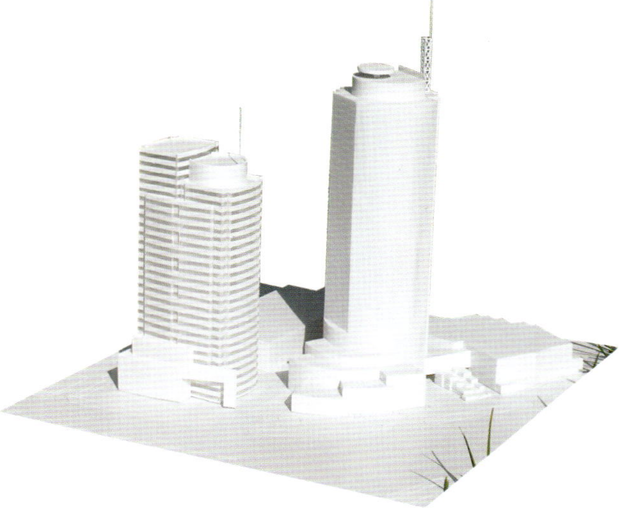

今晚传媒大厦

Jinwan Media Building

项目地点：天津
用地面积：3000 m^2
建筑面积：30 000 m^2

Location: Tianjin
Site Area: 3,000 m^2
Building Area: 30,000 m^2

扫描查看更多信息

天津加尚建元建筑设计有限公司
Tianjin CUN Architectural Design Co., Ltd.

天津加尚建元建筑设计有限公司创立于2009年4月，其与天津建筑设计院合作，真正做到了强强联合。
公司在城市规划、大型地产项目开发和标志性建筑设计领域积累了独特而宝贵的经验，主创设计师均为具有前沿设计思想的资深专家。
公司秉承国际化现代化设计理念，以全新的模式为客户提供品质卓越的服务。公司尤其擅长超高层写字楼、高级公寓、星级商务酒店、城市总体规划、大型公建、住宅等建筑领域的设计。公司先后完成了天津梅江会展中心二期工程、滨江道建筑综合改造工程、重庆市综合办公楼、天津市地铁东南角B地块开发、和平路提升改造等诸多项目的建筑方案设计。
公司锐意进取，目前正处于高速、持续发展阶段。

地址：天津市南开区士英路与云际道交口
东海岸商务中心8层
邮编：300381
电话：+86-22-23669303
邮箱：jianyuanzhaopin@126.com
tianjinjianyuan@126.com
网址： www.cansun-ca.com

Add: 8F,East Coast Business Center, the Link of Shi Ying Road and Yunji Dao,
Nankai District, Tianjin City
P.C.: 300381
Tel: +86-22-23669303
E-mail: jianyuanzhaopin@126.com
tianjinjianyuan@126.com
Web: www.cansun-ca.com

天津梅江国际会展中心二期工程
MJCEC Phase II Project,Tianjin

设 计 师：赵军
项目地点：天津
用地面积：71 900 m²（展览区）
建筑面积：282 300 m²

Designer: Jun Zhao
Location: Tianjin
Site Area: 71,900 m² (exhibition area)
Building Area: 282,300 m²

天津梅江国际会展中心二期紧邻一期项目，总建筑面积为28.23万 m²，约为一期总建筑面积的3倍。
二期工程建筑整体造型犹如张开双翼的飞燕，自然而奔放，令人充满了无尽的想象。其整体建筑主要由三部分构成：中央部分共3层，最高45 m，设计用途为会议办公、餐饮娱乐、小型展厅及相关配套设施；两翼部分为双层结构，共4个展厅，展厅净高38 m。
在建筑基调上以浅色为主，与梅江地区建筑环境总体风格相一致。
设有地下停车场，可容纳1300辆机动车的存放，将缓解大型展会举办时的停车压力。
为方便参展货物运输，通过巧妙设计，设置运货坡道，可将货物直接运送到二层展厅；西面广场地下引入地源热泵系统，可为二期建筑供暖及制冷，充分体现节能环保理念。
这只翱翔在梅江的“飞燕”初展“双翼”，将进一步服务会展经济，提升天津城市形象。

The MJCEC phase II project is close to phase I project. It covers a building area 282,300 m², nearly three times the building area of phase 1 project.
The phase II project is in shape of a flying swallow, natural and energetic, full of endless imaginations. The building includes three parts: the central part has 3 floors, 45 m high (peak height), designed for convention, office, catering & entertainment, some small display halls and supporting facilities; the two wings has two floors, totally 4 display halls, 38 m high (clear height).
The building is keynoted by light colors, according to the general style of Meijiang building conditions.
It has a underground parking lot with 1,300 parking spaces, which will mitigate the parking pressure on occasion of large exhibition.
It has introduced many swift designs to facilitate cargos transport in exhibition, for example, it has set up cargos ramp which can directly transports cargos into the display hall on the second floor; the west square has a underground ground source heat pump system to provide warming and refrigeration for the phase 2 building, fully representing the philosophy of energy-saving and environmental protection.
This “flying swallow” is extending its “two wings” above Meijiang, which will further serve for the development of convention & exhibition economy, and improve the image of Tianjin City.

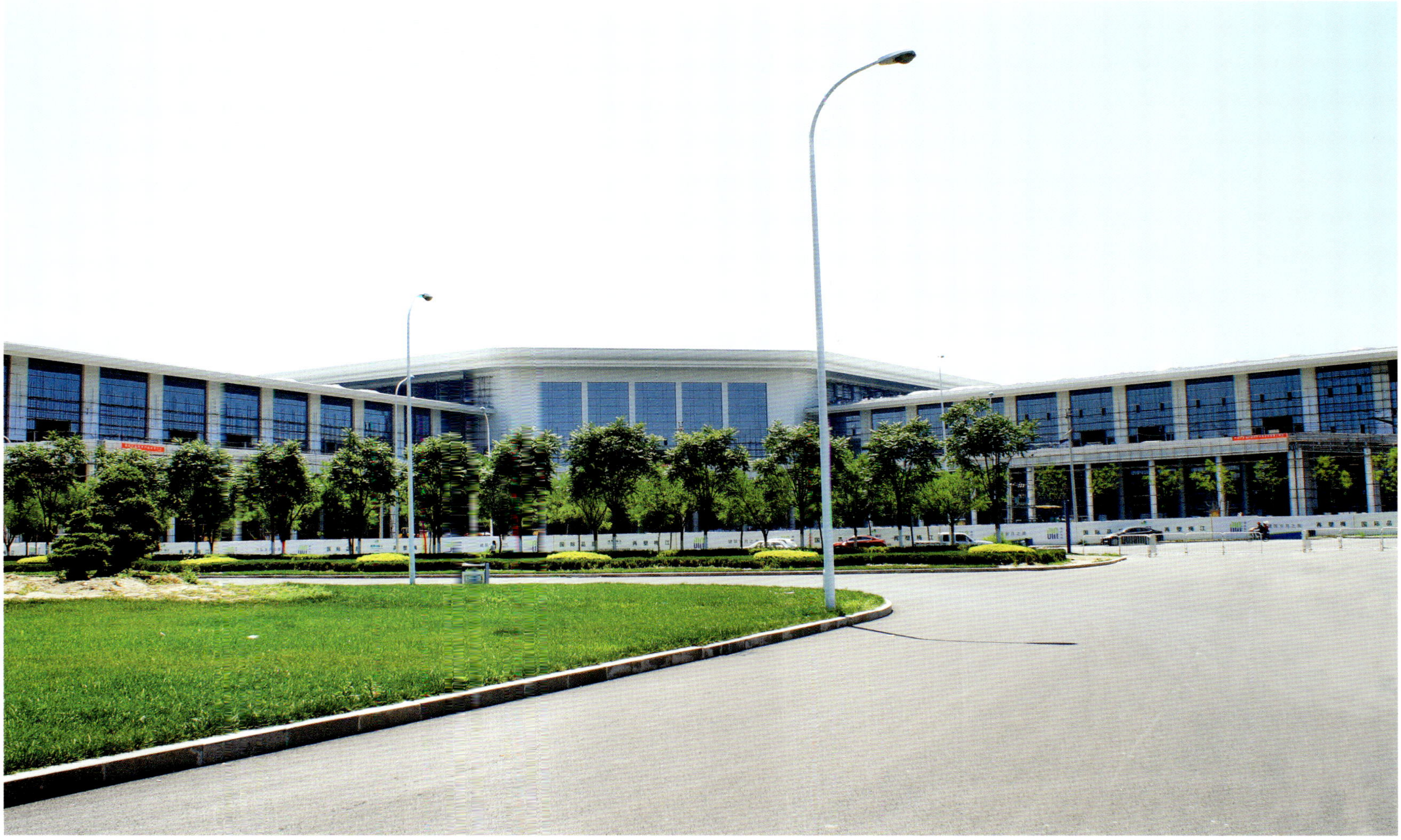

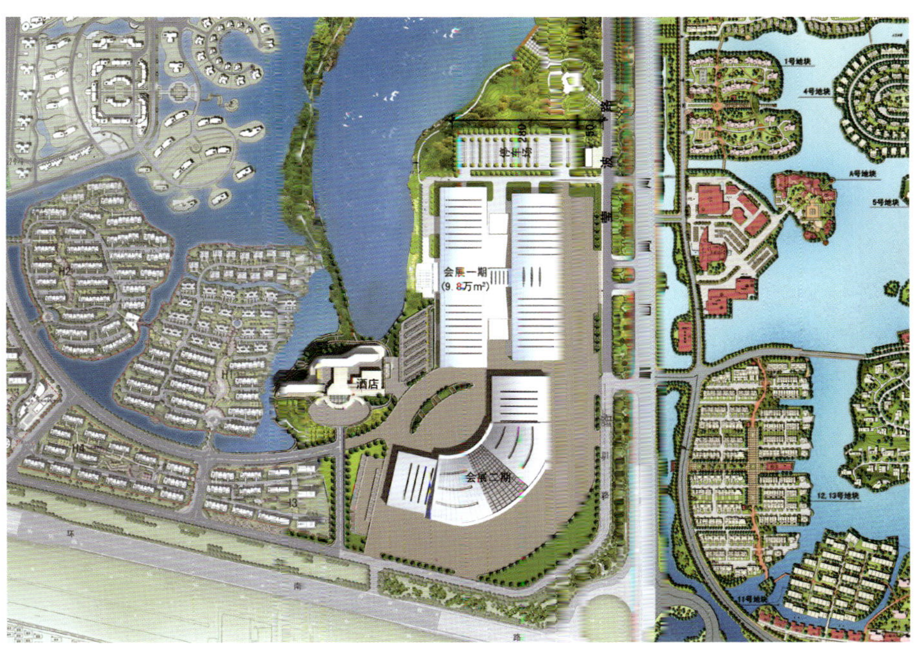
停车场
会展一期
(9.8万m²)
酒店
会展二期

天津梅江会展中心
Tianjin Meijiang Convention Center

天津地铁项目
Tianjin Metro Project

设 计 师：刘志涛 Designer: Zhitao Liu
项目地点：天津市 Location: Tianjin

天津地铁多条线路的相继开通带动了地铁周边楼盘的快速发展，其中地铁2号线B地块东南角项目、地铁2号线咸阳路项目和地铁3号线王顶堤项目均是集商业、办公、酒店、住宅等功能于一体的大型建设项目，建设规模及投资规模相对较大，它们必将进一步使周边经济发展增速并提升区域价值。

The operation of Tianjin Metro lines has driven the rapid development of metro peripheral buildings. Among them, Metro 2 Lot B Southeast Corner Project, Metro 2 Xianyang Road Project, and Metro 3 Wangdingdi Project are large building projects integrating commercial, office, hotel, residential and other functions, with large building scale and investment scale, which will further improve the peripheral economic growth and regional value.

地铁2号线B地块东南角项目
Metro 2 Plot B Southeast Corner Project

用地面积：10 039m^2
建筑面积：79 839m^2
地上建筑面积：55 451m^2
地下建筑面积：24 388m^2

Site Area: 10,039m^2
Building Area: 79,839m^2
Aboveground Building Area: 55,451m^2
Underground Building Area: 24,388 m^2

地铁2号线咸阳路项目
Metro 2 Xianyang Road Project

用地面积：641 000 m^2
建筑面积：399 000 m^2
地上建筑面积：243 000 m^2
地下建筑面积：156 000 m^2

Site Area: 64,100 m^2
Building Area: 39,900 m^2
Aboveground Building Area: 243,000 m^2
Underground Building Area: 156,000 m^2

地铁3号线王顶堤项目
Metro 3 Wangdingdi Projec

用地面积：12 719 m^2
建筑面积：82 530 m^2
地上建筑面积：50 730 m^2
地下建筑面积：31 800 m^2

Site Area: 12,719 m^2
Building Area: 82,530 m^2
Aboveground Building Area: 50,730 m^2
Underground Building Area: 31,800 m^2

唐山市世博二期
Tangshan World Expo Phase II

项目地点：河北 唐山
用地面积：73 855.09 m²
建筑面积：463 071.41 m²
容 积 率：6.27

Location: Tangshan, Hebei
Site Area: 73,855.09 m²
Building Area: 463,071.41 m²
Plot Ratio: 6.27

本项目用地位于唐山市中心区核心地段，北至新华西道、东至建设南路。以营造新世纪的居住环境及高价值的商业环境为目标，从生活方式的角度规划设计，布局规整合理，可使空间被充分利用，达到土地利益最大化。结合总体布局，采用点、线、面结合的方法，合理运用造景手法，引入整体统一、渗透交融的绿化空间概念，在用地东侧设置开阔城市广场，商业内街布置景观小品，为商业增加趣味性，并提供休息场所。裙房屋顶设屋顶花园和球类运动场，丰富景观层次，提升商业区品质与形象。

The project is located in the core of downtown Tangshan City. The land lot extends to Xinhua West Avenue in the north and Construction South Road in the east, with planning land area 20,635 m². The project tries to create new century residential & commercial conditions, in lifestyle design and reasonable layout, to make full use of spaces, without waste of land, and maximize land benefits. According to layout plan, the design introduces greening spaces that integrate with the whole project, in rational landscaping manner, considering a combination of point, line and plane layout. It sets a broad-view city square on the east side of the site, and sets landscaping gadgets on the inner commercial streets, to add some interests to the commercial atmosphere, and provides resting place, including rooftop garden and ball sports court, with rich landscaping layers, to improve local commercial quality and image.

天津市港建建筑设计有限责任公司

Tianjin GangJian Architectural Design Co.,Ltd.

天津市港建建筑设计有限责任公司，拥有30多年的历史，其前身为天津市建工集团一建设计室，后经资产重组，改制为有限责任公司。公司经过艰苦创业，不断学习国际先进的建筑行业管理模式，现已发展成为技术实力雄厚、人才济济的综合性设计单位。

公司工程设计资质为甲级，设有建筑、结构、给排水、暖通、电气五大专业。主要承接民用建筑、工业建筑、城市规划及居住区、住宅小区规划设计，同时也承接民用与工业建筑项目的可行性研究和工程项目咨询等业务。

地址：天津华苑产业园区榕苑路15号8号楼2楼；
天津市北辰区经济开发区双辰中路西双街创业园A座5楼
电话：+86-22-23859697/23859798
传真：+86-22-23859697-8000
邮箱：23859798@86-22.cn

Add: F2, No. 8 Building, 15 Rongyuan Road, Huayuan Industry Park, Tianjin; F5, Tower A, Shuangjie Entrepreneurship Park, West of Shuangchen Middle Road, Economic Development Zone, Beichen District, Tianjin
Tel: +86-22-23859697/23859798
Fax: +86-22-23859697-8000
E-mail: 23859798@86-22.cn

城际美景小区

"Beauty between the Cities" Community

项目地点：天津
建筑面积：205 000 m²

Location: Tianjin
Building Area: 205,000 m²

本案位于天津市北辰科技园区，东侧为双辰中路，南至双川道，西侧为园区景观内河。

设计主要原则：

1．从城市景观与建筑的设计角度，为城市建造一个生态化、本土化、舒适化、品位化的居住小区。

2．从住户的角度考虑，为追求品质生活的人们创造舒适和谐的生活空间。

3．从园林景观塑造角度，力求做到社区与外部环境形成互动关系，能互为景观。

The project is located in State Level Tianjin Beichen Hi-tech Industrial Park. It is beside Shuangchen Middle Road in the east, extending to Shuangchuan Avenue in the west, and beside the park landscaping inner river in the west.

Design principles:

1. To create an ecological, local, comfortable, and high-taste residential community, from the perspective of city landscaping and architecture.

2. To create a comfortable and harmonious living space for the residents who are pursuing quality life, from the perspective of residents.

3. To form interaction between community and external conditions, in an integral landscape, from the perspective of gardening and landscaping.

天津市北辰区人民法院审判综合楼

Judge General Building of Beichen District People's Court, Tianjin

项目地点：天津　　Location: Tianjin

建筑面积：32 000 m²　　Building Area: 32,000 m²

本案位于天津市北辰区京津公路以东，北仓道以北的北辰区中心地段。

外观采用了完全对称设计，线条笔挺大气，立面材质采用浅灰色石材，能够充分展现出建筑严肃，端庄，正气凛然的形象。方案力求创造一个通透、明朗的建筑形象。这体现了新时期法院公正、公开、透明的司法理念，而不是过去的深宅大院，密不透风，深不可测的法院形象。与中国当代司法机构公开、公正、透明、为民的理念和形象又正好贴切，外观庄重、简洁、鲜明、大气，具有现代气息。

The project is located east of Beijing-Tianjin Highway, north of Beicang Avenue, in downtown Beichen District, Tianjin.

The appearance is in absolute symmetric design, with straight generous lines. The façade is made of grayish stone materials, to be solemn, decent, and righteous. The project tries to create a transparent architectural image, to represent the fair, just and transparent judicial philosophy in new epoch, abandoning the past image official deep court which was secret and impermeable and unpredictable. This design idea agrees with Chinese modern juridical system which is open, just, transparent and for the people, and the appearance is solemn, concise, clear, generous and epochal.

天津市公安局北辰分局应急指挥中心、警训基地、刑事科学研究所

Emergency Command Center, Police Training Base, and Criminal Science Research Institute of Tianjin Public Bureau Beichen Branch

项目地点：天津　　Location: Tianjin

建筑面积：43 600 m²　　Building Area: 43,600 m²

本项目位于天津市北辰区北辰道南侧，处于北辰区中心地段，地理位置相当重要，可用地面积为16 117 m²。

北辰区应急指挥中心是北辰区的重要建筑，它的建成将为该区域的空间形态和活动形式的形成发挥重要作用。

1．形态：建筑形态产生于不同角度的思考，设计师试图用最简洁的形式体现公安局的力度，形成一种挺拔的感觉。

2．整体形象意识：建筑群体要和谐统一，强化建筑区的整体形象。突出公安局开放、公正、民主、亲民的形象。

3．鲜明的建筑个性特征：单体外观上追求一种庄重、威严的设计风格。

The project is located on the west side of Beichen Avenue, Beichen District, Tianjin. It is in the downtown of Beichen District, with important geography and available land area 16,117 m².

The emergency command center is an important building of Beichen District, which upon completion plays an important role in the space form and activity form of this district.

1. Form:

The building form comes from our thinking from different perspectives, trying to show the power of local public security bureau in the most concise form.

2. Overall image awareness:

The building cluster shall be harmonious and uniform, to strengthen the overall image of the area, and highlight the "open, fair, democratic, people-oriented" image of local public security bureau.

3. Obvious building character:

The single building appearance seeks for a solemn and mighty design style.

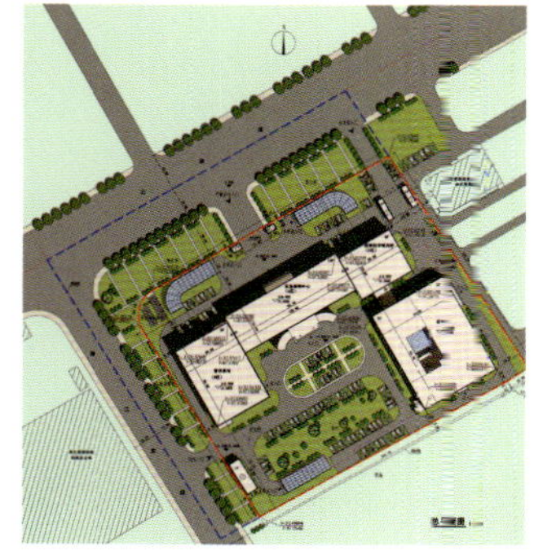

盛世华府小区5号楼（综合商业）

No. 5 Building, Prosperous Manor (Commercial Complex)

项目地点：天津
建筑面积：123 000 m^2

Location: Tianjin
Building Area: 123,000 m^2

项目位于京津公路西侧，北运河以东，北侧为城市绿地，京津城际武清站近在咫尺，交通区域优势明显。
地上首层至五层为商业，五层局部为多厅电影院、游泳馆，地下一层为超市，地下二层平时为车库，战时为人防。

The project is located on the west side of Beijing-Tianjin Highway. It is east of North Canal, and beside urban greenbelt in the north. It is near Beijing-Tianjin Intercity High-speed Railway Wuqing Station, with obvious regional advantages of traffic and transportation.
The aboveground first to fifth floor is for commercial purpose, part of the fifth floor is for multi-hall cinema and natatorium, the underground first floor is for supermarket, and the underground second floor is for garage in ordinary time and for air defense in wartime.

天津港津建筑设计工程有限公司
Tianjin GangJin Architecture Design & Engineering Co.,Ltd.

天津港津建筑设计工程有限公司于1993年7月成立，是天津市第一家获准成立的具有建筑行业甲级设计资质的中外合作综合性设计公司。公司于2001年9月通过了ISO9001质量管理体系认证，建立了完整的科学管理制度和质量管理体系。

公司具有丰富的建筑设计经验，设计范围覆盖建筑行业的各个领域；以"严谨、勤奋、求实、创新"的理念，树立公司的精品意识。近20年来，共设计包括高层、超高层在内的各类工程项目累计2000余项，总建筑面积达数千万平方米，项目遍及国内10余省市。并获得多项部、市级优秀设计奖项。

公司视人才为企业的基础，以培养和造就优秀人才作为公司的核心竞争力。公司现有技术人员90余名，其中高级职称25名、中级职称31名；有国家一、二级注册建筑师8名、国家一、二级注册结构师10名、注册设备工程师6名、注册造价师2名、注册规划师5名；同时聘请数名国内外高水平的工程设计专家为高级顾问。公司在自己的设计团队基础上，走强强联合之路，广泛与国内外知名高校、境外设计公司合作。

在"质量求生存，服务求信誉，经营求发展，管理求效益"的工作方针和"以人为本，服务至上，科学严谨，创新发展"的质量方针指导下，赢得了社会各界的广泛赞誉。

简洁、高效的工作流程，严格的岗位责任制，科学的奖罚激励机制，先进的质量管理体系和雄厚的科研人才队伍，为公司向客户提供高质量的产品和服务夯实了基础，也为公司树立了良好的品牌形象。

地址：天津市空港经济区凤鸣道华盈大厦6层
电话：+86-22-66267758/66267757/66267756
传真：+86-22-66267759
邮箱：tjbhgj@163.com
网址：www.tjgangjin.com

Add: 6th Floor, Huying Building, Fengming Road, Konggang Economic Zone, Tianjin
Tel: +86-22-66267758/66267757/66267756
Fax: +86-22-66267759
E-mail: tjbhgj@163.com
Web: www.tjgangjin.com

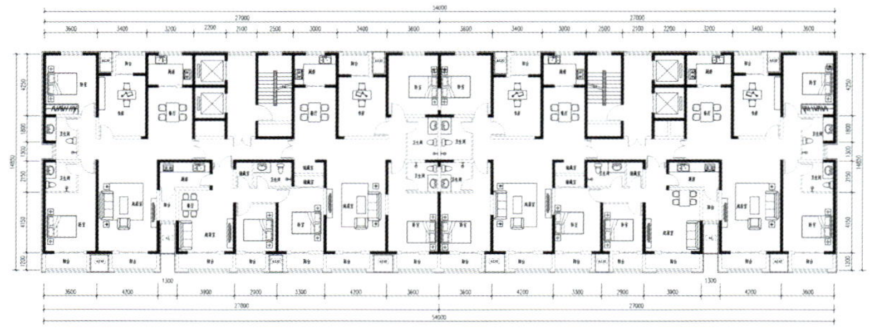

包头日月天地广场
Sun & Moon World Plaza, Baotou

设 计 师：孟力、徐凤富、郭海燕、夏颖、郑文波
用地面积：17 998.83 m²
建筑面积：67 990 m²
建筑密度：49.39%
绿 化 率：31%

Designer: Li Meng, FengfuXu, Haiyan Guo, Ying Xia, Wenbo Zheng
Site Area: 17,998.83 m²
Building Area: 67,990 m²
Building Density: 49.39%
Green Ratio: 31%

项目用地位于包头市稀土高新技术产业开发区新光路以北，日月豪庭小区以东，地块西侧为区间路，该用地临近住宅小区、办公区域、商业中心，区域优势显著。建筑地下一层，主要功能为停车场；地上一层至四层为大型综合商场及独立的二层临街商业；五层及以上为公寓楼。公寓和商业分设独立出入口。

熊岳河商业综合体项目
Xiongyue River Commercial Complex Project

设 计 师：王连文、王可、王扬
用地面积：106 500 m²
建筑面积：340 800 m²
容 积 率：3.2
建筑密度：29%
绿 化 率：35%

Designer: Lianwen Wang, Ke Wang, Yang Wang
Site Area: 106,500 m²
Building Area: 340,800 m²
Plot Ratio: 3.2
Building Density: 29%
Green Ratio: 35%

本项目位于辽宁省营口市鲅鱼圈区平安大街路东侧，规划高铁车站广场北侧，和高铁站相对。项目设计重点研究城市发展总体规划及现实趋势对区域（街区）发展的影响，区域性发展需求对特定建筑的影响和要求。根据周边的影响因素，即高铁车站及站前广场、南侧的自然河流来进行整体设计，力求通过该项目设计，对整个区域完善发展起到推动作用。整体建筑风格采用现代风格和新古典主义相融合的模式，力求清新典雅的建筑特色。

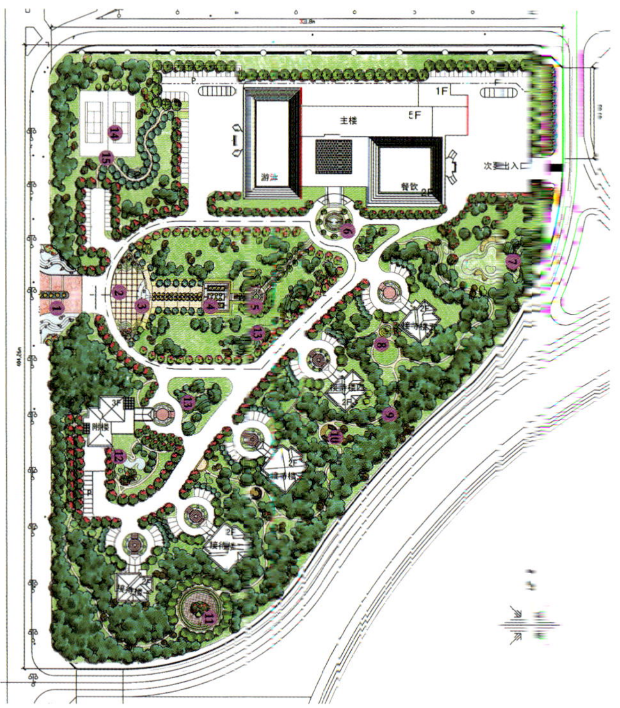

1. 入口景观
2. 入口广场
3. 跌水池
4. 景观水池
5. 观湖亭
6. 酒店入口喷泉
7. 果岭
8. 阳光草坪
9. 休闲林荫小路
10. 休息花架
11. 小型聚会广场
12. 丽日湖
13. 起伏的地形
14. 网球场
15. 景观亭

景观总平面图

营口迎宾馆

Yingkou Welcome Hotel

设 计 师：	孟力、张月雷、李大顺、李琳琳、郭海燕、李轩明	Designer:	Li Meng, Yuelei Zhang, Dashun Li, Linlin Li, Haiyan Guo, Xuanming Li
用地面积：	103 576 m²	Site Area:	103,576 m²
建筑面积：	29 537 m²	Building Area:	29,537 m²
容 积 率：	0.29	Plot Ratio:	0.29
建筑密度：	11.6%	Building Density:	11.6%
绿 化 率：	50%	Green Ratio:	50%

项目位于辽宁省营口市环湖北路以北，新联北大街以南，澄湖西路以东，规划路以西，为高规格礼宾接待的五星宾馆。由1栋主楼、1栋附楼、1栋会议中心楼、5栋独栋接待楼组成。酒店设施完备，有游泳馆、洗浴中心、会议中心、大型宴会厅等系统完善的康体娱乐功能。主建筑群掩映在亭台雕塑、林木草坪之间，是景色宜人的花园式酒店。

营口御景山温泉宾馆

Yujingshan Hot Springs Hotel, Yingkou

设 计 师：	张英、张月雷、梁凤岭、郭海燕、郑文波	Designer:	Ying Zhang, Yuelei Zhang, Fengling Liang, Haiyan Guo, Wenbo Zheng
用地面积：	43 960 m²	Site Area:	43 960 m²
建筑面积：	20 445.28 m²	Building Area:	20 445.23 m²
容 积 率：	0.42	Plot Ratio:	0.42
建筑密度：	18.6%	Building Density:	18.6%
绿 化 率：	50%	Green Ratio:	50%

御景山温泉宾馆位于营口市双台镇境内，南临沙河、北靠归名寺。该项目是依托于当地得天独厚的地热温泉旅游资源建设的四星级温泉休闲旅游度假酒店，由1栋综合楼、1栋服务楼、9栋VIP贵宾洗浴楼、4栋独栋贵宾洗浴接待楼、2栋四合院贵宾洗浴接待楼组成。酒店设施包括有游泳馆、洗浴中心、会议中心、大型宴会厅等系统完善的康体娱乐功能，另外提供特色室内、室外温泉养生疗养康体服务。

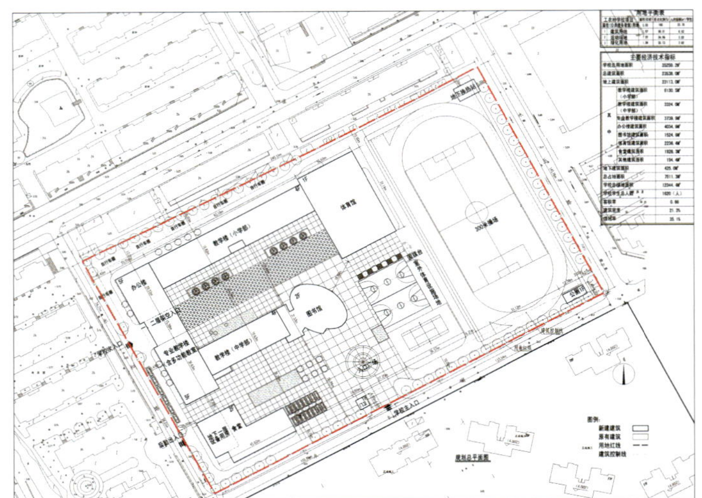

天津塘沽工农村九年一贯制学校

Nine-Year School of Workers & Farmers Village, Tanggu District, Tianjin

设 计 师：张英、马文敏、郭海燕
用地面积：35 250.2 m²
建筑面积：23 538 m²
容 积 率：0.66
建筑密度：20.9%
绿 化 率：35%

Designer: Ying Zhang, Wenmin Ma, Haiyan Guo
Site Area: 35,250.2 m²
Building Area: 23,538 m²
Plot Ratio: 0.66
Building Density: 20.9%
Green Ratio: 35%

本项目位于天津市塘沽区工农村永和路以东、永乐路以西、永兴道以南、永丰道以北。该项目由中、小学部教学楼、专业教学楼、办公楼、图书馆、体育馆、食堂、操场等组成，设施完备，现已投入使用。在校学生约有1600人（走读生），教师员工160人。

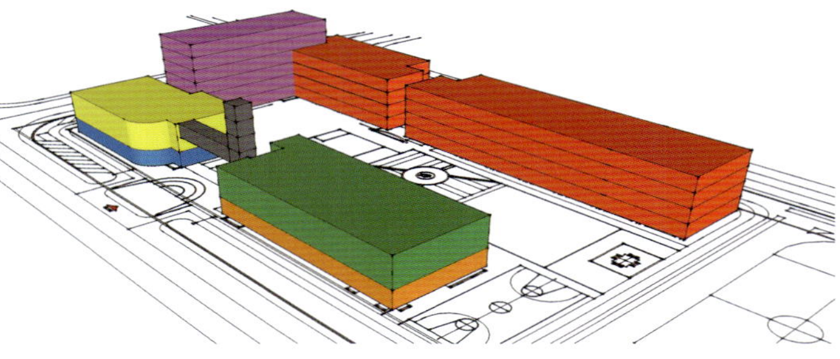

天津天保金海岸D-03地块小学

Primary School on Tianbao Golden Coast Lot D-03, Tianjin

设 计 师：张英、李鑫卉、李博、李琳琳、夏颖、李轩明
用地面积：20 713 m²
建筑面积：18 900 m²
容 积 率：0.91
建筑密度：27.0%
绿 化 率：26.1%

Designer: Ying Zhang, Xinhui Li, Bo Li, Linlin Li, Ying Xia, Xuanming Li
Site Area: 20,713 m²
Building Area: 18,900 m²
Plot Ratio: 0.91
Building Density: 27.0%
Green Ratio: 26.1%

本项目位于天津市开发区宏达街南侧、太湖东路东侧、发达街北侧地块内。该学校包含教学楼、风雨操场、600人综合报告厅、食堂、60人的员工宿舍等。整体布局较为规整，功能合理布局，节约用地。各功能建筑形成半围合空间，景观结合建筑所围合的空间形成了两条主景观轴，景观轴上设四个景观节点，将校园的内外景观相结合，体现了学校的开放性。

鹿鸣山庄

Deers Village

设 计 师：崔洪涛、马立媛、徐凤富、李瑞娟、冯丹、张乐
用地面积：82 835 m²
建筑面积：128 960 m²
建筑密度：25.0%
绿 化 率：26.1%

Designer: Hongtao Cui, Liyuan Ma, Fengfu Xu, Ruiqing Li, Dan Feng, Le Zhang
Site Area: 82,835 m²
Building Area: 128,960 m²
Building Density: 25.0%
Green Ratio: 26.1%

本项目位于包头市达茂旗滨河路东侧、纬三街南侧、纬六街北侧地块内。本项目以营造丰富、自由、灵性、有创意、符合人体尺度的空间为目的。建筑单体间相互渗透、融合，形成多元、神秘、奇异的欧洲风情，彰显豪宅贵气。外立面设计着重突出整体的层次感和空间表情，建筑是一种凝固的艺术，通过空间层次的转变，打破传统立面的单一和呆板，使其节奏、比例、尺度热情奔放，透露着淡淡南加州情调。

山东水岸名都住宅小区

River Shore Honorable City Residential Community, Shandong

设 计 师：刘子侬、李大顺、郭海燕、夏颖、李轩明
用地面积：57 967.79 m²
建筑面积：111 059.7 m²
容 积 率：1.92
建筑密度：21%
绿 化 率：36%

Designer: Zinong Liu, Dashun Li, Haiyan Guo, Ying Xia, Xuanming Li
Site Area: 57,967.79 m²
Building Area: 111,059.7 m²
Plot Ratio: 1.92
Building Density: 21%
Green Ratio: 36%

本项目位于山东省高青县东部偏北的芦湖公园附近，北临青高路，东临东环路，南临青城路，周边配套设施完善、交通便利、环境优美。用地周边布置小高层住宅建筑，中心布置四层洋房。利用建筑层数有节奏的高低变化，形成有韵律的空间变化与丰富的城市天际线。沿街布置集中商业，青城路一侧布置一层底商，促进商业的辐射面积成倍增加，构筑区域性商业圈，既可服务于城市又可服务于社区。

营口御景山温泉山庄规划

Plann of Yujingshan Hot Springs Hotel, Yingkou

设 计 师：张英、王可、张月雷、王扬
用地面积：254 200 m²
建筑面积：211 131 m²
容 积 率：0.77
建筑密度：18%
绿 化 率：40%

Designer: Ying Zhang, Ke Wang, Yuelei Zhang, Yang Wang
Site Area: 254,200 m²
Building Area: 211,131 m²
Plot Ratio: 0.77
Building Density: 18%
Green Ratio: 40%

御景山温泉山庄位于辽宁营口双台镇。本项目为山地项目，包括公寓、酒店、洋房、联排别墅、独栋别墅、商业等，业态较全。整体布局为错列式围合，组团功能及建筑形式结合山势，使住户拥有全景视野。尊重现场地形地貌，划分绿地与活动区域，"S"形行道路、跌水水景观都对地形生态进行了高度保护和恰到好处的营造。

营口仙人岛规划
Plan of Immortal Island, Yingkou

设 计 师：孟力、张英、王连文	Designer: Li Meng, Ying Zhang, Lianwen Wang
用地面积：78 610 000 m²	Site Area: 78,610,000 m²
建筑面积：1 560 000 m²	Building Area: 1,560,000 m²
容 积 率：0.2	Plot Ratio: 0.2
建筑密度：15%	Building Density: 15%
绿 化 率：45%	Green Ratio: 45%

辽宁省营口市仙人岛地区范围北至熊岳河，与鲅鱼圈新城区隔河相望；南抵浮渡河，与大连市域接壤；西临渤海湾，拥有丰富的港口岸线与沙滩资源；东濒沈大高速公路、黑大公路与规划中的沈大高速铁路，与辽宁省域交通干线紧密相连。在制定和实施规划过程中，着力于协调经济、社会、人口、环境的关系，考虑仙人岛地区长远发展的可能性，突出创新，使得规划适应市场经济条件下区域发展的需要。整体用地空间结构可以概括为“两轴、三心、四廊、四片”的总体框架，通过“轴向串联、分片集中，水绿相间、生态可续”的布局构成仙人岛地区未来空间发展的总体结构。

天津华厦建筑设计有限公司

Tianjin Huaxia Architectural Design Co., Ltd.

扫描查看更多信息

天津华厦建筑设计有限公司始创于1992年，拥有建筑工程设计甲级、城乡规划编制乙级、市政行业（燃气轨道交通除外）乙级、房屋建筑工程监理甲级以及房屋建筑施工图审查等多项资质。业务范围涉及建筑设计、市政设计、规划设计、建筑咨询、施工图审查、工程监理等领域的综合性服务，同时扩展到新型建材、混凝土、砂浆预拌、房地产开发、物业管理等多种行业。公司已通过GB/T19001—2008质量体系认证，员工近500人。

董事长兼总经理刘存发先生以诚为本、锐意进取、任贤纳才，旗下汇聚了一批资深专家和业界精英，其管理团队精诚干练、高效务实；设计团队创新求异、精益求精；开发团队熟悉市场，勇于开拓，是一支充满活力、开放而多元的战斗集体。公司在学校、医院、宾馆、展厅、商厦、写字楼、住宅等民用建筑及各类厂区、大跨度钢结构工程等工业建筑的规划、设计、监理等方面业绩丰富、奖项云集，作品遍及华夏大地。

华厦人秉承精心设计、优质服务的宗旨，以先进的设计理念、精湛的专业品质赢得了广大业主的盛赞，公司以“爱岗敬业、忠诚守信、团结协作、以德兴业”的企业精神与各界友人精诚合作，共建美好家园。

地址：天津市南开区华苑产业区榕苑路16号
鑫茂科技园中心楼三层
电话：+86-22-58693336/58693341/23672000
传真：+86-22-58598916
邮箱：tjhxjzsjgs@126.com
网址：www.tj-huaxia.cn

Add: 3 Floor, Center Building, Xinmao Science and Technology Park, No. 16 Rongyuan Road, Huayuan Industry Area, Nankai District, Tianjin
Tel: +86-22-58693336/58693341/23672000
Fax: +86-22-58598916
E-mail: tjhxjzsjgs@126.com
Web: www.tj-huaxia.cn

霞浦县公安局

Xiapu County Public Security Bureau

项目地点：福建 宁德
建筑面积：23 680.3 m²

Location: Ningde, Fujian
Building Area: 23,680.3 m²

库尔勒罗马假日花园
Roman Holidays Garden, Korla

项目地点：新疆 库尔勒
建筑面积：142 888.56 m²

Location: Korla, Xinjiang
Building Area: 142,888.56 m²

荔波国际会议中心酒店

International Convention Center Hotel, Libo

项目地点：贵州都匀
建筑面积：47 420.23 m^2

Location: Duyun, Guizhou
Building Area: 47,420.23 m^2

泰悦经典居住小区
Taiyue Classic Residential Community

项目地点：天津
建筑面积：162 440 m^2

Location: Tianjin
Building Area: 162,440 m^2

静海第六中学
Jinghai No. 6 Middle School

项目地点：天津
建筑面积：46 000 m²

Location: Tianjin
Building Area: 46,000 m²

华厦科技园
Huaxia Science & Technology Park

项目地点：天津
建筑面积：90 000 m²

Location: Tianjin
Building Area: 90,000 m²

河北钢铁集团鑫达钢铁有限公司办公楼
Office Building of Hebei Iron & Steel Group Xinda Iron & Steel Co., Ltd.

项目地点：河北 迁安
建筑面积：15 000 m²

Location: Qian'an, Hebei
Building Area: 15,000 m²

天津宏筑建筑设计有限公司

Tianjin Hongzhu Architectural Design Co., Ltd.

扫描查看更多信息

宏筑是以“宏筑设计”品牌为核心，专业化与综合化并存的设计集团。宏筑缘起于2006年成立的设计工作室，2009年“宏筑”品牌诞生。2012年宏筑公司作为“特邀编委”参与中国第一部《中国建筑设计作品年鉴》的编撰及组织工作。
宏筑拥有具备独立法人资格的天津宏筑建筑设计有限公司、天津宏筑景鸿景观规划设计有限公司、天津鸿腾装饰设计有限公司、天津思景域建筑设计咨询有限公司、天津火域科技有限公司等5个子公司，提供专业的景观设计、建筑规划设计、建筑装饰设计、设计可视化及视觉创意等服务。打造一个横向联合的综合技术服务平台，相互协作的运营与管理支撑体系。

地址：天津市南开区红旗南路588号
仁爱濠景国际大厦D座10F
电话：13702167898/13702167899
邮箱：hz_sjy_cc@126.com
网址：www.hongzhujz.com

Add: 10th Floor, Tower D, Ren Ai Hao Jing International Building,
No.588 Hongqi South Road, Nankai District, Tianjin
Tel: 13702167898/13702167899
E-mail: hz_sjy_cc@126.com
Web: www.hongzhujz.com

天津河东区营门口地块商业综合体概念性规划设计

Concept Plan and Design of Commercial Complex in Yingmenkou Plot, Hedong District, Tianjin

项目地点：天津
用地面积：33 400 m²

Location: Tianjin
Site Area: 33,400 m²

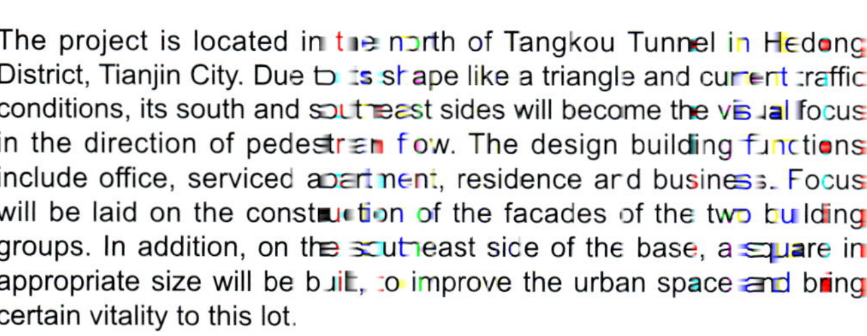

The project is located in the north of Tangkou Tunnel in Hedong District, Tianjin City. Due to its shape like a triangle and current traffic conditions, its south and southeast sides will become the visual focus in the direction of pedestrian flow. The design building functions include office, serviced apartment, residence and business. Focus will be laid on the construction of the facades of the two building groups. In addition, on the southeast side of the base, a square in appropriate size will be built, to improve the urban space and bring certain vitality to this lot.

项目位于天津河东区堂口地道北侧，三角状的基地及交通现状使南侧和东南侧成为人们的视觉焦点。设计建筑使用功能包括办公、酒店式公寓、住宅及商业，着重打造两个建筑群落立面，并积极地在基地的东南侧退让出一个尺度较为舒适的广场，将为该地段的城市空间做出改善并带来一定的活力。

天津河东区李公楼地块方案设计

Concept Design of Ligonglou Plot, Hedong District, Tianjin

项目地点：天津
用地面积：37 300 m²

Location: Tianjin
Site Area: 37,300 m²

规划地块与其北侧的金康园及久福园居住小区仅隔李公楼前街一条街道，对建筑日照的要求较高。
本案虽处于城市核心，毗邻城市商圈，但并非在核心商圈之中，必将面对客源与商户的分流；区域商业地段价值属于B类，商业形象较弱，如“百货”等大型商业的发展空间有限。
项目定位为中高端宜居社区、甲级写字楼以及生活配套型商业等多业态的复合开发。周边以居住型配套商业为主，生活型商业氛围已形成，将是发展重点；另鉴于周边商业中心的经营状况与竞争错位的考虑，应实现定位升级、业态丰富。
综合而言，本案所在区域难以支撑大型百货类商业的发展，应控制集中型商业体量，以丰富周边生活配套型商业业态为发展重点；同时可兼顾具备区域辐射级的餐饮、休闲娱乐的街区式商业。

Jinkang Garden and Jiufu Garden residential communities, on the north side of the planning lot, are only one street away from Ligonglou Front Street, so the design has high requirements for sunshine on the buildings.
Although the project is located in the center of the city and is close to the urban commercial district, it is not in the core commercial district, so it must face the diversion of customers and merchants. With a commercial district value of Class B, the region has a weak business image; therefore, large business like "department store" has a limited development space here.
The building products are positioned as multi-ownership composite developments, including agreeable mid- and high-end communities, class-A office buildings and life supporting commercial buildings. The surroundings are mainly residential supporting commercial buildings, with established lifestyle business atmosphere, which will be the focus of the development. In addition, considering the operating status and misplaced competition of peripheral commercial centers, the project shall have positioning upgrade and rich ownership forms.
In a word, the location is unable to support the development of large department stores, so it is necessary to control the volume of centralized business and focus on enriching peripheral life supporting commercial ownership forms.

乌兰察布集宁一中新校区概念性规划设计方案

Concept Plan and Design of the New Campus of Jining No. 1 Middle School, Ulanqab

项目地点：内蒙古 乌兰察布
用地面积：南区208 000 m²、北区152 400 m²
建筑面积：南区98 000 m²、北区273 300 m²

基地位于集宁新区霸王河东、满达街南，教育园区西，布拉格街北。整体规划分为两部分：

一、南区教学部分是“一轴一线五点”的景观结构。两条主、次景观轴形成了开敞与密闭、色彩与浓绿的鲜明对比。宿舍间五个不同的广场以“德育、智育、体育、美育、劳育”为主题，并通过不同的参与性景观构筑物表现各自的主题；

二、北区教师公寓部分是“一心两轴”的景观结构。主题水景广场为区域核心，展示“人与水”的主题。东西、南北两条主景观轴依托宅自然式绿化来体现“人与绿”的景观意境。整体景观结构力求通过“从容行进间感受来自天然的收获”的空间环境打造真正适合现代人居住的空间。

Location: Ulanqab, Inner Mongolia
Site Area: 208,000 m² for the south area, 152,400 m² for the north area
Building Area: 98,000 m² for the south area, 273,300 m² for the north area

The site is located east of Bawang River, north of Manda Street, west of Education Park, north of Bulage Street, in Jining New District. The whole planning consists of two parts:

1. The teaching section in the south area is landscaping structure comprising one axis, one line and five points. The main landscaping axis and the secondary landscaping axis form the sharp contrast between open and close, a diversity of color and a purity of dark green. The five deformed squares among dormitories are themed as "Moral Education, Intellectual Education, Physical Education, Aesthetic Education, Labor Education" respectively, and displayed with different participant landscaping structures.

2. Teachers apartments in the north area are landscaping structures comprising one center and two axes. The theme waterscape square is the area center with the theme of "man and water". The two landscaping axes east-to-west and south-to-north present the landscaping conception of "human and greening" relying on the natural greening between buildings. The overall landscaping structure is designed to tailor-make modern residential space, where “people can walk in leisure to feel and enjoy the harvest from nature”.

乌兰察布集宁区盛世新城居住小区方案设计

Schematic Design of Flourishing New Town Residential Community, Jining District, Ulanqab

项目地点：内蒙古 乌兰察布
用地面积：138 200 m²
建筑面积：331 500 m²
建筑密度：15.9%

地块位于集宁区东部工农大街东侧，泰昌北路西，北接学府路，南临110国道。区位交通优势十分突出。地块西侧为白泉山生态公园，北侧为集宁师范高等专科学校，人文、自然生活环境优势明显，是连接集宁南部区域与市政府所在地的东部节点。
本方案属于混合型居住小区，包括廉租房，公租房和经济适用房。经济适用房和公租房采用多层、高层组合的建筑方式。廉租房采用小高层的建筑方式。三种住宅产品统一规划，整体考虑，相互交融，和谐共生。

Location: Ulanqab, Inner Mongolia
Site Area: 138,200 m²
Building Area: 331,500 m²
Building Density: 15.9%

The land lot is located east of Workers & Farmers Street, west of Taichang North Road, north to Xuefu Road, south to No. 110 National Highway, in eastern Jining District, Ulanqab City. Convenient transportation is an outstanding advantage. With Baiquan Mountain Eco-Park on the west side, and Jining Teachers College on the north side, it has obvious advantage of human and natural living environment, and is the east joint between southern Jining District and local government office premises.
This scheme is a mixed residential community, including low-rent houses, public-rent houses and economically affordable houses. Economically affordable houses and public-rent houses will be built as a combination of multi-story buildings and high rise buildings. Low-rent houses will be built as small high rise buildings. All the three residential products are consistently planned, generally integrated, closely interconnected, and harmoniously coexistent.

▲ 方案一

乌兰察布客运枢纽中心规划方案设计

Plan and Concept of Passenger Hub, Ulanqab

项目地点：内蒙古 乌兰察布
用地面积：东部83 300 m²、西部55 000 m²

Location: Ulanqab, Inner Mongolia
Site Area: 83,300 m² in the east, 55,000 m² in the west

▲ 方案二

基地所在区域距离市中心直线距离大约10 km。位于110国道附近，交通优势明显。
规划用地分为东西两部分，东部为客运枢纽中心建设用地，用地面积为8.33 hm²，设计建设客运站房及配套服务用房3.1万 m²。西侧为商业用地，用地面积为5.5 hm²，设计建设配套高端商业8万 m²。

The linear distance from the base area to the city center is about 10km. It is nearby No. 110 National Highway, thus convenient transportation is quite an advantage.
The planned land is divided into east part and west part; the east part is the construction land for the passenger hub, covering a land area of 83,300 m², of which, 31,000 m² area is planned for passenger station and supporting service station. The west part is commercial land, covering a land area of 55,000 m², of which, 80,000 m² area is planned for supporting hi-end commercial buildings.

▲ 方案三　　▼ 方案四

天津三五互联移动通讯有限公司设计方案

Design Proposal of Tianjin 35 Technology Co., Ltd.

项目地点：天津
用地面积：65 000 m^2
建筑面积：一期27 424 m^2、二期64 027 m^2、三期34 282 m^2

Location: Tianjin
Site Area: 65,000 m²
Building Area: Phase I: 27,424 m²; Phase II: 64,027 m²; Phase III: 34,282 m²

项目位于天津市西南部高新区的核心区域——华苑产业区（环外），地理位置优越，是天津第一个"无燃煤区"和"国家ISO14000环保示范区"。
本地块东至海泰大道，南至亚安电子，西至华科七路，北至华科五路，周边多为高科技工业产业区，基础设施配套完善。项目分为三期建设，自北向南，自西向东，先建办公大楼，再建科研实验区；先开发沿街地带，外部带动内部经济发展，为内部建设做好前期准备。

Located in a superior geological location in the core area in southwest Tianjin Binhai High-tech Industrial Development Area - Huayuan Industrial Zone (outside the ring), the project is Tianjin's first "no coal combustion area" and "National ISO14000 Environmental Protection Demonstration Area".
The lot arrives at Haitai Avenue in the east, YAAN Tech headquarters in the south, 7th Huake Road in the west and 5th Huake Road in the north, surrounded mostly by high-tech industrial zones, with complete supporting infrastructures. The project is constructed in three phases and starts from north to south and west to east. Office buildings will be built first, and then the scientific research and experiment area. The land along the street will be first developed to let the external areas drive the economic development of the internal areas and make good preparations for the internal construction.

滨海（天津）全球汽车零部件采购中心

Coastal (Tianjin) Global Auto Parts Purchase & Trading Center

项目地点：天津
用地面积：121 300 m^2
建筑面积：244 048 m^2
容 积 率：2.0

Location: Tianjin
Site Area: 121,300 m²
Building Area: 244,048 m²
Plot Ratio: 2.0

规划用地位于天津市津南区葛沽镇南部，毗邻天津大道，交通非常便利。在功能布局上：根据行业特点、性质进行明确的分区设计。建筑呈围合式布局，将整个基地划分为四个功能区。
在建筑设计上：整体建筑风格采用简洁大方的现代建筑风格，园区整体采用框架式建筑结构。在色彩上：以米黄色和冷灰色为主，配合少量的玻璃幕墙形成园区典雅大方的风格。在环境设计上：注重环境的营造，各片区建筑围合形成独立的绿化空间，通过巧妙的景观营造手法，使环境景观与园区和谐的融合在一起，中心大型的公共绿化广场为园区释放出难得的休息空间。

The planning land is located in the south of Gegu Town, Jinnan District, Tianjin City. Adjoining Tianjin Avenue, it enjoys a very convenient traffic. In functional layout, a clear zoning design is made according to industrial characteristics and nature. The buildings are arranged in an enclosure style, and the whole base is divided into four functional areas.
Architectural design: The buildings adopt the simple and openhanded modern architectural style, while the whole site is a frame structure. Beige and cold gray are the dominant colors, and together with a small number of glass curtain walls, they give the site a sense of elegance and liberality.
Environmental design: The design tries to create an environment, where all buildings are enclosing an independent green space in each zone. Through ingenious landscaping techniques, the ambient landscape is integrated with the site harmoniously, and the large public green square in the center provides a scarce rest space for the site.

沙河一、二期项目概况

Overview of Phase I & II of Shahe River

本次设计范围西至怀远大街，东至康宁路（氧化塘），北至沿河建筑及解放大街，南至沿河建筑及虎山路。几个重要的现有城市道路和规划道路纵贯沙河，现状道路包括光明路、建设路、幸福路、工农北路，规划道路包括虎山一纬路、长征街、以及规划三经路、四经路、五经路、六经路、八经路、九经路等。
以“山水相依、生态宜居、宜业宜游”的城市总体规划目标为依托，努力打造注重文化品位，强化商业氛围，提升居住环境，并与地区经济发展同步进行的高档滨河景观。以概括的手法，将内蒙文化的元素融入设计，以体现当地特有的历史文化形式和内涵。

This design extends to Huaiyuan Street in the west, Kangning Road (Oxidation Pond) in the east, riverside buildings and Liberation Street in the north, and to riverside buildings and Hushan Road in the south. Some important existing urban roads and planning roads spread across Shahe River. The existing roads includes Bright Road, Construction Road, Happiness Road, Workers & Farmers North Road; the planning roads includes Hushan First South-North Road, Planning Third East-West Road, Fourth East-West Road, Fifth East-West Road, March Street, Planning Eighth East-West Road and Ninth East-West Road.
Based on the general planning target of “an ecological and livable city with waters and mountains, suitable for enterprises and tourism”, this project strives to build a high-end waterfront landscape that emphasizes cultural tastes, strengthens commercial atmosphere, increases residential conditions, and synchronizes with local economic development. Besides, the design, through generalization, integrates Mongolian rock scripts and paintings development with Mongolian cultural elements, to represent the unique form and substance of local history and culture.

沙河三期

Phase III of Shahe River

三期设计突出“沙河源”的概念，意在延续一期、二期的设计风格，将“岩文化”融会贯通。沙河三期划分三个区段：区段一、游赏体验区：与沙河二期相接，并有京包铁路穿过。设计遵循以人为本，生态宜居，注重人与自然和谐共存；驳岸形式由硬质驳岸到自然式转变，以自然式驳岸为主。区段二、休闲品味区：由于分布商业建筑和大面积硬质广场，所以驳岸形式采用复式硬质驳岸和复式生态驳岸。区段三，益智娱乐区：以自然驳岸来打造山水相依，亲近自然的生态宜游公园。

The Phase III design highlights the concept of “the origin of Shahe River”, to continue the Phase I & II designing style and integrates “Rock Culture”.
There are three sections for the phase III of Nansha River Project. The first section is the touring experience area, which connects the phase I project, with Beijing-Baotou Railway passing through it. The design follows the idea of human orientation to build an ecological livable place. It emphasizes the harmonious coexistence between human and nature. The revetment transforms from hard revetment to natural revetment which is as the main type. The second section is the leisure taste area. Due to the distribution of commercial buildings the large-scale hard squares, the revetment type is compound hard revetment and compound ecological revetment. The third section is the intelligent entertainment area, Connecting waters and mountains by natural revetment, it is to build an eco-park which is close to nature and suitable for touring.

盛世新城居住小区景观设计

Landscape Design of Flourishing New Town Residential Community

项目地点：内蒙古 乌兰察布
用地面积：138 000 m^2
绿 化 率：40%

Location: Ulanqab, Inner Mongolia
Site Area: 138,000 m^2
Green Ratio: 40%

项目位于内蒙古乌兰察布市集宁区，110国道北侧，泰昌北路西侧。
景观设计遵循并延续现代简约的建筑风格，创造亲切的居住小环境；景观方案设计将居住区整体地形以台地式呈现，从而体现尊重现状、因地制宜，寻求与场地和周边环境密切联系的设计构思；因项目容积率和建筑密度相对较高，楼宇间的景观可控面积相对较小，这就要求设计师必须珍惜每一寸土地，将居民活动空间与植被生长空间合理结合，既保证方案的完整性，又突出局部景观细节；设计关注项目的“地域性”，即地区自然景观与历史文化的特点。景观设计着眼于地域、着手于场地，营建具有当地特色的园林景观类型和满足当地人们活动需求的空间场所。

This project is located in Jining District of Ulanqab City in Inner Mongolia, which is on the north side of No. 110 National Highway, and on the west side of Taichang North Road.
The landscape design follows and continues the modern minimalist architectural style to create a small cozy living environment. It presents the integral landform of the residential community as a tableland, to represent the designing conception of respecting the present situation, following the landform and seeking for close connections between the site and the surrounding environment. Given high plot ratio and building density of this project, the controllable area for the landscape between buildings is relatively small. Therefore, the designer must treasure every inch of the land and reasonably combine the residential space with the vegetation space, to guarantee the integrity of project and also highlight the details of partial landscape. This design pays great attentions on the "locality" of project, namely the characteristics of local natural landscape and local historical culture. Focusing on the region and the site, the project is to build a gardening landscape with local characteristics and a space or place meeting the activity requirements of local residents.

中金汽车产业园区

Zhongjin Automobile Industrial Park

项目地点：天津
用地面积：113 200 m^2
建筑面积：181 900 m^2
容 积 率：1.47

Location: Tianjin
Site Area: 113,200 m^2
Building Area: 181,900 m^2
Plot Ratio: 1.47

规划用地位于天津市津南区海河工业园内，交通便利，具有优越的地理位置和区域优势。依托海河工业产业园区，定位为现代化工业产业园。
园区利用建筑围合的特点，设置多个共享空间，并设置标志景观区及重要空间节点。整体布局以中心为轴，采用规则的对称方式布置。
大型工业办公建筑采用简洁、明快的玻璃幕墙结构，局部用棕咖啡的色系点缀，既突显大气、稳重，又不失现代感。

This project is located in Haihe River Industrial Park of Jinnan District in Tianjin city, with convenient transportation, favorable geographic location and regional advantages. Due to the support of Haihe River industrial park, it is positioned as a modern industrial park.
The park sets several public areas by means of building enclosure, and also set landmark landscaping areas and important space joints. The general layout is symmetric regularly along the center line.
The large industrial office buildings adopt concise, bright glass curtain wall structure. Some parts are dotted with brown and coffee colors, to highlight the grandness, steadiness as well as the modern sense.

新疆四方建筑设计院有限公司

XINJIANG SIFANG INSTITUTE OF ARCHITECTURAL DESIGN co.,Ltd.

新疆四方建筑设计院有限公司成立于1993年1月1日，其前身为乌鲁木齐经济技术开发区建筑勘察设计院有限责任公司，具有建筑工程设计甲级、规划设计乙级、岩土勘察设计及施工乙级、市政设计乙级、咨询设计丙级等设计资质。现有职工151人，其中国家一级注册建筑师5人、一级注册结构师5人、二级注册建筑师6人、注册城市规划师3人、注册咨询师4人；高级工程师40人，工程师51人、助理工程师32人，具有较强的高端人才优势。

地址：新疆乌鲁木齐市水磨沟区安居南路70号
中国万向招商大厦13层
电话：+86-991-4697637
传真：+86-991-4697637
邮箱：xjsf-_d@sina.cn

Add: Floor 13, China Universal Merchants Mansion, 70 Anju South Road,
Shuimogou District, Urumqi City, Xinjiang Region
Tel: +86-991-4697637
Fax: +86-991-4697637
E-mail: xjsf-_d@sina.cn

“康乐名居”综合楼

"Kangle Fame" Multipurpose Building

项目地点：新疆 乌鲁木齐
用地面积：2142.61m²
建筑面积：地上面积 38 788.73m²，
地下面积 13 586.37m²
建筑密度：25.77%

Location: Urumqi, Xinjiang
Site Area: 2142.61 m²
Building Area: Aboveground 38,788.73 m²,
underground part 13,586.37 m²
Building Density: 25.77%

康都时代花园小区 1 号—19 号楼

Buildings No. 1-19, Kangdu Times Garden Community

项目地点：新疆 库尔勒	Location: Korla, Xinjiang
用地面积：153 000 m^2	Site Area: 153,000 m^2
建筑面积：400 000 m^2	Building Area: 400,000 m^2
容 积 率：2.69	Plot Ratio: 2.69
建筑密度：20.6%	Building Density: 20.6%

南湖尚苑高层住宅小区

High-rise Residential Community of South Lake Modern Garden

项目地点：新疆 乌鲁木齐
用地面积：5904 m^2
建筑面积：37 551 m^2
建筑密度：26.2%

Location: Urumqi, Xinjiang
Site Area: 5,904 m^2
Building Area: 37 551 m^2
Building Density: 26.2%

天和广场高层商住楼

Tianhe Plaza High-rise Commercial & Residential Buildings

项目地点：新疆 乌鲁木齐
用地面积：4020.93 m^2
建筑面积：73 866.97 m^2
建筑密度：25%

Location: Urumqi, Xinjiang
Site Area: 4,020.93 m^2
Building Area: 73,866.97 m^2
Building Density: 25%

库车市民服务中心

Public Service Center, Kuche

项目地点：新疆 库车
用地面积：3669.46 m²
建筑面积：34 146.24 m²
建筑密度：16.3%

Location: Kuche, Xinjiang
Site Area: 3669.46 m²
Building Area: 34,146.24 m²
Building Density: 16.3%

米东区残联

Disabled Persons' Federation, Midong District

项目地点：新疆 乌鲁木齐
用地面积：5085 m²
建筑面积：4303.54 m²
建筑密度：16.8%

Location: Urumqi, Xinjiang
Site Area: 5,085 m²
Building Area: 4,303.54 m²
Building Density: 16.8%

乌鲁木齐烈士陵园

Martyrs' Cemetery, Urumqi

项目地点：新疆 乌鲁木齐
用地面积：2323.88 m²
建筑面积：3942.54 m²

Location: Urumqi, Xinjiang
Site Area: 2,323.88 m²
Building Area: 3,942.54 m²

达坂城青少年活动中心
Youth Activity Center, Daban Town

项目地点：新疆 达坂城
用地面积：1946.98 m^2
建筑面积：3599.8 m^2
建筑密度：17.45%

Location: Daban Town, Xinjiang
Site Area: 1,946.98 m^2
Building Area: 3,599.8 m^2
Building Density: 17.45%

维吾尔医院
Uyghur Hospital

项目地点：新疆 乌鲁木齐
用地面积：2229.47m^2
建筑面积：36 768.69 m^2
建筑密度：18.98%

Location: Urumqi, Xinjiang
Site Area: 2,229.47 m^2
Building Area: 36,768.69 m^2
Building Density: 18.98%

新疆印象建设规划设计研究院（有限公司）

Xinjiang impression Construction Planning and Design Institute (Limited)

* 集文化思考、城市导演、工程设计于一体的特色设计研究机构。
* 具有中国环境艺术设计甲级、城市规划、旅游规划、建筑设计、室内设计、风景园林乙级资质的综合设计团队。

领导印象

高级建筑师、中国策划20年十大策划专家、中国环境艺术委员会专家委员会委员、全国首批注册高级环境艺术师、首届全国百位优秀环境艺术师、清华大学城市导演课题客座教授、新疆城市规划协会常务理事、新疆优秀中青年勘察设计工作者、董事长王元新；一级注册结构师、总工程师张晓民；一级注册建筑师、国家注册规划师、副院长张士鹏；一级注册建筑师、总建筑师苗新育、李萱、徐乃钦；副院长、高级工程师时卫国、王冀东、王洋等多位资深专家。

团队印象

该院设有建筑、规划、市政、景观、装饰5个设计所，王陆易农教授规划工作室，阎顺研究员旅游工作室，王洋创作事务所，徐乃钦方案室，王红霞影视动漫工作室、袁志君、伊莎雕塑家工作室，新疆印象建设沙龙等。现拥有教授级高级工程师、注册建筑师、注册结构师、注册城市规划师、注册环境艺术师、雕塑家等各类专业人才80余人。

成果印象

近几年来，该院作品有110余入选《中国建筑设计作品年鉴》，10件入选《中国创意优秀建筑设计作品集》。有30余项规划、建筑、景观设计成果获国家建设部、新疆维吾尔自治区、新疆生产建设兵团、乌鲁木齐市、昌吉市等各级优秀设计奖。其中：达坂城王洛宾音乐艺术城获2008年“中国策划20年经典策划”金奖，昌吉回民小吃街和莎车十二木卡姆农家乡园获“2009年中国人居典范规划设计方案”金奖，沙湾大盘美食文化城、阜康市瑶池园、昌吉滨湖河东“2010年中国人居典范规划设计方案”金奖。

文化印象

“创新与研究是设计院第一生产力”，董事长王元新一直要求设计团队在创作过程中做到“七分产品研究，三分作品设计”，要高举“现代地域文化”大旗，用优秀作品来阐释“新疆印象，不同创意”的境界。

印象2011

2011捷报频传，该院的昌吉市水岸林居获中国建筑学会2011年“全国人居经典建筑规划设计方案”规划、环境双金奖；昌吉回民小吃街获住房和城乡建设部环境艺术委员会“中国环境艺术奖最佳范例奖”；库尔勒梨园民街“华夏第一包”获“中国环境艺术奖最佳创作奖”，同时院长王元新、副院长王洋双双获得“全国百位优秀环境艺术师”称号。新疆印象建设规划设计研究院（有限公司）被国家住房与城乡建设部中国建筑文化中心《中国建筑设计作品年鉴》编委会授予“中国最具创新力建筑设计机构”荣誉称号，同时授予院长王元新“中国建筑设计行业杰出贡献人物”称号，授予总建筑师李萱“中国建筑设计行业卓越贡献人物”荣誉称号。该院还获得了中国民族建筑研究会授予的“2011年度中国民族建筑事业杰出贡献奖”，院长王元新获得“2011年度中国民族建筑事业杰出贡献奖”。

地址：乌鲁木齐高新区苏州东街568号
金邦大厦15楼、11楼、10楼、7楼
电话：+86-991-7819196
传真：+86-991-7819189
邮箱：irismuse_shao@163.com
网址：www.xjyinxiang.com

Add: 15 Floor, 11 Floor,10 Floor,7Floor Jinbang Building,No.568 Suzhou East street, Hign-tech District,Urumqi,Xinjiang
Tel: +86-991-7819196
Fax: +86-991-7819189
E-mail: irismuse_shao@163.com
Web: www.xjyinxiang.com

昌吉滨湖河花间集休闲街建设工程

Lakeshore River With Flowers Leisure Street Construction Project, Changji

设 计 师：王元新、徐乃钦、韩思雯、戴金梅、张世鹏、韦世武、梁爽、邵丽伟
项目地点：新疆 昌吉
用地面积：39 000 m²
建筑面积：50 000 m²

Designer: Yuanxin Wang, Naiqin Xu, Siwen Han, Jinmei Dai, Shipeng Zhang, Shiwu Wei, Shuang Liang, Liwei Shao
Location: Changji, Xinjiang
Site Area: 39,000 m²
Building Area: 50,000 m²

根据滨湖河的区域位置、场地现状和周边环境，在规划设计之前对基地的用地适宜性、使用者人群、滨湖河对外交通和外部视觉景观等作出了详细的分析，以便使滨湖河规划功能合理、健全，景观优美并与城市环境有机结合。分析目前滨湖河的道路系统基本完善、缺乏景观性的现状，本次设计主要完善滨湖河的景观效果，遵循整体结构承上启下、左右衔接，空间设计张弛有度、舒适宜人，道路系统层次分析、功能清晰，单体形象大方得体、清新自然的规划目标，使得这里有人们所需的物质生活品质，以及精神生活得以满足。

According to the geography, site conditions and surroundings of Lakeshore River, we have made detailed analysis on the applicability, users group, external traffic and external visual landscape of the project base before planning & design, so as to integrate reasonable functions with beautiful landscapes and urban environment. At present, the road system of Lakeshore River is generally complete, lacking landscape, so this design is mainly to improve the landscaping effect of Lakeshore River, following the integral structure continuation and linkage; the space design is flexible, agreeable; the road system has distinct levels and clear functions; every building is generous and decent, with natural planning objective, enabling material life and spiritual life quality for residents who can enjoy life here.

屯垦博物馆
Farming Defense Museum

设 计 师：王元新、袁赵君
项目地点：新疆 乌鲁木齐
用地面积：153 000 m²
建筑面积：5000 m²

Designer: Yuanxin Wang, Zhaojun Yuan
Location: Urumqi, Xinjiang
Site Area: 153,000 m^2
Building Area: 5,000 m^2

本工程立于104团新七连连部南侧。整体布局遵循中华文化理念，东西南北中风水精神，选择对称、均衡、互补、阴阳、虚实的建城原理，构筑东方建筑文化特色。

整体构成结合内部路网形成“中”字与“车”字的结合、构成文化形象。主馆分馆围合布置，形成独特的屯垦文化印迹。

建筑场地与外部路网形成“鼎”字，意为屯垦戍边，鼎盛千秋。采用中国传统花纹样式的各类条田分布周边，穿插景观绿化带在其中，有序又不失灵性。

七馆重复形成序列，又以中轴对称布置。馆与馆之间的连廊，设置汉阙。建筑连廊的外侧墙壁均为浮雕，展现出各时期屯垦戍边的历史场面。

The project is located on the south side of the camp of New 7th Company of No. 104 Regiment. The layout plan follows Chinese cultural ideas, and the four directions represent traditional Chinese geotechnical spirits, to be symmetric, balanced, complementary, downsun-sunward, unreal-real, to construct the featured Oriental architectural culture.

The structure combines with inside road network to form Chinese characters “中” and “车”, serving as cultural images. The main hall and branch hall enclose a cultural impression of special farming defense.

The building site and outside road network form the Chinese character “鼎”, implying the farming defense policy supporting the country for ever. Chinese traditional strips surround the architecture, interlaid with landscaping greenbelts. The sequence is also flexible.

The seven halls form into a sequence of museum, symmetric by the central axis. The corridor between the halls has palatial watchtowers in Han style. The external walls are all relief sculptures, displaying the historic scenes of farming defense in different historic periods.

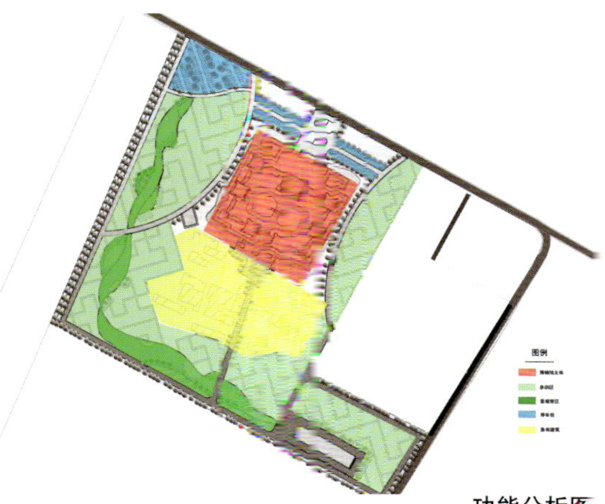

功能分析图

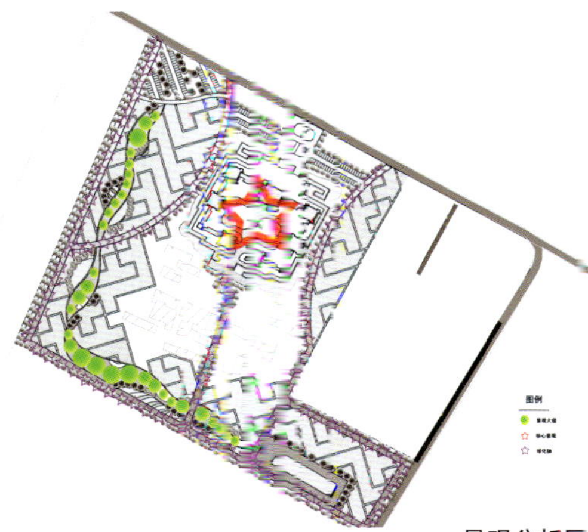

景观分析图

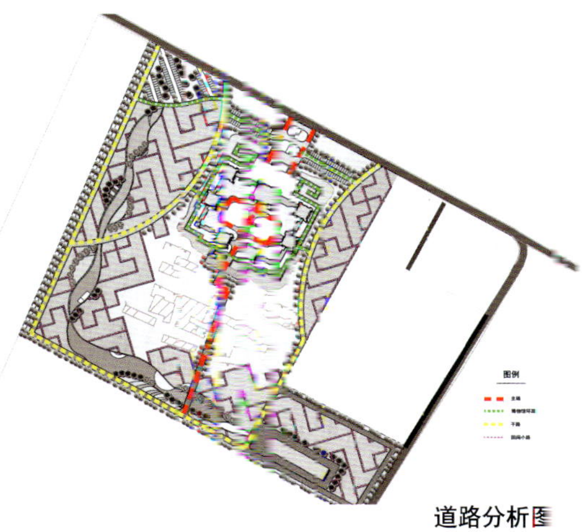

道路分析图

新疆维吾尔自治区建筑设计研究院
Xinjiang Architectural Design Institute

新疆维吾尔自治区建筑设计研究院成立于1956年，是新疆最大的国家甲级建筑勘察设计单位之一，技术实力和业务规模居于全国同行前列，主要从事建筑工程设计、城市规划、市政工程设计、工程勘察、工程咨询、工程监理、工程概预算等业务。

该院现有正式职工570多人，集中了新疆建筑设计行业的主要设计力量和优秀专家，拥有中国工程院院士1人、全国勘察设计大师1人、享受国务院政府特殊津贴的专家9人。

该院设计创作了一大批新疆不同历史时期的代表作品，如：20世纪的人民剧场、昆仑宾馆、新疆人民会堂、新疆科技馆、新疆迎宾馆；21世纪至今的海德酒店、中银广场、新疆国际大巴扎、新疆博物馆、新疆最高建筑中天大厦、新疆体育中心、乌鲁木齐国际机场T1、T2、T3航站楼等。

地址：新疆乌鲁木齐市光明路125号
电话：+86-991-8869192（总机）
传真：+86-991-8865683
邮箱：xadi@vip.163.com

Add: No.125 Guangming Road, Urumqi City, Xinjiang
Tel: +86-991-8869192
Fax: +86-991-8865683
E-mail: xadi@vip.163.com

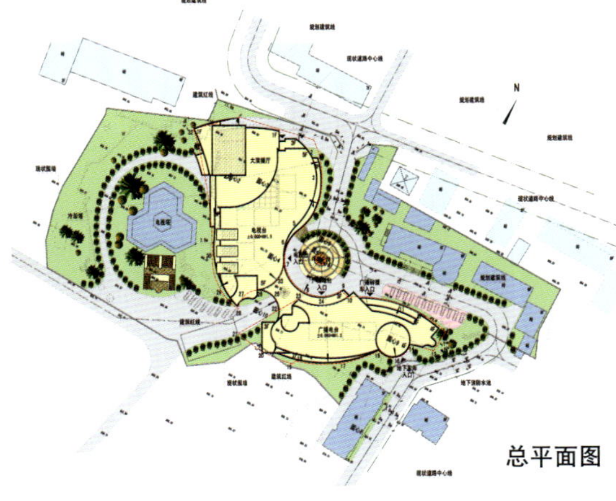

总平面图

乌鲁木齐广播电视技术中心
Radio & TV Technology Center, Urumqi

项目地点：新疆 乌鲁木齐
建筑面积：18 350.84 m^2

Location: Urumqi, Xinjiang
Building Area: 18,350.84 m^2

该工程为集广播电台、电视台两大传媒机构的技术办公、节目制作与播出、大型节目演播交流于一体的综合性公共建筑。坐落于乌鲁木齐红山顶，处于“四山三塔”生态景观带，显著的地理环境特征，特殊的功能，独特的建筑形式与空间形象，突出了其标志性。

设计功能完善、布局合理、流线清晰。建筑形态反映了建筑特点并与地理环境特征相协调。

乌鲁木齐人民广播电台

新疆人民广播电台播控译制楼

Broadcasting Control & Translation Building of Xinjiang People's Radio Station

项目地点：新疆 乌鲁木齐
建筑面积：22 000 m^2

Location: Urumqi, Xinjiang
Building Area: 22,000 m^2

该项目位于乌鲁木齐市团结路新疆维吾尔自治区广电局院内，主要功能为播音室、演播厅、录音棚、主控机房及附属办公。
设计注重挖掘独特的文化元素与建筑功能、造型巧妙结合，创造出具有独特场地精神的现代广播建筑。本项目设计获“建设部优秀工程设计”三等奖。

新疆广汇企业集团中天广场

Zhongtian Plaza, Guanghui Group, Xinjiang

项目地点：新疆 乌鲁木齐　Location: Urumqi, Xinjiang
建筑面积：110 600 m^2　Building Area: 110,600 m^2

中天广场是新疆维吾尔自治区首府乌鲁木齐市地标性建筑之一，建筑主体高211.5 m，至塔尖229 m。
中天广场外观为简约、挺拔的塔形，主楼顶部斜向收进的玻璃晶体，寓意天山重重雪峰，同时象征广汇集团蒸蒸日上的事业，将现代感与地域性较完美地结合在一起。

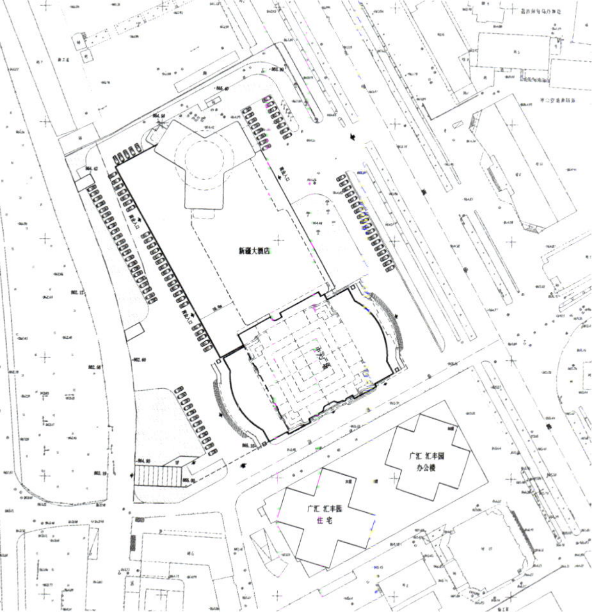
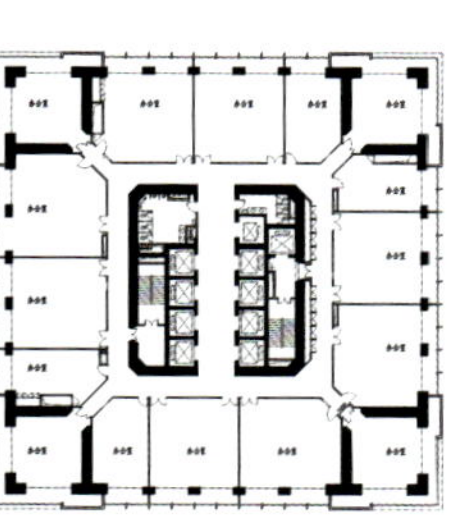
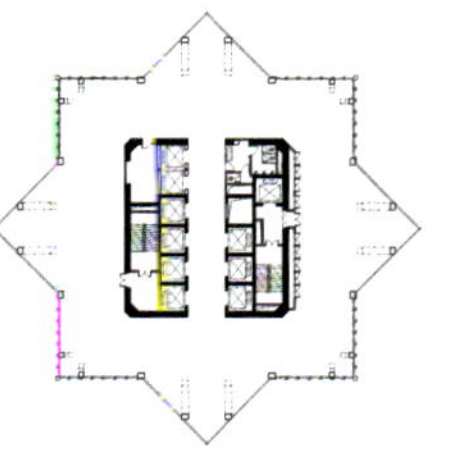
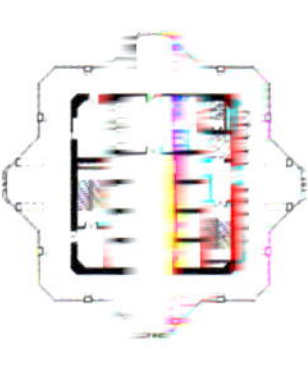
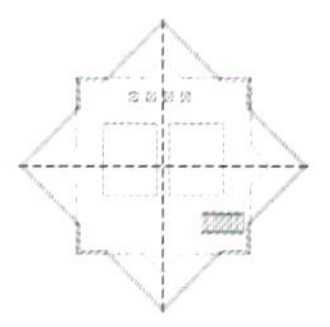

扫描查看更多信息

云南华凌建筑设计有限公司
YUN NAN HUA LING BUILDING DESIGN CO.,LTD.

云南华凌建筑设计有限公司，成立于2005年，具有建筑工程设计、咨询及装饰设计甲级资质。公司立足于昆明，面向全国，设计项目遍及云南省内外多个城市，是昆明地区较有影响力的综合型建筑设计公司之一。公司经过多年艰苦创业，建立并完善了适应市场竞争的自我激励、自我约束和自我发展的内部机制，已逐步形成规模化经营与管理。

公司业务以建筑设计为主，服务领域涉及城市规划、景观设计、室内外设计等。一批优秀的设计项目已相继建成，并得到业主及社会各界的广泛好评。为提高设计师的业务水平和设计作品的质量，增强自身竞争力，公司先后与国内外多家知名设计公司合作，相互学习，使公司保持比较先进的设计理念和手法。

总公司
地址：昆明市官渡区关上中心区关兴路239号8层
电话：+86-871-7197609
传真：+86-871-5711559
邮箱：809733532@qq.com

分公司
地址：昆明市二环东路685号6楼
电话：+86-871-3306883/3306433
传真：+86-871-3136725
邮箱：554148156@qq.com

Company
Add: 8th Floor,Number 239 Guanxing Road,Guandu District, Kunming
Tel: +86-871-7197609
Fax: +86-871-5711559
E-mail: 809733532@qq.com

Branch
Add: 6th Floor, No.685 East second Ring Road. Kunming.
Tel: +86-871-3306883/3306433
Fax: +86-871-3136725
E-mail: 554148156@qq.com

百盛财富中心
Parkson Center of Wealth

项目地点：云南 个旧
建筑面积：120 243.37 m²

Location: Gejiu, Yunnan
Building Area:120,243.37 m²

某公安分局行政办公楼
A Branch of Public Security Administrative Office Building

项目地点：云南
建筑面积：28 343 m²

Location: Yunnan
Building area: 28,343 m²

某县行政中心拆除重建项目

A County Administration Center Demolition and Reconstruction Project

项目地点：云南
建筑面积：31 815.56 m²

Location: Yunnan
Building area: 31,815.56 m²

曲靖汽配城

Qujing Auto Parts City

项目地点：云南 曲靖
建筑面积：20 281.83 m²

Locatoin: Qujing, Yunnan
Building area: 20 28[illegible] m²

石林创富广场

Stone Forest Wealth Square

项目地点：云南 石林
建筑面积：37 400.82 m²

Location: Stone forest, Yunnan
Building area:37,400.82 m²

文庙群艺楼

The Confucious Temple Arts Group Building

项目地点：云南 昆明
建筑面积：2000 m²

Location: Kunming, Yunnan
Building area: 2,000 m²

扫描查看更多信息

昆明兰德设计有限公司

Kunming LAND Design Co.,Ltd.

昆明兰德设计有限公司是中国资深民营勘察设计企业。成立26年来，现已拥有建筑、化工、医药行业设计及装饰装潢、建设工程咨询、工程总承包等资质，已先后承接并出色地完成了二千余项工程设计、项目管理工作，是中国同行业中唯一荣获联合国奖旗的优秀企业。
本公司拥有以12位国际注册高级经理人为核心、28位高级工程师及43位注册工程师为技术骨干组成的复合型专业人才队伍，建成了涵盖国际技术合作投资咨询、建筑、化工石化医药、生物工程等行业的20余个分公司（设计室）。长期以来，公司在国内外工程建设中做出了重要贡献，所取得的业绩闻名遐迩，诸如在大理鹤庆建成了极具民族风格的现代商务中心，在丽江展开了新兴旅游城市战略规划（投资71亿元），云南橡胶基地建设，闻名国内外的昆明关上中心贸易区的开发，滇黔桂石油基地的建成，西南最大的纺织厂的建设，大批制药厂GMP的达标，曲靖富源大型煤化工基地的策划、国内外知名顶级抗癌天然药物紫杉醇基地的建成投产，五星级版纳、万象大酒店的建设，越南制药厂的建成投产，国内大型工业磷酸基地的建设，国内最大实木加工厂的建成、国内最大超临界液态CO_2生物药厂的建成投产，以及数百万平方米房地产开发等，其优良的业绩均得到了社会各界的较高评价。曾荣获"设计精巧誉满春城"、"认真负责 服务优良"、"信得过单位"及"设计之神"等多项奖旗；特别自豪的是承接了联合国援建昆明知名东南亚酒楼建设工程，并以出色成果荣获联合国"设计精巧 独创风格"奖旗。
公司是云南省勘察设计协会副理事长单位、专家组成员。

地址：云南省昆明市东风东路47号建业商务中心25楼
邮编：650041
电话：+86-871-3200488
传真：+86-871-3114758
邮箱：design@land-design.cn
网址：www.land-design.cn

Add: 25Floor, Jianye Busniess Center, No.47 East Dongfeng Road, Kunming, Yunnan
P.C.: 650041
Tel: +86-871-3200488
Fax: +86-871-3114758
E-mail: design@land-design.cn
Web: www.land-design.cn

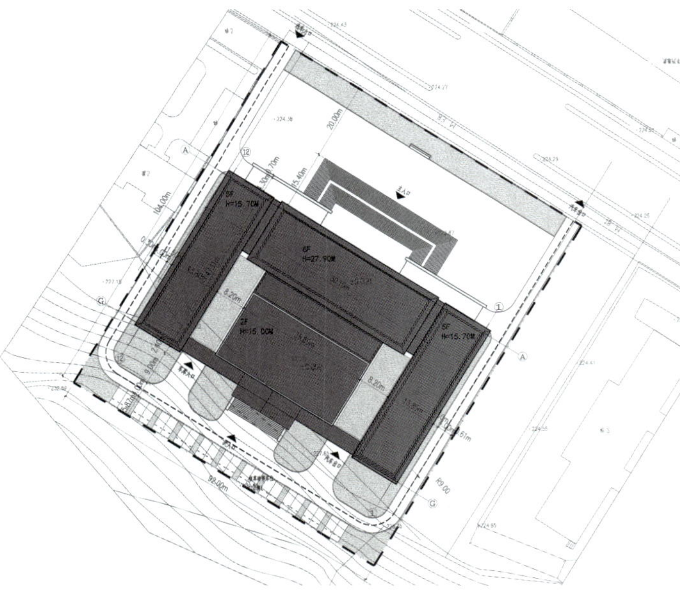

蒙城法院

Mengchen Court

设 计 师：段云龙 Designer: Yunlong Duan
项目地点：云南 Location: Yunnan
建筑面积：10 572 m² Building Area: 10,572 m²

项目为云南某市中级法院审判庭及行政办公大楼，地处该县中心景观大道一侧。立面风格采用了中式屋顶挑檐以迎合地方整体风貌，结合西式建筑石柱元素作为经典司法建筑的符号，以简洁的线条和均衡对称的外观体现司法机关的廉洁公正。建筑立面重点突出了石材柱的挺拔贯通和转角部分石材墙体厚重稳固的视觉效果，以此产生明显的立面阴影和大体块元素，突出司法建筑的庄严、神圣感。
大楼沿街设置办公日常用入口，独立的审判法庭设有独立的入口及大台阶，在办公楼三面围合的空间中营造司法的威严、庄重气氛。外墙凸凹的部分不但可以为空调机等设备提供空间，同时产生更加厚重、稳固的形象。

丽江市市直机关办公楼及会议中心

Public Services Office Building & Conference Center of Lijiang City

项目地点：云南 丽江　　Location: Lijiang, Yunnan

建筑面积：13 336.92 m²　　Building Area: 13,336.92 m²

项目位于丽江市古城区祥和片区香格里大道南段西侧，会议中心以庄严宏伟为设计主旨，同时突显流畅的线条，让人眼前为之一亮。

The project is located on the west side of Shangrila Avenue Southern Section, Xianghe Block, Old Town District, Lijiang City. The conference center is designed to be solemn and magnificent, highlighting the smooth lines, very eye-catching.

扫描查看更多信息

GOA 绿城东方

GOA绿城东方自1998年创立以来，伴随着中国经济的迅速崛起，现已成为中国最具实力的建筑设计机构之一，具有国家建筑工程甲级资质。目前已分别在杭州、上海、北京、南京、宁波等地设立办公室，共拥有专业技术人员近600人，是涵盖都市规划、建筑设计、结构工程、给排水工程、机电设备、暖通设备、室内设计等多领域的专业设计团队。

公司恪守合伙人负责制的项目管理方式，使得每一个项目都由一个或多个合伙人全程控制，从而保证项目品质。在长期的执业实践中，公司与国内外优秀的规划建筑设计、室内设计、景观设计等相关设计及咨询机构建立了良好的业务合作关系，有能力为客户提供从项目前期策划、规划设计直至建筑设计等全过程的高水准设计及综合咨询服务，并通过协调与整合各相关专业的设计服务以进一步提升项目的最终品质。

在10余年的专业服务实践中，公司已为社会奉献了许多优秀的设计作品，其中桂花城、春江花月、北京御园等的设计创意均被业界视为“现象”来关注，这也使公司成为设计潮流的引领者。

Greentown Oriental Architects(GOA), founded in 1998, is a professional design team specialized in Urban Planning, Architectural Design, Structural Engineering, Plumbing Engineering, Electrical Engineering, HVAC Engineering. Riding China's rapid economic growths, it has become one of the best architectural design firms in China and is accredited with National Construction Class A Qualification. With the support of nearly 600 professional staff, the company has set offices in Hangzhou, Shanghai, Beijing, Nanjing and Ningbo.

We always follow partner responsible system in our project management. Every project is directly supervised by at least one partner in order to guarantee our highest service quality. During our past long term design practice, we have established excellent professional relationships with domestic and foreign firms which are outstanding in urban planning, architectural design, interior design, landscape design, etc. We are capable of providing high quality professional services from project strategy planning, master planning to architectural design. We also always coordinate with other related professions to further enhance the ultimate quality of the project.

In over ten years of professional practice, we have provided a lot of excellent design work, such as Osmanthus Town, Chun Jiang Hua Yue, Beijing Majestic Mansion and etc. The design concepts of these projects have been considered as a "phenomenon" by the peers. It also makes us become the trendsetter of architectural design.

地址：浙江省杭州市古墩路389号
电话：+86-571-88366180
传真：+86-571-88366080
邮箱：goa@goa.com.cn
网址：www.goa.com.cn

Add: 389 Gudun Road, Hangzhou, Zhejiang, China
Tel: +86-571-88366180
Fax: +86-571-88366080
Email: goa@goa.com.cn
Web: www.goa.com.cn

宁波科技研发基地

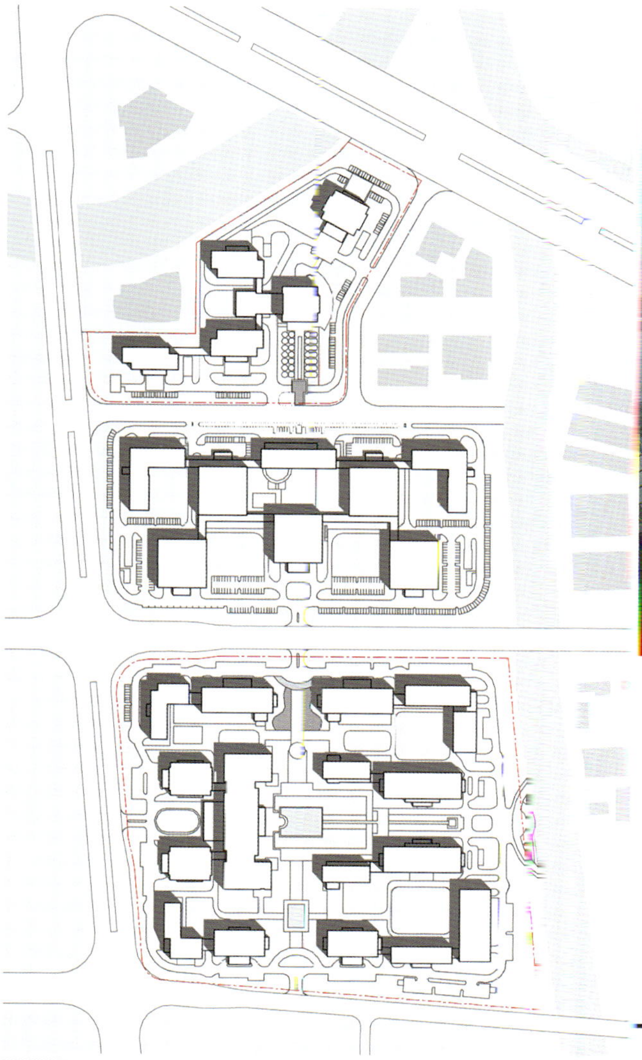

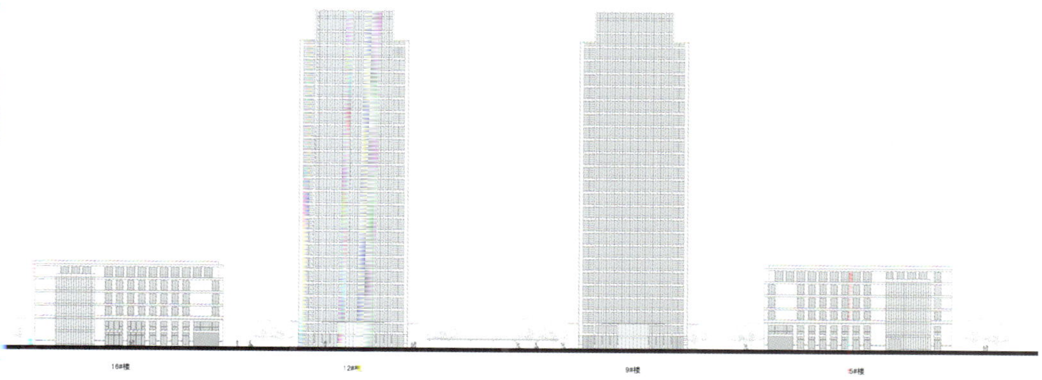

宁波科技研发基地

Science and Technology R&D Base, Ningbo

项目地点：浙江 宁波
建筑规模：620 000 m²

Location: Ningbo, Zhejiang
Building Area: 620,000 m²

项目基地分三期开发设计，周期较长，但良好的对位和轴线关系，同中求异的细部设计，使得项目获得了稳定而丰富的空间效果。项目采用高层和多层相结合的群体关系，同时引入庭院空间的概念将每个用地分成若干个庭院空间，点式或板式的高层建筑作为庭院的控制性体量，多层建筑则与高层建筑相互对位呼应，最终围合形成组团庭院。并通过高层建筑之间的互相对位，使组团与组团之间也能形成良好关系。这样围合形成的半封闭的庭院空间，结合景观处理，创造出宁静、稳定的办公环境。

This project base is developed and designed in three phases with a long term. However, the project has gained stable and rich space effects by its good contraposition and axial relations, as well as details designed in difference and general resemblance. The project combines high-rise and multi-floor buildings together. Meanwhile, it introduces the concept of courtyard space, to divide each land into several courtyard spaces with high-rise buildings of point type or plate type as the controlled volume of courtyard, while the multi-floor buildings echo with high-rise buildings to finally enclose a group courtyard. The mutual contraposition of high-rise buildings enables the group courtyards to form good relations with each other. This semi-enclosed courtyard space integrates with the treated landscape to create a tranquil and stable office environment.

苏州御园
Suzhou Majestic Mansion

项目地点：江苏 苏州
建筑面积：220 000 m²

Location: Suzhou, Jiangsu
Building Area: 220,000 m²

整体住宅布局分为合院住宅和平层住宅两大片区：基地西南部的合院住宅旨在打造尊贵大气的社区核心空间，通过围合的建筑组合营造安定、宜人的邻里空间；沿社区东侧和北侧的城市道路，布置五排平层住宅，结合社区主入口、会所和社区围墙，打造连续变化的城市界面和气派、精致的社区对外形象。
项目力求将住宅的庭院真正打造成为私家自由生活的天地，尤其通过合院住宅这一产品的研发和呈现，巧妙地解决了相邻住户间视线的相互干扰问题，一系列建筑灰空间和组合庭院的设置，使得住宅室内外得以密切对话和无缝相融，营造舒适、宜人的庭院生活空间。

The general residential layout is divided into two big zones, namely courtyard residential zone and flat residential zone. The courtyard residences in the southwestern base create a elegant and grand core space of community, and the enclosing architecture creates stable and pleasant neighborhood space. Along the urban roads on the east and north sides of the community set five rows of flat residences, together with the main community entrance, clubs and community enclosing walls, to create a continuous varying city interface as well as gorgeous and delicate external image of community.
This project strives to build the residential courtyard into a real private world of free life. In particular, the development and presentation of courtyard residences have skillfully resolved the mutual interference of sight to courtyard between the neighboring residents. A series of gray spaces and combined courtyards are set to achieve the close dialogue and seamless integration between the interior and exterior of residence, which has built a comfortable and pleasant courtyard living space.

桐庐桂花园
Sweet Osmanthus, Tonglu

项目地点：浙江 桐庐
建筑面积：160 000 m²

Location: Tonglu, Zhejiang
Building Area: 160,000 m²

项目地处桐庐江南新区核心区域，四周青山环抱，内拥5700 m²自然湖泊，地理位置得天独厚，景观优势明显。规划采用人车分流的道路系统，以一条贯穿地块南北的车行主干道为骨架串联起各组团，机动车于组团外就被引入地库，组团内部为纯粹的人行步道系统；园区空间布局灵活，在强调各组团独立性的同时，也在组团内部的围合感以及进入组团的仪式感方面有所加强。住宅类型以多层、高层和平层官邸为主，辅以少量排屋和商铺，力求营造低密度、低容积率、高绿化率的大型现代生活园区。

This project is located in the core area of Jiangnan New District of Tonglu County. It is surrounded by green mountains and contains 5,700 m² natural lakes. The location enjoys advantageous geography and landscape. The planning is to build a road system diverting pedestrian flow from vehicular flow, with an arterial driveway penetrating the land lot from north to south as the frame to connect each group together. The motor vehicles enter into the basement from the group entrance and the group interior is purely a pedestrian system. The layout of park space is flexible, which strengthens the enclosure sense of group interior and the ritual sense of entry into the group, while emphasizing the independence of each group. The residential buildings are mainly multi-storey buildings, high-rise buildings and flat mansions with a few townhouses and shops as auxiliaries. It strives to build a large and modern life park with low density, low plot ratio and high green ratio.

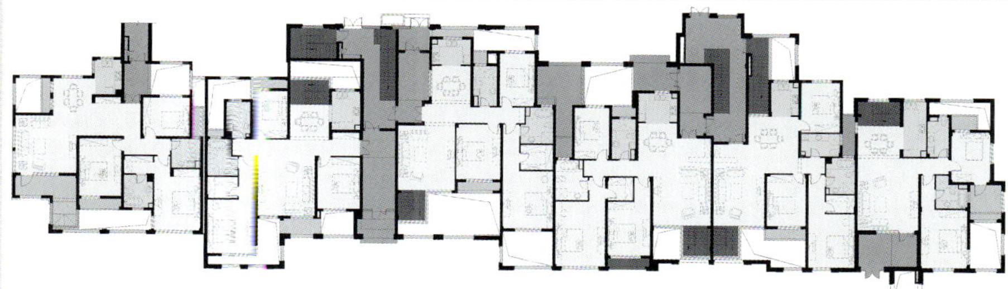

一层平面图

杭州九里松度假酒店
Jiulisong Holiday Hotel, Hangzhou

项目地点：浙江 杭州　Location: Hangzhou, Zhejiang
建筑面积：5500 m²　Building Area: 5,500 m²

九里松酒店是一座始建于20世纪后期的砖混结构建筑，经过几次改建，建筑形体复杂，高低错落。项目业主对于改建的期望是使其成为一座具有中国传统意味空间的当代商务酒店。所以此次建筑的更新并非一次建筑外观和样式的改变，而是在现有的环境和建筑结构基础上，重新安排建筑内部功能及空间，并建立建筑内部与外部的统一的空间关系与形式秩序。设计着重关注：整理进入建筑的空间序列，建立内外部空间的使用与视线关系，以及建筑外墙的更新。在此次改建中，严谨的现状结构测评和加固设计，周到的酒店功能配置和服务设计均得以贯彻，同时业主对于当代中式的诉求，也在材料运用，形态比例和空间序列的设计中得以含蓄表达。

Murraya pine hotel is a built in last century later brickconcrete building, after a few times alteration, building complex shape, the floor height strewn at random project owner for rebuilding expectations is to become a space that has Chinese traditional means of contemporary business hotel so the building update is not a architectural appearance and style change, but in the current environment and building structure basis, rearranging building internal function and space, and establish building internal and external uniform spatial relationship and form order design focuses on: finish into the building space sequence, the establishment of internal and external space use and line of sight relationship, and building exterior wall update in the reconstruction, the rigorous situation structure evaluation and reinforcement design, good hotel function configuration and service design are to be implemented, and at the same time the owner for contemporary Chinese style.

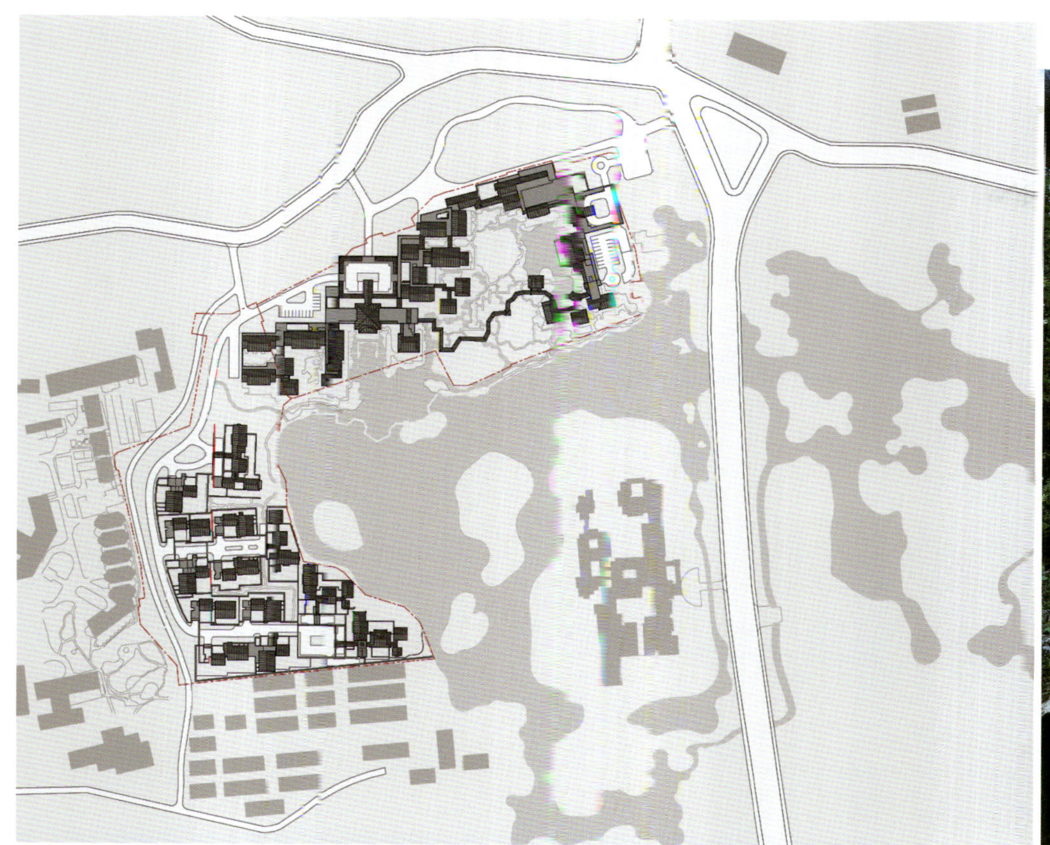

杭州西子湖四季酒店
Four Seasons Hotel Hangzhou at West Lake

项目地点：浙江 杭州
建筑面积：43 000 m²

Location: Hangzhou, Zhejiang
Building Area: 43,000 m²

酒店位于有着悠久历史文化和景观特性的西湖景区。设计将强化西湖“如画”的意境、营造西湖新风景作为努力方向，并将发掘景观因素、延续自然意境作为酒店主题和营造重点。整个建筑群体采用传统江南庭院的建筑风格，用传统建筑的语言和方法，对建筑整体关系、庭院空间营造进行了精心思考与设计。设计将“入口庭院—大堂—泳池—跌水—湖面—别墅区客房—远山”作为酒店的景观主轴和中心意向，并做了层次丰富的营造，使得建筑与周边湖景结合，具有长久的吸引力和感染力。

This hotel is located in West Lake of long history and culture, with landscape characteristics. The design strives to strengthen the 'picturesque' imagery of West Lake and build a new landscape in West Lake. Meanwhile, it is to explore the landscape factors and continue the natural conception, which are the main sceneries and construction emphasis of hotel. The entire building cluster and landscape are built in traditional South Chinese courtyard style. The integrity of buildings and construction of courtyard space are carefully thought and designed with traditional architectural language and methods. With "entrance courtyard – lobby – swimming pool – drop water – lake surface – hotel rooms of villa area – distant hills" as principal axis and central intention of hotel landscape, the design is to construct various layers to integrate the buildings with the surrounding lakes sceneries which has lasting attractiveness and appeals.

嵊州市新医院
New Hospital of Shengzhou

项目地点：浙江 嵊州
建筑规模：一期149 000 m²

Location: Shengzhou, Zhejiang
Building Area: Phase I 149,000 m²

为了在保证医院使用功能的前提下建设一座分散的园林式医院，将原本的医疗街作了一些小小的改动，使它的形状变成一个围绕中心庭园弯折的"L"形。医院核心部分的门诊、急诊和医技分别分散于几栋尺度相近的4层建筑中，"挂在""L"形医疗街的外侧，它们之间仅在一层相连通。结合医院的标志系统，希望通过简单明确的目标位置，使医务人员和病患在绝对运动距离因分散布局而增加的情况下，尽量减少无效的运动。同时，作为人流量最大的"L"形医疗街，因为单侧向庭园开放而变得生动有趣，在心理上降低病患在就医过程中的紧张和不适感。

In order to build a scattered garden hospital without affecting the hospital functions, the project makes some minor changes to the original Medical Street, to bend it into a "L" shape surrounding the central courtyard, the inside of which is directly faced toward the courtyard. As the core parts of hospital, the outpatient department, emergency department and medical technology department are separately scattered in several four-storey buildings of similar scale and are 'hung' on the outside of medical street, with only one storey connected with each other. Considering the existing hospital marking system, the design uses simple and clear target locations to minimize useless movement, given medical staff and patients' absolute motion distance is increasing due to the scattered layout. Meanwhile, the L-shaped Medical Street with the greatest pedestrian flow becomes lively and interesting due to its one-sided openness to the garden, which can reduce the nervousness and discomfort of patients in the process of medical treatment.

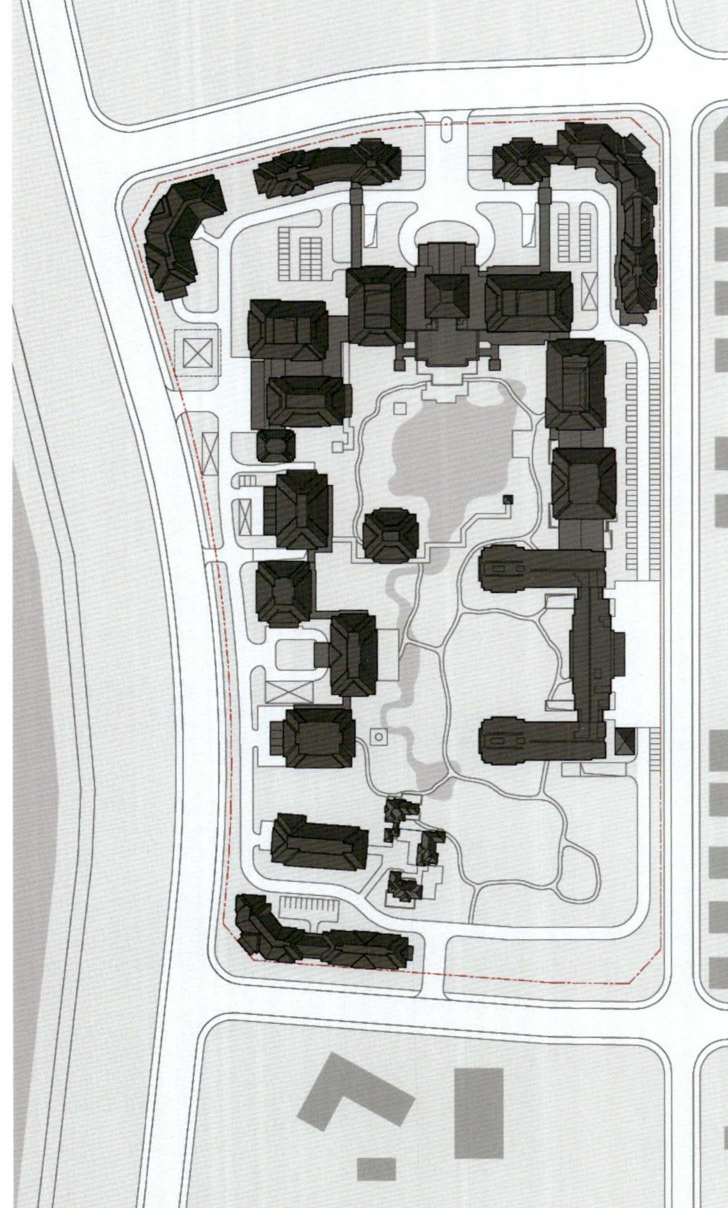

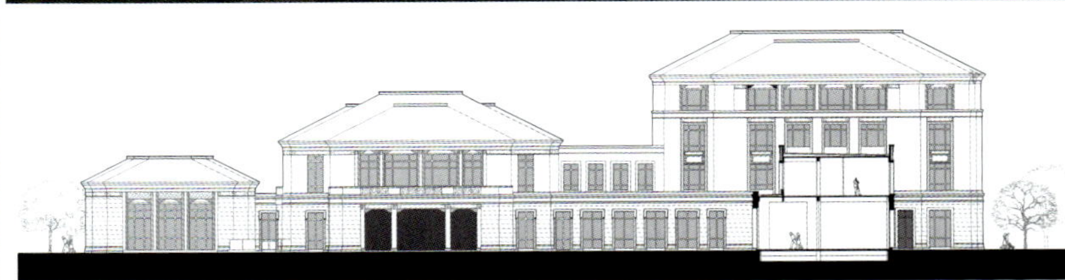

东立面

百大绿城·西子国际

Baida Greentown · Xizi International Center

项目地点：浙江 杭州　　Location: Hangzhou, Zhejiang

建筑规模：276 000 m²　　Building Area: 276,000 m²

项目处于杭州庆春广场商圈，是集居住、商业、娱乐、办公、购物等多功能于一体的复合高品质城市综合体。建筑师将设计理念一直贯彻到景观设计及室内设计中，有非常高的完成度。在设计阶段，建筑师就提出了和建筑概念相吻合的景观、室内空间、灯光、商业店招牌甚至标志设计的概念方案，这就给其他顾问设计公司提出了更准确的任务书。建筑不是孤立的体量，在各功能建成、投入使用之后，会有各种元素强加于建筑之上，以致失控。所以不仅需要在设计上考虑概念的贯彻，还需要在方案阶段考虑建成后可能遇到的种种状况。

This project, located in Qingchun Plaza business area, is a multifunctional and compound high-quality city complex integrating residence, commerce, entertainment, office, shopping and other functions. The architects have implemented the designing ideas into landscape design and interior design with high degree of completion. In the design stage, the architects put forward the conceptual plan of landscape, interior space, lighting, logo and even signage design that coincide with the architectural concept, which has given more accurate task requirements to other design consulting firms. The building is not isolated and upon completion and operation, various elements will be imposed to the building out of control. Therefore, it not only needs to consider concept implementation in design, but also needs to consider the various possible post-completion situations even in the conception stage.

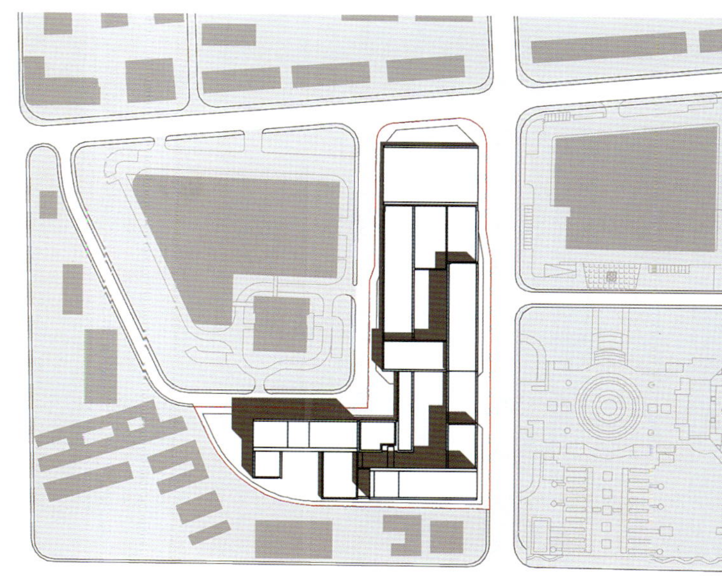

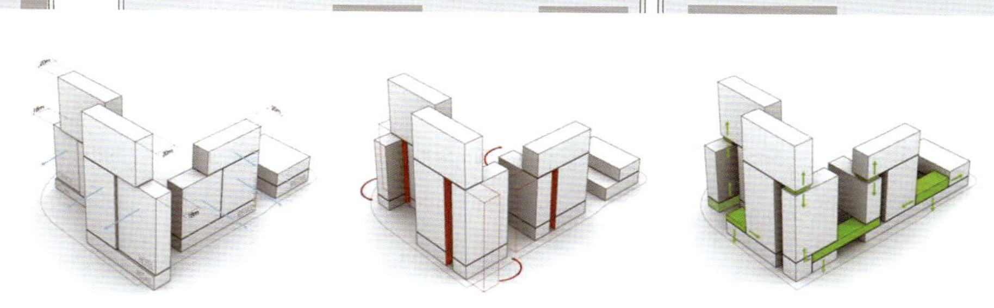

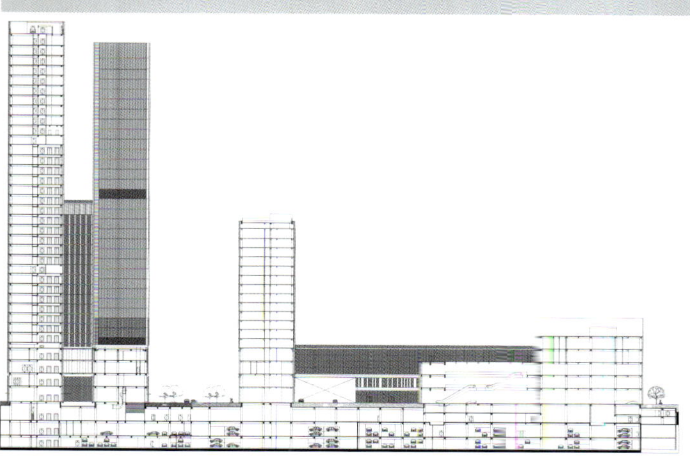

剖面图

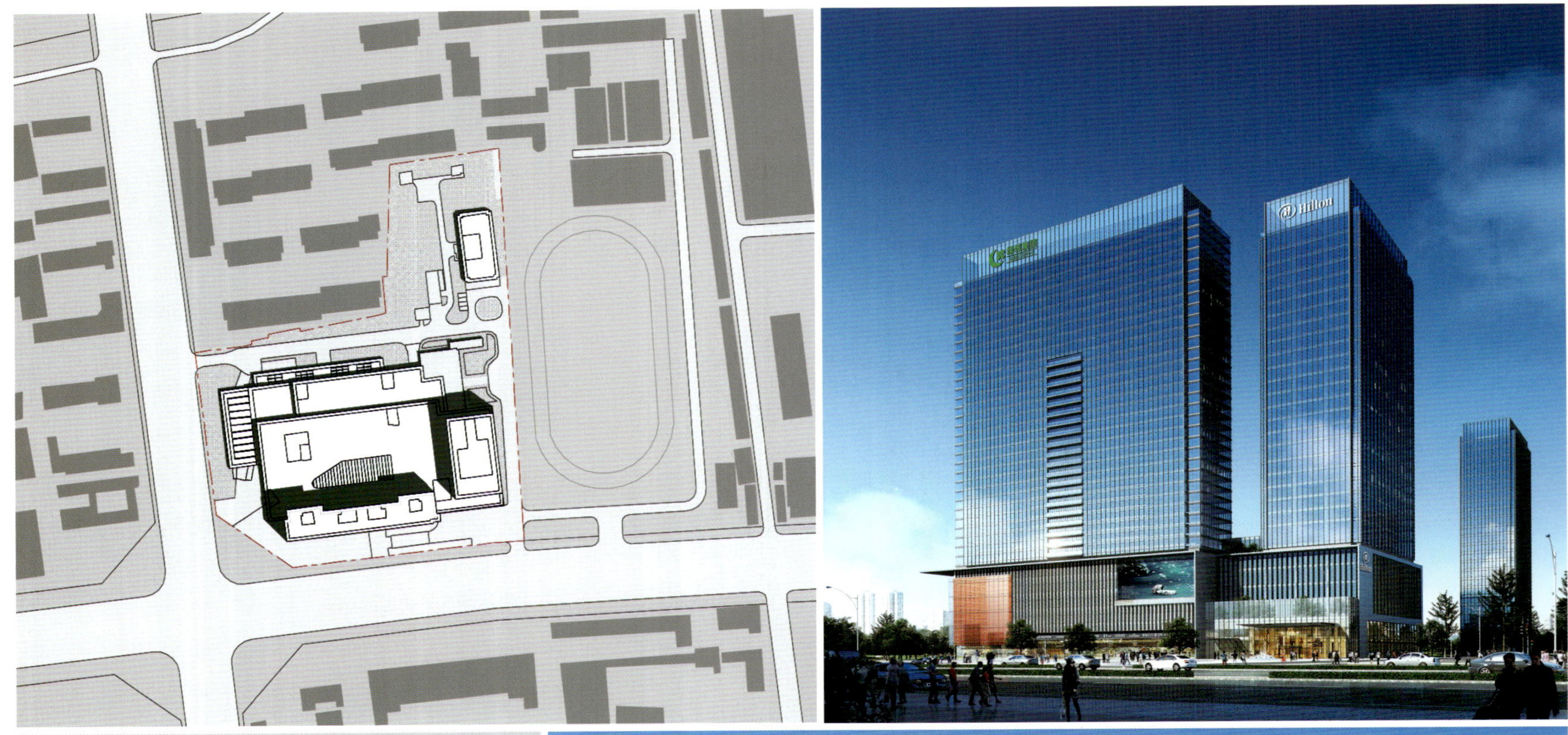

京杭广场
Jinghang Plaza

项目地址：北京　Location: Beijing
建筑面积：278 000 m²　Building Area: 278,000 m²

项目位于通州区的核心地段，是通州新城的启动项目。设计通过对人流、车流的精心组织，使得整个综合体繁杂的流线得以有序地解决。为实现各种不同的功能，一方面用玻璃幕墙作为最主要的立面要素来保证项目的整体感，使之在周围的凌乱秩序中突显项目的整体气势；另一方面，创造出不同品类的空间使所有的使用者能够各得其所。安静优美的屋顶花园是住宅的私享领地，生动的六层中厅给商业注入生气和活力，热闹的下沉广场是城市的舞台。建筑群力图以积极而开放的姿态，在城市生活中扮演一个充满感染力的角色。

This project, located in the core area of Tongzhou District, is the initial project of Tongzhou New Town. The design carefully organizes the pedestrian flow and vehicular flow to streamline the complicated traffic in order. To meet different demands, the façade with glass curtain wall as the major element ensures the project integrity standing out of peripheral disorder; on the other hand, a variety of spaces satisfy different users. The tranquil rooftop garden is the private residential realm to enjoy. The lively atrium of six floors brings vitality and vigor to the business atmosphere, and the noisy sinking plaza is a city platform. The building cluster seeks to play an appealing role in city life with a positive and open attitude.

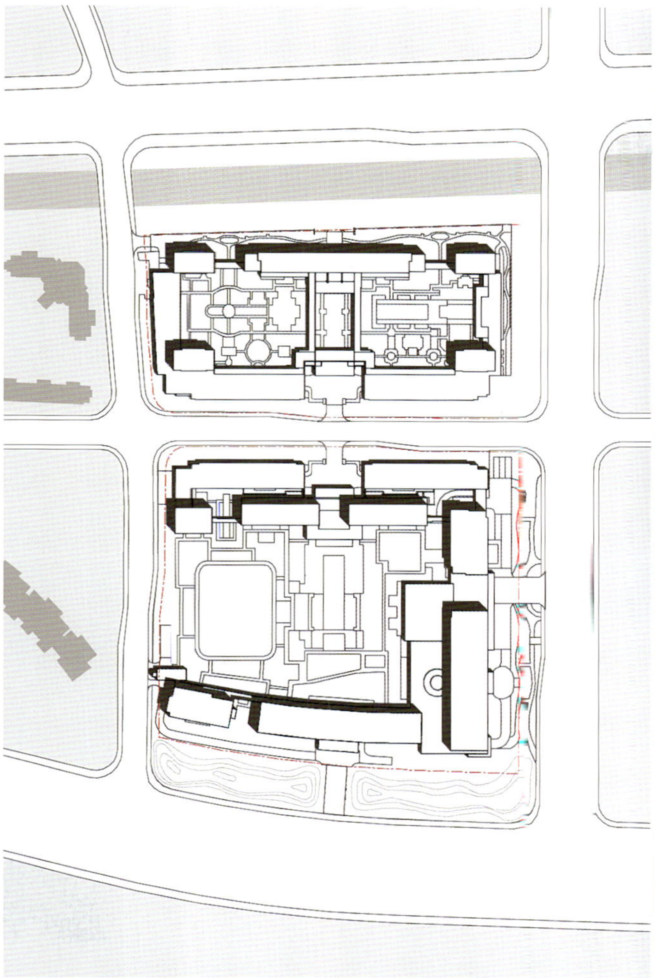

蓝色钱江

Sapphire Mansion

项目地点：浙江 杭州
用地面积：8.4 hm^2
建筑面积：406 800 m^2

Location: Hangzhou, Zhejiang
Site Area: 84,000 m^2
Building Area: 406,800 m^2

设计试图探寻一条最佳途径以诠释城市综合体在城市特定区域所应承担的引领作用。经过多次调整，总体布局更注重建筑组群间的主次关系及与城市的联系：住宅组群横跨南北两个地块，沿南北向中轴线呈对称展开，让出南地块的东侧布置酒店办公综合楼，以便与两地块间的商业街形成连续的商业界面，并呼应东侧的办公及社区公园，为整个区块注入了活力。设计将玻璃、石材与金属等元素进行巧妙组合，在追求简洁大气的建筑形态的同时，塑造出层次丰富的立面细节，使整个项目焕发出与钱江新城区域现代化氛围相协调的时代气息。

The design seeks to explore the best way to interpret the leading role of building complex in specific areas of city. The general layout, through many adjustments, pays more attentions to the primary and secondary relations between building clusters and the connections with city. The residential building groups stretch across the north lot and the south lot, and spread along the north-south symmetric axes. The land on the east side of south lot set the multipurpose building for hotels and offices to form a continuous commercial interface together with the commercial street between two lots. Meanwhile, it is to echo with the office and community park on the east side and bring vitality to the whole lot. The design skillfully combines glass, stone materials, metal and other elements, seek for concise and grand architecture, and build the façade details of rich layers, which enable the entire project to reveal the modern spirit coordinating with the regional modernized atmosphere of Qianjiang New Town.

新昌玫瑰园
Rose Garden, Xinchang County

项目地点：浙江 绍兴 Location: Shaoxing, Zhejiang
用地面积：25.3 hm² Site Area: 25.3 hm²

项目分为南北两片，北区为酒店用地，南区为住宅用地，将开发建设一个以酒店和低密度住宅相结合的高端社区。在本项目的规划设计中，基地拥有优美的自然景观条件，而要强化居住与景观的联系，创造一种富有诗意的栖居环境，就必须采用整体设计的方法。本案力求结合基地特征，最大限度地保持基地原有环境，在此基础上以酒店和低层住宅作为主要建筑类型，力求使其与环境结合形成不同的特色，并通过环境将其整合成为一个有机而丰富的整体。

This project includes the north area and the south area. The north area is for hotels while the south area is for residence. It is to build a high-end community integrating hotels and low-density residences. In the planning design of this project, the base has beautiful natural landscape. However, it must adopt the integral design method to strengthen the connections between residences and landscape and to create a poetic dwelling environment. This project strives to combine the base characteristics and tries to best to retain its original environment, and on this basis, integrate the principal building types, namely the hotels and low-storey residences, into the environment with different characteristics and rich contents.

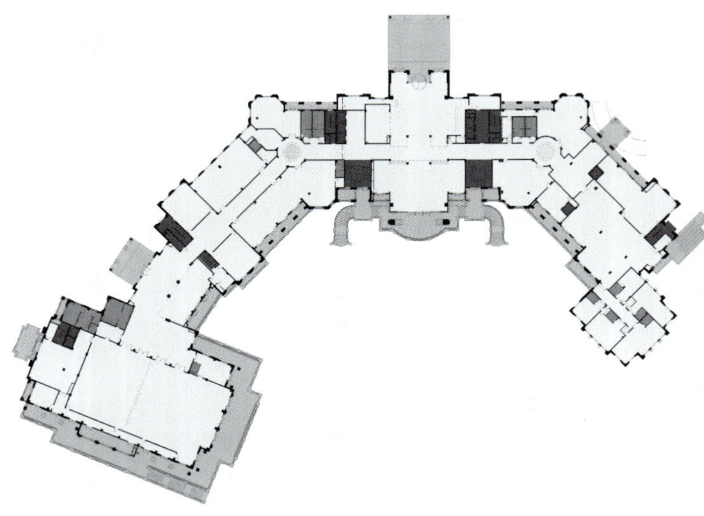

杭州普立策建筑设计有限公司
Hangzhou PLC Architectural Design Co.,Ltd.

杭州普立策建筑设计有限公司（香港PLC国际设计集团）公司成立于2008年，参照国际化的设计服务模式，为客户提供房产开发咨询、城市规划咨询和建筑设计等专业服务。
普立策设计（PLCD）拥有众多对建筑充满热情的优秀建筑师、结构和设备工程师，具有丰富的项目运作经验。目前公司能提供高端别墅、豪华公寓、城市综合体、办公建筑、学校建筑等各类造型项目从规划设计、概念设计到施工阶段的全过程服务。
普立策设计（PLCD）自成立之日起，就致力于将优秀的建筑理念、技术手段和设计手法融为一体，追求高质量设计。坚持为业主提供完善的、高水准的专业服务，坚持环境效益，社会效益和业主的经济利益相结合，真诚沟通，相互尊重，长期合作，价值共享。
普立策设计（PLCD）始终将企业的生存发展与个人的事业前途紧密结合，坚持企业和员工共同成长，同心协力形成一个有朝气、有理想、有创造力、有共同目标的战斗团体。

设计创造价值

普立策设计（PLCD）始终坚持设计创造价值为公司的第一目标，只有设计为客户创造了价值才能赢得市场，这也是公司能在目前严峻的市场竞争中得以生存并不断壮大的根本原因，公司始终坚持为客户设计更具价值的规划，筑建的更理想空间，实现自身最佳的盈利模式——和客户双赢。

全程化设计

普立策设计（PLCD）专业从事项目从拿地前的可行性研究分析、项目定位分析、经济技术成本测算到拿地后项目的前期策划、项目业态分析、盈利模式分析、产品定位、建筑方案设计、施工图设计、各专项设计配合和工程实施配合的全过程服务，公司对每个承接的任务都开展全过程跟踪，以便于优秀的设计方案能在现实工程中得到最大程度地实现。

集体创作

普立策设计（PLCD）深知每个设计方案都代表着公司品牌，所以公司坚持集体创作，坚持每个项目的设计方案都必须经过集体讨论，严格把关，多个方案比较后才能推出，这是对客户负责，也是对自己负责。

地址：杭州市拱墅区湖墅南路186-1号丽阳国际商务中心6幢
电话：+86-571-88103291（前台）/+86-571-88102636（业务）
+86-571-88103373（投诉）
传真：+86-571-88103291
邮箱：plcd@plcdesign.cn
网址：www.plcdesign.cn

Add: Tower 6, SunShine International Business Centre, No.186-1 Hushu South Road, Gongshu District,Hangzhou
Tel: +86-571-88103291/88102636/88103373
Fax: +86-571-88103291
E-mail: plcd@plcdesign.cn
Web: www.plcdesign.cn

石家庄西山御园
Western Noble Garden, Shijiazhuang

设 计 师：王巍、黄森、俞越泷、石苏芹
项目地点：河北 石家庄
用地面积：800 000 m^2
建筑面积：77 600 m^2 （一期）

Designer: Wei Wang, Sen Huang, Yuelong Yu, Suqin Shi
Location: Shijiazhuang, Hebei
Site Area: 800,000 m^2
Building Area: 77,600 m^2 (phase 1)

石家庄西山御园项目位于石家庄西山国宾馆北侧，包括小院、合院、排屋、叠排、高层住宅等物业类型，总用地面积达80 hm^2，其中一期总建筑面积为7.76万m^2，现已成为石家庄高端住宅领域的标杆。
以人为本的理念结合当地人文、气候、市场等条件，运用新型设计手法，合理将各建筑单体融合在大自然中。

The project is located north of Xishan Guesthouse, Shijiazhuang City. It includes small courtyards, enclosed courtyards, townhouses, overlaid townhouses, high-rise residential and other buildings, with total land area 800,000 m^2, of which, phase 1 total floor area 77,600 m^2, now completed, benchmarking high-end residential domain of Shijiazhuang City.
Human-based philosophy combined with local culture, climate and market conditions, in novel design manner, has organically integrated all buildings in the Nature.

南京九月森林
September Forests, Nanjing

设 计 师：王巍、陈滋达、俞越泷、石苏芹
项目地点：江苏 南京
建筑面积：110 000 m^2

Designer: Wei Wang, Zida Chen, Yuelong Yu, Suqin Shi
Location: Nanjing, Jiangsu
Building Area: 110,000 m^2

南京九月森林项目位于南京浦口区老山森林公园区块，总建筑面积11万m^2，由小院、合院、叠排等高端住宅组成，目前一期、二期已经建成。
结合地形变化，使建筑天际线变化丰富，总体布局规则。建筑局部边角和沿绿化带处因地制宜，定制扩大户型或地下室打开户型，打造最优居住品质和最大化利用土地资源。法式、意式、西班牙式等经典欧式风格建筑错落布置，结合材质和色彩上的变化，使小区风格丰富多样。

The project is located in Laoshan Forest Park area, Pukou District, Nanjing City. With total land area 110,000 m^2, the project includes small courtyards, enclosed courtyards, townhouses and other high-end residential buildings. Now phase 1 and phase 2 are completed.
The landform enriches the skyline changes in front of the buildings. The general layout is regular, with corners and greening zones appropriate, featuring custom extended house type or basement open house type, to create the highest quality and maximal land efficiency. French-style, Italian-style, Spanish-style and other classical European-style buildings are deployed in order, combining the changes in materials and colors, diversifying the flavors of the community.

舟山朱家尖集散中心宾馆

Tourism Terminal Center Hotel in Zhujiajian,Zhoushan

设 计 师：张敏军、王巍、许培、俞越泷、石苏芹
项目地点：浙江 舟山
用地面积：19 626 m²
建筑面积：34 759.61 m²

Designer: Minjun Zhang, Wei Wang, Pei Xu, Yuelong Yu, Suqin Shi
Location: Zhoushan, Zhejiang
Site Area: 19,626 m²
Building Area: 34,759.61 m²

本项目位于舟山市朱家尖，总用地面积19 626 m²，总建筑面积34 759.61 m²，容积率1.0，在2012年5月招投标中获得第一名，目前正在实施图纸设计中。
在"尊重环境和地域文脉"核心理念上，处理好旅游集散中心整体和局部之间的相互关系，满足酒店各功能板块的合理布局，同时注意整体的风格协调，做到既有统一又有变化，从而打造成一个既和周边环境相协调，又有自己特色的高品质酒店。

The project is located in Zhujiajian Town, Zhoushan city. With total land area 19,626 m², total floor area 34,759.61 m², the project ranked first in the May 2012 biddings, and now it is under design of drawings.
Based on the core philosophy "to respect environmental and regional culture", the project will rationalize the interrelationship between tourism terminal center on the whole and its locality, to satisfy all functions of hotel, while paying attentions to integral style harmony. Its inheritance with changes will build a high quality hotel featured by natural harmony.

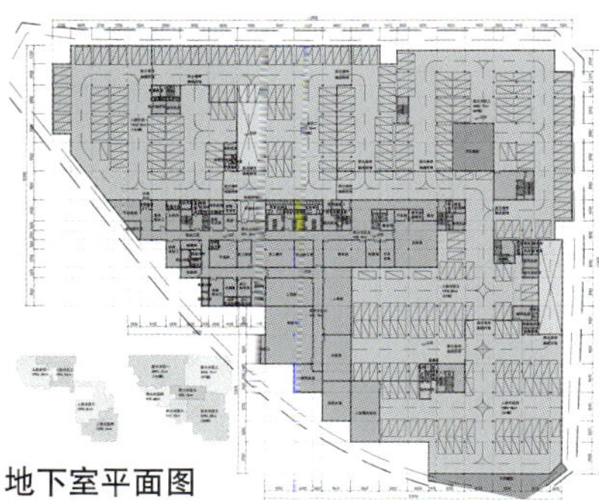

地下室平面图

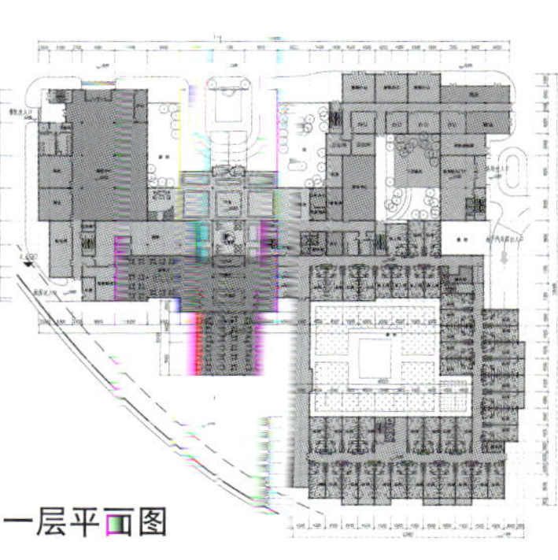

一层平面图

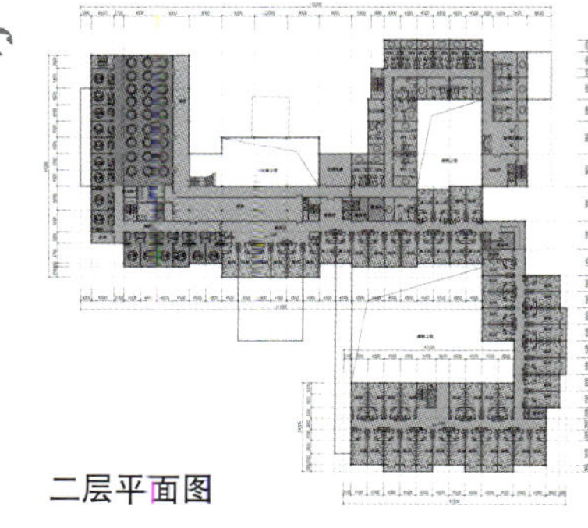

二层平面图

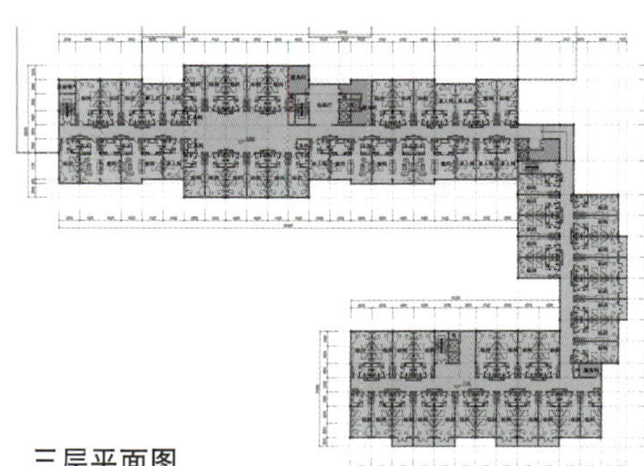

三层平面图

杭州晓城天地
Sunrise World, Hangzhou

设 计 师：黄森、王巍、俞越泷、石苏芹
项目地点：浙江 杭州
用地面积：30 446 m^2
建筑面积：161 444 m^2

Designer: Sen Huang, Wei Wang, Yuelong Yu, Suqin Shi
Location: Hangzhou, Zhejiang
Site Area: 30,446 m^2
Building Area: 161,444 m^2

杭州晓城天地项目位于杭州下沙街道上沙社区，总用地面积30 446 m^2，总建筑面积161 444 m^2。该项目由2幢高达百米的写字楼和1幢高50 m的板式酒店以及商业裙房组成，为杭州城东最重要城市综合体之一，目前正在建设中。
充分考虑狭长地形和周边日照等限制条件，挖掘地块价值，因地制宜地布局主楼和各种商业空间形态，形成高低起伏的城市视觉界面，营造出一个富有动感的城市综合体建筑。

The project is located in Shangsha community, Xiasha Sub-district, Hangzhou City. With total land area 30,446 m^2, total floor area 161,444 m^2, this project includes two 100 m high office buildings and one 50 m high plate hotel with commercial skirt buildings, marking one of the important city complexes in eastern Hangzhou city, and now the project is under construction.
Limited by stripe landform and surrounding sunlight conditions, the design explores land value, puts the main buildings and various commercial spaces on the streamline of city vision, to create a dynamic urban architectural complex.

经济技术指标

指标	数值	指标	数值
总用地面积(平方米)	30446	办公楼B	92.98
总建筑面积(平方米)	161444	酒店	48.06
其中 地上计入容积率建筑面积(平方米)	109605	裙楼高度(米)	24.28
地下总建筑面积(平方米)	49006	机动车停车位(个)	1144
地上不计入容积率建筑面积(平方米)	2833	其中 地上停车位	75
建筑密度(%)	40%	地下停车位	1069
容积率	3.6	非机动车停车位(个)	4065
绿地面积(平方米)	6090	其中 地上停车位	519
绿地率(%)	20%	地下停车位	3546
建筑总高度(米) 办公楼A	98.26	残疾人停车位(个)	20
备注	1、不计容积率建筑面积为酒店转换层，层高2180。 2、绿地面积为屋顶绿化、植草砖绿化折算后面积。		

新疆蓝领公寓
Blue Collar Apartment, Xinjiang

设 计 师：张敏军、胡建宏、俞越泷、石苏芹
项目地点：新疆 乌鲁木齐
用地面积：54 972.14 m^2
建筑面积：158 775.95 m^2
容 积 率：2.9

Designer: Minjun Zhang, Jianhong Hu, Yuelong Yu, Suqin Shi
Location: Urumqi, Xinjiang
Site Area: 54,972.14 m^2
Building Area: 158,775.95 m^2
Plot Ratio: 2.9

新疆蓝领公寓项目位于乌鲁木齐，总用地面积54 972.14 m^2，总建筑面积158 775.95 m^2，容积率2.9，由11幢高层住宅和商业建筑组成，2011年开始设计，项目正在建设中。
合理考虑居住、商业两种业态布局，组合出三个不同形态与氛围的景观空间，形成一个建筑与景观相得益彰的住区环境。

The project is located in Urumqi City, with total land area 54,972.14 m^2, and total floor area 158,775.95 m^2, plot ratio: 2.9, including 11 high-rise residential and commercial buildings. The project started design in 2011 and now it is under construction.
In reasonable consideration of residential and commercial patterns, the project includes three different forms of landscape spaces, shaping a residential community perfected by the landscape.

杭州北城天地

Beicheng World, Hangzhou

设 计 师：张敏军、陈滋达、王巍、俞越泷、石苏芹
项目地点：浙江 杭州
建筑面积：130 000 m^2

Designer: Minjun Zhang, Zida Chen, Wei Wang, Yuelong Yu, Suqin Shi
Location: Hangzhou, Zhejiang
Building Area: 130,000 m^2

杭州北城天地项目位于杭州市城北中心地带的湖州街汽车城旁，总建筑面积13万m^2，是由写字楼、星级酒店、酒店式公寓、商业街等大型商业组成的城市综合体，配套齐全，定位高端，预计2013年初建成使用。
项目充分挖掘各类商业空间价值，合理布置各类物业形态，建筑采用ART–DECO风格，比例优美，造型精致，塑造杭州城北区域性高端城市综合体形象，为杭州十大城市综合体优秀设计蓝本。

This project is located beside Huzhou Street Automotive Town in the north core of Hangzhou City. With total floor area 130,000 m^2, this city complex includes office buildings, star hotels, apartment hotels, and other buildings, with complete auxiliary facilities, high-end orientation, to be completed and put into use by early 2013.
It makes full use of all commercial space values, rationalizes all forms of properties, and its buildings are in ART-DECO style, in beautiful proportion, elegant shape, marking a high-end city complex in northern Hangzhou City, and serving as one of the top 10 excellent complex designs in Hangzhou City.

台州东方美地

Orient Beauty Land, Taizhou

设 计 师：沈凡、陈滋达、王巍、俞越泷、石苏芹
项目地点：浙江 台州
用地面积：450 492m^2

Designer: Fan Shen, Zida Chen, Wei Wang, Yuelong Yu, Suqin Shi
Location: Taizhou, Zhejiang
Site Area: 450,492 m^2

项目位于台州黄岩区院桥三块岩水库东南侧，总用地面积为450 492 m^2，分为居住用地和文体用地两部分，由低层休闲住宅、高层住宅、企业家运动会所、体育训练中心、运动会所、文化中心及服务用房、休闲度假中心、五星级酒店、酒店式公寓等多种建筑业态组成，2011年开始设计，目前正在建设中。
充分考虑变化复杂的地形，在总图设计时因地制宜，精选适建场地，同时配合局部的土方平整；充分利用地形的变化定制户型，使每一幢单体都能争取到最合理的朝向和最大的景观。整个园区建筑形体错落有致，绿树掩映，成功塑造了山地别墅和度假休闲的高端物业形象。

The project is located southeast of Three Rocks Reservoir, Yuanqiao Town, Huangyan District, Taizhou City. With total land area 450,492m^2, comprising of residential land and cultural/sports land, the project includes low-rise leisure residential, high-rise residential, entrepreneurial sports club, physical exercise center, sports club, cultural center and service rooms, leisure & resort center, 5-star hotel, apartment hotel and other buildings. The design commenced in 2011 and the project is now under construction.
Considering the landform complexity, and steep topography, the general plan selects the appropriate site, while customizing house types by landform changes, in combination of local earthwork leveling, so every single building has maximal direction and landscape. The entire buildings are different but in orderly place, shaded by green trees, making a high-end image of hilly villas and resort properties.

扫描查看更多信息

宁波中鼎建筑设计研究院
Ningbo Zhongding Architecture Design & Research Institute

宁波中鼎建筑设计研究院是一家以建筑设计与研究为主的股份合作制企业，于2002年成立，旨在培养与造就一支高素质的建筑设计队伍并潜心于城市与建筑的研究与创作，不断为社会奉献精品工程；在工程设计方面充分发挥技术优势，专长于建筑功能、形式与经济三方面的综合效益分析与研究；在科题研究方面能积级探索具有学术前瞻性的基础理论，推进建筑艺术与技术的进步。
中鼎建筑设计研究院致力于应用先进的理念、技术和设备进行工程设计研究，依靠技术人员的创造力、丰富的实践经验和现代信息技术等手段，能应对各种复杂的建设条件和不同客户的要求，并提供先进可靠、经济安全的解决方案。
研究院将秉承不断为社会奉献精品工程、推进建筑技术进步为宗旨，坚持“诚信为本，追求卓越”的质量方针，为客户提供先进的设计技术和优质的工程服务。

Founded in 2002, Ningbo Zhongding Architecture Design & Research Institute is a joint-stock corporation and has been developing a high quality architectural force team, focusing on architectural design and research,. We have intended to create and research urban and architecture as well as devoting collected projects. In design, the institute exerts fully technical superiority, specializes in providing a combination of synthesis benefits among architectural function, form and economy by analysing and researching. In research, the institute zealously explores fundamental theory toward academic perspective.
Ningbo Zhongding Architecture Design & Research Institute dedicates to applying advanced conceptions, technologies, equipments for research of project design. Depending on talent of technicians' creation, abundant practical experience and modern informational techniques, the institute not only could deal with complicated constructional troubles and feed different clients' demand, but also could provide safely and economical settlements, which pay attention to the design between structure and space, make use of natural resource, pursuit harmony developing between human and environment.
By constantly offering collected projects, by promoting architectural technology as principle, the institute insists on a quality policy of pursuiting honesty and excellence in order to offering advanced design technology and prominent project service.

地址：浙江省宁波市环城西路南段695号
电话：+86-574-87495078
传真：+86-574-87497098
邮箱：zddesign@zddesign.cn
网址：www.nbzd.cn

Add: No. 695 the southern section of Ring West, Ningbo City, Zhejiang Province
Tel: +86-574-87495078
Fax: +86-574-87497098
E-mail: zddesign@zddesign.cn
Web: www.nbzd.cn

宁波诺丁汉大学三期工程
University of Nottingham Ningbo Phase III

项目地点：浙江 宁波
用地面积：6267 m²
建筑面积：59 600 m²

Location: Ningbo, Zhejiang
Site Area: 6,267 m²
Building Area: 59,600 m²

西安工业大学附属学校
Xi'an Technological University Affiliated School

项目地点：陕西 西安
用地面积：387 000 m²
建筑面积：269 616 m²
容 积 率：0.6

Location: Xi'an, Shan'xi
Site Area: 387,000 m²
Building Area: 269,616 m²
Plot Ratio: 0.6

西安工业大学附属学校的规划核心为：一个中心、两条主道及多个院落空间。是一座大规模的中小学校，学生人数达 13 000 人，胜过一般大学的规模，因此主张建筑应该像文化一样多样化。在设计中，采用了建筑分类的差异设计，如：行政图书中心大楼强调个性的同时兼顾城市与校区的主要景观效果，中小学幼儿园各阶段的教学楼细部设计与材料使用也有一定的差异，甚至学生宿舍的功能配置、居住生活模式、层数及立面设计也有一定的差异。各教学区的报告厅作为点缀，其设计偏向时尚。

The planning core of Xi'an Technological University Affiliated School: one center, two arterial roads and courtyard space. The design will build an immortal school with diverse ideas. This is a large-scale middle and primary school with 13,000 students, which is above average university scale. More students mean more buildings, as well as much more differentiations among them. Thus I propose that architectures should be diverse just as culture, otherwise the buildings in campus seem to be all in the same pattern, which is too tedious. The differentiation design for building classification will be applied in the project design, for example: the Administrative Books Centre Building emphasizes individualization, considering the main landscaping effects of the city and campus, and there are some differences among the detail design and materials of teaching buildings for each stage of middle school, primary school and kindergarten, including functional configurations of students dormitories, life mode, number of floors, and façade design. As ornament, the lecture halls of each teaching area are designed to be stylish.

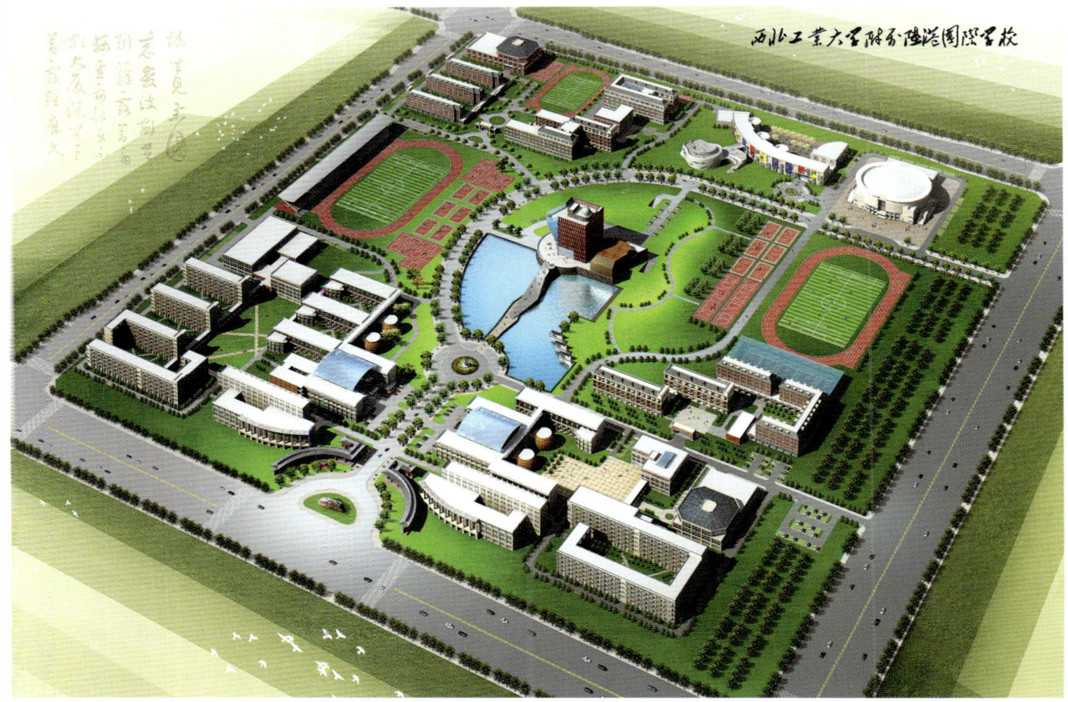

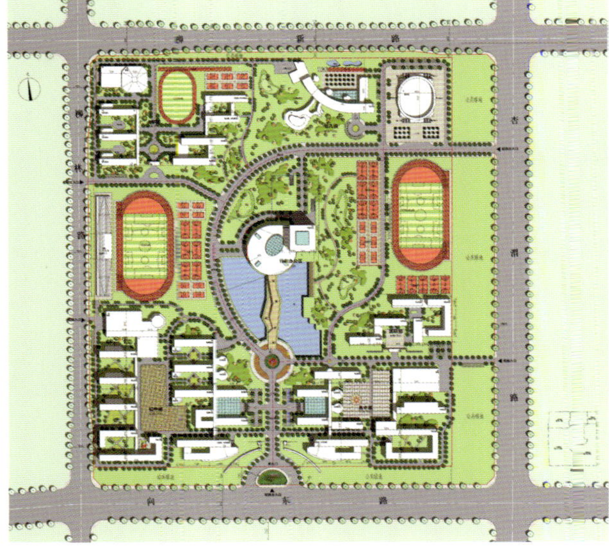

梁祝公园改造及博物馆
Liangzhu Park Reconstruction and Museum

项目地点：浙江 宁波
用地面积：1200 m²
建筑面积：5376 m²

Location: Ningbo, Zhejiang
Site Area: 1,200 m²
Building Area: 5,376 m²

项目位于宁波梁祝公园西南角，东侧为梁山伯墓，西侧为姚江，设计必须严控体量，以求得与公园内建筑尺度之间的协调，因此将博物馆的主要展厅均安排在地下，地面以上部分的空间主要用于入口、序厅、庭院、茶室及管理用房，同时力求分解体量。设计试图从汉晋的石刻艺术中探索当时建筑的一些具有代表性的特征。率直的石瓦当、粗犷的屋脊、设有举梁的屋面等，仅在入口廊檐部得到现代化的体现，而非复古搬抄，从而在保留传统建筑语言的同时赋予建筑时代特征。

The project is located on the southwestern corner of Ningbo Liangzhu Park, beside the Tomb of Liang Shanbo in the east, and Yao River in the west. The architectural mass should be strictly controlled in order to make the project in harmony with the architecture scales in the park. Therefore the main exhibition halls are all arranged in basement, and the aboveground spaces are used for entrance and lobby, courtyard, teahouse and management space, at the same time the breakdown of the mass also ought to be done; the project design tries to explore from the stone carving art of Han and Jin dynasties some representative features of then architecture; the modern sense of the artless stone eave tiles, grand ridge and the roof with beams and etc. are reflected only in the eaves of veranda of the entrance, not to copy the ancient ways, and the architecture is featured by the times.

宁波市二纵二横主干道整治
Rectification of Two North-South and Two West-East Arterial Roads, Ningbo

项目地点：浙江 宁波

Location: Ningbo, Zhejiang

鉴于国际普遍存在的中心城区街道交通拥堵、形象零乱以及缺乏特色的现象，宁波市政府决定对中心城区二纵二横主干道（药行街、通途路、人民路及环城西路）进行整治，整治目的是改善交通秩序、提升街景形象、展现街道特色、优化步行环境。

Considering traffic jamming, sparse image and lack of features in downtown area as an international problem, Ningbo People's Government was determined in 2010 to rectify the two south-north and two east-west arterial roads (Yaohang Street, Tongtu Street, People's Street and Ring West Road), for the purpose of: better traffic order, better street image, better street features and better pedestrian conditions.

宁波高新区科贸中心大厦
Technology & Trade Center Building, National Hi-tech Industrial Development Zone, Ningbo

项目地点：浙江 宁波
用地面积：4643 m^2
建筑面积：82 863 m^2

Location: Ningbo, Zhejiang
Site Area: 4,643 m^2
Building Area: 82,863 m^2

总平图设计中，建筑靠近基地北侧，沿城市道路一字排开。正前方布置近 7000 m^2 的绿化广场，以美化整个办公场所，并促建筑与景观形成互补关系。在满足自身的要求下，也给周边城市空间带来宜人的室外开放空间。
建筑双塔的设计思路，首先旨在表现“双纸鹤”的象征含义，同时满足了规划提出的容积率等要求；另一方面也产生了大面积的室外绿化空间，对于营造人性化、生态化的环境提供了更大的选择空间。与此同时，从类型学的角度出发，对主体建筑进行了折变，使建筑显得更加有生气有活力；主楼自身设计之中又产生了“一高一低，一虚一实”的变化，使建筑在周边环境中显的更加谦逊，能更好的融入周边建筑群中。

In the general layout, the project is near the north of the base, and lined along the city road. Right in front of the building is a greening square of nearly 7,000 m^2, which beautifies the whole office space and integrates the building and the landscape. Fitting in the project design, it also brings a comfortable outdoor open space to the surrounding urban space.
The twin towers design, symbolizing "twin paper cranes" meets the planned plot ratio and other requirements, and also produces large outdoor spaces, offering more choices for building humanistic and ecological environment. In the meantime, viewed from typology, the design refracts the appearance of the main building to make it look more lively and energetic. On the other hand, the contrast between high and low, real and virtual in the main building design stands it out of the surrounding environment, appearing more modest, and more integrated in the peripheral building clusters.

宁波大宗商业交易中心
Ningbo Commodity Exchange Center

项目地点：浙江 宁波
用地面积：9002 m^2
建筑面积：50 200 m^2

Location: Ningbo, Zhejiang
Site Area: 9,002 m^2
Building Area: 50,200 m^2

宁波大宗商业交易中心主要追求中国要素与时代感的体现。使开放的城市空间与静谧的办公环境、高效的交易空间与情感化的交易环境和谐统一。以生生不息的细部力量铸就建筑的永恒魅力。设计中尽量采用新材料、新技术、新设备、新工艺，做到节电、节水、节能。设计合理的道路体系，妥善处理好交通流线，形成一个完整的运行系统。以适度超前为原则，经济合理地利用土地与空间，充分考虑今后使用、管理及未来发展的需求。

Ningbo Commodity Exchange pursues the expression of Chinese elements and epochal sense. Open urban spaces and tranquil office conditions, efficient trading space and emotional trading environment are in a harmonious unity. The undying detail power builds the eternal building charm. The design make optimal use of new materials, new technologies, new devices, and new processes, for the purpose of electricity saving, water saving and energy saving. The design has reasonable general plan road system, with good traffic flows, to form a complete running system. In the moderate avant-garde principle, the design make reasonable and cost-effective use of land and space, in full consideration of current and future use, management and the requirements for future development.

杭州友慧润城建筑设计有限公司

Hangzhou Youhui Runcheng Architectural Design Co., Ltd.

杭州友慧润城建筑设计有限公司成立于2009年，具有建筑行业（建筑工程）甲级资质。自公司成立以来，凭借自身的实力，以独特新颖的视觉角度、全新的设计理念，在城市规划、建筑设计、景观设计等领域取得了优秀的成绩，公司成立一年内签订合同产值达2000多万人民币。
公司拥有一支精英团队，队员们满载梦想、执著追求、专业进取、创新拼搏，使得友慧润城与国内诸多房产公司、策划顾问公司等有着许多良好的合作项目。目前公司的设计、规划业务遍及浙江、江苏、安徽、湖北、广西、山东、贵州、江西等省市。

地址：杭州市拱墅区湖墅南路271号中环大厦405室
电话：+86-571-88387906
传真：+86-571-88388172
邮箱：yhrc2011@126.com

Add: Room 405, Zhonghuan Building, Hushu South Road No.271, Gongshu District, Hangzhou
Tel: +86-571-88387906
Fax: +86-571-88388172
E-mail: yhrc2011@126.com

咸宁贺胜路地块居住小区方案设计

Concept Design of Hesheng Rd. Plot Residential Community, Xianning

项目地点：湖北 咸宁
用地面积：163 333 m^2
建筑面积：500 000 m^2

Location: Xianning, Hubei
Site Area: 163,333 m^2
Building Area: 500,000 m^2

设计中沿袭法式人文基调，以大气、沉稳的规划布局，浪漫气质、精雕细琢的建筑风格，自然生态的景观环境，着力打造咸宁高端别样的“浪漫之城”。
在规划布局上强调建筑与环境交融一体，充分利用地块的高低变化合理布局。运用动静、商住分离的设计手法，以气势磅礴的主体建筑围合出城廓的形式。严谨对称的格局、壮观华丽的轴线、丰富气派的空间序列无不渲染着豪华尊贵的古典气氛，在不同时间通过光影的交错体现出极丰富的变化。建筑由法式花园、电梯洋房、园景小高层公寓、宽景高层公寓、配套商业及会所组成。多层次的面积配比，满足城市不同人群对于舒适性生活的美好愿望。

The design is keynoted by French culture, planned in generous and stable layout, built in romantic and elaborate character, and surrounded by scenic ecology, with all efforts to building a high-end unique “romantic town” in Xianning.
The layout integrates architecture with environment, makes full use of landform, design in dynamic and static manner, commercial and residential manner separately. The magnificent main buildings make a circle in form of the outer walls of a castle. The symmetric layout, magnificent axis and rich spatial sequence are full of luxury and noble classicism. There are rich changes in light and shadow at different times. There are French-style garden, houses with elevator, landscaped high-rise apartment, panoramic high-rise apartment, auxiliary commercial building and club. Multi-level area ratio can satisfy different people’s aspiration for a comfortable life.

荆门尚海明珠居住小区方案设计

Concept Design of Marine Pearl Residential Community, Jingmen

项目地点：湖北 荆门
用地面积：163 333.3 m^2
建筑面积：320 000 m^2
容 积 率：2.0
建筑密度：18.5%
绿 化 率：35.1%

Location: Jingmen, Hubei
Site Area: 163,333.3 m^2
Building Area: 320,000 m^2
Plot Ratio: 2.0
Building Density: 18.5%
Green Ratio: 35.1%

本规划设计从既有的水系资源出发，以异域情调的威尼斯水城作为设计出发点，结合双泉公园现状的水资源，力求达到人与水，人与自然的和谐共存。

规划布局注重动静分离，各种类型单体与环境的关系有别。在平衡用地经济性与高尚社区舒适性方面，体现疏密有致的布局特色，在紧凑中发挥灵活性。别出心裁的路网形态和规划中富有变化的建筑排列，在保证户户南向的同时打破阵列式的格局，满足经济性的同时体现出高尚居住区的休闲气息。小区道路与环境设计体现步移景异的景观效果，绿地集中与分散的均衡布局，追求景观环境的共享性与私密性，形成舒适的欧式园林居住空间。

Based on existing water resource in Double-Fountain Park, this design is inspired by exotic Venice, in seek for harmonious coexistence between human and water, human and nature.

The layout is planned in dynamic and static manner separately, where all kinds of single buildings are classed in order with the surroundings. The balance of economy and noble community comfort is featured by considerable flexibility in pleasant density. Unique road network and various building array ensure every house is southward while breaking out the array pattern, and satisfy the requirements for economy while showing noble community leisure. The roads and surroundings are designed in rich landscapes ensuring sightseeing effect at each step, greening concentration and diverse balance, landscape sharing and privacy assuring form a comfortable European-style garden living space.

主要经济技术指标

总用地面积			64429	m^2
总建筑面积			376540	m^2
其中	地上建筑面积		316540	m^2
	其中	酒店	50400	m^2
		公寓	22000	m^2
		商业	75000	m^2
		住宅	167340	m^2
		会所	1000	m^2
		物管及经营用房	800	m^2
	地下建筑面积		60000	m^2
容积率			4.92	
绿地率			20.0	%
建筑密度			35.8	%
住宅总户数			1395	户
机动车总车位数			2395	辆
其中	住宅车位数		1395	辆
	商业车位数		1000	辆

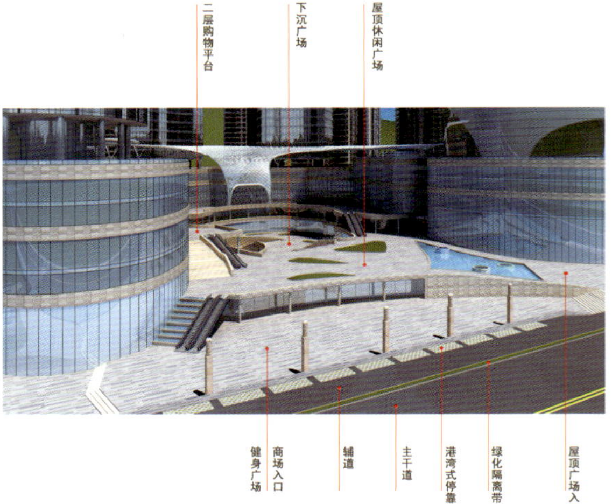

荆门锦辉广场商住综合体概念设计

Conceptual Design of SSG Square Commercial & Residential Complex, Jingmen

项目地点：湖北 荆门
用地面积：约64 666.7 m^2
建筑面积：320 000 m^2
容 积 率：4.92
建筑密度：35.8%
绿 化 率：20%

Location: Jingmen, Hubei
Site Area: about 64,666.7 m^2
Building Area: 320,000 m^2
Plot Ratio: 4.92
Building Density: 35.8%
Green Ratio: 20%

根据场地西低东高，北低南高的特点，将商业及广场沿白云大道展开设置，采取内凹的建筑形式，形象犹如跃起的“江中鲤鱼”。通过对场地综合分析处理，设计多个广场平台，合理地处理场地高差，形成室内丰富的购物空间和室外广场空间，给单调的商业流线赋予感性的生命。入口两侧有一栋超高层五星酒店，带空中旋转餐厅；1栋2.2万 m^2的高端公寓；2栋高端精装住宅。通过复杂而又简约的空间设计手法，与白云大道的沿街形象引起深刻的共鸣，从多个方面提升城市的形象。

The site is low in the west and high in the east, low in the north and high in the south, the commercial buildings and square are extended along Baiyun Avenue, and the concave façade looks like “a carp leaping up from the river”. The design includes many square platforms, and the height difference of this site shapes indoor shopping space and outdoor square space, giving sensational life to commercial streamline. At the entrance are an overhigh-rise 5-star hotel with a revolving restaurant, a high-end apartment with gross floor area 22,000 m^2, and two high-end fine decorated residential buildings. Responsible and concise space design along Baiyun Avenue has promoted the image of this city in many respects.

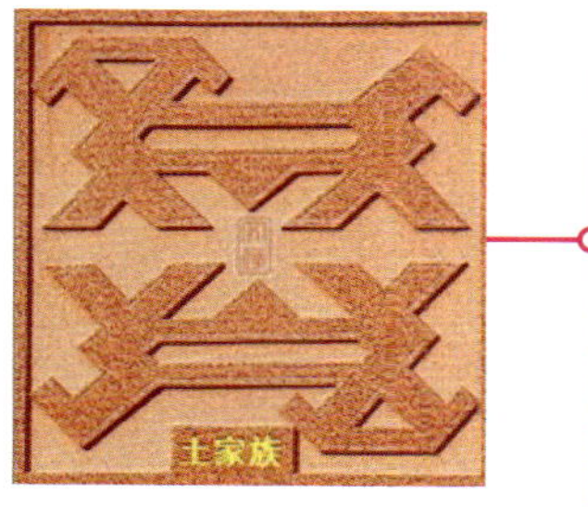

●总体定位

清江画廊青山绿水

城市形象门户空间

和谐共生的生态环境

“一站式”旅游服务

●形象定位

徜徉巴邑意何稠？

九曲清江风景幽。

水是回廊山是画，

车来船往画中游。

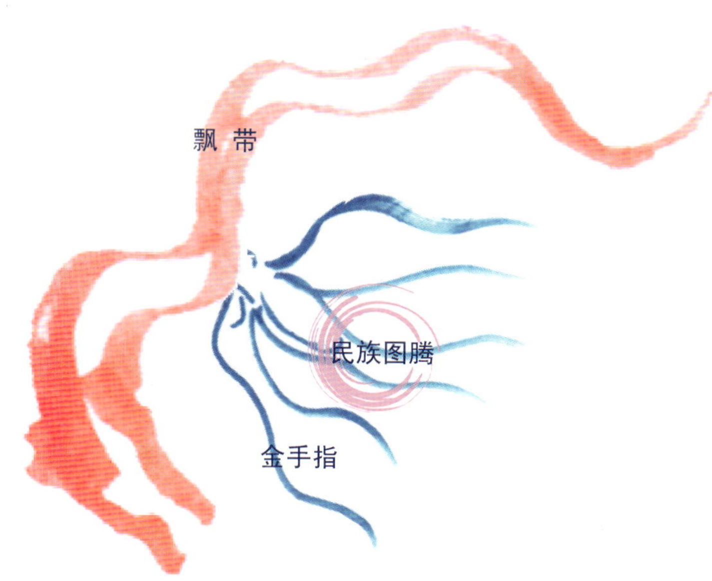

长阳清江画廊旅游服务中心概念设计

Concept Design of Clearwater Gallery Tourist Service Center, Changyang

项目地点：湖北 宜昌	Location: Yichang, Hubei
用地面积：17 000 m^2	Site Area: 17,000 m^2
建筑面积：31 000 m^2	Building Area: 31,000 m^2
容 积 率：1.82	Plot Ratio: 1.82
建筑密度：34%	Building Density: 34%
绿 化 率：17%	Green Ratio: 17%

本次规划立足于长阳历史的发展历程，在长阳土家族文化的影响下，既具有本土基础，又呈现出开放与多元的特色，通过整体布局充分体现旅游服务特色。整体规划以长阳舞蹈为主题，建筑群体用“飘带”连接，与清江遥遥相对，寓意长阳文化源远流长。入口大厅周边“飘带”及树池组成一双“金手掌”，将其民族图腾敬于掌心。设计中充分体现长阳舞蹈文化和土家族的图腾文化、窗文化、建筑文化等。

The conceptual design is based on the historical development of Changyang County (Yichang, Hubei), under the influence of Tujia ethnic culture, and extended to tourist service features through overall layout, with open and diverse characteristics. The overall planning is themed by local dance, and buildings are linked with “ribbons”, corresponding to Clearwater (Qing river), implying its cultural origin. The ribbons and trees surrounding the entrance hall form a pair of “golden fingers”, holding ethnic totem in the palm. The design has fully expressed local dance culture, ethnic totem culture, windows culture and building culture etc.

扫描查看更多信息

浙江新中环建筑设计有限公司
Zhejiang New China-Jaicn Architectural Design Co., Ltd.

浙江新中环建筑设计有限公司，是具有甲级建筑设计资质、乙级市政公用（排水、给水、道路、桥梁、风景园林）设计资质和乙级勘察专业资质、甲级监理资质、杭州市代建资质的独立法人企业。经过10多年的发展，公司已发展成为内部组织机构健全、专业结构配套合理、技术水平一流、市场竞争实力雄厚和社会信誉良好的大型设计公司。公司现有职工300多人，其中，国家一级注册建筑师11人、国家一级注册结构师12人、注册公用设备工程师5人、国家注册监理工程师15人、国家注册造价工程师3人、国家一级注册建造师4人，公司高级技术职称28人、中级技术职称98人。

近年来，公司在建立健全内部管理、完善企业规章制度的同时，坚持以人为本，对外引进高端人才，对内培养新生力量，狠抓建筑设计这一龙头业务，努力拓宽市政公用、勘察、弱电、室内设计等相关业务领域，并积极向高、精、尖技术挑战，取得了累累硕果。

公司推崇人性化管理体制，追求“和谐环境、创新技术、追求完美”的设计理念，通过优化、调整、激励等措施，大大地提高了员工的工作热情和积极性，充分地挖掘和发挥了员工的聪明才智，使公司获得了多个“钱江杯”优质工程奖和全国人居建筑环境双金奖等综合大奖。面对越来越激烈的竞争市场，公司将适时调整未来企业的发展方向，把握时代的脉搏，紧跟时代的步伐，构筑良好的工作氛围，搭建更宽阔的发展平台，让员工与企业携手共赢，用更多更好的作品，为这个城市乃至整个社会增光、添彩。

地址：杭州市天目山路217号江南电子大厦6-10楼
邮编：310013
电话：+86-571-88228015
邮箱：1061740422@qq.com
网址：www.zjxzh.com

Add: Floors 6-10, Jiangnan Electronic Building, 217 Tianmushan Road, Hangzhou City
P.C.: 310013
Tel:+86-571-88228015
E-mail: 1061740422@qq.com
Web: www.zjxzh.com

湖州星际广场
Huzhou Star Plaza

项目地点：浙江 湖州
用地面积：29 078 m²
建筑高度：162 m

Location: Huzhou, Zhejiang
Site Area: 29,078 m²
Building Height: 162 m

城市性：符合城市规划要求，考虑周边建筑、道路、环境的关系。建筑群自身形态完整、优美，与周边建筑相互呼应，更添魅力。

文化性：采用富有东方文化韵味的群体组合关系（组合曲线）造型，生动活泼，标识性突出。

综合性：建筑群由超高层塔楼、商业裙房和三幢公寓组成，内部功能复杂丰富。设计整合各类功能形态，处理好内部各项功能关系，并以现代的外立面处理手法将建筑群组成为整体。

杭州钱江新城金融城综合体

New City Financial Complex, Qianjiang Economical Development Zone, Hangzhou

项目地点：浙江 杭州
建筑面积：1 000 000 m²

Location: Hangzhou, Zhejiang
Building Area: 1,000,000m²

市民中心大尺度的敞开界面和绿色坡道，犹如一场戏的序幕，提示人们继续感受序列中的市民中心、杭州大剧院，使孤立的金融城地块与市民公园地块合二为一，创造丰富的办公空间。同时最大化地保留原有建筑，并进行部分立面改造，打造统一的钱江金融城。

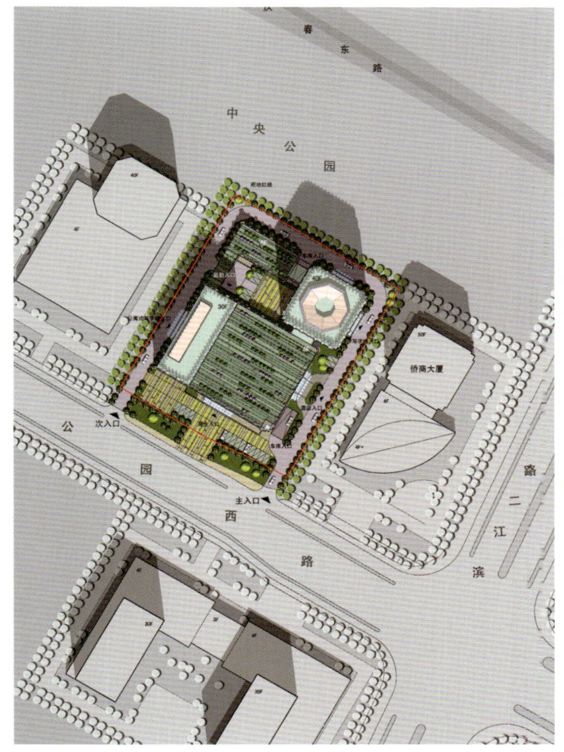

杭州宝盛大厦

Hangzhou Baosheng Mansion

项目地点：浙江 杭州　　Location: Hangzhou, Zhejiang
建筑面积：110 635 m²　　Building Area: 110,635 m²

杭州宝盛大厦位于钱江南岸，钱江世纪城内（G-04-01和G-04-02地块），与杭州钱江新城隔江相望。
建筑布局以突出超高层的塔楼为核心，同时在高容积率下寻求体量的消解，以及创造丰富的多重空间关系。裙房围绕主塔楼展开布置，烘托出塔楼挺拔向上的形象。
造型上寻求时代特色的同时结合杭州丰富的历史底蕴，创造出独特的建筑形象。主塔楼四个角上采用多面切削的手法结合向上收分的形态，成就玲珑宝塔的意境。竖向线条的重复排列，成就塔楼的高耸挺拔感。

浙江时代大厦

Zhejiang Times Mansion

项目地点：浙江 杭州
建筑面积：45 000 m²

Location: Hangzhou, Zhejiang
Building Area: 45,000 m²

本工程位于杭州市中心武林广场附近，设计遵循“建筑形体应是内部功能和其性格特征的反映”这一原则，主、副楼均采用简洁的长方体，既利于办公使用，也有利于结构和经济性。但设计并不满足于简单地做两个平淡的长方体，而是通过精心考虑建筑物的长宽高比例和顶部造型，使本建筑在与周围环境融合相处的基础上，铸造出一种昂扬向上、生机勃勃的时代精神，也使本大厦在三维空间得到形象的体现。

外墙采用全景式的落地大玻璃幕墙给使用者以360°全景式景观体验，同时又使建筑如水晶般晶莹剔透，在白天的阳光与夜晚的灯光下展现出迷人风采，凸现其卓尔不凡的独特气质。

杭州龙禧福朋·喜来登酒店

Four Points By Sheraton Hotel, Hangzhou

项目地点：浙江 杭州
建筑面积：100 000 m^2

Location: Hangzhou, Zhejiang
Building Area: 100,000 m^2

本建筑为综合性建筑，位于杭州市滨江区高新技术产业开发区。本建筑为强调广场入口的轴线关系，沿轴线布置一发散的圆形、扇形娱乐中心，入口广场与城市空间形成对话关系。二幢塔楼围绕一圆形娱乐、休闲中心而设立，遥相呼应，分别为18层酒店及22层专家楼，中间圆形建筑为商业、娱乐、会展等综合性配套用房。整个项目按现代化，智能化高层建筑标准建设，进一步提升了该地段的城市商业、休闲价值。

浙江广播电视大学教工路校区

Faculty & Staff Road Campus, Zhejiang Radio Broadcasting and Television University

项目地点：浙江 杭州 Location: Hangzhou, Zhejiang
建筑面积：55 000 m² Building Area: 55,000 m²

浙江广播电视大学教工路校区教学用房改造工程位于杭州市教工路，世贸丽晶城北侧。设计力求在高密度的城市空间中营建一个“现代书院”，一个具鲜明个性的“知识之门”，成就勃勃雄心的超现代作品。

“现代书院”：整个“书院”空间可以解构为三大部分：校前广场空间、轴线序列、庭院空间。

“知识之门”：通过拓扑美学的应用法则，将“数字信息”的特征运用于建筑立面表皮上，从而创造了一栋有鲜明特色的建筑。

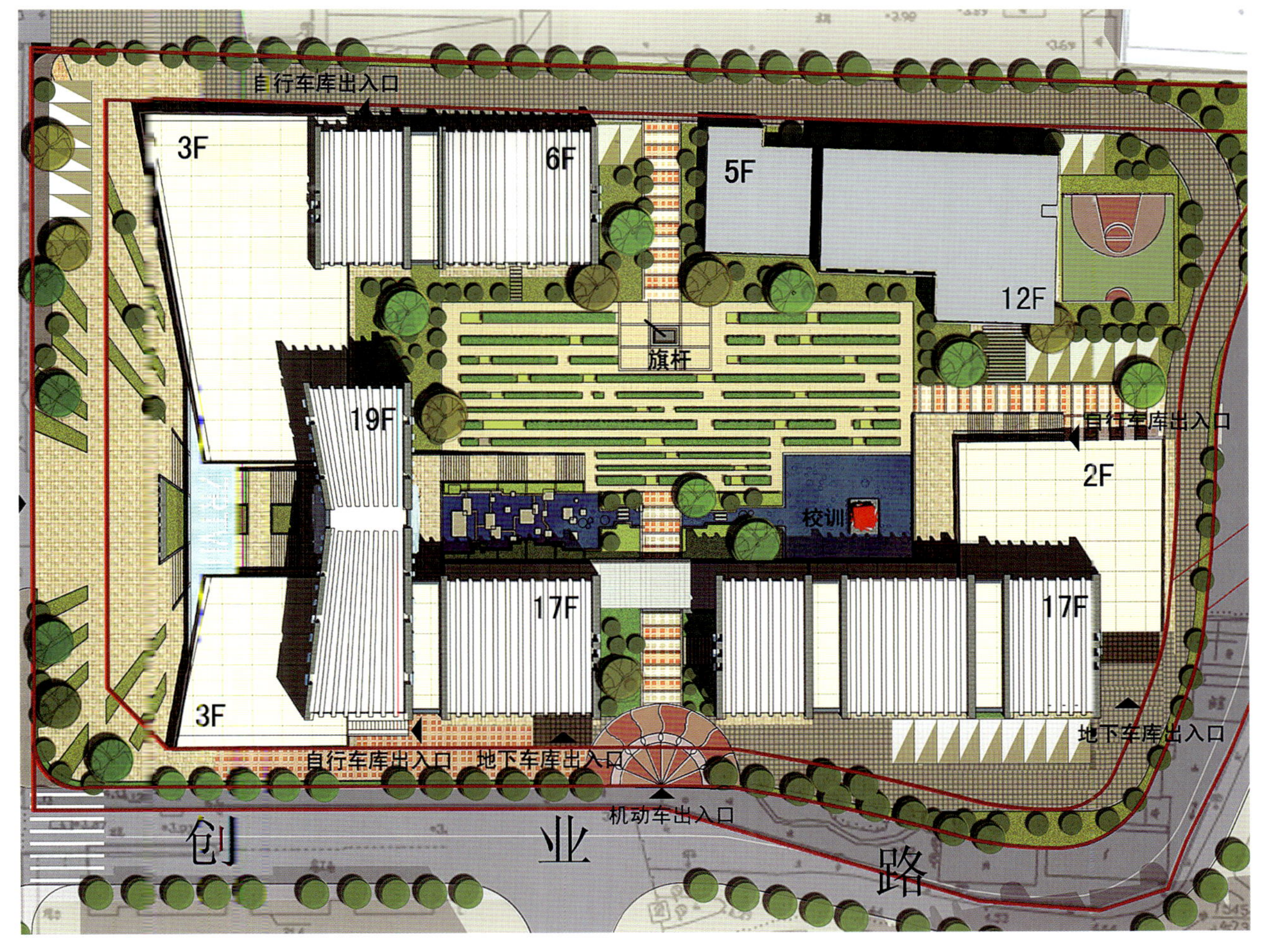

自行车库出入口
3F
6F
5F
12F
旗杆
19F
自行车库出入口
2F
校训
17F
17F
3F
地下车库出入口
自行车库出入口
地下车库出入口
机动车出入口
创
业
路

浙江广播电视大学

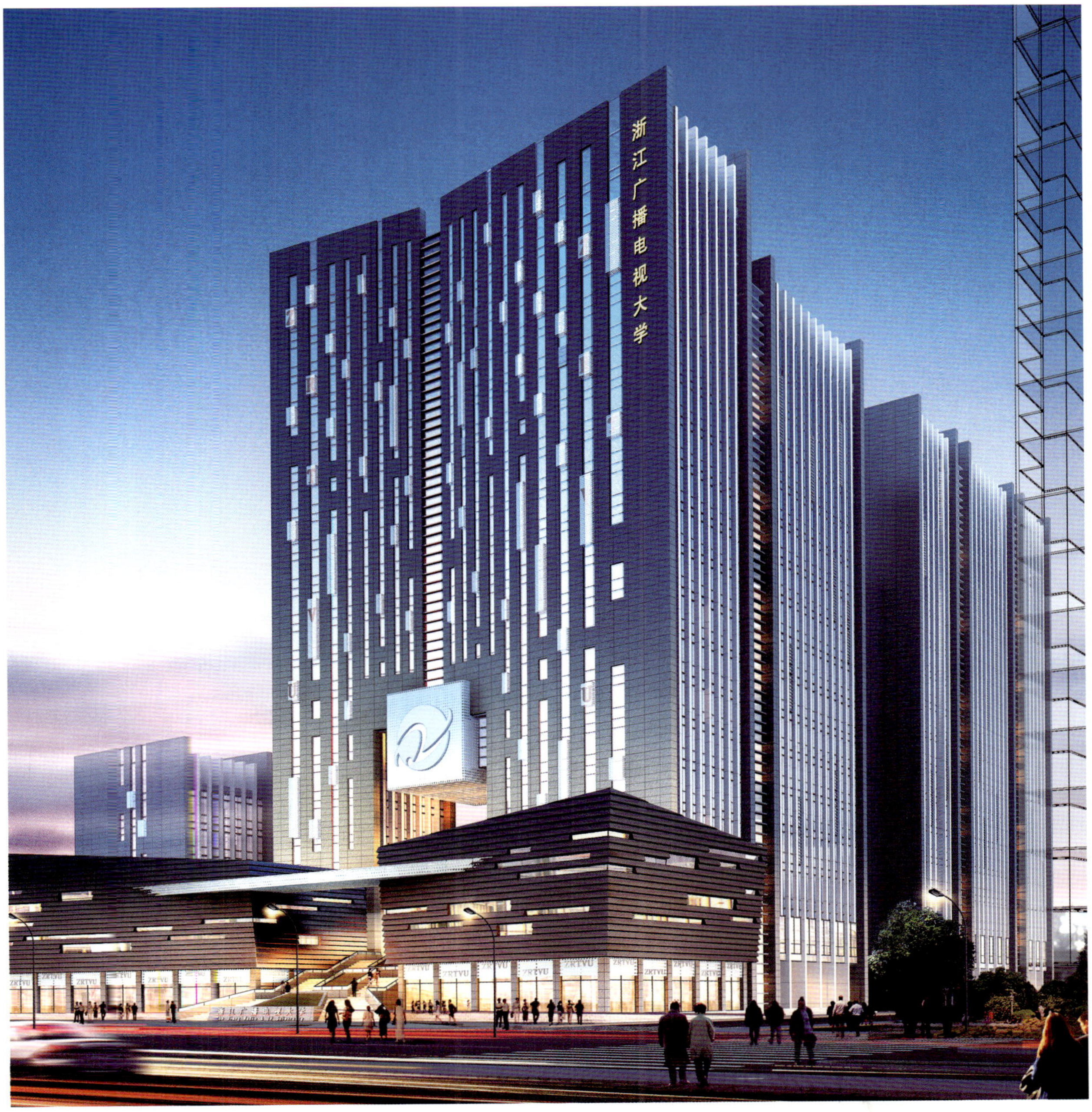

浙江安居建筑设计有限公司
Zhejiang Anju Architectural Design Co., Ltd.

浙江安居建筑设计有限公司成立于1997年12月，具备国家建设部认证的建筑工程设计甲级资质。多年来主要承接国内各省市的厂区、住宅区、商业综合体、学校、星级酒店及各城区的旧城改造等项目。
公司的业务领域涵盖了规划、建筑、室内、景观、智能化等专业设计，其中尤以居住社区、五星级酒店和城市商业综合体地产的设计最为擅长。我公司一贯专注于工业与民用建筑等诸多领域的研究，提供从概念策划、定位方案、初设优化到施工图的整体设计服务，并一直在为创造超越客户期望的产品而努力。
公司目前拥有各类技术设计人员70余人，其中高级职称17人、中级职称28人，技术骨干人员占员工总数的95%以上。现设有5个设计业务部门，具有GB/T19001–2008标准的设计质量保证体系与经营管理制度。
公司成立至今，历经创业阶段，稳步发展阶段及质量全面提高阶段，坚持以创新求实的思路抓住企业发展机遇，昂首迎接挑战，阔步向前迈进，并继续本着"没有最好，只有更好"的服务宗旨，竭诚为客户提供最满意的服务。

近年主要项目介绍：

项目名称	**建筑类型**	**总建筑面积**
加格达奇·金马饭店	五星级宾馆建筑	总建筑面积5.7万m^2
舟山云厦住宅小区	居住建筑	总建筑面积9.8万m^2
下沙头格住宅小区	居住建筑	总建筑面积20.5万m^2
杭州下沙东方社区农转居多层公寓	居住建筑	总建筑面积13.23万m^2
杭州闲林省直经济房	居住建筑工程	总建筑面积约45万m^2
安吉浒畔居别墅	居住建筑工程	总建筑面积约8万m^2
杭州"天城银座"综合大厦	居住及公建	总建筑面积为6.1万m^2
大洋彼岸二期	居住建筑	总建筑面积为10.7万m^2
江冠·现代英冠商务中心	高层办公建筑	总建筑面积为12.2万m^2
萧储[2009]12号地块	超高层办公建筑	总建筑面积为14.5万m^2

地址：杭州市西湖区文一路68号坤和商务楼9楼
电话：+86-571-89710598
传真：+86-571-85117797
邮箱：wawawa7797@163.com
214263452@qq.com
66090247@qq.com

Add: 9th Floor Kunhe Business Building, Wenyi Road No.68, Xihu District, Hangzhou City
Tel: +86-571-89710598
Fax: +86-571-85117797
E-mail: wawawa7797@163.com
214263452@qq.com
66090247@qq.com

杭州下沙东方社区农转居公寓

Farmers-to-Citizens Apartment, Oriental Community, Xiasha Economic Development Zone, Hangzhou

项目地点：浙江 杭州
用地面积：55 401 m^2
建筑面积：132 300 m^2

Location: Hangzhou, Zhejiang
Site Area: 55,401 m^2
Building Area: 132,300 m^2

项目位于杭州经济开发区内，地块北侧为元成南路，东侧为文渊北路，西侧为东方路，南侧为纬一路。采用"小区入口—小区景观中心—景观节点—单元入门"系列景观空间组织，使住户从进入小区开始，便能感受到强烈的"家"的气氛，享受到一系列的交往、活动、休憩的空间。景观由主入口开始通过集中景观绿地向各个方面发散，入口景观视线开阔，景观环境优美，整个小区的景观连贯，让人们由景而入，顺景而出。

小区内分别布置了儿童游乐、老人活动、运动健身等大型居民活动场地，并结合绿化景观布置，创造出宜人的活动空间。

This project is located in Xiasha Economic Development Zone of Hangzhou City, Yuancheng South Road in the north, Wenyuan North Road in the east, Oriental Road in the west, and Weiyi Road in the south.

The landscaping sequence of "community entrance – community landscaping center – landscaping node – unit entrance" will give residents a strong "home" atmosphere at the first moment of entering the community, who can enjoy various spaces for interactions, activities and recreation. The landscape is diversified from the main entrance to all directions through the collective greening belt, with borad visions at the entrance, beautiful environment, and the whole community is set in a scene where people come into landscape and go out of landscape.

In the community set large outdoor spaces for children entertainment, elder recreation, sports and fitness etc. Combining with greening landscape, the community creates an agreeable space.

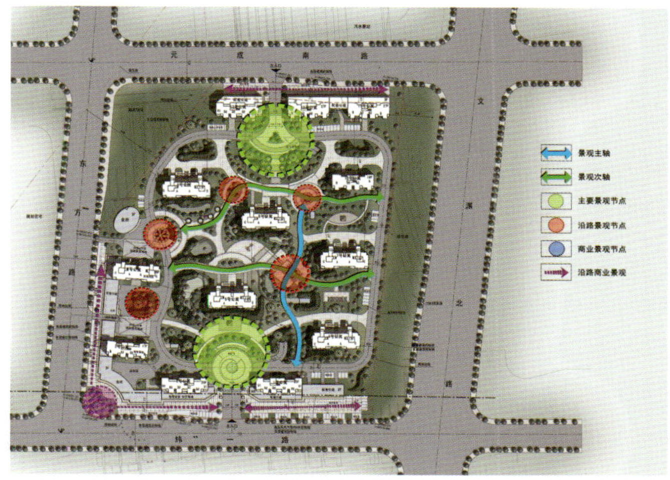

扫描查看更多信息

华东勘测设计研究院（简称华东院）是中国水电工程顾问集团公司直属的大型综合性国家甲级勘测设计研究单位，于1954年在上海建院，现总部设在杭州，在四川、重庆、福建、广东、江西、云南等地设有分支机构，在越南、土耳其、尼日利亚等国设有驻外办事机构。
华东院现有1600多名员工，85%以上是专业技术人员，其中，持有国家注册各类执业资格的有500余人。华东院是一个多专业、跨领域、综合性的勘测设计研究院，先后承担了大批大型公共建筑、高层建筑、花园住宅小区、工业建筑等各类工程项目设计，承担完成了多个城乡区域发展规划。
华东院的创作遵循最生态、最环保、最个性的思路，显独特于细节，立新意于经典，用换位思考的方式实现建筑与城市的共鸣，体现个性与共性的融合。
华东院坚信：只有人文的、绿色的才是持久的、经典的。
华东院的建筑设计理念是以人为本，崇尚建筑与自然的和谐，为社会提供独特的、节能的、环保的绿色建筑。

地址：浙江省杭州市潮王路22号
电话：+86-571-56738888/56737418
传真：+86-571-88392805
邮箱：zhao_wb@ecidi.com
网址：www.ecidi.com

Add: No.22, Chaowang Road, Hangzhou City, Zhejiang Province
Tel: +86-571-56738888/56737418
Fax: +86-571-88392805
E-mail: zhao_wb@ecidi.com
Web: www.ecidi.com

华东院三墩基地复建改建工程
Sandun Base Rehabilitation and Reconstruction Project, Huadong Engineering Corporation

设 计 师：赵文冰
项目地点：浙江 杭州
用地面积：6876 m²
建筑面积：85 000 m²

Designer: Wenbing Zhao
Location: Hangzhou, Zhejiang
Site Area: 6,876 m²
Building Area: 85,000 m²

山、水是构成自然形态的重要元素，在人类日常生活中扮演着重要的角色。而华东院的主业为水电站工程，和山、水也是息息相关的因此建筑设计及总图设计由山、水引申而来，通过对象形文字“山”“水”的变异与提炼，反映到建筑立面上及景观设计中，这种山、水元素的特性，与华东院的企业文化、企业精神产生共鸣，从而形成一种区域环境、建筑、使用者三位一体的效果。
作为能源的开发者，同样也应该是环保的提倡者和领导者。节能不光要做到节约能源，利用洁净能源，减少能耗，同样还应该是节约资源。在大楼设计的时候应该处处考虑到节能的要求。

Mountain and water are two important landforms, playing important roles in people's daily life. So the business is hydropower station engineering, closely related to mountain and water. Therefore, our architectural design and general design are extended from mountain and water, and the icons of mountain and water are reflected in the facades and landscape designs through the deformation and extraction of Chinese pictograms for mountain and water. Such features echo with our corporate culture and spirits, and form a trinity of regional environment, architecture and users.
Energy-saving: As an explorer of energy resources, our corporation is also pioneer and leader of environmental protection. Energy saving is not only energy saving, use of clean energy, and reduction of energy consumption, but also resource saving. In the design, we consider the requirements for energy saving in all aspects.

桐乡电力调度大楼

Tongxiang Power Dispatching Building

设 计 师：吴登国
项目地点：浙江 桐乡
用地面积：3500 m^2
建筑面积：25 000 m^2

Designer: Dengguo Wu
Location: Tongxiang, Zhejiang
Site Area: 3,500 m^2
Building Area: 25,000 m^2

本项目设计追求桐乡产杭白菊“茶一般境界”的意境，突出表现在以下三个方面：1．高效的功能，功能被融汇在一个逻辑性的平面形态中，空间骨骼犹如迷人的“蜷体玉龙”。2．独特的气质，整个形体挺拔、干净、大气，所有生成元素都遵循一致的逻辑，层层相扣，充分体现了这个城市的文化胸怀，茶一般清新。 3．入味的空间，“转、承、启、和”的空间节奏给人以似曾相识的感觉，向当地传统的“江南院落”空间情结致敬的同时白描一帧“落花无言，楼淡如菊”，一茶一筑！

This project design is to follow the "Tea Realm" conception of white chrysanthemum produced in Tongxiang city, and it is outstandingly expressed in three ways:
1. The high efficient function is performed in a logical surface, and the space is like a charming "Curled Jade Dragon".
2. Unique feature. The entire figure is straight, neat and glamorous. All generated elements follow a consistent logic with layers connecting with each other, and fully embodies the cultural vision of the city as a fresh tea path.
3. The "Turning, Developing, Changing and Concluding" space makes people feel similar. It plays tribute to the local traditional "South Chinese Courtyard" space complex and at the same time draws a picture of "silently the flowers are falling, and the building is as slight as the chrysanthemum ". Tasting tea in the building!

景洪流沙河片区规划

Planning of Liusha River Area Jinghong

设 计 师：王晓庆
项目地点：云南 景洪
用地规模：13 400 000 m^2

Designer: Xiaoqing Wang
Location: Jinghong, Yunnan
Site Area: 13,400,000 m^2

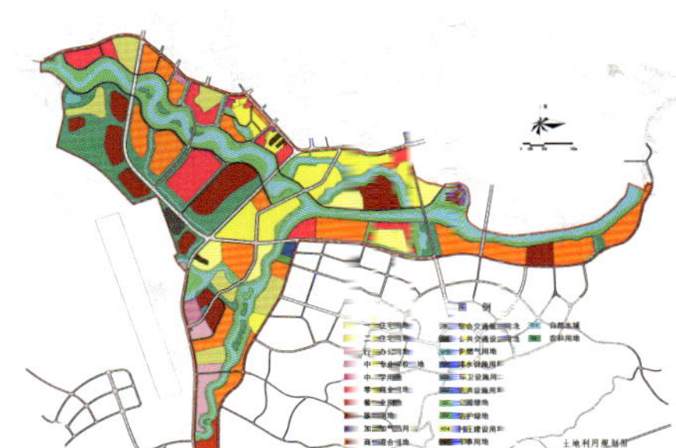

水上城市——多功能的水岸空间。
整合水体自然资源，在城市内部开凿河道，联通流沙河—南凹河水系，打造集民族风情、旅游观光、商业集会等功能为一体的水岸空间，塑造“水上傣家”城市形象。
宜居城市——生态和谐的人居环境。
将自然山水及传统文化元素融入城市，塑造景洪城市“风貌新区”，将生态能源及生态建筑技术引入城市，建立优雅、舒适、生态、富有民族特色的人居环境。
文化城市——充满诗意的文化家园。
傣家文化与传统农耕生活的现代演绎，傣乡传统文化与高尚文化场所相结合突显城市品位和生活品质，人文与自然景观结合，构建“特色城市”。
旅游城市——无以伦比的度假生活。
引入集会庆典，创造生活之余的多元化生活，重现傣族民族非物质传统文化生活，便捷的出行条件、多阶层的购物环境、丰富的民族文化、宜人的自然风光，打造“灵韵景洪”。

City on Water – a multifunctional waterside space
In order to integrate water body resources, the project digs canals within the city, to connect Liusha River – Nan'ao River water system, build a waterside space integrating ethnic flavors, tourism & sightseeing, commerce & meeting and other functions, and create a city identity of “Dai home on water”.
Habitat City – an ecologic habitat in harmony
The project integrates natural landscapes and traditional cultural elements into the city, to create a “scenic new area” of Jinghong City, which will introduce eco-energy resources and eco-building technologies into the city, to create a habitat featuring elegance, comfort, ecology and ethnic customs.
Cultural City – a cultural home full of poetic senses
It is modern performance of ethnic culture and traditional farming life. Ethnic traditional culture coincides with noble culture to highlight city taste and life quality, and human culture integrates with natural landscape, to build a “featured city”.
City of Tourism – unrivalled resort
It introduces meetings and ceremonies to create diverse lifestyles in free time, and reproduce ethnic nonmaterial traditional culture and life. Convenient travel conditions, shopping environment for every social class, rich ethnic culture, and agreeable natural scenes will build an “inspiring Jinghong City”

福建省档案馆新馆
The New Archives of Fujian

设 计 师：吴登国 赵文冰
项目地点：福建 福州
用地面积：6800 m^2
建筑面积：42 000 m^2

Designer: Dengguo Wu, Wenbing Zhao
Location: Fuzhou, Fujian
Site Area: 6,800 m²
Building Area: 42,000 m²

在该项目的创作中，着力于形、意、理三个层面，力求将文化性、开放性和可持续性融于一体，推陈出新，“章”法天成，构筑符合时代需求的现代公共档案馆。
文化性：方案最大程度保留了基地的水面，使“水”成为环境设计的主角，彰显福建独特的“海峡西岸”地文特征；新馆设计方案的基本形体接近正方形，犹如一方收藏印。
开放性：档案馆开始走向历史的前台，以其内容丰富、价值珍贵的馆藏，在资政的同时服务于民。档案馆本身所拥有的珍贵馆藏以及所强调对外服务功能的双重特色与百宝格的特质完全契合，于是在建筑的外观采用百宝格独特韵味的处理手法。
可持续性：方案注重环境的生态节能性设计，借用传统的“借景”手法，同时将建筑东南角打开，可以引导常年的主导风将新鲜空气源源不断地带入建筑内部，达到了利用自然力加强通风的目的。

This project puts forth effect on three levels on form, meaning and texture, integrating the culture, openness and sustainability. It brings forth the new through the old in orderly ways to create a modern public archives by meeting the epochal demand.
Culture: The project retains the base surface to a maximum extent so that "water" leads in the environmental design to embody the unique "West Coast of the Strait" physiographic feature of Fujian province. The basic shape of the new archives is close to a square shape like a collection seal.
Openness is heading forward to the front of history by serving both the politics as well as his people with rich content and precious collections. The dual characteristics of precious collections and emphasized foreign service functions of the archives coincide perfectly with the specialties of "treasure display racks". So it adopts unique treatment on the appearance of the building.
3. Sustainability. This project focuses on ecological energy-saving design of the environment by using the traditional "landscape-borrowing" method. It has an open in southeast corner to guide fresh air of the perennial prevailing wind into the inner side of the building constantly for the purpose of enhancing ventilation by using the natural force.

丽水市滨水公园
Waterfront Park, Lishui

设 计 师：赵文冰、王晓庆
项目地点：浙江 丽水
用地面积：85 000 m^2
建筑面积：27 600 m^2

Designer: Wenbing Zhao, Xiaoqing Wang
Location: Lishui, Zhejiang
Site Area: 85,000 m^2
Building Area: 27,600 m^2

本项目是集城市展览、历史文化、科普教育、休憩娱乐于一体的滨水公共活动中心。由滨水广场、生态公园、博物馆、规划展示馆和青少年科技活动中心组成。本项目紧扣公共开放空间的主题，通过建筑物精心布局，保证绿化带的延续性。建筑物、自然环境和历史文化和谐共生，情景交融，展现出一幅优美的人文活动和自然生态交织的滨水画卷。

This project is a waterside public center integrating urban exhibition, history & culture, science generalization, rest and entertainment. It includes waterside park, eco-park, museum, display hall and youth science center. The project is themed by public open space, to ensure a continual riverside greening belt for citizens, whose continuity is ensured through elaborate layout of buildings. The buildings coincide with natural conditions, history and culture, to show a beautiful waterside picture of human cultural activities and natural ecologies.

厦门市天铜山综合性度假区酒店
Tiantongshan Integrated Resort Hotel of Xiamen

设 计 师：徐堃
项目地点：福建 厦门
用地面积：39 587 m^2
建筑面积：42 000 m^2

Designer: Kun Xu
Location: Xiamen, Fujian
Site Area: 39,587 m^2
Building Area: 42,000 m^2

项目位于福建省长泰县天铜山综合度假区内，位于整个天铜山一期项目用地的东北侧，项目着力打造一个热带休闲山地滨水度假休闲区。
整体设计以景观为核心塑造“度假天堂”，整个建筑群以入口大堂为形体中心，层层展开，形成标志性制高点。从入口大堂到会所，再到客房，公共空间均最大程度地面向水景庭院内，既可形成对内庭环境的围合，又可更好地与环境融合。
丰富、灵动、错落有致的建筑布置通过不同庭院组合起来，形成前庭、中间半围合庭院以及滨水庭院，充分利用基地有限的用地现状，并通过加强景观的设计来保证客房休闲、自然的舒适性。同时，较分散的建筑布局符合亚热带的气候特征，每个客房都能保证良好的采光通风，房间和景观结合紧密。

This project is located in Tiantongshan Comprehensive Resort Zone, Changtai County, Fujian Province. It is in the northeast side of Tiantongshan Phase 1 land lot. It pays great efforts in creating a tropical leisure mountainous waterfront resort
The overall design is to shape a "vacation paradise" taking landscape as the core. The building cluster, centering the entrance lobby, is extending layer by layer, to form a marking highest point. From the lobby to the club, to the guestrooms, all public spaces are facing to the waterscape inner courtyard to a maximal extent. This can not only enclose the inner courtyard but also better integrate with the environment.
Variable, agile and orderly architectural layout is made up of different courtyards with front yard, middle semi-enclosed yard and waterfront yard. It makes the best of the limited land resources in the base, and strengthens the landscape design to ensure the leisure natural comfort of the rooms. At the same time, a more decentralized architectural layout fits the subtropical climate conditions so that each guestroom can guarantee good light and ventilation and combine closely with the landscape.

杭州市城东新城 0901 人防工程
0901 Civil Air Defence of East New Town in Hangzhou

设 计 师：陈楠
项目地点：浙江 杭州
用地面积：93 518 m²
建筑面积：63 880 m²

Designer: Nan Chen
Location: Hangzhou, Zhejiang
Site Area: 93,518 m²
Building Area: 63,880 m²

基于本项目位于道路下方，并与周边的道路及工程相接的特点，以超前的眼光实现地上、地下一体化发展，实现地下系统与地上系统的有机联系；同时与周边各地块未来的地下空间统一考虑，将地下各项设施进行系统整合、统筹安排，着力构建城东新城地下空间大系统。设计中做到了与周边工程，如商业地下空间、地铁等的无缝衔接，做到了地面景观、地下空间的顺利过渡和功能互补，做到了将空气、阳光引入到地下空间等绿色理念，推进了城东新城"地下城"的宏伟规划。

Base on the project is located underground, joining the surrounding roads and engineering, relate the underground system with the aboveground system, looking forward to the integration of the aboveground and underground developments. At the same time, it takes full account of the fugure underground integration with peripheral lots, to systematically integrate all underground facilities, and make collective arrangement, focusing on the larger system of east new town underground spaces. The design realizes seamless joint with peripheral projects, such as commercial underground space and the metro. It also realizes smooth transition and functional complementation of the ground landscape and underground space, and realizes the green idea of leading natural air and sunlight into the underground space, to advance the ambitious plan of an "underground city" of east new town.

华东勘测设计研究院闲林办公楼项目
Xianlin Office Building Project of Huadong Engineering Corporation

设 计 师：徐堃
项目地点：浙江 杭州
用地面积：49 711 m²
建筑面积：70 609 m²

Designer: Kun Xu
Location: Hangzhou, Zhejiang
Site Area: 49,711 m²
Building Area: 70,609 m²

项目位于常余路南侧，东靠高教路，四周临高档住宅区。项目包含办公主楼和餐饮会议中心、健身中心。
办公主楼强调整体的严谨大气，通长的竖向构件丰富建筑立面的韵律感，营造简洁、明快的办公风格。设计以严谨而又不失灵性的立面组合契合行业特质，以金属、玻璃的挺阔与光感体现现代的时尚和高效，以石材的内敛与凝重反映传统的历史与发展，创造出庄重、简洁的办公建筑氛围。
东西辅楼的设计吸取办公大楼的立面风格元素，以简洁、明快为主，与办公大楼相辅相成。
综合运用现代建筑的设计语言，高技术的理念，多层次的文化内涵，表达出设计研究院知识密集型、富有创新精神和前瞻性的科研单位的特色

The project is on the south side of Changyu Road, near Gaojiao Road in the east, surrounded by high-grade residential communities. It includes main office building, catering and convention center and fitness center.
The office building emphasizes the overall prudence and generosity, creating a simple and neat office style through long vertical structures to enrich the rhythm of façade. In the design, a combination of prudent yet not less agile façades complies with the industrial character, the broadness and brightness of metal and glass structures represent modern fashion and efficiency, and the modesty and weight of stone materials reflect traditional history and development, to create solemn and concise office building conditions.
The east and west auxiliary buildings adopts façade elements from the office building, concise and clear, complementing the office building. It combines modern architectural design language, high-tech ideas, and multilevel cultural significance, to express the knowledge-intensive, innovative and forward-looking features of this corporation as a scientific research organization.

钱江创新创业产业园二期
Qianjiang Innovation & Entrepreneurship IndustryPark Phase II

设 计 师：陈楠　　Designer: Nan Chen
项目地点：浙江 杭州　　Location: Hangzhou, Zhejiang
用地面积：180 000 m^2　　Site Area: 180,000 m^2
建筑面积：450 000 m^2　　Building Area: 450,000 m^2

钱江创新创业园二期项目位于杭州钱江经济技术开发区核心地块内，是集产业创新、企业孵化、办公生产及配套设施的综合建筑群体。以标志性的总部经济办公楼为主体，与孵化楼、研发楼、独栋产业用房、创业综合楼及部分裙房组成。
本项目分三个区块，中间区块为工业用地，占地约 11 万 m^2，以为独幢产业用房、孵化楼、研发楼为主，总建筑面积约 18 万 m^2；东西两个区块为商业办公用地，东侧由 4 幢高层办公楼及其裙房组成，总建筑面积约 20 万 m^2，西侧地块位于胡家文河以西，以酒店住宿为主，总建筑面积约 7 万 m^2；地下室均为车库及设备用房。南侧的 7 幢楼为钱江创新创业园一期项目，总建筑面积约 5 万 m^2，已投入使用。建成后，一、二期将共同成为一个服务于高新产业从创业到孵化、再到走向成熟的一系列的配套完整、环境优美的孵化—生产办公中心。

The phase 2 project is located in the core lot of Qianjiang Economic Development Area, Hangzhou City. It is a comprehensive building cluster of industrial innovation, corporate incubation, office & production and supporting facilities. It includes marking HQ economical office building as the main body, as well as incubation building, R&D building, independent industrial houses, entrepreneurship comprehensive buildings and some skirt buildings.
This project has three land lots. The middle lot is industrial lot, with land area approx. 110,000 m², mainly developed into independent industrial houses, incubation buildings and R&D buildings, with total floor area approx. 180,000 m². The east and west land lots are commercial lots, of which, the east lot includes 4 high-rise office buildings and skirts with total floor area approx. 200,000 m² and the west lot is west of Hujiawen River, mainly developed into hotels and lodges, with total floor area approx. 70,000 m². The underground are garages and equipment rooms. The south 7 buildings are phase 1 works of Qianjiang Innovation & Entrepreneurship Park, with total floor area approx. 50,000 m², have been in service. Upon completion, the two phases of works will become a beautiful mature center of incubation, production and office center, serving for high and new technologies from entrepreneurship to incubation and to maturity.

西昌市一环路历史风貌核心区
First Ring Road Historical Style Core Area, Xichang City

设 计 师：王晓庆　　Designer: Xiaoqing Wang
项目地点：四川 西昌　　Location: Xichang, Sichuan
用地规模：123 000 m^2　　Site Area: 123,000 m^2
建筑规模：99 100 m^2　　Building Area: 99,100 m^2

项目因西昌唐嶲州古城墙遗址而得来，所以总体定位是唐代风格的文化休闲公园。设计中将盛唐遗风的景观设计及建筑风格贯穿整个地块。在展现大唐盛世恢弘气势的同时，也展示出浓郁的西昌风情，使其拥有强烈地域特色。一园"五彩唐诗园"及一湖"烟雨印月湖"的设计，将古代山水园林"效法自然"表现得淋漓尽致。

This project originates from the wall ruins of Xizhou ancient town of Tang dynasty, remained in today's Xichang City, and the general positioning is cultural leisure park in Tang style. In the design, the landscape design and architectural style of prosperous Tang legacy penetrates the entire area. It shows the magnificent Tang as well as rich Xichang customs, showing strong local features. The design of "Colorful Tang Poetic Park" and "Rainy Moon Lake" perfectly reproduces the ancient landscaping garden by "Following the Nature".